THE
DYNAMICS
OF
INDUSTRIAL
LOCATION

THE DYNAMICS OF INDUSTRIAL LOCATION

The Factory, the Firm and the Production System

ROGER HAYTER

Simon Fraser University
Burnaby, Canada

JOHN WILEY & SONS

Chichester · New York · Weinheim · Brisbane · Singapore · Toronto

Copyright © 1997 by John Wiley & Sons Ltd,
Baffins Lane, Chichester,
West Sussex PO19 1UD, England

National   01243 779777
International (+44) 1243 779777
e-mail (for orders and customer service enquiries): cs-books@wiley.co.uk
Visit our Home Page on http://www.wiley.co.uk
or http://www.wiley.com

Other Wiley Editorial Offices

John Wiley & Sons, Inc., 605 Third Avenue,
New York, NY 10158-0012, USA

VCH Verlagsgesellschaft mbh, Pappelallee 3,
D-69469 Weinheim, Germany

Jacaranda Wiley Ltd, 33 Park Road, Milton,
Queensland 4064, Australia

John Wiley & Sons (Asia) Pte Ltd, 2 Clementi Loop #02-01,
Jin Xing Distripark, Singapore 129809

John Wiley & Sons (Canada) Ltd, 22 Worcester Road,
Rexdale, Ontario M9W 1L1, Canada

Library of Congress Cataloging-in-Publication Data

Hayter, Roger, 1947–
    The dynamics of industrial location: the factory, the firm, and
the production system / Roger Hayter.
        p.   cm.
    Includes bibliographical references and index.
    ISBN 0-471-97083-2 (cloth). – ISBN 0-471-97119-7 (paper)
    1. Industrial location.   I. Title.
    HC79.D5H39   1997   96-44231
    338.09 – dc20   CIP

British Library Cataloguing in Publication Data

A catalogue record for this book is available from the British Library
ISBN 0-471-97083-2 (cloth)
ISBN 0-471-97119-7 (paper)

Typeset in 10 on 12pt Times from the author's disks by Dobbie Typesetting, Tavistock, Devon
Printed and bound in Great Britain by Bookcraft (Bath) Ltd, Midsomer Norton, Somerset
This book is printed on acid-free paper responsibly manufactured from sustainable forestation,
for which at least two trees are planted for each one used for paper production.

For Mum, Dad and Max

CONTENTS

PREFACE

Industrialization is an historically significant process which has contributed much to the distinctive nature of national, regional and local economies and to geographically uneven trajectories of growth and decline or, more generally, to the geographically uneven process of 'creative destruction'. Industrialization adds wealth and jobs to places directly and indirectly, brings places together economically in a variety of ways, serves and creates consumer demands, and is a profoundly cumulative process. Moreover, since at least the Industrial Revolution, the world's hegemonic powers have been the world's industrial leaders. In the 19th century, *Pax Britannica* and Britain as the world's workshop were two sides of the same coin. As the 21st century is approached, hegemonic rivalry features the world's three leading industrial powers, the US, Japan and Germany, striving for industrial and political supremacy. Yet, industrialization in one place can also threaten and undermine wealth and jobs elsewhere, including even the longest established industrial places, and it is relentlessly demanding of the environment. In short, industrialization is important to geography and development.

Industrial geography's distinctive contribution to understanding the process of industrialization originates from the emphasis placed on the perspective of location. Thus, from its beginnings the central tasks of industrial geography have been, and remain, to explain the location of industrial activity (from global to local scales of reference), changes in the location of industrial activity (from historical to annual time-scales), and the implications of changing patterns of industrial location for national, regional and local economic development. As a sub-discipline of human geography, industrial ('manufacturing') geography traditionally has been preoccupied with manufacturing activities and, as such, developed rapidly in the 1950s when the first textbooks were written. The importance of manufacturing industry to local economies within advanced countries and the role of industry in shaping the 'character' of places underlay this development. The sub-discipline of industrial geography was further stimulated at this time by the innovation of a form of regional policy, notably in the United Kingdom but rapidly emulated elsewhere, which emphasized the role of manufacturing industry in providing jobs and income to regions suffering from high levels of unemployment, low per capita incomes and high rates of out-migration. To put the matter simply, this policy, in its original formulation, argued that to reduce these problems 'work must be taken to the workers' and that only manufacturing activities could be readily dispersed. The geography of primary and service activities, it was assumed, could not be altered since the former depended upon nature and the latter on population. Manufacturing activities, on the other hand, were considered to be more locationally flexible and would also generate 'spin-offs' in 'designated' regions.

If regional policy, as originally conceived, is now largely defunct, scholarly interest and the social relevance of industrial geography remains vital. The opening up of 'new industrial spaces' in many parts of the globe in both developed and developing countries is intimately interrelated, albeit in no simple cause and effect way, with adjustment and diversification in some established industrial spaces and with the deindustrialization of others. The problem of industrial transformation is global in scope and the process of creative destruction is evident to all. It is the assumption of this book that industrial geography can contribute towards an understanding of this problem and this process.

This book seeks to present a systematic understanding of the dynamics of industrial location as researched and developed by industrial geographers and related scholars. Following the two introductory chapters of Part I which briefly outline the concerns of industrial geography and place the 'industrial transformation problem' in geographical and historical perspective, the book progressively explores industrial location dynamics at the level of individual factories (Part II), individual firms (Part III), and finally with respect to production systems which comprise bundles of inter-firm (and inter-factory) relations in the manufacture of particular products (Part IV). While this sequential focus on the factory, the firm and the production system (inter-firm relations) reflects the chronology of industrial geography as a sub-discipline, the primary rationale for this organization is analytical. Given the historical and geographical context of industrialization outlined in Part I, the factory, the firm and the production system represent progressively more complex units of investigation (and location structures) and principles and issues in the early parts of the discussion are incorporated and elaborated in later parts of the discussion.

The book is targeted primarily for senior level undergraduates and represents the way I have been teaching my Geography of Manufacturing course at Simon Fraser for several years. In this course, I have found H. D. Watts' (1987) *Industrial Geography* and K. Chapman's and D. Walker's (1987) *Industrial Location* to be extremely useful and much appreciated by my students. Another good textbook is E. J. Malecki's (1991) *Technology and Economic Development*, although as its title implies this book has a specific focus on technological change. My own book, I believe, offers a distinctive organization which explicitly links industrial geography with the 'industrial transformation problem' (Part I) and progressively builds on the three most theoretically important categories of analysis used in industrial geography, namely the factory, the firm and the production system (Parts II–IV). In addition, in comparison to existing texts, there is more emphasis placed on production systems, and the related topic of industrial districts, since they have been such an important thrust of research within industrial geography over the past decade. In any event I hope that this book will continue to stimulate interest in industrial geography and encourage students to continue to think about the relations between industrialization and regional development.

A textbook, by definition, is supposedly comprehensive and, given the particular structure of the book and length limitations, I have tried to be as comprehensive as possible. There has been an explosion of research (information) within industrial geography over the last 25 years, however, and I have no doubt that I have excluded many important contributions by various authors. All I can say in this regard is that I apologize (while shamelessly inviting those excluded to send me their reprints!). I have also had to be selective in my examples. While I have tried to choose a range of cases from various contexts, my preference for discussing these cases in some detail inevitably limits the number of examples cited. Hopefully, the cases selected demonstrate important issues and at least provide a basis for comparison of examples better known and more relevant to readers.

I have been thinking about writing a book on industrial geography for many years and perhaps I should have spent less time watching soccer and hockey. After all, my teams (Sheffield Wednesday and the Vancouver Canucks) have only rarely been inspirational. In the end, I owe a debt to many people in the evolution of this book. Brian Fullerton and Ken Warren (at the University of Newcastle upon Tyne) and Bruce Proudfoot and Geoff Ironside (at the University of Alberta) initially encouraged my interest in economic geography, while my thinking about industrial geography was profoundly affected by my doctoral alma mater, the University of Washington, especially through the help and direction provided by Gunter Krumme and the teachings of Morgan Thomas, Bill Beyers and Richard Morrill. Subsequently, I have gained much from being a part of a vibrant international community of industrial geographers and in recent years I have benefited considerably from my local partnerships with Trevor Barnes and David Edgington.

With respect to the book itself, Trevor Barnes, Doug Watts, Bill Lever, Kevin Rees and Jerry Patchell offered valuable comments on earlier drafts for which I am most thankful. In fact, Jerry had the fortitude to closely read and criticize an early draft, and his assistance has been invaluable in trying to come to grips with the industrial geography of Japan. Over the years, several of my graduate students have provided all kinds of help in teaching my course in industrial geography, notably Ben Ofori-Amoah, Eric Grass, Kevin Rees, Tanya Behrisch, Jon Skulason and Robyn Adamache and I gladly tip my hat to my students at Simon Fraser who have shown considerable patience and under-standing as guinea pigs in my efforts to get my course and this book properly organized. Thanks to Dieter Soyez and Johannes Hamhaber, summer school in 1995 at the University of Cologne, provided a relaxing and supportive environment in which to re-write a draft of this book, and Iain Stevenson, Louise Portsmouth and Claire Walker of John Wiley have been most helpful in keeping this book on track. I am also extremely grateful for the help and good humour provided by our front (and back?) offices in the Geography Department at Simon Fraser, including the typing help provided by Mary Ward, Marcia Crease and Diane ('tables') Sherry and the cartographic expertise of Ray ('I suppose you want this yesterday') Squirrell and Paul ('sure, that's easy') Degrace. I cannot divulge, however, the support received from Bill, my fellow departmental co-conspirator. Bill, does Hilary know what Ida knew?

My final words of gratitude are to my family. Jacquie, Alison, Lynn and Megan have been a bedrock of encouragement and support while simultaneously striving mightily to enrich my understanding of a wider world. Thanks and don't stop! Finally, I would like to dedicate this book to my parents, Dorothy and Eric ('Ernie') Hayter. My privileged life owes much to them and stands in contrast to their own sacrifices. They left school at age 14. In Dad's case, between 1942 and 1949, he soon travelled to (present day) Myanmar, India, Bangladesh, Egypt, Cyprus and Israel, as a British soldier. After I left school, in somewhat more comfortable circumstances, I collected degrees in England, Canada and the US. His education may have been more radical but I have been luckier. My thanks to him must be posthumous.

ACKNOWLEDGEMENTS

We are grateful to the following for permission to reproduce copyright material:

Academic Press for Table 5.2 from Table viii, p. 196 (Tillman 1985)

Addison Wesley Longman for Figure 3.8 from Figure 1.3, p. 4 (Watts 1987), for Figure 2.8 from Figure 2 (Hayter 1996)

American Geographical Society for Table 8.3 from Table II, p. 227 (Hayter 1976); for Figure 8.10 from Figure 3, p. 218 (Hayter 1976)

Association of American Geographers Figure 17.1 from Figure 1, p. 484 (Patchell 1996)

Bernie Hayter for Exhibit 16.1

Blackwell for Figure 4.5(a) and (b) from Figure 1, p. 439 (Florida and Smith 1993); Figure 13.8 from Figure 2, p. 208 (Patchell and Hayter 1992)

Blackwell/Polity for Table 14.1 from Table 1.1, p. 16 (Morales 1994)

Canadian Association of Geographers for Table 6.1 from Table 10, p. 275 (Steed and De Genova 1983); for Table 4.6 from Table 3, p. 224 (Bathelt and Hecht 1990); for Table 11.4 from Table 1, p. 136 (Soyez 1988); for Table 13.5 from Table 2 (Rees and Hayter in press)

Cassell Academic for Table 2.3 and Figure 2.4 from Table 1, pp. 68–75 (Freeman 1987)

Croom Helm for Figure 8.5 from Figure 9.5 (Watts 1980); for Table 16.2 from Table 11.1, p. 280 (Noble 1987)

Department of Social and Economic Geography, Lund University for Figure 6.3 from Figure 11, p. 92 (Pred 1967)

Economic Geography for Table 4.4 from Table 2, p. 133 (Ley and Hutton 1987); Figure 5.8 from Figure 5, p. 106 (Smith 1966)

Elsevier Science for Figure 7.8 from Figure 3, p. 169 (Hayter 1979)

Richard Florida for Table 16.2 from Table 2 (Florida 1994)

The Geographical Association for Figure 14.6 from Figure 1, p. 356 (Peck 1990)

Geographical Journal for Figure 3.4 from Figure 1 (Taylor et al 1938)

Guilford Press for Figure 13.10 from Table 2.2, p. 47 (Angel 1994); for Table 15.2 from Table 3.4, p. 55 (Dicken 1992); Table 12.2 from Table 16.5, p. 186 (Dicken 1992); for Figure 3.1 from Figure 2.5, p. 21 (Dicken 1992); for Figure 4.4 from Figure 6.11, p. 182 (Dicken 1992)

Peter Gould for Figure 7.11 from Figure 9.5, p. 302 (Abler et al 1970)

The Controller of Her Majesty's Stationery Office for Table 3.11 from Table 16.2, p. 321 (Healey and Ilberry 1990)

Harvard Business School for Table 10.2 from unnumbered Table, p. 119 (Simon 1992) and for Table 17.1 from unnumbered tables, pp. 117 and 119 (Johnston 1991)

Institute of Developing Economies for Figure 15.3 from Figure 1, p. 472 (Nakajo 1980)

Institute of Geoscience, University of Tsukuba for Figure 13.5(a) and (b) from Figures 3(b) and (c), pp. 69 and 78 (Suyama 1996)

Johns Hopkins University Press for Figure 7.6 from Figure 2, p. 51 (Gregerson and Contreras 1975)

Journal of European Economic History for Table 3.1 from Table 2, p. 275 (Bairoch 1982), and for Table 3.2 from Tables 10 and 13, pp. 296 and 304 (Bairoch 1982)

Macmillan for Figure 9.5 from Figure 3.3, p. 41 (Burns 1989)

The McGraw-Hill Companies for Table 5.1 from Table 3.3, p. 47 (Hoover 1971), for Table A2.1 from Table 8.7, p. 226 (Hoover 1971), for Table A2.2 from Table 8.8, p. 228 (Hoover 1971), for Table A2.3 from Table 8.9, p. 229 (Hoover 1971), for Table A2.4 from unnumbered Table, p. 247 (Hoover 1971), for Table A2.5 from unnumbered Table, p. 248 (Hoover 1971), and for Table 10.1 from unnumbered Table, p. 44 (Business Week 1993).

Y. Miyakawa for Figure 14.4 from Figure 2, p. 30 (Miyakawa 1980)

Oxford University Press for Figure 3.5 from an unnumbered figure, p. xv (S. Pollard 1981); for Table 3.13 from Table 1.1, pp. 6–7 (Madison 1991); Table 3.14 from Table C.5, pp. 248–9 (Madison 1991)

Pion for Figure 8.3 from Figure 1, p. 1605 (Taylor and Thrift 1982); for Figure 11.1 from Figure 1, p. 790 (Hayter 1986); for Figure 2.2 from Figure 1, p. 994 (Barnes 1990); for Figure 4.7 from Figure 1, p. 38 (Watts 1991); for Figure 14.7 from Figure 3, p. 813 (Patchell 1993a); for Figure 14.3 from Figure 2, p. 811 (Patchell 1993a); for Table 19.8 from Table 3, p. 1354 (Healey and Clark 1985); for Table 16.6 from Table 1, p. 43 (Watts 1991); for Table 16.7 from Table 2, p. 44 (Watts 1991); for Table 4.8 from Tables 5 and 6, pp. 49 and 51 (Watts 1991)

Prentice Hall for Figure 3.7 from unnumbered figure, p. 24 (Alexandersson 1967); for Table 6.2 from Tables 3.3 and 3.4(b), pp. 89 and 94 (Schmenner 1982)

The Regional Studies Association for Figure 13.6 from Figure 5, p. 336 (Scott 1992); for Figure 9.4 from Figure 2(b), p. 399 (Olofsson 1994); for Figure 9.3 from Figure 5, p. 417 (Keeble and Walker 1994); for Table 16.9 from Table 1, p. 307 (Healey and Clark 1984); for Table 4.5 from Tables 1 and 2, pp. 128–9 (Malecki and Bradbury 1992), for Table 9.5 from Table 3, p. 451 (Reynolds et al 1994)

Routledge for Figure 7.10 from Figure 9.2, p. 187 (Alvstam and Ellegard 1990); for Figure 8.7 from Figure 3.1, p. 67 (de Smidt 1990); for Figure 8.8 from Figure 4.1, p. 83 (Fuchs and Schamp 1990); for Figure 8.9 from Figure 10.5(d), p. 225 (Park 1990); for Figure 5.9 from Figure 2.4, p. 28 (Bloomfield 1991); for Table 14.7 from Table 5.7, p. 127 (Rubenstein 1991); for Figure 6.5 from Figure 5.2, p. 131 and Figure 5.3, p. 132 (Rubenstein 1991); for Table 9.9 from Table 2.8, p. 31 (Massey et al 1992)

The Royal Dutch Geographical Society for Figure 12.5 from Figure 1, p 345 (Patchell and Hayter 1995)

The Royal Geographical Society (with the Institute of British Geographers) for Table 4.6 from Tables 12.1 and 12.2, pp. 235 and 246 (Christy and Ironside 1987); for Table 3.12 from Table III, p. 211 (Townsend 1993); for Figure 3.10 from Figure 3, p. 397 (Martin 1988)

M.E. Sharpe for Table 13.2 from Table 1, p. 9 (Yaginuma 1993)

A. Takeuchi for Figure 15.1 from Figure 5, pp. 204 and 207; and Figure 15.2 from Figure 10, pp. 212 and 214 (Takeuchi 1994)

Wadsworth Publishing Co. for Table 13.3 from Table 5.01, p. 97 (R. L. Morrill 1974)

Karel Williams, Glen Haslam, John Williams, Sukdhev Johal for Table 2.4 from Table 7.2, p. 99 (Williams et al 1994)

Mira Wilkins for Table 15.7 from Tables 1–4, pp. 20–23 (Wilkins 1994); for Table 15.4 from Table 8, p. 33 (Wilkins 1994); for Table 15.6 from Table 5, p. 30 (Wilkins 1994)

While every effort has been made to contact copyright holders some enquiries remained unanswered at the time of going to press; it is hoped that any further acknowledgements may be added in future editions.

PART I

THE PROBLEM OF INDUSTRIAL TRANSFORMATION

Industrial and industrializing societies are dynamic societies which become increasingly interdependent within the global economy. From this perspective, recent attention to 'globalization' refers to deepening forms of interdependence that have for centuries been gaining momentum. Part I offers a broad overview of industrial dynamics and the perspective of industrial geography in analysing these dynamics. Chapter 1 identifies the scope and nature of industrial geography, especially contemporary industrial geography. Chapters 2 and 3 then outline the broad historic and geographic dimensions of change of global manufacturing activities, especially since the Industrial Revolution.

The problem of industrial transformation defined

In the latter part of the 18th century, particularly in the British Isles, the pace of economic life accelerated and 'revolutions' in transportation and industry began, while that in agriculture continued. The generally accepted summary signature assigned to this period, however, is 'The Industrial Revolution', a label which reflects the emergence of the manufacturing sector as a major employer and economic driving force within modernizing societies. The modern problem of industrial transformation had been established (Chapter 2).

The problem of industrial transformation may be defined as the ability of nations, regions or communities to initiate, rejuvenate and retain viable industrial structures which meet long-term goals, however these goals may be articulated. If progressive increases in per capita income can be said to significantly capture the economic purpose of national and regional development, then in relatively abstract terms the problem of industrial transformation depends upon increasing the productivity of processes and/or innovating products with high-income elasticities of demand (Thomas 1964). In more concrete terms, industrial transformation is often expressed as the need for nations and regions to shift from low-value to high-value-added activities or from old, stagnant, even declining industries into new, fast-growing ones that can provide jobs and higher incomes. In this view, industrial competitiveness increasingly depends on research, development and innovation (Freeman et al 1982).

However defined, the problem of industrial transformation varies across time and space. In the late 18th century, the Industrial Revolution posed a unique situation for the UK, namely how to shift from an agricultural, rural based society to an urban, industrial one, in the absence of any available 'model' to copy or from which to learn. In subsequent decades, the US and European countries such as Germany and France had to face the challenge of industrializing a rural economy and, if the UK offered an example and source of expertise, it also provided powerful and highly efficient competition. Indeed, to an important degree industrial strategy for all 'late' developing countries, from Canada, Japan and Sweden in the late 1880s to China in the 1980s, has involved how to play catch-up with established industrial powers. Once established, in large part because of the enormous size of investments in fixed capital and the accumulation of labour pools and skills, industrialization generates its own powerful momentum. Yet, such is the dynamism of the industrialization process that even this momentum can be halted and reversed. Thus 200 years after the start of the Industrial Revolution, the UK faced another unique problem of industrial transformation as the first established industrial power to face, during peacetime conditions, large-scale de-industrialization as measured by an absolute decline in the number of manufacturing jobs (Singh 1977; Thirlwell 1982). The UK's experience in this regard was soon emulated by large regions in the US containing populations larger than that of the UK (Bluestone and Harrison 1982) while the German Ruhr and other older industrial regions of western Europe have suffered a similar fate. Indeed, within the past decade no industrialized country has been immune to profound forces of rationalization and restructuring.

In the 1990s (and surely beyond), the problem of industrial transformation poses a formidable challenge to nations and regions, whether they are industrializing for the first time or trying to rejuvenate deindustrialized structures. This challenge is underlined by Barnet's (1993) recent argument that in the globe today existing industrial capacity already substantially outstrips effective demand (see Barnet and Cavanagh 1994). In such a situation it is almost inevitable that new industrial growth will undermine existing industrial centres. Yet, there are strong arguments for continuing to believe that industrial success is important for job generation, increasing income and political power. The reasons for studying the relationships between industrial location, industrial organization and national, regional and local development are all the more pressing. Industrial geography is central to this task.

1

INDUSTRIAL GEOGRAPHY

The purpose of this chapter is to briefly introduce the subject of industrial geography by outlining the principal approaches that define it and the activities that are central to its concerns.

Approaches to industrial geography

The fundamental rationale for industrial geography, commonly also referred to as manufacturing geography, rests on the geographic unevenness and constantly changing geography of manufacturing activity. Following several pioneering studies which sought to explain the geographic distribution of particular manufacturing activities (Hartshorne 1928, 1929) or the concentration of activities in specific localities (Wise 1949), specialized textbooks in manufacturing geography were first published (in English) in the 1960s (Miller 1961; Alexandersson 1967; Estall and Buchanan 1980). These books revealed the growing interest in industrial (manufacturing) geography and a 'systematic' approach based on the classification of location conditions and factors. A mushrooming and vibrant research literature soon led to new theoretical perspectives and research designs, and raised new questions. A *few* well-known, relatively comprehensive books which reflect the development of the literature since 1970 include Chapman and Humphrys (1987), Collins and Walker (1975),

Hamilton (1974), Hamilton and Linge (1979), Massey (1984), Rees et al (1981), Scott and Storper (1986), Sayer and Walker (1992), Storper and Scott (1992), Storper and Walker (1989) and Taylor and Thrift (1984). In turn, mounting research and shifts in the research frontier encouraged the publication of additional textbooks on industrial geography, notably those by Smith (1971), Bale (1976), Webber (1984), Chapman and Walker (1987), Watts (1987), Malecki (1991) and, most recently, Harrington and Warf (1995).

Despite considerable debate over approach and forms of analysis within manufacturing geography, however, location, location change and implications for local development, have been constant themes in the development of the subject. As Watts (1987: 1) says:

'The central task of contemporary industrial geography is to describe and explain changes in the spatial pattern of industrial activity . . . The emphasis in industrial geography is on explaining where and why changes in location of industrial activity have taken place and on trying to understand why some areas experience industrial growth and other areas experience industrial decline.'

Simply put, industrial geography seeks to explain the location dynamics of manufacturing activity and the local development implications of these dynamics. In a broader context, if economic geography explains geographical variations in how people gain a livelihood and 'how place makes a difference to the economic process' (Barnes 1987: 307), industrial geography is an integral part of this task. From an interdisciplinary perspective, the concerns of industrial geography are clearly at the core of the industrial transformation problem defined above as 'the ability of nations, regions and communities to initiate, rejuvenate and retain viable industrial structures which meet long term goals, however these goals may be articulated.'

From the idiographic to the nomothetic

As contemporary introductory textbooks suggest, it is common to recognize that a watershed in approaches to explanation within economic geography occurred around the late 1950s and 1960s (Healey and Ilbery 1990; Berry et al 1993). Before the watershed, the various traditional approaches to economic geography, including industrial geography, are summarized as 'idiographic' in contrast to the 'nomothetic' or theoretical approaches of contemporary times. This book is clearly rooted in contemporary research and a note on traditional industrial geography sets the stage for brief comments on the various theoretical approaches that inform industrial geography.

Within industrial geography, the idiographic tradition is explicitly (and impressively) exemplified by Warren's (1970) meticulously detailed analysis of the British iron and steel industry from 1740 to 1968. For him, theories and principles may define how industry should be located but they cannot come to grips with understanding actual location dynamics. In Warren's (1970: viii) view, 'analysis in economic geography requires a resolution to probe as far as possible into the reasons for location decisions, to master the rudiments of technology and to look into the contingencies of industry and company development.' Such a statement indicates that idiographic approaches go beyond

description and are interested in explanation and the causal processes underlying location outcomes. In practice, Warren's search for 'the reasons for location' is fundamentally historical; his understanding of the various steel districts of Britain is based on an extraordinarily detailed historical account of decisions by individuals and companies in the steel industry in the particular (historical, geographical and technological) circumstances they were made and of how the (intended and unintended) outcomes of one decision shape subsequent behaviour. In this account there are no universal notions of rationality. For Warren, the economic geography of the British steel industry can only be understood by going back to its origins and connecting the present with the past through an intricate maze of decisions. Ultimately, British steel regions are each unique and each region should be considered idiographically, that is, as a separate case study.

There was a 'systematic' (generalizing) side to traditional industrial geography which primarily took the form of inductive classifications of location factors and of industrial regions (Estall and Buchanan 1980). By and large, however, these classifications were seen more as the summary outcome of idiographic studies than as a basis for theorizing about location behaviour and regional industrial change.

For a variety of reasons, during geography's watershed years (and since), idiographic approaches were widely criticized; indeed Warren's (1970) book, if not *the* last, is one of the last substantive studies in industrial geography in this tradition. The criticisms are profound (see Johnston 1987). Suffice to say that, preoccupied with uniqueness, many idiographic studies seemed overwhelmed by detail, were deemed to be overly descriptive and ignored, or at the very least, underestimated broader economic, political and social processes which in important ways encourage tendencies towards similarities, as well as differences, between places. Idiographic studies did not offer clear yardsticks by which to judge what is important for explanation (say, regarding industrial location) and the lack of theories created growing difficulties in articulating ideas with other disciplines, participating in general (but not local!) policy debates and communicating skills and critical thinking among students. Indeed, as industrial geography's information base has exploded the problems underlying an idiographic approach have mounted. After all, theories and models seek to present complex realities in ways that can be 'readily' or at least parsimoniously understood and critically evaluated. From this perspective, theories, models and frameworks seek to emphasize 'important' causal relationships and processes and in so doing provide ways of looking for, structuring and evaluating evidence. Moreover, theories make explicit the values by which evidence is assessed.

Given that what is theoretically important is judgemental and value based, several theoretical perspectives have emerged in industrial geography over the last 30 years or so. Just how to classify these perspectives is open to some debate. The most common starting point is to distinguish neoclassical, behavioural and structuralist theories and to further characterize neoclassical and behavioural as 'conventional' or mainstream and the latter as 'radical' (Lever 1985; Watts 1987; Healey and Ilberry 1990). Yet this trichotomony of approaches omits reference to the geography of enterprise, a major theme of industrial geography since the mid-1960s, whose enduring legacy has been to explicitly incorporate organization, especially (large) business organizations, as a key dimension of industrial location change (Krumme 1969b). Indeed, Storper (1981) argues that industrial geography's radical tradition comprises a structuralist view and what he terms an open system view which overlaps substantially with the geography of enterprise. In addition, over the past decade the influence of structuralism has waned and has been modified, or perhaps replaced, by such radical approaches as critical realism and regulation theory which have gained currency in industrial geography.

The following overview of theoretical approaches makes the primary distinction between conventional and radical theoretical approaches in which the geography of enterprise is placed in the latter category. Four prefatory remarks need to be made. First, any classification of theoretical approaches to industrial geography is judgemental. For example, while Storper (1981) sees open systems/enterprise geography approaches as radical, Walker (1989) does not. Second, distinctions in theory are often less obvious in the practice of industrial geography. Indeed, alternative approaches have become increasingly blurred as concepts, terms and methods once associated with one approach are exchanged, borrowed and modified by other approaches. Third, theorizing is a largely derivative exercise in geography, and in industrial geography the most important source of conventional and radical theories has been economics (although in recent years sociology and anthropology have become more important). Fourth, only an introduction to alternative (theoretical) approaches is attempted here to provide some general perspective on the recent evolution of industrial geography and on how this book is organized.

Conventional theoretical approaches

Conventional theoretical approaches to industrial geography are rooted in neoclassical models and theories (Smith 1966, 1971). These theories are conventional in at least two general senses. First, they represent established doctrine within economics and form the basis for the first wave of location and regional development theories in industrial geography. Second, they are conventional by providing widely prevalent rationales (and accepted yardsticks of performance) for the economic operation of capitalist societies. In neoclassical thinking, competition is the great regulator of economic behaviour that maximizes individual efficiency and social welfare in the long run. If conventional now, however, it is worth noting that neoclassical theories evolved from classical theory pioneered by Adam Smith (1776) at the end of the 18th century, then a truly radical idea suggesting that free markets, rather than custom, tradition and manipulation, should govern economic behaviour.

Several general characteristics of the neoclassical explanation of industrial location can be noted here before its fuller discussion in Chapter 5 (Smith 1971). First, it focuses solely on 'economic' variables (transportation costs, labour costs, etc.) with history, political economy and social processes ignored or interpreted as 'complications' to the basic economic forces. Second, neoclassical location theory analyses economic factors in an abstract, deductive manner to derive generalizations as to where industry should locate. The theory so derived provides a 'normative' yardstick to compare with actual behaviour (and specific hypotheses by which to empirically examine theory). Third, neoclassical models assume 'universal' economic laws, based on 'universal' notions of rationality, that govern behaviour. In other words, it is the iron laws of economics that govern behaviour, rather than the idiosyncrasies of individual agents. In neoclassical perspectives where 'free' competition prevails, the characteristics of individual firms ('agents') are essentially irrelevant as the competitive process itself sorts out the most efficient (or profitable) location pattern. That is, whether production and markets are organized by small firms or giant firms ultimately depends on (universal) rational notions of efficiency (Williamson 1975).

In summary, neoclassical theory constituted a completely different (and in this sense radical!) form of explanation of industrial location than that offered by idiographic studies and the antithesis of the idiographic view that places are unique, only to be understood in terms of their own history. In practice, neoclassical theory encouraged a 'positive' approach to explanation, in which theory provided a basis for specific hypotheses that searched (tested) for generalized statistical relationships among representative samples of factories which linked industrial location to measurable economic variables (e.g. Stafford 1960; Watts 1971; Auty, 1975).

Within industrial geography, neoclassical location theory was quickly criticized from a behavioural perspective (Pred 1967, 1969). The behaviouralists emphasize that in the real world, decision-makers differ in terms of goals, preferences, knowledge, abilities and rationality. In an important sense, behavioural models may be seen as attempts to make neoclassical theories more realistic by incorporating issues of location preference and organizational structure in explanations of industrial location patterns. Thus, the behaviouralists sought to develop theories of the (location) decision-making process to improve understanding of location choices, especially in advanced societies where there are many location options even within the economic constraints imposed by competition. In this regard, behaviouralists recognize that location behaviour varied considerably between large and small firms and so helped reinforce industrial organization as an important variable in explanations of industrial location. If behavioural and neoclassical theories are complementary they are nevertheless inspired by different approaches to explanation. In particular, while the starting point for neoclassical location theory takes the form of spatial variations in economic costs and revenues, that for behavioural studies is the spatially biased information horizons of decision-makers.

Radical theoretical approaches

Neoclassical and behavioural theories overwhelmed the idiographic tradition and quickly formed the basis of a new pedagogy in economic geography. Indeed, these theories became the 'conventional' theories of the subject. With gathering momentum, however, the conventional approaches themselves have been subject to considerable criticism by so-called radical theories (Barnes 1987). Within industrial geography, radical approaches were initially associated with the geography of enterprise (Krumme 1969a; Hayter and Watts 1983), and, with greater force, by structuralist theories of industrial location (Massey 1979; Storper 1981). Radical theories drew from institutional and Marxian economics respectively and are considered as 'radical' primarily because they offer different 'world views' of the capitalist process compared to neoclassicism. Thus, in contrast to mainstream neoclassical thinking, radical theories explicitly recognize contradictory tendencies within capitalism in which competitive processes do not automatically guarantee socially desirable outcomes. Whereas in neoclassical theories, competition generates stable and fair outcomes, in radical theories competition, unless regulated, generates unstable and unfair outcomes. Economic growth both generates great wealth and is crisis ridden.

It might also be noted that radicals are often critical of neoclassical theory's reliance on research methods which favour a 'linear' way of thinking (statement of theory, formulation of hypotheses, the collection of data, the testing of hypotheses, and the re-evaluation of theory) and for statistical explanations of industrial location patterns which narrowly focus on measurable economic variables, isolate location from underlying ('deeper') processes and fail to assess the influence of contingent circumstances (see Sayer's 1982a reply to Keeble 1980; Sayer and Morgan 1985). Instead, radicals often suggest case studies as a way of revealing the complex and dynamic interplay of local and global, tangible and intangible factors and context which influence industrial locations (Krumme 1969a; Schoenberger 1994). However, as will be noted, it is too facile to allocate statistical and case study research methods respectively to conventional and radical perspectives.

In radical perspectives, economic behaviour, including location behaviour, has to be understood in terms of the forces of political economy. For enterprise geography, the starting point of analysis is the giant multi-product, multi-plant corporation. Such corporations, it is argued, have considerable power which means they have abilities to modify and even manipulate the locations (and markets) in which they operate. In contrast to neoclassical assumptions, corporate giants do not simply passively respond to broader forces of competition but actively seek to control, as far as possible, their own destiny. In this view, multinational corporations in particular enjoy considerable discretion over the location of investment and employment opportunities, and this discretion in turn enhances their bargaining power in relation to local and even national governments. For enterprise geography, industrial location, to be properly understood, had to be seen as part of the strategy and structure of firms.

The structuralist criticism of neoclassical theory went deeper (Bradbury 1979; Massey 1979, 1984; Peet 1983). Drawing explicitly on Marx, structuralists emphasize that economic growth under capitalism is a crisis-ridden, disequilibrium process at the heart of which is capital's *inherent* tendency to exploit labour, that is to pay workers less than the value extracted from their labour, and to create class divisions. Simply put, the interests of capital and labour are diametrically opposed. In the structuralist view, explanations of industrial location have to be placed within the context of broader global forces and broader relations of production in which the key element is the labour process. In taking this view it might be noted that structuralists are dismissive of enterprise geography (Walker 1989). In general terms, structuralists criticize enterprise geography for not articulating with broader theories of development and for its focus on individual corporations which deflects from the structuralist analytical priority, the labour process.

There are also profound criticisms of structuralism. Critics have noted that structuralist explanations are often opaque, lack clear forms of evaluation, and are prone to economic determinism by over-emphasizing macro-economic forces and underplaying the experience and power of individuals and institutions (Duncan and Ley 1982; Taylor 1984; Johnston 1987). Moreover, structuralism's criticism of neoclassical theory that it is essentialist, by its attempt to explain or 'reduce' economic activity to the effect of universal, abstract economic laws of economic rationality (Barnes 1987), can also be levelled at structuralism. Indeed, in structuralist accounts of location patterns and regional industrial change (e.g. Bradbury 1979; Peet 1983), capital and labour (and government) are often treated in a standardized, largely undifferentiated way in which, for example, labour is essentially treated as dupes, capital as relentlessly hyper-mobile and placeless, and states serving only to support capital. The critics of structuralism stress that agencies such as business, labour and governments are not passive players orchestrated by forces totally beyond their control but pursue strategies that can and do make a difference (Walmsley and Lewis 1984). Moreover, these strategies, including the ways in which institutions relate to one another, develop differently in different places. In this regard, Walker's (1989) recent requiem for enterprise geography might have been better directed towards structuralism.

Indeed, recent developments in radical arguments, notably critical realism and regulation theory, are giving much greater stress to the role of agency than previously admitted under structuralism (Sayer 1982b; Lipietz 1986; Storper and Scott 1992; Tickell and Peck 1992; see also Barnes 1987). From this perspective, explanations of industrial location dynamics give priority to the policies of nation states, the structure of labour markets and business strategies and structures, as well as to the relationships between them.

Places and institutions matter   In the old, for the most part forgotten, industrial geography, idiographic studies stressed the role of individuals and organizations (agents) in the creation of areally differentiated landscapes in which individual regions are unique (Warren 1970). But such studies were highly detailed and failed to systematically link regions and industrial locations within a wider world of competition. Neoclassical and structuralist theory rejected the idea of regional uniqueness and the importance of agents and gave priority to this wider world of competition. But such a rejection denies the richness (if not the existence) of economic geography and the role played by agents in creating

differences among regions. As theoretical positions have evolved, however, from within both conventional and radical vantage points there has been widespread acceptance that institutions, not only of business, but also of labour and the state, do matter in the creation of industrial geographies (Peck 1992, 1993). Similarly, it turns out that regions, after all, are unique (Johnston 1984) and geography (still) matters (Massey and Allen 1984; Wolch and Dear 1989).

As Johnston (1984) stresses, however, regions may be unique but they are not singular, by which he means they are not isolated, closed systems but (increasingly) tied to wider forces of globalization. The theoretical challenge is how to conceptualize the interplay of local and global forces. In Barnes' (1987, 1988) view, no one theory can account for the variability and dynamism of industrial geography. Rather, 'There is no Industrial [Economic] Geography, only industrial [economic] geographies' (Barnes 1988: 349). In this view, different 'local' models have to be developed to explain different local situations. If recent traditions are instructive, any pursuit of local models is likely to emphasize, among other considerations such as those of physical geography and relative location, institutional structures and the distinctive ways these structures unfold in particular places.

Places matter because people matter. People in different places have distinct attitudes and beliefs and organize their lives and economies in distinctive ways. Industrialization both threatens and creates these differences. On the one hand, the spread of industrialization contains powerful tendencies to standardize forces of demand and supply. On the other hand, industrialization is a geographically uneven process and its geographical spread is continually modified in light of local circumstances. The development and diffusion of industrialization is intimately shaped by the actions, policies and institutions of people.

This book's structure

Ultimately, industrial location patterns and regional industrial change needs to be under-

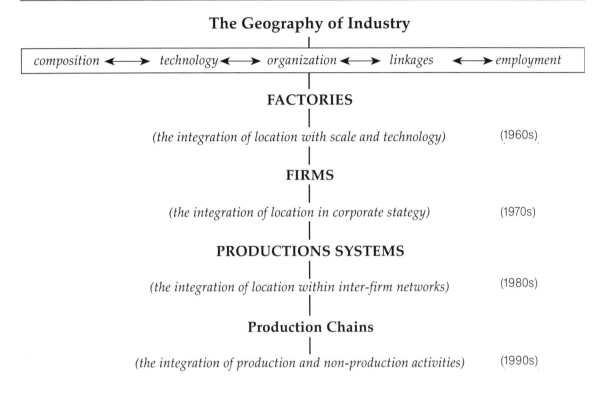

Figure 1.1   Industrial Geography: The Main Units of Investigation

stood in terms of prevailing global forces and how economic agents respond to, modify and even lead such forces within particular places. Within industrial geography, the business firm is widely recognized as the main 'agent' of change. While enterprise geography first gave explicit reference to the role of the firm, especially in the form of the large corporation, analyses of the locational implications of corporate strategies and structures, and more generally of the relationships between forms of industrial organization and regional development, are now widely incorporated throughout the subject and in both conventional and radical traditions (e.g. Watts 1980b, 1981; Sayer 1986; Scott and Storper 1986; Schoenberger 1987, 1994; LeHeron, 1990; Clark 1993; de Smidt and Wever 1990). In recent years, there has been rapidly growing interest in the geography of labour market institutions (Clark 1989; Peck 1996) and this geography is not neglected in this book. Nevertheless, within

capitalist societies, it is the individual firm that is the principal decision-making unit and the principal dynamic of change. The substantive core of this book largely 'sees' industrial location and industrial location change from the perspective of the organization.

The remainder of Part I provides the broad context for analysis of industrial location dynamics by reviewing the broad temporal (Chapter 2) and geographical dimensions (Chapter 3) of the industrialization process since the Industrial Revolution. In this discussion, emphasis is given to institutional as well as to technological change. Subsequently, Parts II, III and IV provide an interpretation of industrial geography progressively based on theories and principles which have been developed to explain the location of factories, the geography of firms and the geographical structure of production systems. The factory, the firm and the production system have provided three principal units of investigation for much conceptual thinking and empirical work in manufacturing or industrial geography over the last several decades (Figure 1.1). Each unit has a strong rationale in theory and practice and each progressively complements the other, while it is the firm that provides the organizational and decision-making framework in which to understand the location of factories and the location structure of production systems

First, factories, commonly referred to as plants or establishments, are discrete physical structures or buildings in which manufacturing or production actually occurs. While manufacturing continues to take place in households, global production is dominated by factories in industrialized and especially developed countries. Factories are specific production and work places which exist at specific sites as part of the built environment of communities. Second, manufacturing firms, commonly referred to as enterprises or corporations, are decision-making and legal entities which own, control and organize the manufacturing activities that occur in factories (and sometimes in households). Firms may comprise a single factory and so constitute single-plant firms. Firms can also comprise two or more plants and so constitute multi-plant firms. If firms control plants in more than one country they are international, multinational or transnational firms. In practice, the pattern of ownership and control among firms is extremely complex and the location of firms cannot be pinpointed in the same way as factories. Third, production systems comprise networks of firms that are linked to one another through flows of goods and services in order to manufacture a particular product such as a car or computer.

It might be noted that the factory–firm–production system organization is reflected in the chronology of industrial geography. The initial focus of neoclassical theorizing, for example, was the individual factory; enterprise geography then focused on the firm and in recent years, production systems have been the centre of attention. Most recently, there have been several studies urging for a greater consideration of 'production, commodity or value-added chains'

(Gereffi 1989; Porter 1990; Storper 1992: Dicken 1994; Gereffi and Korzeniewicz 1990, 1994). While closely related to the idea of production systems, value-added chains appear to be more explicitly concerned with networks of firms that include producer services, retail, wholesale and financial services, as well as manufacturing functions, and which are international in scope (Edgington and Hayter in press). Since the idea of value-added chains is still at an embryonic stage and because the basic conceptual issues of their organization are raised in the more established discussion of production systems, in this book discussion is restricted to production systems.

Research designs in industrial geography Research designs (i.e. the way in which empirical information is collected and analysed) vary considerably in the practice of industrial geography. As a way of illustrating these differences, useful, closely related distinctions are made between extensive and intensive research methods (Sayer and Morgan 1985) and standardized and non-standardized interviewing techniques (Healey and Rawlinson 1993). Brief comments on these approaches serve as an introduction to the main ways industrial geographers have conducted empirical research and thereby to the main sources of information for this book.

Extensive research methods principally involve obtaining highly structured information from a large sample of respondents, such as firms which are chosen in a way that ensures, as far as possible, the sample is representative of the population from which it is chosen. Highly structured information on a sample of firms may be obtained from standardized interviews that are based on questionnaires which as far as possible emphasize factual and pre-coded answers. Randomness is typically an important design feature of extensive surveys in order to ensure that samples are (statistically) representative, which in turn allows the data so collected to be analysed by a wide range of statistical techniques. Samples may be stratified in some way (e.g. by firm size or ownership status) in order to facilitate comparisons (and remove complications). Samples may also be chosen from one particular type of industry ('industry surveys') or from across several industries ('cross-sectional surveys') and from one or more regions. If the data collected are for one point in time they are labelled 'static' surveys; if for two points in time they are labelled 'comparative static', and if for many, sequential points in time, the surveys are said to be 'longitudinal.' Extensive surveys, it might be noted, may also draw on the structured information provided in government censuses, which may be based on information derived from an entire population of firms or from a sample of firms.

The major empirical advantages of extensive surveys are twofold. First, they generate statistically valid general characteristics and trends which are representative of the population as a whole. Second, they generate a consistently collected database which can be analysed by a battery of statistical techniques to test for hypotheses; for example, regarding factors that are purported to explain the location pattern of a particular industry. Indeed, questionnaires ('survey instruments') are normally structured to collect information to test specific hypotheses which are directing the research. Typically, extensive research is associated with quantitative or statistical explanations which test for 'order', or 'generalizations' in location patterns and behaviour. Since the shift from idiographic to nomothetic approaches, extensive research designs continue to be extremely important to the practice of industrial geography and in recent years illustrative examples include O'Farrell and Hitchens (1988), Peck (1985), and Keeble and Walker (1994).

Intensive research designs principally involve in-depth analysis of individual cases and primarily comprise non-standardized interviews of one or more representatives of individual firms. Case studies can be selected for various reasons but often because in some important way they are 'distinctive'. Thus individual firms may be chosen because they are the biggest and most powerful firm in an industry, or because they represent some kind(s) of leading edge or best practice behaviour, or because they help illuminate particular kinds of processes. In some senses, individual cases may also be thought of as representative but the extent to which this can be argued will inevitably be a matter of judgement. The principal source of information is the non-standardized interview which is conducted in an interactive way with respondents and which explores lines of inquiry in a flexible manner. The kind of information generated by this sort of interview is often 'qualitative', although quantitative data, for example on the costs of operations, may also be provided. It might be noted that newspaper and magazine interviews provide second-hand sources of case study interviews.

The empirical strength of intensive research designs, and related case study interviews, first, lies in revealing insights into complex processes which evolve over time within changing internal and external constraints (Healey and Rawlinson 1993; Schoenberger 1991, 1992; McDowell 1992; Markusen 1994). Second, they can document distinctive and unusually important processes which cannot be effectively subjected to extensive research and can illuminate arguments, reasons and debates over alternatives that would otherwise remain uncovered. Although not as prevalent as extensive research, non-standardized surveys have a long lineage in industrial geography. Idiographic approaches, including Warren (1970), frequently relied on such surveys, while the emergence of the geography of enterprise and behavioural geography signalled a growing commitment to these methods (Krumme 1969a; Stafford 1974; Hayter 1976). Since then there have been a growing number of corporate case studies and some recent examples include Clark (1993), Patchell (1993b) and Schoenberger (1994).

It is tempting to suggest that extensive and intensive research designs can be classified according to theoretical perspective (Schoenberger 1991; Healey and Rawlinson 1993). There is some substance to this suggestion in that conventional theorizing, especially from a neoclassical perspective, is closely allied to a positive methodology and the quantitative testing of precisely stated hypotheses. Moreover, structuralists were strongly critical of such approaches and suggested that statistical models provided only description and did not address the deeper causes of behaviour which are rooted in the capitalist system itself (Massey 1984). On closer inspection, however, it is extremely difficult, possibly misleading, to simply classify research methods by theoretical perspective. The methods themselves are not mutually exclusive (or inherently ideological). After all, the strengths and weaknesses of extensive and intensive research are largely mirror images and, where resources have permitted, both approaches have been effectively combined. A recent study of the geography of labour markets combined an extremely large-scale survey (of over 700 firms) with case studies of individual plants (Hayter and Barnes 1992; Barnes and Hayter 1992; see also Barnes et al 1990). More generally, there is much conventional theorizing that has employed case studies: the Harvard Business School was an important pioneer in this respect. Similarly, some radical analyses have been rigorously quantitative (e.g. Webber and Rigby 1986; Rigby 1991). Finally, it might be noted that there has been much discussion of the pros and cons of alternative research designs from within particular theoretical perspectives.

It is a fact that industrial geography is an empirically grounded subject which has traditionally employed a variety of research designs. This tendency is likely to continue. What does need emphasizing is that data and forms of data analysis do not speak for themselves. Rather, information needs to be evaluated from particular conceptual perspectives. This book reviews a large number of such perspectives, models or frameworks that provide this kind of evaluation.

Manufacturing industry

In the English language literature, the term industry refers to all economic activities such as the mining industry, the transportation industry, the hotel industry and the retail industry, as well as manufacturing industry which is sometimes referred to as secondary industry. It is also common within the English language literature to interpret industry as manufacturing! Within (English language) geography, industry and manufacturing are often used as synonyms; textbooks which focus on manufacturing, for example, have frequently used 'industrial geography' or 'industrial change' within their titles (e.g. Estall and Buchanan 1980; Watts 1987). This book follows this somewhat ambiguous tradition! As a further semantic matter, it might be noted that the term 'sector' is often used as a synonym for industry, especially when the latter refers to industries of a somewhat aggregate nature, e.g. the manufacturing sector.

At the core of what is meant by manufacturing (or secondary) industry are those activities in which raw materials or already manufactured materials are fabricated, assembled, processed or transformed by mechanical, electrical or chemical means into more valuable products. Actual manufacturing activities occur in a bewildering variety of forms. These include spinning, weaving, stitching, knitting, turning, plating, machining, moulding, forging, hammering, screwing, stamping, pressing, bolting, installing, cutting, edging, heating, melting, mashing, soldering, sorting, polishing, cooking, drying, blowing, packaging, inserting, painting and welding. Over time, these and other activities in a wide variety of manufacturing industries have been increasingly performed by machines of one kind or another. However performed, manufacturing activities also have to be designed, controlled, co-ordinated, financed, watched, checked and protected, inputs purchased and outputs sold and distributed. All of these activities, in one way or another, add value and, in essence, manufacturing is a value-adding process.

Manufacturing activities are classified in a variety of ways. For example, crude dichotomies are made between primary manufacturing (activities which manufacture inputs from the primary

Table 1.1 Canada: Employment by Major Sector, 1971 and 1991

Sector	Employment (000s)	
	1971	1991
Agriculture	514	448
Other primary	221	280
Manufacturing	1776	1865
Construction	489	695
Transportation	707	916
Trade	1335	2169
Finance, insurance and real estate	399	760
Service	2128	4376
Public administration	545	832
Total	8114	11 392

Source: Canada Yearbook, Ministry of Industry, Science and Technology, Ottawa: *Statistics Canada* (1993: 208).

sector) and secondary manufacturing (activities which manufacture already manufactured components); durable (e.g. fridges, cars) and non-durable goods (e.g. food); consumer goods (purchased by households) and capital goods (machinery and equipment purchased by firms to manufacture other goods); high and low tech industries (research and development employees and budgets are important in the former but not in the latter); and heavy goods (e.g. iron and steel) and light goods (e.g. electronics). In addition, for census purposes, most countries have developed more detailed classes of manufacturing activity as part of 'standard industrial classifications' (SICs). Indeed, it is national SICs that provide the principal source of aggregate data on industries in general and manufacturing industry in particular.

In the case of Canada, for example, the government publishes data on nine major sectors of the economy (Table 1.1). According to these data, employment in manufacturing in 1971 accounted for 21.9% of the workforce and, in terms of total number of jobs, was the second most important sector after services. In 1991, the

employment level in manufacturing was greater than in 1971 but its share had fallen to 16.4% of the workforce, and (in terms of total number of jobs) it was the third most important sector behind service and trade. To some extent, however, how jobs (or some other variable) are allocated among sectors is arbitrary. For example, independent lawyers whose services are contracted by manufacturing firms are in the service sector while lawyers on the staff of manufacturing firms to provide the same legal services are part of manufacturing industry. This point is not unimportant at the present time. Downsizing by large corporations, for example, often involves contracting out services, ranging from computing and legal services to cleaning, security and maintenance services, which were formerly provided internally.

Even more fundamentally, it is important to appreciate the interdependent nature of the entire economy. Manufacturing industry does not exist in isolation but is closely integrated with several other industries, notably the primary, construction, utility, wholesale trade and transportation industries, while important links exist with business service industries. Indeed, there are usually whole departments of governments whose primary function is to service and regulate manufacturing and primary activities. In this regard, the so-called 'goods producing sector' includes all manufacturing activities plus activities in other industries which provide inputs and services to the production and distribution of material goods. Manufacturing industries are at the heart of the goods producing sector, which continues to dominate the employment base of advanced societies. According to Britton and Gilmour (1978: 71), for example, in 1971 about 65% of the Canadian (and the US economy) could be considered as part of the goods producing sector.

The manufacturing sector is itself a highly aggregated mix of activities which can be further broken down into separate industrial categories. SIC schemes vary among countries and in the Canadian case, for example, in 1993 the manufacturing sector was broken down or 'disaggregated' into 22 two-digit industries (Table 1.2). Although not always the case, the idea behind this classification scheme is that activities within an industry share similar characteristics (e.g. in terms of markets, technology and/or inputs). Each of these two-digit industries can be further disaggregated into three- and four-digit industries (Table 1.3). For example, the wood industries comprise six three-digit industries including sawmills, plywood mills, shingle and shake mills and miscellaneous activities, and several of these industries are further disaggregated. The electrical and electronics products industries are broken down into even more categories.

The Canadian census, as do other censuses, provides various kinds of information on individual manufacturing, including value added, number of establishments, and employment (Tables 1.2 and 1.3). In fact, the Canadian census also breaks down employment into production and administrative employees, and male and female, as well as documenting information on total wages, cost of materials and supplies, cost of energy, and value of shipments. These variables provide aggregate indicators of the size and characteristics of manufacturing industries which are useful in comparisons over time, space and between industries. In practice, geographical studies of manufacturing activity often focus on employment change, and numerous statistical measures have been devised to describe and summarize the extent to which local employment is specialized, diversified and participates in national trends. Some of the better known of these descriptive statistics are summarized in Appendix 1.

Conclusion

Industrial geography is a dynamic sub-discipline which has pursued a variety of theoretical

Table 1.2   Canada: Selected Characteristics of Manufacturing Industries, 1993

Industry	No. of establishments	Total employment	Value added ($M)
1. Food	3 008	189 499	15 898
2. Beverages	194	26 602	4 209
3. Tobacco	17	4 778	1 221
4. Rubber	173	22 964	2 142
5. Plastic	1 153	50 410	3 080
6. Leather and allied	230	12 818	495
7. Primary textiles	184	18 346	1 377
8. Textile products	744	27 646	1 374
9. Clothing	1 921	82 737	3 131
10. Wood	2 894	109 961	8 344
11. Furniture and fixtures	1 331	44 654	2 139
12. Paper and allied	664	101 926	8 081
13. Printing and publishing	4 655	124 867	8 505
14. Primary metal	409	84 416	7 770
15. Fabricated metal	5 117	132 606	7 906
16. Machinery	1 855	74 379	5 441
17. Transportation	1 349	209 879	21 139
18. Electrical and electronic products	1 365	118 629	9 602
19. Non-metallic mineral products	1 519	42 661	3 401
20. Refined petroleum and coal	157	14 084	2 368
21. Chemical	1 248	90 490	12 317
22. Other	2 756	63 050	3 842
Total	32 943	1 647 432	133 789

Source: *Statistics Canada* cat 31–203.

Table 1.3 Canada: Total Employment in the Wood Industries and the Electrical and Electronic Products Industries

Electrical/electronics	Total employment	Wood	Total employment
1. Small electrical appliances	2 785	1. Sawmills, planning and shingle mills	58 043
2. Major electrical appliances	6 266	– shingle and shake units	1 954
		– sawmills, planning mills	56 091
3. Electrical lighting	5 767		
– lighting fixtures	3 701	2. Veneer and plywood mills	8 099
– lamps and shades	1 013	– hardwood veneer and plywood	3 593
– bulbs and tubes	1 053	– softwood veneer and plywood	4 506
4. Record players, TVs, radios	1 433	3. Sash, door and other millwork	32 545
		– prefabricated wooden buildings	2 522
5. Electronic equipment	53 333	– kitchen cabinets, bathroom vanities	9 977
– telecommunications	18 704		
– parts and components	13 065	– wooden doors and windows	10 842
– other	21 564	– other millwork	9 174
6. Office and business machinery	15 427	4. Wooden box and pallet industry	2 390
– computers and peripherals	10 983		
– other	4 444	5. Coffin and casket industry	725
7. Electrical industrial equipment	191 187	6. Other wood industry	8 175
– transformers	4 779	– wood preservation	1 201
– switch gear	7 117	– particle board	1 867
– other	7 291	– waferboard	2 092
		– others	2 997
8. Communications wire and cable	7 361		
9. Other electrical products	7 070	Total	109 961
– batteries	1 441		
– non-current carrying devices	1 293		
– other	4 336		
Total	118 629		

Source: *Statistics Canada* 31–203.

approaches and implemented a variety of research designs. While there is much evidence of cross-fertilization between approaches, there remains a tension between explanations which stress universal tendencies in location behaviour and those which assign greater priority to local context and contingency. From both starting points, however, there is increasing interest in the organization of industrial activity. This interest is reflected throughout this book.

2

MANUFACTURING CHANGE IN HISTORICAL PERSPECTIVE

A plethora of manufacturing industries developed in many parts of the world prior to the 18th century. These industries include branches of metal and textile manufacture, food and beverage and wood processing as well as papermaking, printing, glass-making, candlestick-making and numerous others. Traditional industries existed largely as small-scale labour-intensive activites by workers who typically owned their own tools. Traditional industries frequently enjoyed strong, complementary links with agriculture and were widely dispersed among rural areas and towns of all sizes. Traditional industry could be found in households and workshops supplied by water power. Prior to the 18th century, there were a few factory-like operations while waged labour, and industrial slavery, existed. In the late 18th century, however, the Industrial Revolution and the factory system set in motion unprecedented changes in the nature, scale and growth rates of manufacturing activities. These changes have scarcely abated since.

This chapter summarizes the evolution of manufacturing industry from a broad historical perspective. In particular, the chapter traces the evolving nature of manufacturing industry before, but particularly since, the Industrial Revolution. Prior to the Industrial Revolution, traditional industries did experience change,

albeit slowly, and there were alternative forms of organization. Since the Industrial Revolution, the dynamism of manufacturing has become a defining characteristic; the structure of manufacturing has become increasingly complex; and the industrialization process has become one of 'creative destruction' of global magnitude.

The evolution of manufacturing industry

The evolution of manufacturing industry over historical time is often summarily expressed in the form of 'stages of development' models. Berg (1985) reviews two such 'models of manufacture' relevant to British and West European experience in which industrial capitalism developed indigenously in long-established societies. In particular, she distinguishes a Marxian model of 'primitive accumulation and manufactures' and a 'proto-industrial' model (Figure 2.1). Both models recognize the important early connections between agriculture and industry, the role played by commercial interests and the dispersed growth of handicraft production in rural cottages and urban workshops. Both models also portray a linear sequence of manufacturing evolution. Thus, the primitive accumulation model emphasizes a three-stage transition featuring handicraft

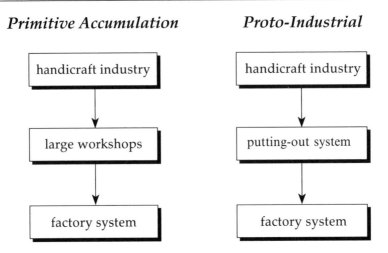

Figure 2.1   The Evolution of Industry: Two Stages of Development Models. Source: based on Berg (1985)

industry, manufacturing in large workshops, and the factory system and full-scale industrialization (Berg 1985: 70–77). In this scheme, the large workshop, located particularly but not exclusively in towns, plays a key role in promoting the division of labour, separating workers from the means of production and setting the stage for large-scale industrial concentration in factories which further extend the accumulation process and labour exploitation by virtue of their size, more sophisticated machinery and by removing worker control over working conditions.

In the case of the proto-industrial model, the crucial second stage in the linear sequence features the 'putting-out' system rather than the large workshop as the dominant way of organizing manufacture prior to the factory system (Berg 1985: 77–86). In this model, rural putting-out systems pave the way for large-scale industrialization, not by refining a division of labour (rural households might just as well develop polyvalent skills as specialized ones), but by the capacity of rural areas to increase levels of production (not

necessarily productivity) at competitive prices. Thus the advantages of putting-out systems to serve growing demands for manufactured goods from the late 16th century related to their relative freedom from guild restrictions; access to labour which was cheap and, with agricultural change, increasingly plentiful; reliance on dispersed rural workers with a tradition of low wages and who faced difficulties in forming unions; reliance on rural workers who had access to agricultural work and to subsistence levels of food; and domination by merchants who controlled market access, wage levels, the materials required in production, and who accumulated the profits and capital which in turn provided funds for industrialization.

In reality, as Berg (1985) notes, pathways towards industrialization are more complex than anticipated by either of these two models. Apart from workshops and putting-out systems, traditional industry was organized according to artisan and co-operative principles, and different forms of organization existed side by side or within the same industry in different places. Moreover, if the large factory has become the dominant unit of organization since the Industrial Revolution, the small firm, workshop, putting-out systems and artisan production have never been relegated to the pages of history. As Piore and Sabel (1984) argue, industrialization is not a simple linear process and the "pathways to industrialization and regional development" remain varied (Storper and Scott 1992).

Traditional manufacturing industry

Traditional manufacturing industries developed as 'handicrafts'. Indeed, the Latin roots of the word 'manufacturing' imply 'making by hand' and traditional manufacturing processes were intimately and fundamentally associated with sentient labour skills. While the precise origins of many manufacturing activities cannot be stated categorically, the diffusion of particular manufacturing skills from regions of origin to other parts of the globe often required centuries. Over historical time, changes in organization and improvements in methods of production were made to traditional industries, but such developments tended to be sporadic and limited in impact.

The evolution of paper-making   A brief history of paper-making illustrates several key aspects of the diffusion of a traditional industry (Table 2.1). The development of calligraphy and the adoption of the animal-hair brush around 250 BC created a need for a writing material that was cheaper and more practical than silk (Library of Congress 1968). Although not entirely clear, it is thought that Ts'ai Lun, an official in the Emperor of China's court, invented the process of paper-making, using old rags, hemp, tree bark and fish nets. He related his findings to the Emperor in AD 105 who became Patron of Paper-making, and paper-making gradually spread throughout China over the next several hundred years. Subsequently, in association with the travels of monks, merchants, conquering armies, explorers,

the sponsorship of kings and princes, and the kidnapping of skilled artisans, paper-making slowly spread to Japan and Korea before AD 610, various parts of the Middle East around AD 750–900, North Africa (AD 1100), Spain (AD 1150), various parts of western Europe (AD 1190–1586) and from Europe to Mexico in 1580 and the US in 1690 (Library of Congress 1968). During this slow evolution, experiments were made with different materials and, for example, the Spanish introduced a water-powered 'stamping' mill (to reduce raw materials, mainly cloth rags, to fibres) in the 12th century. About 500 years later, the Dutch developed the Hollander Beater which required less power and was four times faster than the stamping mill.

After 2000 years of development, paper-making in 1800 existed as a workshop-based activity located adjacent to streams for water power and close to towns which provided supplies of rags and markets. At this time, the workshops varied in size although most were small and paper-making remained a labour-intensive process, more art than science (Roberge 1972). Within just 100 years, however, as part of the Industrial Revolution, paper-making was transformed into a capital-intensive activity organized in large factories by radical technological innovations in the paper machine and in wood fibre-based pulping processes. In this industry, as in others, the process of creative destruction had begun. In geographical terms, as pulp and paper manufacture migrated to remote rural locations, specifically those accessible to the coniferous forests of North America, northern Europe and Russia, craft forms of production declined and paper-making in metropolitan centres became more specialized (Hunter 1955). Moreover, the Industrial Revolution established technological change as a continuous process in the pulp and paper industry. While numerous small-scale ('incremental') innovations have collectively significantly enhanced productivity (Cohen 1984; Davie 1984), occasionally there have been radical innovations including, for example, in wood-pulping technologies and paper-machine processes (Ofori-Amoah 1993,

Table 2.1   The Spread of Traditional Paper-Making

Location	Year	Comment
China	105	Ts'ai Lun, an official of Emperor Ho-Ti's court credited with inventing paper-making. Oldest archaeological discovery: AD 109.
Korea	600?	Probably transferred by Buddhist monks.
Japan	610?	Probably transferred by Buddhist monks. Considerable experimentation. Hand paper-making still important.
Samarkand	750	Two Chinese paper-makers captured by Arabs in war and taken to Samarkand.
Baghdad	795	Chinese paper-makers brought by Harun al Rashid to start second factory. Spread of paper-making through Arab world including Cairo by 1040.
Spain	1056–1151	Introduced by the conquering army of Moors. Spain exporting paper by 1150.
Italy	1255	Introduced by Moors. Soon became exporter.
France	1326–38	A legend claims a Frenchman, who was captured by Saracens in Second Crusade and worked in a Damascas paper mill, returned with the skill in 1150s.
Germany	1320	
Netherlands	1428	Paper-making did not become important until after 1586 when war with France cut off French supplies of paper. Hollander Beater invented in Amsterdam in late 17th century.
Switzerland	1450?	
England	1550s	After failure of a mill established in 1490s, Bishop of Ely sponsored Spanish paper-makers and Royal Court sponsored German paper-makers.
Mexico	1580	Introduced by Spanish.
US	1688	Introduced by a German immigrant, William Rittenhouse, who built first mill in Pennsylvania; second and third mills built in 1710 and 1728.

Source: Library of Congress (1968), Studley (1977) and Hills (1988).

1995). In turn, the new wood-pulping technologies have permitted the dispersal of the pulp and paper industry to tropical and subtropical forest regions, to access hardwoods, and its relocation to urban areas, to access recycled paper. In turn, pulp and paper production in the 'old' coniferous regions has been forced to restructure or disappear (Mather 1990; Marchak 1995)

Alternative forms of organization: the English wool industry   Many traditional industries were embedded in agricultural regions and exhibited different forms of organization to paper-making. An example is provided by the English woollen industry, England's most important industry, its chief source of wealth prior to the Industrial Revolution, and found in every region of the country. Forms of organization, however, varied considerably, including between the two leading producing regions, namely the West Riding of Yorkshire, where the domestic system dominated, and the West Country (south-west England), where the putting-out system dominated. There were some workshops in both regions. The two regions also experienced different transitions with the factory system (Table 2.2).

In the domestic system of the West Riding, spinning, weaving and carding were organized among many small land-holding families: women and children did most of the spinning and men the weaving; their time was further divided between manufacture and farming which principally involved cattle, pigs and chickens. These

Table 2.2 The Organization of Traditional Industry: The English Woollen Industry

Characteristic	Domestic system	Putting-out system
Dominant region	West Riding of Yorkshire	West Country
Principals	Independent manufacturers, small land owners	Merchants
Control of materials	Manufacturers and land-owners	Merchants who distribute (put-out) materials to home workers
Control of workshops	Manufacturers and workers	Workers originally but increasingly merchants
Agricultural connection	Part-time, some livestock, gardening	Farming important and increasingly so
Division of labour	Within family, live-in workers and community	Within family and community
Exchange	Exchange relations controlled by manufacturers	Relations controlled by merchants
Class distinction	Weak; live-in workers can start their own business	Strong; growing sense of alienation
Marketing	Manufacturers sell their own cloth in markets and to merchants	Merchants control all materials
Impact of Industrial Revolution	Co-operative mills set up to house machinery until 1850s when factory system dominates	Factory system imposed in face of strong worker resistance. Industry fails

Source: adapted from text discussion in Berg (1985) and Mantoux (1966).

artisan manufacturers bought and owned cloth and equipment while the journeymen that were hired typically received free board and lodgings in the master's house plus a wage. Such workers often stayed with the same master for a long time, occasionally leaving to create their own business. There were limited class differences between masters and men, and a strong sense of community spirit and values existed. As production exceeded local needs the cloth would be taken to market to be sold to a merchant who might also "meet the expenses of certain minor details of manufacture" such as dying and finishing (Mantoux 1966: 62). It was this kind of domestic system that underlay subsequent romanticizing of traditional industry in terms of health, stability, family and spiritual values – characteristics that were contrasted to the dark, satanic mills of the factory system.

In contrast to the domestic system, which was dominated by small independent producers, the putting-out system in the woollen industry was organized by merchant manufacturers. Merchant manufacturers, who became important in the West Country at an early stage in the industry's development, owned the wool, cloth, and ultimately even the equipment, as well as being responsible for marketing. The merchants 'put-out' the wool to workers who specialized in a particular process for a wage from the merchants. These workers typically relied more on farming than the Yorkshire artisans, either as land-owners or increasingly as farm labourers. Indeed, as harvests failed or agricultural wages were reduced these workers were often forced to borrow from the merchants, using their equipment, such as a loom, as collateral. Consequently, over time, merchants gained control over the entire production process, which in turn encouraged class alienation between capitalists and workers. As workers became more dependent on the merchant their wage levels became more vulnerable, a situation particularly acute for town dwellers. Not surprisingly, worker complaints and violent disputes were more common in the West Country than in the West Riding prior to the Industrial Revolution (Mantoux 1966; Berg 1985).

During the 18th and 19th centuries the evolution of the woollen industry in the two regions varied in a manner reflecting the general and rapid shift of industrial activity in Britain to the North. Bearing in mind that wool remained the country's dominant export during the 18th century, a massive switch in the industry occurred in this period as Yorkshire increased its share of the country's output from 20% in 1700 to 60% in 1800 (Berg 1985: 125). Moreover, the industry's evolution at this critical time is neither consistent with the primitive accumulation model, since workshops were not a dominating feature, nor the proto-industrial model, since the region (the West Country) where the putting-out system dominated did not become the centre of factory production. While some large factories were established in the West Country, these too failed. In contrast, in Yorkshire, the growing centralization of woollen production in factories was achieved by both artisan and capitalist forms of organization. Thus woollen factories in this region were created by artisans in the form of 'co-operative mills' containing machinery each could utilize, a form of organization which lasted until the 1850s, and by merchants who built wholly owned factories employing wage labour (Berg 1985: 222). In this same period, the related worsted industry similarly concentrated in the West Riding of Yorkshire. In its case, however, the process of concentration and transformation into a factory system more closely resembled the proto-industrialization model; the industry had been organized according to putting-out networks, rather than on an artisan basis, and it was the former putting-out merchants who built the new factories.

The Industrial Revolution and the factory system

From the late 18th century onwards, the artisan systems, putting-out networks, workhouses and early factories that comprised traditional industries were increasingly challenged by the factory system, initially in the UK. This challenge was led by the cotton textile industry and Richard Arkwright's new factories in Derbyshire and Lancashire which contained spinning machines known as the water frame, which he patented in 1769 and 1775, even if he did not invent it (Mantoux 1966: 220–270). Arkwright's factories were large, multistoried buildings located on streams to obtain water power; these factories employed 200 to 800 workers, including women and children. They provide the model for the factory system.

Mantoux (1966: 39) quotes an early (1836) definition of the factory system as that which "designates the combined operations of many orders of work people, adult and young, in tending with assiduous skill a series of productive machines, continuously impelled by a central power." More recent definitions of the factory system similarly emphasize the manufacture and concentration of large outputs of standardized products in large buildings housing numerous, mechanically powered machines operated by large numbers of waged employees performing highly specialized tasks under the strict supervision and control of specialized management and the clock (Chapman 1972; Watts 1987: 37). Some of these features were evident in some traditional industries, especially those comprising workshops, 'mills' or early factories which utilized water power to produce relatively standard products with waged labour. After the 1760s, however, the factory system increasingly dominated existing and new industries and the biggest factories operated on an unprecedented scale in terms of size and number of machines. Initially, the factories relied on water power. Subsequently, the innovation of the steam engine by James Watt in 1769 (the first patent date), made practical at Matthew Boulton's Soho works in Birmingham, opened "the final and most decisive stage of the industrial revolution" by providing reliable power at any location (Mantoux 1966: 337).

The factory system did not suddenly replace traditional forms of organization as an inevitable consequence of 'machine' imperatives which required large investments, power sources and centralized production. Traditional and factory systems, in the cotton as well as woollen and other industries, existed side by side for decades. In the case of Arkwright's factories, which contained over 1000 spindles, the water frame had originally been designed as a small machine for home use and operated by hand. Arkwright, however, had the business acumen to realize the possibility of large-scale production, and profit, of the water frame. "It was Arkwright's patent which enclosed the machine within a factory, had it built only to large-scale specifications and henceforth refused the use of it to anyone without a thousand spindle mill" (Berg 1985: 243). By restricting its use to 1000-spindle mills, the water frame only became economic if supplied by water power, and subsequently steam-powered mills. In other branches of cotton manufacture, key innovations such as Hargreaves' spinning jenny (patented 1770) were widely used in domestic and workshop settings. Moreover, the size and structure of the new factories themselves varied considerably. In Manchester, for example, as late as 1841 the average size of cotton mills in primary processing was 260 employees and 25% of the factories employed less than 100, so that "Extremely small firms fitted in besides the giants. Some were single process firms; others combined several processes. Some were multistoried mills with an assembly line type of organization. Others were a combination of shacks and workshops" (Berg 1985: 230). In activities with limited markets, such as the framework knitting industry of the East Midlands, the factory system was not the 'end point' of the Industrial Revolution but a degraded putting-out system featuring sweated labour.

Nevertheless, the factory system represented fundamental changes which were recognized by contemporary observers; none more so than the workers and artisans who engaged in the factory burning and machine breaking activities of the Luddite riots of 1811 in Yorkshire, Lancashire and Nottinghamshire. Notwithstanding such opposition and alternative forms of work organization, the factory system grew in importance. The new factories could manufacture goods in much greater quantities and at significantly lower cost than traditional industry. For capitalists, the motivations to establish factories were rooted in profits, prestige, status and greater control over workforces (Marglin 1983).

For Mantoux (1966: 42), the critical changes comprising the Industrial Revolution in the UK occurred between 1760 and 1820 or 1830. By this time:

"the great technical inventions, including the most important invention of all, the steam engine, had all become practical realities. Many factories were already at work, which apart from certain details as to tools, were identical with those of today (1927). Great centres of industry had begun to grow up, a factory proletariat made its appearance, the old trade regulations, already more than half destroyed, made way for the system of *laissez faire,* itself even then doomed . . . The law which inaugurated factory legislation was passed in 1802. The stage was ready set . . ."

Perhaps 'doomed' is too strong a word to use in relation to *laissez faire,* that great radical idea of economic thought developed by Adam Smith (1776) to extol the virtues of an economic system driven by principles of competition and profit with minimalist government intervention (and no regard for custom and tradition). Mantoux's point is that the relatively raw version of *laissez faire* was soon modified by an increasingly complex set of social and institutional initiatives, beginning with government legislation in the form of the factory acts which 'interfered' with business behaviour in terms of such basic issues as a minimum age for employment and length of hours worked in a week. As Polanyi (1944) articulates, capitalist societies, as others, are regulated, and in the 19th century the purpose of these regulations was to ensure that an economic system motivated by *laissez faire* remained a society. Nevertheless, whether *laissez faire* was doomed or regulated, at the end of the 18th century "the stage was ready set". The Industrial Revolution, in association with the agricultural and transportation revolutions, established capitalism's most striking characteristic, that for self-generated change (Heilbroner 1992: 25).

Industrialization as a process of creative destruction

Industrialization as the engine of capitalism has long provoked different interpretations, beginning with the classical economists, notably Adam Smith (1986) and Karl Marx (1978) and their contemporaries writing in the later decades of the 18th century and the first half of the 19th century. While both Smith and Marx recognized labour as the source of value and wealth, they offered distinctive visions of the implications of the capitalist-based industrialization process for labour and economic development. According to Smith, the basic means by which production is increased is through an increasing division (specialization) of labour and the introduction of machinery. As Smith anticipated, industrialization within the framework of capitalist society has been a major source of 'the wealth of nations'. In general terms, the use of more specialized, productive labour created larger outputs, which in turn required an expansion of markets and for markets to operate efficiently. In Smith's view, markets work best (most efficiently) when they are regulated by freely competitive processes, i.e. the principles of *laissez-faire*. So long as governments could ensure that capitalists themselves did not restrict competition, a natural expression of self-interest according to Smith, the market forces of demand and supply would continually stimulate an efficient allocation of resources. Competition thus simultaneously encourages individuals to pursue economic self-interests while eliminating surplus profits, wages and rents. In this way, the wealth of nations is enhanced and capital and labour get fair returns for their efforts. Capitalism and industrialization are potentially expansive, liberalizing 'open-ended' forces.

In contrast, Marx (1978) interpreted industrialization as a fundamentally exploitative process in which capitalists sever labour from the means of production and are able to drive down wages to provide the source of profit or surplus value, at least some of which could be reinvested to reproduce capital and to continue to exploit labour. Marx (1978: 109–199) modelled this process in terms of interrelated circuits of money, productive and commodity capital (see also Bradbury 1979; Healey and Ilberry 1990: 185–187). In the case of the circuit for money capital, for example, the formula is (Marx 1978: 109):

$$M—C....P....C'—M'$$

where capitalists start the process by using ('transforming') money (M) to buy commodities (C) and these commodities are then further transformed in a productive process (P) which creates ('transforms') new, more valuable commodities (C') which in turn can be exchanged for a larger sum of money (M') that began the circuit. The difference between C and C', and M and M', is the surplus value created by exploiting workers who are always exploited because they inevitably receive less in wages than they provide in value of output (Barnes 1990: 995). Marx offered similar formula for productive and commodity capital. In general terms, Marx's vision of industrialization is dominated by a circular process in which the metamorphoses of capital are based on the realization of surplus value (Figure 2.2). As Marx emphasized, the circuit of capital faces inherent contradictions which are ultimately expressed in massive crises as capitalists over-invest to create excess capacity, and rely more and more on exploiting labour to extract surplus value thereby immiserating the working classes. For Marx, economic injustices can only be resolved when workers regain the means of production.

Industrialization under capitalism, confounding Marx's own predictions, has proven to be extremely resilient in coping with its internal contradictions. Industrial structures have been continually transformed by technological and

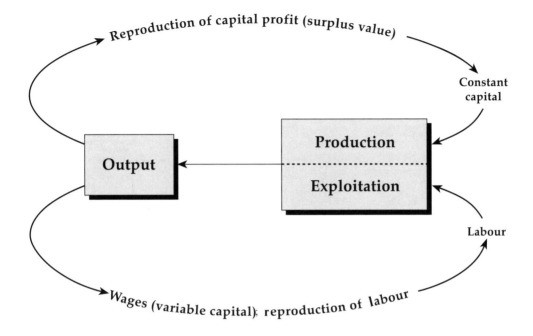

Figure 2.2 Marx's Model of Reproduction. Source: Barnes (1990: 994)

institutional changes; labour has not passively waited to be exploited but has developed abilities to cope with capitalism on its own terms, and governments, in different ways in different societies, have introduced batteries of regulations to regulate as well as to stimulate industry. To use Schumpeter's (1942) phrase, industrialization under capitalism is a process of 'creative destruction'. Industrial capitalism is crisis-ridden and for many people a source of sustained injustice; it is also enormously wealth-generating and for many people a source of sustained improvements in standards of living.

Within the long shadows cast by Smith and Marx, there have been various attempts to conceptualize industrialization processes since the Industrial Revolution. Many of these studies have sought to categorize change in terms of discrete stages of development (Storper and Walker 1989: 204; Berry 1992). In recent years, these approaches have been strongly influenced by the idea that industrialization occurs in the form of long cycles or waves of growth which are periodically interrupted by recessionary crisis. While the particular timing and length of each cycle or wave, and the number of severe recessionary crises varies among the models, Kondratieff (1978) provides a widely cited model.

Kondratieff cycles For Kondratieff, industrial evolution since the late 18th century has occurred in terms of a series of long waves or cycles lasting about 50 years, before being terminated by a major crisis or depression. So far, four complete waves have been identified, approximately beginning in the 1770s, 1820s, 1880s and 1940s, and a fifth wave supposedly began in the 1980s. On the basis of the wholesale price index, US experience provides a good illustration for the existence of Kondratieff cycles (Figure 2.3). Each wave comprises periods of recovery, prosperity and recession and is terminated by severe depression which in this model occurred around the 1820s, 1880s, 1930s and 1980s. The first Kondratieff

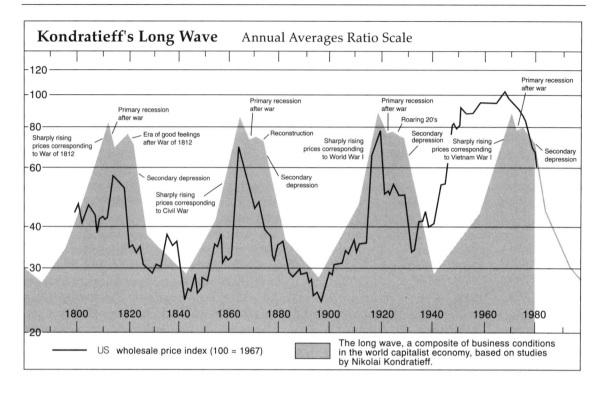

Figure 2.3 Kondratieff's Long-Wave Cycle and American Experience. Source: Knox and Agnew (1994: 114)

wave therefore more or less overlaps with Mantoux's (1966) periodization of the Industrial Revolution in England.

There is empirical support for the idea of long waves of industrialization, although somewhat different periodizations do exist (Mensch 1979; Mandel 1980; Freeman 1982; Freeman et al 1982; Gordon et al 1982). Explanations for Kondratieff waves vary nevertheless. According to Mensch (1979), each new wave is created by the clustering of basic innovations which stimulate massive opportunities for investment and employment in new branches of industry (see Abernathy and Utterbach 1978). In the initial phases of each Kondratieff wave, technological changes focus on employment and market-expanding product innovations. Over time, as markets for the new goods become saturated and as investment

increasingly favours process technology to reduce labour and material costs, a combination of excess capacity and decreasing demand creates a crisis Mensch labels 'technological stalemate'. The way out is another cluster of innovations. From this perspective, the first clustering of innovations occurred in the iron and textile industries, the second clustering in steam power and railways, the third in electric power and chemicals, the fourth in petrochemicals, electronics, autos and aerospace, and the fifth in micro-electronics.

Mensch's view, however, has been criticized for its technological determinism, because available evidence does not support his idea of the clustering of innovations and because the evolution of leading industries is more complicated than anticipated by this model (Freeman 1982;

Freeman et al 1982; Chapman and Humphrys 1987; McArthur 1987). Freeman (1987) outlines an alternative approach, based on recognizing shifts in techno-economic paradigms, which suggests that long waves of economic activity are more broadly based and embedded within society (Perez 1983; Marshall 1987; Freeman and Perez 1988). This model has several key features. First, economic development is generated by technological and institutional changes which form the basis for each long wave. Second, industrialization is a secular process which becomes increasingly complicated over time. Third, industrialization is characterized by economic crises which in turn help stimulate transformation. Fourth, economic transformation is led by particular 'leading edge' economies and industries. Finally, fundamental economic changes occur to realize productivity advantages that could no longer be obtained by previous arrangements.

Shifts in techno-economic paradigm

To help understand technological change and its wider impacts on society, Freeman and Perez (1988: 45–47; Freeman 1982) distinguish incremental innovations, radical innovations, new technology systems and techno-economic paradigms. Incremental innovations occur more or less continuously within industry and typically occur on the initiative of engineers and workers directly engaged in production. While no single incremental innovation has a dramatic effect, over a period of time the cumulative effects of incremental innovations on productivity are extremely important (Hollander 1965; Cohen 1984). Radical innovations, on the other hand, occur unevenly over time, space and sectors, and have dramatic impacts which create new markets and the basis for investment booms which support the growth of new products. Even so, the impacts may be localized around these new products. Changes in technology systems, which combine radical and incremental technological innovations with organizational and managerial innovations, have broader impacts on several

branches of the economy and create new industries. The cluster of innovations in synthetic materials, petrochemicals, injection moulding and extrusion machinery and related applications that occurred from the 1920s to the 1950s provides an example (Freeman et al 1982).

Changes in techno-economic paradigms occur when new technology systems exercise *pervasive* effects throughout the entire economy. New techno-economic paradigms include new product and process technologies which in themselves form new industries and also affect the input cost structure and conditions of production and distribution in the economy as a whole. While new techno-economic paradigms evolve out of the downswing phase of the previous Kondratieff wave because of some decisive advantage, the shift from one paradigm to the next inevitably involves structural crisis and institutional as well as technological innovation. For Freeman (1987), Kondratieff waves are created by shifts in 'techno-economic paradigm', successively defined as the early mechanization, steam power and railway, electrical and heavy engineering, fordist mass production, and information and communication techno-economic paradigms. Each techno-economic paradigm is associated with major industrial and infrastructural innovations, new principles of productivity (and engineering 'common sense') and by the emergence of new 'ideal' forms of business organization (Table 2.3). Changes in business organization are further related to evolving labour relations and sources of innovation (Table 2.5). Each paradigm is also characterized by innovations in international and national systems of regulation, and broad shifts in industrial and technological leadership (Figure 2.4).

Techno-economic paradigms do not imply that industrialization is a simple teleological process. New techno-economic paradigms occur for particular reasons because they offer decisive economic advantages, which are most fully incorporated by the leading firms of the new main branch industries, over existing paradigms. With each new paradigm, new industries, types of infrastructure, sources of energy and 'ideal'

Table 2.3 Characteristics of Techno-economic Paradigms

Long wave (key factor industries)	Main carrier branches and infrastructure	Other industries growing rapidly from a small base	Source of productivity improvement	Key organizational features	Typical entrepreneur
1. Early mechanization 1780s–1830s (cotton, pig iron)	Textiles, textile chemicals, textile machinery, iron working and castings, water power, potteries. Canals, turnpike roads	Steam engines, machinery	Mechanization and factories create scale economies not possible in traditional industry	Entrepreneurs and small firms. Local capital	Arkwright
2. Steam power and railway 1830s–1880s (coal, transport)	Steam engines, steamships, machine tools, iron, railway equipment. Railways, world shipping	Heavy engineering	Steam engine and new transportation systems enhance mechanization possibilities	Limited liability joint stock companies	Stephenson
3. Electrical and heavy engineering 1890s–1930s (steel)	Electrical engineering, electrical machinery, cable and wire heavy engineering, armaments, steel strips, heavy chemicals, synthetic dyestuffs. Electricity supply and distribution	Autos, aircraft, telecommunications, radio, aluminium consumer durables, oil, plastics	Cheap steel superior engineering material to iron. Belts, pulleys driven by one steam engine replaced by more flexible unit and group drive for electrical machinery, overhead cranes and power tools allowing improved layout. Standardization facilitates world-wide operations	Giant cartels, trucks. Middle management. Stable utilities	Krupp
4. Fordist mass production 1930s–1980s (energy-oil)	Autos, trucks, tractors, tanks, armaments, aircraft, consumer durables, process plant, synthetic materials, petrochemicals. Highways, airports, airlines	Computers, radar, NC machine tools, drugs, nuclear weapons and power, missiles, microelectronics	Flow processes and assembly line extend scale economies beyond batch production. Standardization of components. New industrial spaces	Oligopolies, MNCs. Subcontracting based on arms-length vertical integration	Ford
5. Information and communication 1980s– ('chips')	Computers, electronic capital goods, software, telecommunications, optical fibres, robotics, FMS, ceramics, data banks, information services, digital telecommunications network. Satellites	Third generation biotechnology products and processes, space activities. Fine chemicals, SDI	FMS, networking and economies of scope. Electronic control systems. Networking of design, production and marketing. Systemation	Networks of large and small firms. Quality, control, training. JIT	Kobayashi

Source: based on Freeman (1987: 68–75).

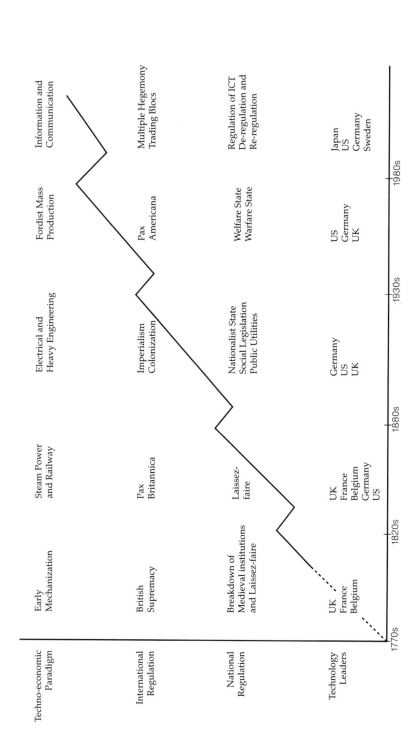

Figure 2.4 Kondratieff Long Waves: Selected Features of National and International Regulatory Regimes. Source: based on Freeman (1987)

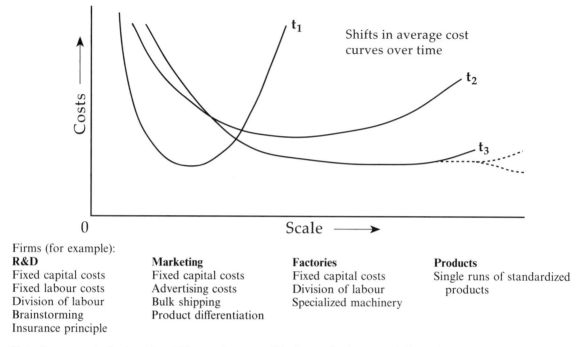

Firms (for example):

R&D	**Marketing**	**Factories**	**Products**
Fixed capital costs	Fixed capital costs	Fixed capital costs	Single runs of standardized
Fixed labour costs	Advertising costs	Division of labour	products
Division of labour	Bulk shipping	Specialized machinery	
Brainstorming	Product differentiation		
Insurance principle			

Note: Average cost = fixed cost + variable cost ÷ by output (Total cost = fixed cost + variable cost)

Figure 2.5 Sources of Internal Economies of Scale (Along a Given Curve)

types of business organization become the 'leading edge' models of the period. Yet, existing industries remain, to greater or lesser degrees, affected by the effects of the new paradigm. Similarly, not all places are equally affected by new paradigms. Moreover, even as one paradigm becomes dominant, the seeds of the next are evident.

Main industries and infrastructure Each paradigm is associated with specific mixes of dominant industries ('main carrier branches') and expansions of particular forms of infrastructure. In addition, there are new small industries which grow rapidly and become the main carrier branches of the next paradigm. Thus, in the early mechanization techno-economic paradigm, the textile and iron industries were the main carrier branches of the new technologies and new forms of organization (especially factories) while transportation was facilitated by canals and turnpike roads. The steam engine and machinery industries also grew rapidly and became the dominant industries of the next long wave. With each new long wave, the leading industries change and others are created while existing industries adapt and grow, sometimes rapidly, often by adopting features and attitudes of the main carrier branches. In other words, each wave adds layers of new activity and infrastructure while simultaneously forcing changes in existing structures.

It might also be noted that the diversification of industry over successive long waves extends to resource manufacturing as well as secondary manufacturing industry. Thus, entirely new branches of resource-based industries, other than those associated with power generation, have been created since the Industrial Revolution. The aluminium industry, for example, dates from around the beginning of the second Kondratieff, and the modern pulp and paper industry dates from around the beginning of the third Kondratieff.

Interdependencies among industries (and other sectors of the economy) is a vital feature of the industrialization process. In the early mechanization techno-economic paradigm, textile factories stimulated the textile machinery industry which was fed by the developing iron industries, while both textile and iron factories provided markets for the newly developing steam engines. In the steam power and railway techno-economic paradigm, the close connections that developed among the coal, iron and steel, railroad, and heavy engineering industries provided the core of the industrial agglomerations that developed in the UK, the US and Germany. As Freeman and Perez (1988: 47–49) emphasize, so-called 'key factor' industries play a vital role in the creation of new techno-economic paradigms and the establishment of broadly based industry interdependencies.

Sources of productivity improvement The success of new paradigms is based on 'decisive advantages' which for individual firms are in the form of productivity improvements. The factory system and mechanization of the early mechanization techno-economic paradigm, for example, created internal economies of scale (i.e. declines in the average cost with increasing size of factory) not possible in traditional industry. As Adam Smith (1986) noted, specialized workers performing the same task on the same machine on a continual basis provided a basis for scale economies by working faster, learning tasks more readily and encouraging use of specialized machinery (Figure 2.5). The fixed costs of new plant and equipment (i.e. costs that do not vary with output) also encouraged scale economies since the spreading of large fixed investments reduce average costs as output increases. In subsequent techno-economic paradigms, possibilities for internal scale economies have been further enhanced; for example, by the development of new sources of power, materials, machinery and power tools, factory organization and transportation systems which allowed

cheaper access to larger markets (Table 2.3). In addition, increasingly larger firms realized firm-level internal economies of scale related to marketing, finance, purchasing, computing and research and development (Figure 2.5).

Economies of scale as a source of productivity improvement was pronounced under fordism. While fordism became the dominant model for industrialization after World War II, its origins can be traced to the 'mass production system' developed by Henry Ford at the turn of the 20th century, and the improvements made by Alfred Sloan at General Motors. Ford's innovations, notably with respect to the interchangeability of parts, the simplicity of attaching parts to each other and the moving or continuous assembly line, significantly enhanced scale economies to the production of complex products (Womack et al 1990: 27). These innovations extended the principle of scale economies beyond batch production.

In the specific case of autos, prior to the innovations introduced by Ford between 1908 and 1913, manufacture was still in the age of craft production, each one being custom built. In craft production, there was no standard gauging system and machine tools could not cut hardened steel so that part specifications were typically approximate and often warped as a result of an oven-based hardening process. Auto assembly in practice involved extensive fitting; there were no economies of scale and no two cars were exactly the same (Womack et al 1990: 22). Beginning in 1908 with the introduction of the Ford T, Ford's adoption of the same gauging system throughout the manufacturing process and of newly developed machine tools which cut prehardened metals, allowed complete interchangeability of parts. In addition, Ford introduced designs that reduced the number of parts and made them easy to attach. Finally, in 1913 he introduced the moving assembly line so that individual workers could perform single, simple tasks without moving.

Ford's innovations greatly reduced the need for skilled workers (fitters) and the efficiencies

Table 2.4   Ford Motor Company: Selected Production Characteristics, 1909–1916

Year	Production (000 autos)	No. of Employees	Autos per employee per annum	Labour hours per auto
1909	14	1 665	8.5	357
1910	21	2 773	7.6	400
1915	369	18 892	19.5	123
1916	585	32 702	17.9	134

Source: Williams et al (1994: 99).

were enormous. Thus, in Ford's craft production factories of the early 1900s, an assembler's average task cycle (i.e. the time worked on one task before repeating it) was almost nine hours (514 minutes) and a whole car took about a month to complete (Watts 1987: 40; Womack et al 1990: 27). Much of the workers' time was spent filing and fitting parts that were not perfectly interchangeable or cut to precise specifications. After the perfection of part interchangeability in 1908, task cycle times were dramatically reduced to 2.13 minutes by August 1913 and, later in the same year, following Ford's introduction of the moving assembly line (for little capital cost) at the new Highland Park factory in Detroit, workers' task cycle times were reduced to 1.19 minutes (Womack et al 1990: 28). It now took about 93 minutes to assemble a car. By 1916, Ford's improvements to the mass production system created enormous efficiencies and sources of corporate growth (Table 2.4). In the same time period, while the price of a Ford T dropped by more than half, Ford's profits escalated enormously from US$3 062 000 to US$57 157 000 (Williams et al 1994: 98).

In addition to autos, other industries became the main carrier branches for assembly line and continuous flow technology under fordism. But there were also diseconomies of scale rooted in the boring, repetitive work of highly specialized labour, bureaucratic management structures heavily oriented to labour control, and expensive and inflexible machinery. The productivity advances of fordism began to peter out and in the dynamic and uncertain times of the 1970s more flexible manufacturing systems (FMSs), in

which economies of scale are complimented by economies of scope, promised much greater productivity improvements. In this respect, 'economies of scope' define the ability of firms to shift among a range of closely differentiated products or processes, thereby maintaining continuous operation of fixed investments even in dynamic market conditions. Of course, versatile, general-purpose machinery is not a new phenomenon. To serve new dynamic markets, however, prototype flexible firms in leading industries invest in multipurpose machinery featuring robots and especially computer numerically controlled machines (CNC) capable of "producing a variety of new products, or old products in new ways" (Gertler 1988: 420), while the reprogrammable nature of flexible technology potentially implies an ability to adapt to several model changes thus extending machine life (Schoenberger 1987: 205).

Key factor industries Each new techno-economic paradigm involves a transformation of established engineering and managerial *common sense norms* underlying what constitutes efficient and profitable practices throughout virtually all industries. For Freeman and Perez (1988), each techno-economic paradigm is associated with a 'key factor', which is a particular input or set of inputs that fundamentally changes relative cost structures throughout the economy. In general, key factors fulfil three conditions. First, the average cost (and price) of key factors falls rapidly. Second, key factors are in almost unlimited supply for long periods and any bottlenecks are perceived as short term. Third,

Table 2.5 Long Waves: Selected Characteristics of Business-related Institutions

Techno-economic paradigm	Decision-making structure	Labour relations	Sources of innovation
Traditional industry	Artisans, capitalists, merchant manufacturers	Custom and tradition	Individual
1. Early mechanization	Entrepreneurs	Foreman. Drive system	Engineer and inventor-entrepreneurs
2. Steam power and railway	Entrepreneurs	Foreman. Drive system	Engineer and inventor-entrepreneurs
3. Electrical and heavy engineering	Captains of industry	Taylorism	Inventors plus in-house R&D
4. Fordist mass production	Technostructures	Industry unions and collective bargaining	In-house R&D and government R&D
5. Information and communication	Collaborative networks	Union/non-union, factory/ firm level bargains	In-house R&D, factories as labs, collaborative R&D

Sources: columns 1 and 4 based on Freeman and Perez (1988); column 3 based on Gordon et al (1982).

key factors have potential for incorporation in many products and processes. On the basis of these combinations of characteristics, Freeman and Perez suggest that pig iron and cotton, coal and transportation, steel, oil, and micro-electronics are, respectively, the key factors associated with each techno-economic paradigm (Table 2.3).

In the fordist techno-economic paradigm, for example, oil became the cheapest form of energy that was readily available and widely applicable as a feedstock for energy-intensive industries such as petrochemicals and synthetics; a source of heat for industries and households throughout the economy; and the source of fuel for autos, the dominant carrier branch of the paradigm. Oil also became a major industry in its own right, providing the product for some of the largest corporations in the world. Massive public and private investment in motorways, service stations, airports, oil and petrol distribution systems provided the necessary infrastructural investment. As a key factor, oil required radical rethinking of cost structures in many industries and the economy as a whole. Similarly, at the present time, cheap and abundantly available micro-electronics have fuelled the growth of new industries, transformed existing industries and created demands for new workplace skills while undermining old ones. The infrastructure of the 1990s is the information highway. Computers are

now routinely associated with all types of production in the form of NC (numerically controlled) machine tools, robots, process control instrumentation, and CAD (computer-assisted design) systems, while computerized data transmission and processing systems provide the basis for administrative control and integration. Computerized operations have become a major part of fixed investment in plant and equipment. At the same time, public policy has become increasingly involved with providing telecommunications infrastructure and associated regulations regarding the processing and exchange of information for which space and time has become significantly compressed (Amirahmadi and Wallace 1995).

Institutional innovations In addition to technological change, each paradigm is implicated with new forms of international and national systems of regulation (Figure 2.4). In terms of national regulation, for example, in the latter part of the 18th century *laissez-faire* constituted an intellectual attack on feudal and medieval restrictions on trade, whether in the form of guilds, tolls, monopolies, privileges and restrictions on apprentices and worker movement. Instead, *laissez-faire* sought the liberalization of market forces as the guide to public policy. Subsequently, the social inequities arising from market forces and the inefficiencies of markets, especially with regard to the under-provision of public goods such as education and the costly over-provision of infrastructure such as railways, led to more government involvement in the economy. In the electrical and engineering techno-economic paradigm, governments increasingly passed legislation to govern health and safety, and education, and to establish public utilities. Under fordism, governments became widely committed to some form of welfare state and to Keynesian economics which required policies to stimulate demand in recessions and to reduce inflation during recoveries.

Each techno-economic paradigm is also associated with institutional innovations affecting business organization, labour relations and systems of innovation (Table 2.5). In terms of business organization, for example, successive paradigms feature the entrepreneur, the limited liability company, cartels and monopolies, multinational corporations and closely integrated networks of small and large firms. Alternatively put, Adam Smith's (1986) entrepreneurs of the first two paradigms were replaced by Veblen's (1932) 'Captains of Industry' in the electrical and heavy engineering techno-economic paradigm, and under fordism by Galbraith's (1967) 'technostructures' in which ownership is divorced from control. While Veblen's captains exercised decision-making control over large firms, and often retained controlling equity, they were supported by a growing bureaucracy of middle managers and administrative departments, including a growing army of secretaries and typists. In Galbraith's technostructures, decision-making functions are separated from ownership and are themselves decentralized along specialized lines in terms of function ('vice-presidents of finance, production, R&D, marketing'), product group ('vice-presidents of product A, B and C') and geography ('vice-presidents of Asia, Europe, North America') (see Chandler 1962). In the present paradigm, more flexible forms of business organization are seeking to integrate entrepreneurial and corporate forms.

Similarly, forms of labour control have evolved since the emergence of the factory system (Table 2.5). Prior to the Industrial Revolution, to admittedly varying degrees, the pace and rhythm of work in traditional industries had been affected by tradition and custom, and workers enjoyed some degree of self-supervision. The factory system required more disciplined workforces in terms of the pacing and length of work. In the US, the new factory workforce often comprised unskilled immigrants and, initially, the dominant mode of labour control was known as the 'drive system', which featured the use of 'foremen' to 'drive' workers, including by threat, force and cunning, to increased productivity (Gordon et al 1982; see also Marshall and Tucker 1992). Under this system, wages were low, non-wage benefits were virtually non-existent, workers could be fired without notice and rights of appeal, and ethnic and gender differences among workers were often exploited (e.g. in order to break strikes). According to Gordon et al (1982), the productivity gains achieved by this system of labour control were soon limited and during the electrical and heavy engineering paradigm, in the US, Taylor developed the principles of scientific management. According to these principles ('Taylorism'), engineers apply 'time and motion' principles to break work down into the simplest possible tasks, each of which is performed by specialist workers. As such, scientific management sought to reduce the control of skilled workers over production and to effectively use poorly educated labour. Under fordism, Taylorism became an accepted part of industry-wide collective bargains between management and unions which formally (and legally) represented worker interests. In the present paradigm, however, fordism's collective bargains are under attack as firms seek more flexibility in labour markets.

The innovative process has also been reorganized since the Industrial Revolution (Table 2.5). In the first two techno-economic paradigms, innovations were developed by individual inventors who were often practical people skilled in some way in the use of the machinery then available. The history of the cotton textile, iron, steel and pottery industries, steam engines and railways in the first two long waves, for example, can be written as the biographies of individual inventors (Mantoux 1966). The beginning of the third Kondratieff witnessed a significant innovation in the process of invention and innovation, namely the creation of 'research and development' (R&D) laboratories by industry. In Freeman's (1982) terms, led by the new chemical and electrical industries in the US and Germany, R&D became 'professionalized' as firms invested in large, expensive laboratories employing scientists and engineers organized in teams designed to apply formally learned scientific principles to whole systems. In-house R&D laboratories soon developed in many industries to become part of broader 'R&D systems', combining in-house R&D (performed by individual firms) with industry association R&D (sponsored by industry association members), government R&D and university R&D, as well as some 'co-operative' forms of R&D involving more than one of these institutions. The nature of R&D systems varies considerably among nations, even within the same industry (Nelson 1988: 325). In general, the technological leaders associated with each long wave (Figure 2.4) define the most effective national systems of innovation.

The crisis of structural adjustment Historically, shifts from old to new techno-economic paradigms have been gradual, difficult processes which have typically gathered momentum during increasingly severe recessions and depressions. Thus, old paradigms offer considerable resistance to change as a result of the inertia of existing capital investments and of human attitudes, capabilities and relationships which to some extent are enshrined in law as well as in tradition and prevailing notions of 'common sense'. Moreover, the implementation of new paradigms is slowed by the expense of new capital projects, uncertainty over appropriate choices and by the effort required to develop new skills and relationships. From this perspective, increasingly severe downturns in the latter part of long waves, in which unemployment and business failure become increasingly serious, progressively under-

line the declining potential of old paradigms while encouraging fundamental re-thinking of economic structures. Thus, it is through the stress of recessions that the 'decisive advantages', in terms of productivity and profitability gains, offered by the new paradigm, and not available to the old paradigm, become more widely recognized, and new forms of engineering 'common sense' established (Table 2.3).

In turn, rethinking within business implicates extensive changes in transportation, communication and community infrastructure and in economic and social policies. Thus, as the economic and social problems of the old paradigm become more apparent, the advocates of the new paradigm are likely to have a growing influence on public attitudes and policy. For the same reasons, it is in periods of crisis that governments are more able and willing to implement new policies.

Industrialization as a debate between flexible specialization and mass production

Although they appreciate its sophistication, Hirst and Zeitlin (1991: 15–16) criticize Freeman and Perez's (1988) model of techno-economic paradigms because they claim it over-emphasizes technology as a mechanism of change and fails to incorporate more explicitly the role of local agents and institutions in making choices about industrial trajectories. In this view, there can be no "parsimonious account of the paths to mechanization" (Sabel and Zeitlin 1985: 141). Rather, the geographical argument noted in the preface that "there is no economic geography, only economic geographies" (Barnes 1988: 349) is paralleled by an historical argument which emphasizes the "many-world history of modern industry" (Sabel and Zeitlin 1985: 169). For Hirst, Piore and Sabel, stages of development models inevitably distort local history (see Hiebert 1990).

According to Piore and Sabel (1984), the process of industrialization since the Industrial Revolution has occurred as a confrontation between systems of flexible specialization, also

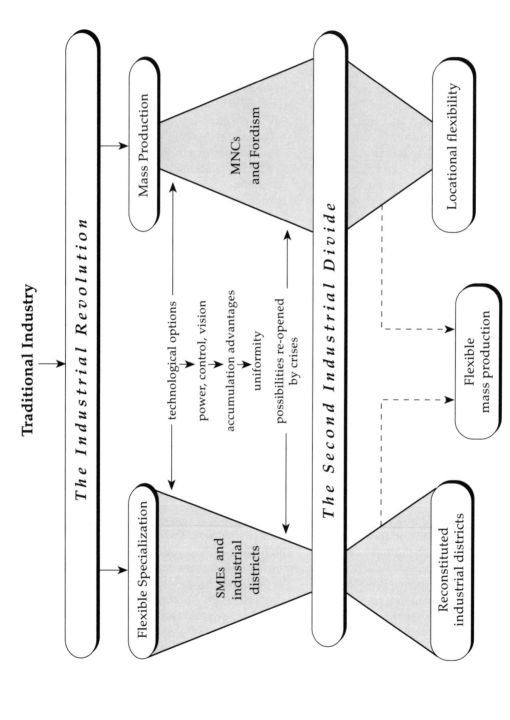

Figure 2.6 Industrialization Since the Industrial Revolution: Flexible Specialization versus Mass Production. (MNCs = multinational corporations; SMEs = small and medium-sized enterprises)

labelled craft production, and systems of mass production (Figure 2.6). On the one hand, flexible specialization is most notably illustrated by small and medium size firms in industrial districts in which a wide and changing array of products are manufactured by skilled workers using multipurpose machinery. On the other hand, mass production is most notably illustrated by multinational 'fordist' corporations manufacturing high volumes of standardized products by primarily unskilled labour using special-purpose machinery (Hirst and Zeitlin 1991: 2).

In this view, various forms of flexible specialization have provided alternatives to the mass production model since the Industrial Revolution (Figure 2.6). The chosen technological options, however, do not depend primarily on efficiency but on the interests and 'visions' of those organizations and individuals who control investments and resources in society (Sabel and Zeitlin 1985: 161–164). Moreover, once chosen and supported by infrastructure, a particular technological trajectory develops, accumulating advantages which limits other possibilities. For Piore and Sabel (1984), especially in the US, it was the mass production option that became the dominating vision, culminating in the development of the assembly line and related techniques by Ford. Even in the US, however, SMEs remained important in the economy, and in the 1970s a "Second Industrial Divide" (Piore and Sabel 1984) marked a major transformation that signalled the re-emergence of systems of flexible specialization, including flexible mass production in which high volumes are combined with product differentiation and MNCs rely extensively on SMEs for components and services. Globally, however, the history of the articulation between flexible specialization and mass production varies considerably.

The flexible specialization thesis has had a considerable influence on current debates on industrial location and regional development and it does offer a more detailed and variable account of economic history than that of the techno-economic paradigm model. Yet, there are

significant points of overlap between these two models. Both suggest that there are transformations or turning points in history even if one has more than the other. Both stress the interweaving of technological and institutional change even if relative explanatory emphasis differs, and both consider history vital in understanding economic behaviour. Finally, both models emphasize that since the 1970s, industrial countries have experienced a fundamental change from models dominated by mass production to ones dominated by flexibility.

From fordist mass production to flexible or lean production

According to the techno-economic paradigm framework, since the 1970s, the present period of industrial evolution features a transition from the fordist to the information and communication technology (ICT) paradigm. Related discussions have variously summarized the transition as a shift from fordism to flexible specialization, post-fordism, flexible accumulation and lean production (see Piore and Sabel 1984; Harvey 1988; Womack et al 1990). It is worth re-emphasizing that the shift is not a simple one. Thus, fordism was developed most intensively in the mass production industries of the US and Canada; in other countries, including European countries where US-controlled branch plants are important, fordism was modified in various ways in these same industries (Jessop 1992). At the same time, even in the US, other industries remained non-fordist or developed both fordist and fordist features. As fordism reached its zenith in key US industries, flexible or lean production principles were being developed elsewhere, especially in Japan. There are also historical precedents to the present transition from fordist to flexible production. Hiebert's (1990) discussion of the struggle between the emergence of standardized mass production and more flexible niche production strategies in the Canadian clothing industry in the first few decades of the 20th century is an example, and

(a) Planning System Firms under Fordism

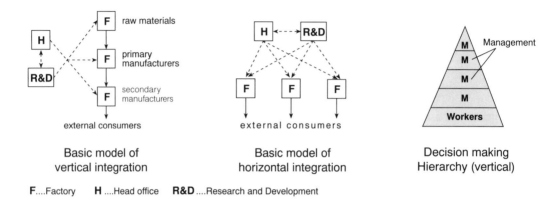

Basic model of vertical integration   Basic model of horizontal integration   Decision making Hierarchy (vertical)

F....Factory   H ....Head office   R&D ....Research and Development

(b) Flexible Business Structures (selected basic types)

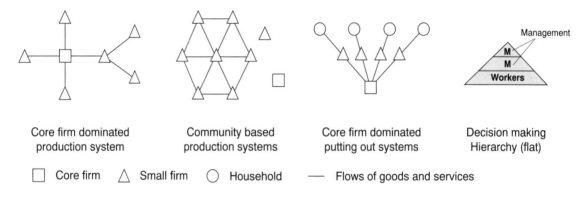

Core firm dominated production system   Community based production systems   Core firm dominated putting out systems   Decision making Hierarchy (flat)

☐ Core firm   △ Small firm   ○ Household   — Flows of goods and services

Figure 2.7   Models of Business Integration. (a) Planning System Firms Under Fordism; (b) Flexible Business Structures

reinforces Piore and Sable's (1984) point that alternative forms of industrial organization to the mass production model have long existed.

Nevertheless, a fundamental theme of current restructuring is the transition from fordist mass production to flexible production (Tables 2.3 and 2.5; Figure 2.4). In outline, the defining characteristics of fordism are the mass production of standardized goods on large-scale assembly lines (or in large-scale continuous flow processes) in big factories, with an emphasis on cost minimization. In terms of organization, the core of the fordist system comprises (a) the horizontal and vertically integrated corporation, especially multinational corporations (MNCs); (b) a strongly hierarchical ('many layered') and demarcated employment structure with professionals in the 'upper' layers and a unionized workforce controlled according to principles of scientific management in the lower layers; and (c) large-scale in-house R&D conducted as a linear sequence of separate functions (consistent with the principles of scientific management). Flexible or lean production, on the other hand, is more

(a) Linear R&D, Production and Marketing

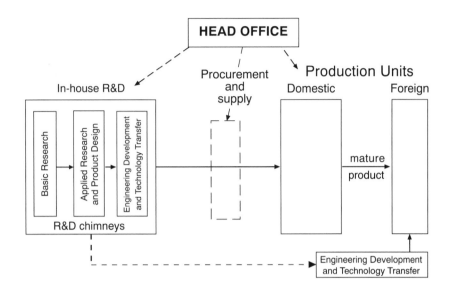

(b) Loopy R&D, Production and Marketing

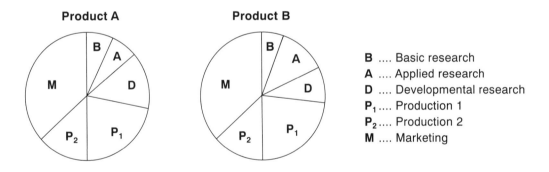

Figure 2.8   The Nature of R&D Systems. Source: adapted from Lutz (1994)

efficient in the use of materials, space and workers, while giving more emphasis to value maximization in the production of a more differentiated range of products in big and small factories. In terms of organization, (a) MNCs remain vital but more specialized than their fordist counterparts and more reliant on small firm suppliers; (b) decision-making structures are 'flatter' as more flexible workforces are smaller, require less supervision, and are more participatory (and unions are often less powerful); and (c) in-house R&D is more closely integrated with production and organized on 'loopy' lines. The next section introduces each of these organizational shifts (Figures 2.7 and 2.8), which are dealt with in more detail in subsequent chapters.

From internal to social divisions of labour

In broad terms, fordism created a 'dual economy' in which business structures (and labour markets) are segmented into two different types. In terms of business structures, for example, Averitt (1968), Galbraith (1967) and Taylor and Thrift (1983) distinguish between planning system, centre or corporate sector firms and market system or peripheral firms (see Chapter 8). As Edwards (1979: 73) notes, market system firms are "under the actual dominance or the long shadow of the big firms". Planning system firms are corporate giants that are controlled by technostructures within various forms of multi-divisional structure to organize dispersed manufacturing operations, typically located in several countries. Planning system firms dominate investment, production and employment in the industries in which they operate, and to varying degrees these firms are able to manipulate market forces in their interests.

Under fordism, the typical planning system firm is horizontally and vertically integrated (Figure 2.7(a)): these firms (horizontally) produce similar products in different factories and they (vertically) manufacture and supply the inputs associated with technically linked operations (Figure 7.1). That is, production is dominated by *internal divisions of labour* (Chapter 13). Ford, for example, sought to pursue principles of integration, and the internal division of labour, throughout the corporate system and in individual factories, the best example of which is Ford's River Rouge plant built in the 1920s at Dearborn, Michigan. This massive factory employed 80 000 workers and in the 1930s came close to producing all the requirements in the manufacture of cars made at the plant. Even after World War II, when it was 'disintegrated', this plant still produced 50% of its needs (Womack et al 1990: 58). Both GM and Ford consistently ranked among the major steel producers of the US.

Under fordism, market system firms comprise small and medium-sized enterprises (SMEs), and while some ('loyal opposition firms') provide localized opposition to planning system firms, they are largely seen as peripheral to the industrial system, serving small market niches which do not offer economies of scale for corporate giants to exploit.

In contrast to fordism and its emphasis on internal control, in flexible or lean production, SMEs play a more central role in production (Figure 2.7(b)). Collectively, SMEs form a *social division of labour* in which individual SMEs perform highly specialized roles in the manufacture of some 'final' product. Thus, among SMEs, the advantages associated with specialized labour are reinforced by the advantages of specialist entrepreneurs who, motivated by self-interest, seek ways to reduce costs and/or new opportunities to define and refine market niches (Patchell 1993a). In this view, the benefits obtained from the highly focused strategic mission of SMEs is complemented by the flexibilities generated by the ability of individual entrepreneurs to act quickly, including with respect to the use of machinery and the deployment of labour. Flexibly specialized production systems can be organized in different ways (Figure 2.7(b)). Core firms, for example, can organize SMEs through subcontracting relations to create different kinds of flexible mass production systems (Sheard 1983; Sabel 1989), or the entire production system can be organized among a population of small interacting firms producing low volumes of high-quality processes and goods (Bagnasco and Sabel 1995). In addition, communication and transportation improvements have permitted a renewal of putting-out systems to re-incorporate households back into the production process. In fact, these putting-out systems can occur on an international basis and reach the most isolated places, often in search of cheap and flexible female and child labour.

Ideally, in flexible production, the role of large corporations and factories tends to be relatively specialized and closely tied to the specialist activities of myriad SMEs. One of the best known complexes which provides a contemporary model of this philosophy is provided by Toyota's assembly plants and supplier firms located in Toyota City. By the 1980s, Toyota was widely recognized as the most efficient auto producer in the world and its assembly factories were typically relying on external sources, mainly SMEs, for 75–85% of their supplies (Chapter 14). To use Sheard's (1983) phrase, this complex represents "a factory without walls".

It might be noted that there is controversy over flexible production and the social division of labour in terms of its social implications. Thus, Piore and Sabel (1984) interpret flexible specialization strategies as a desirable vehicle for local development by stressing the mutual benefits, including profits and wages, arising from co-operation among groups of small, high-quality producers employing a highly skilled workforce that may be union or non-union. Others note the competitive nature of vertical disintegration strategies in which inter-firm relations feature possibilities for labour exploitation, particularly in situations which arise from 'subcontracting' or 'putting-out' to smaller firms or homes where labour is non-unionized, frequently female and poor, unprotected and operating in highly competitive conditions (e.g. Christopherson 1989). In practice, flexible labour can take on very different characteristics (Storper and Christopherson 1987; Hayter and Patchell 1993; Patchell and Hayter 1995).

Towards more flexible workforces

Under fordism, the typical large corporation, and its subsidiaries and factories, structure work done by professional 'white collar', production 'blue collar' and secretarial 'pink collar' workers along strongly hierarchical lines (Figure 2.7(a)). The technostructure occupies the top part of the hierarchy. At the pinnacle itself is the chief executive officer (CEO) and/or corporate president, below which are executive vice-presidents (VPs) representing particular functions (e.g. finance, sales, R&D and production), products and geographic areas. Below the VPs are further layers of management, with each layer reporting directly to the layer immediately above it in the hierarchy. These layers may include assistant

VPs, directors, senior plant managers, maintenance managers, assistant managers, superintendents and assistant superintendents. Each of these professionals supervises secretarial pools of various sizes. In turn, foremen and group leaders often occupy positions at the interface of the technostructure and the (unionized) blue collar workers, who in turn are further layered by seniority and department, where the most senior member of the group exercises a degree of control over the others.

A defining feature of fordist hierarchies is the highly specialized nature of work tasks among professionals, blue collars and pink collars. Indeed, Taylor's ideas of scientific management (or Taylorism) exercised a pervasive effect on work organization in planning system firms under fordism (Table 2.5). As noted, the thrust of Taylorism is to sharply demarcate the job tasks of production workers to realize productivity gains from the specialization on simple, repetitive tasks, which simultaneously increases the chances for automation (Urry 1986). As such, Taylorism required a substantial supervisory layer to ensure that work tasks were carried out and to provide the 'thinking' if decisions had to be made. According to Taylor, the costs of supervisory bureaucracies were more than offset by the benefits of worker specialization. Thus, the jobs of workers and professionals were sharply separated and 'segmented' (Doeringer and Piore 1971; Edwards 1979; Gordon et al 1982; see also Kerr 1954). Within each segment, particular job tasks were typically highly compartmentalized. Moreover, under fordism, collective bargaining between management and unions institutionalized this structure.

The bureaucracies and work structures of fordism started to break down in the 1970s and 1980s. One of the key advantages of fordist corporations seemed to be their stability. By the 1980s, fordist stability was seen as rigidity as fordist corporations encountered increasing difficulty in coping with the dynamism of the ICT techno-economic paradigm. Atkinson (1987), for example, argues that in response to recession, increasing uncertainty and technical change, an increasing number of firms have sought to become more 'flexible' with respect to employment (Chapter 12). At the same time, firms have sought flatter decision-making structures (Figure 2.7(b)). Having fewer workers reduces the need for supervisors, while flexibility potentially implies a widening of job tasks by workers to include decision-making responsibilities. Once-specialized middle management are also expected to be more flexible. On the other hand, a flexible workforce can also mean that workers are hired and fired as needed, or that highly paid workers are replaced by lesser paid workers (Best 1990).

In practice, the search for flexible labour is multifaceted and contentious (Chapter 12). Flexibility, for example, has important implications for unions. In fordist firms, unions are powerful and can bargain with management in developing segmentation. In flexible firms, however, flexibility is both the cause and effect of the unions' weakened power and importance. Thus, firms may be able to introduce new flexible working practices because of the threat of unemployment. At the same time, once the power of unions is weakened, the introduction of flexible working practices may become easier. Clearly, from society's point of view, there is a great deal of difference if communities are dominated by functionally or numerically flexible workers.

From linear to loopy R&D

Under fordism, intellectual skills are concentrated among white-collar workers, including R&D employees who have the responsibility of supplying technology in the form of new products and processes to meet the specific production and marketing needs of the firm. Moreover, under fordism, R&D itself is a linear sequence of specialized (and separate) processes involving basic research, applied research, development research and technology transfer (Figure 2.8(a)). Lutz (1994), for example, illustrates R&D (and production) in the US auto industry from the 1950s until the 1980s as a set of chimney stacks, each of which provided a self-contained department responsible for a specialized range of

tasks which when completed were passed on to the next chimney, with virtually no formally planned interactions between chimneys. From this perspective, R&D, like production, benefits from specialization and economies of scale (Vernon 1970).

In flexible firms, in-house R&D is organized differently (Figure 2.8(b)). In particular, flexible firms seek to develop 'loopy' forms of R&D which plans on the close, ongoing integration of different aspects of the R&D process and between the R&D process and production and marketing. Lutz (1994), for example, illustrates R&D (and production) in the contemporary US auto industry in the form of 'platforms' which integrate all the processes from R&D to marketing for each product. In this model, which is based on Japanese experience, feedback is expected and production workers are also expected to play a role in the innovation process. Indeed, dominant firms are information-intensive and seek to link "design, management, production and management into one integrated system – a process which may be described as 'systemation' and which goes far beyond the earlier concepts of mechanization and automation" (Freeman and Perez 1988: 60). In this view, systemation is an umbrella term for the substantial productivity improvements made possible by flexible manufacturing systems (FMSs), economies of scope, electronic control systems and networking between R&D, design, production and marketing (Table 2.3).

The fundamental change in production from fordist to flexible or lean production methods are well illustrated by the auto industry (Womack et al 1990; Morales 1994; Chapter 14). As noted, Ford's innovations, and associated institutional changes (notably the Taylorization of the workforce and Sloan's development of the divisional structure at GM), had established major sources of productivity improvements and the basis for fordism. As a factory exemplar of this system, in 1950 the Rouge River plant, the largest and most efficient auto factory in the world, was producing 7000 vehicles a day (Womack et al 1990: 48). But observers of this system, notably engineers within

Toyota, also identified sources of waste and began to develop alternative, more flexible and 'leaner' forms of production in the very heyday of fordism. Criticisms of the fordist assembly line and form of organization emphasized the large and costly inventories of materials and space that were maintained, the lack of incentive among workers to develop skills, the wasted time and resources involved in adversarial labour relations, the lengthy time involved in shifting from one product line to another, and the high level of defects that result because quality control is just another specialized task only performed at the end of the assembly line.

At Toyota, the thrust of the development of lean and flexible production was to manufacture cars more efficiently and with a greater emphasis on improving quality. Thus, simpler methods of making dies were developed to allow for much faster changes on the product line and a faster response to consumer demand. The pursuit of just-in-time principles, involving extensive use of suppliers providing small batches of components, eliminated the inventory problem and facilitated quality control at the beginning of the assembly line. More flexible workers became better skilled, more fully employed, more interested in their jobs and more likely to suggest improvements. In addition, the flexibilities permitted by the newly emerging electronic control systems of the 1970s could be readily integrated within a system of manufacture already oriented to lean production and flexibility. It might also be added that in more general terms Toyota represents trends occurring throughout many Japanese industries and it is Japan that has emerged as the technology leader of the ICT techno-economic paradigm. Such a shift, unthinkable in the West two decades ago, confirms again the dynamism of the manufacturing sector.

Conclusion

The institutional and technological changes that have driven the process of economic transformation typically have strong geographic roots: they develop in particular places and societies at particular times. In turn, the institutional and technological changes underlying industrialization alter the geographic conditions of production everywhere. The present ICT paradigm is once again exerting powerful effects on location. Yet these effects are not easy to summarize (Amirahmadi and Wallace 1995). On the one hand, the profound liberalization of communication and information flows has released strong forces of dispersion. On the other hand, the need to control, co-ordinate and mutually develop information among related specialist services and functions still supports forces of concentration. Even in the 1990s, industrialization realizes substantial efficiencies from geographic concentration. The next chapter turns to the geography of industrialization.

3

THE GEOGRAPHY OF MANUFACTURING

Allen Scott (1988a) coined the phrase 'new industrial spaces' to denote recent concentrations of manufacturing in regions that are situated well beyond the long-established industrial regions of western Europe and North America. These new industrial spaces are found in both advanced countries, as in the cases of the Emilia–Romagna area of northern Italy (Third Italy) and Silicon Valley, northern California, and in developing countries, as in the cases of Mexico's Maquiladora exporting processing zone and China's Shenzhen special economic zone. The term 'new industrial spaces' provides an effective image for the dynamism of contemporary processes of industrialization which appear unlimited by geographical boundaries or established notions of where industry 'should' locate. Yet, industrialization has always created new geographies. At one time, the UK's industrial conurbations, especially those within the so-called Axial Belt of industry, the Manufacturing Belt of North America and the German Ruhr, which in the 19th century and much of the 20th century defined the world's pre-eminent industrial agglomerations, were also 'new industrial spaces'. Indeed, even in the UK and Germany, industry colonized what were effectively wilderness areas with sparse populations.

Geographical evolution is a problematical process, with its own tensions and expressions of creative destruction. The technological and institutional innovations of each long wave of industrialization and techno-economic paradigm are implemented by investments in new facilities which have constantly modified the geography of manufacturing. Once established, centres of industry offer tremendous advantages to further industrial growth in terms of existing pools of skilled labour, infrastructure and know-how. Indeed, the 'law' of circular and cumulative causation stresses the tendency of industrial growth to focus on already established areas (Myrdal 1957; Pred 1967). On the other hand, it is typically easier to introduce new forms of work organization, new employment conditions, new technologies and new plant configurations in new industrial spaces. Historically, new rounds of industrialization have created new industrial spaces while reinforcing some existing centres and undermining others. In this way, industrialization incorporates a 'tension' within and between the relative advantages (and disadvantages) of old and new industrial areas. Although stronger in the present information and communication technology (ICT) techno-economic paradigm, because of the increased geographic mobility of capital, this tension has been a continuous force since the onset of the factory system.

In this chapter, the broad features of the evolving global geography of manufacturing since the Industrial Revolution are described and contemporary patterns of manufacturing production and trade that are associated with the ICT techno-economic paradigm are summarized. In terms of regional patterns, particular attention is given to the US, the industrial colossus of the 20th century. The final section of the chapter briefly addresses the role of the manufacturing sector in economic development.

The geographic unevenness of manufacturing

The most significant enduring feature of the geography of manufacturing is its unevenness. The process of industrialization has been geographically selective, resulting in highly variable rates of manufacturing growth among countries. Even within the regions that pioneered the Industrial Revolution in the UK, industrialization was highly localized and isolated. Moreover, as Pollard (1981: 14) notes in a European context, within countries, industrialization has been a profoundly regional phenomenon, an observation which applies equally to North America and those parts of the periphery where industrialization has penetrated. That is, as industrialization has spread across Europe and elsewhere, investment in manufacturing activities in particular time periods has always been geographically selective. A similar point may be made regarding de-industrialization (Chapter 16). However, if industrialization has been strongly regionalized in terms of investment, industrialization has typically exerted far-reaching, often global impacts in terms of trade. Geographical concentrations have necessarily relied upon interregional and international exports, and exports in turn have been a source of competitive stimulus and tension between places, i.e. a way of creating global interdependence. In this regard, it might be noted that, historically, manufacturing firms in the US have been able to rely more on the domestic market, by virtue of its huge size, than manufacturing firms in other countries (Hayter 1986a). Even so, in recent years, in the US there is increasing appreciation for the growing impact of trade on

local economies and for the need for US firms to export more in the increasingly competitive environment of the contemporary global economy (McConnell 1986; Erickson 1989; Erickson and Hayward 1991; Howes and Markusen 1993).

The global distribution of manufacturing industry since 1750: an overview

In a pioneering study, Bairoch (1982, 1993) estimates the evolution of global manufacturing production since 1750 by major world region and selected individual countries (Table 3.1). While the reliability of calculations inevitably decreases as we go back in time, Bairoch's estimates are the most comprehensive available and represent a systematic attempt to comparatively understand global trends in manufacturing distributions, including reference to developing or Third World countries. Clearly, it is to facilitate comparisons that Bairoch maintains the same definition of developed and Third World countries throughout the study period since individual countries do change in development status. Singapore, for example, is not a Third World country and the living standards of South Korea and Hong Kong are significantly greater than in much of Africa and Latin America.

According to Bairoch, on the eve of the Industrial Revolution, the distribution of manufacturing production roughly corresponded to population and almost three-quarters of manufacturing production was located in what he refers to as 'Third World' countries. As the UK industrialized in the first long wave, the Third World's share of global manufacturing declined, modestly at first, such that by 1830 the Third World still accounted for 60% of manufacturing production. During the second wave of industrialization, however, which saw the rapid development of the factory system in the UK, Europe and the US, the Third World's share of manufacturing dropped rapidly and by 1913 was just 7.5% of the world total. While these relative shifts in manufacturing between developed and Third World countries reflect

Table 3.1 Development of Manufacturing Production Among Developed and Third World Countries 1750–1990

Year	Developed countries		Third World		World volume
	Volume	Share (%)	Volume	Share (%)	
1750	34	27.0	93	73.0	127
1800	47	32.3	99	67.7	147
1830	73	39.5	112	60.5	184
1860	143	63.4	83	36.6	226
1880	253	79.1	67	20.9	320
1900	481	89.0	60	11.0	541
1913	863	92.5	70	7.5	933
1928	1 258	92.8	98	7.2	1 356
1938	1 562	92.8	122	7.2	1 684
1953	2 870	93.5	200	6.5	3 070
1963	4 699	91.5	439	8.5	5 138
1973	8 432	90.1	927	9.9	9 359
1980	9 910	88.2	1 320	11.8	11 230
1990	12 090	83.4	2 480	16.6	14 570

Source: Bairoch (1982: 275) and updated in Bairoch (1993: 91).
Notes: The volumes are indexed with UK = 100 in 1900. The estimates for 1750 and 1800 are very approximate. Manufacturing refers to the conventionally defined 'secondary' sector. The definition of developed countries is that traditionally adopted by UN and includes the capitalist countries of the OECD and the planned economies of Europe. All others including Spain and Portugal are classified as Third World.

Table 3.2 Relative Share of World Manufacturing Output by Country and Region 1750–1980

	1750	1830	1860	1880	1913	1938	1953	1963	1980
Developed Countries	27.0	39.5	63.4	79.1	92.5	92.8	93.5	91.5	88.0
Europe									
France	4.0	5.2	7.9	7.8	6.1	4.4	3.2	3.8	3.3
Germany	2.9	3.5	4.9	8.5	14.8	12.7	5.9	6.4	5.3
Italy	2.3	2.3	2.5	2.5	2.4	2.8	2.3	2.9	2.9
Russia	5.0	5.6	7.0	7.6	8.2	9.0	10.7	14.2	14.8
Spain	1.2	1.5	1.8	1.8	1.2	0.8	0.7	0.8	1.4
Sweden	0.3	0.4	0.6	0.8	1.0	1.2	0.9	0.9	0.8
Switzerland	0.1	0.4	0.7	0.8	0.9	0.5	0.7	0.7	0.5
UK	1.9	9.5	19.9	22.9	13.6	10.7	8.4	6.4	4.0
US	0.1	2.4	7.2	14.7	32.0	31.4	44.7	35.1	31.5
Canada		0.1	0.3	0.4	0.9	1.4	2.2	2.1	2.0
Japan	3.8	2.8	2.6	2.4	2.7	5.2	2.9	5.1	9.1
Third World	73.0	60.5	36.6	20.9	7.5	7.2	6.5	8.5	12.0
China	32.8	29.8	19.7	12.5	3.6	3.1	2.3	3.5	5.0
India	24.5	17.6	8.6	2.8	1.4	2.4	1.7	1.8	2.3
Brazil			0.4	0.3	0.5	0.6	0.6	0.8	1.4
Mexico			0.4	0.3	0.3	0.2	0.3	0.4	0.6
World	127	184	226	320	933	1684	3070	5138	1104

Source: Bairoch (1982: 296, 304)
Notes: Figures are based on UK = 100 in 1913. Prior to 1938, India included Pakistan. The 1938 figures are three-year averages.

differential rates of growth, manufactures in the developed countries, such as cotton textiles, directly replaced local production in countries such as India (Singh 1977). In fact, the Third World's share of global manufacturing continued to decline and reached its low point in the 1950s. By the 1960s, manufacturing in the Third World started to become relatively more important and since the onset of the fifth long wave and the ICT techno-economic paradigm, its share of manufacturing has increased rapidly. By 1990, according to Bairoch's estimates, Third World countries accounted for almost 17% of world production.

With respect to individual countries, Bairoch (1982) estimates that in 1750 the largest shares of global manufacturing output were concentrated in China and India/Pakistan (Table 3.2). According to Bairoch's data, the UK established itself as the leading industrial power in Europe between 1800 and 1830 and by 1860 was the world's largest industrial power, with about one-fifth of global production, slightly ahead of China. The relative importance of Britain's share of industrial output peaked around 1880 (at 22.9%), by which time the US was the globe's second most important industrial power. The last two decades of the 19th century witnessed booming industrial growth and by 1900 the US was industrially bigger than Britain. Germany also expanded and in 1900 was the world's third most important industrial power. Japan, on the other hand, according to these figures, was relatively less important in 1900 than in 1750. While Japan had a well-established network of traditional industries, its commitment to modern industrialization did not occur until the Meiji period of the 1860s.

By 1913, the US, Germany and Britain were the established world-leading manufacturing countries, although Japan's importance, despite its 'late start', was evident. In fact, the US, UK and Germany accounted for fully 60.4% of world industrial production at this time. Indeed, the US alone manufactured almost one-third of world industrial output in 1913. The First and Second World Wars arrested the growing importance of Germany and Japan but during the long boom of the fourth Kondratieff and the fordist techno-

economic paradigm they rapidly re-established their industrial strength. Russia also grew rapidly at this time to become the world's second most important industrial producer. Unequivocally, however, the outstanding industrial power of the 20th century has been the US. From the early years of the century until the 1980s, the US has manufactured at least 32% of world industrial production, and in 1953, during the heyday of the fordist techno-industrial paradigm, the US accounted for a remarkable 44.7% of production, an historically unprecedented level of concentration of industrial production. Even by 1963, when European and Japanese industrial growth was in full swing, the US still accounted for over 35% of production, a level that is unlikely to be ever realized again by a single country. Given the overwhelming importance of US-based multinational corporations (MNCs) in patterns of direct foreign investment (DFI) in this period (i.e. the significant US ownership of industrial production outside the US), it is hard to overestimate the dominance of the US during the fordist techno-economic paradigm (Chapter 11).

The German and especially the Japanese industrial economies were also growing rapidly in this period. Indeed, by 1987, following the onset of the fifth Kondratieff long wave, Japan's industrial production, remarkably, was almost as great as that of the US. Germany retained its relative share of world manufacturing output during the transition from the fourth to the fifth Kondratieff long wave or from fordism to the ICT paradigm. For the most part, other west European countries became less important, most notably the UK, while numerous other countries around the globe were starting to account for small but noteworthy shares of manufacturing production. Russia also maintained a substantial industrial base during the fordist paradigm although this base evolved within protected markets lacking internal competition. The underlying weakness of Russian industry became more apparent during the fifth long wave and was ultimately confirmed by Perestroika and the creation of the Commonwealth of Independent States (CIS).

Contemporary global patterns in manufacturing production

In the late 1980s and early 1990s, manufacturing production remained unevenly distributed throughout the world, and was still dominated, albeit less so than in previous decades, by the established industrialized countries of western and north-western Europe, North America and Japan. These countries, along with Australia and New Zealand, comprise the Organization of Economic and Co-operative Development (OECD) and the concentration of manufacturing within the OECD is particularly evident when manufacturing is measured in terms of value added, i.e. the difference between the value of inputs on arrival at the factory and the value of outputs at the factory gate. On the basis of this criterion, in 1990, for example, the US, Japan, CIS and Germany, stand out as the world's biggest industrial powers, while France, the UK and Italy are significant (Figure 3.1). Excluding CIS, these six countries accounted for 62.6% of world value added. Another seven traditionally important industrial countries (Canada, Netherlands, Sweden, Belgium, Australia, Finland and Austria) account for a further 6.2% of manufacturing value added in 1990. Outside of the OECD, in 1990 the CIS was easily the most important industrial power (11.1% of the global value-added total), followed by China (2.4%), Brazil (2.4%), Spain (1.3%) and South Korea (1.1%). Given that the most important 20 countries accounted for almost 85% of global value added in manufacturing, the majority of countries in the world have a very small manufacturing output.

In employment terms, as of the early 1990s, available statistics indicate a somewhat different distribution of manufacturing activity. Thus, the two nations with the most manufacturing employees are China and CIS, followed by the US, Japan and Germany and then India (Figure 3.2). Several former centrally planned economies (CPEs) of eastern Europe, notably Czechoslovakia, Poland and Romania, are also relatively important. Globally, however, there are significant variations in labour efficiency within the manufacturing sector and the importance of the CIS and China in employment terms does not reflect their global competitiveness. Labour efficiency is significantly higher among OECD countries which, generally speaking, over the past decade have increased production with more or less the same or even smaller workforces.

Within the context of the continuing importance of the long-established industrialized countries of the OECD, several important national shifts in global manufacturing are worth noting. First, since World War II the most dramatic change has been the rise of Japan to its position as the world's second largest industrial power, overtaking France, the UK, CIS and Germany in the 1970s and rapidly catching up with the US. As of the 1990s, the relative sizes of the US and Japanese industrial economies are roughly comparable and both are more than twice the size of German industry, the third largest OECD country. Second, in association with the first point, the US has lost its undisputed industrial hegemony. Third, the most dramatic industrial decline has occurred in the UK, which between 1963 and 1990 went from being the second to the fifth largest OECD industrial power.

Among the OECD group of countries, employment trends have varied. The two most significant alternative trajectories are the UK and Japan. On the basis of the OECD's definition of industry, between 1966 and 1981 the UK shed 3.2 million jobs while Japan added 4.1 million jobs (Table 3.3.). The US also added over 4.2 million industrial jobs between 1966 and 1979 although some were shed in the severe recession of 1980 and 1981. Among the other countries employment change was less sharp during this period. Since 1983 and until 1994 the different trends between the UK and Japan are again noticeable (Figure 3.3; Table 3.3). While the UK showed a consistent decline, the number of manufacturing jobs in Japan expanded each year except in 1986 and 1987. In total, between 1966 and 1994 the UK recorded a net loss of manufacturing jobs of over five million, while Japan increased by over six million (Table 3.3). Employment change in the US has been more erratic over the past

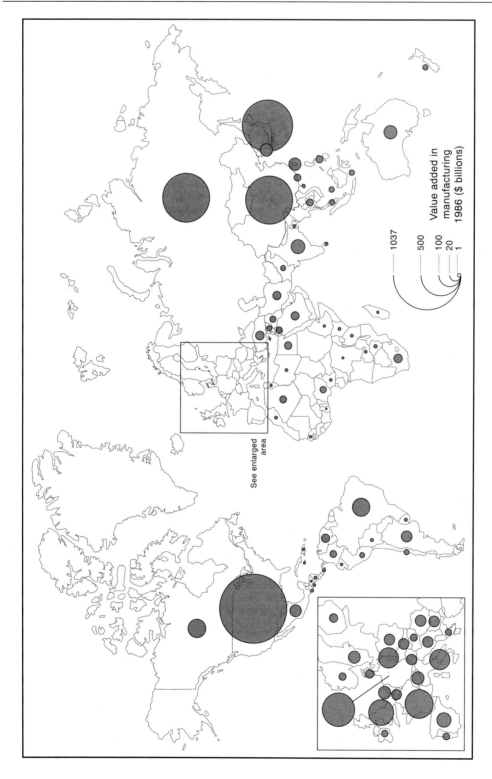

Figure 3.1   The Global Map of Manufacturing Production, 1986. Source: UNIDO (1988) and Dicken (1992: 21)

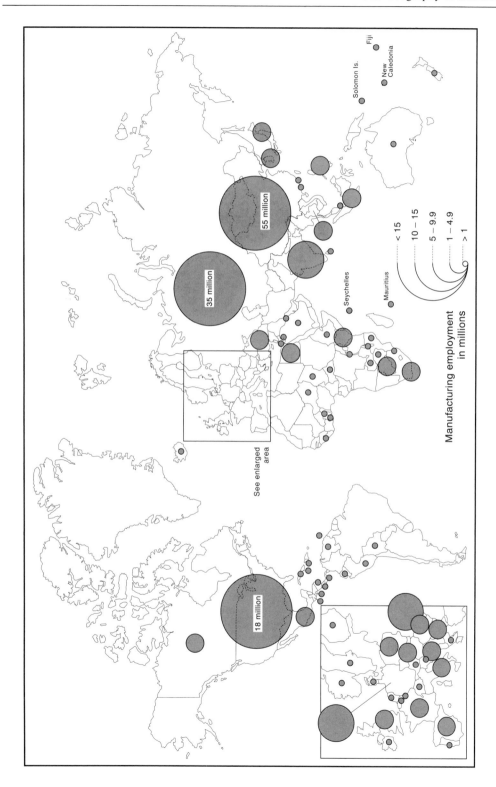

Figure 3.2 The Global Map of Manufacturing Employment, 1992. Source: UNIDO (1993)

Table 3.3 Industrial Employment (Thousands) in OECD Countries, 1966–1994

Country	1966	1973	1979	1980	1981	1986	1994
United Kingdom	11 559	10 484	9 669	9 270	8 354	7 453	6 479
Canada	2 438	2 885	2 994	3 040	3 114	2 898	2 898
United States	26 247	28 045	30 402	29 754	29 995	30 339	29 538
Japan	15 800	19 570	19 140	19 560	19 817	20 180	21 970
Australia	1 805	2 042	1 895	1 937	1 972	1 858	1 861
New Zealand	389	394	398	427	406	444	390
Austria	1 328	1 222	1 233	1 236	1 238	1 241	
Belgium	1 592	1 553	1 334	1 269	1 194	1 048	
Denmark	824	806	750		695	742	
Finland	707	769	731	754	783	774	
France	7 853	8 266	7 670	7 603	7 459	6 574	5 425
Germany	12 656	12 448	11 233	11 327	11 084	10 771	10 067
Greece	709	881	994				
Ireland	293	324	369				
Italy	7 063	7 470	7 646	7 772	7 748	6 821	6 420
Luxembourg	63	67	62				
Netherlands	1 784	1 658	1 481				
Norway	531	560	564	568	573		
Portugal	999	1 137	1 348				
Spain	4 142	4 715	4 303	4 058	3 959		
Sweden	1 556	1 427	1 360	1 363	1 327		
Switzerland	1 430	1 414	1 165				

Source: Labour Force Statistics, OECD, Paris.
Note: Industry is defined to incorporate primary and construction activities and utilities.

decade as manufacturing jobs first expanded rapidly and then declined rapidly after 1989. Even so, there are three million more manufacturing jobs in the US in 1994 than there were in 1966. In Germany, job growth was finally stalled in 1991, while in Australia and Canada manufacturing jobs remained more or less the same during the 1980s although a downward trend is evident in Canada after 1987. All of these countries experienced major recessions in the early 1990s, however, and the number of manufacturing jobs is likely to have declined or, at best, remained at 1990 levels. It might also be noted that there has been a significant net decline in manufacturing jobs in Germany, especially since the mid-1970s.

According to Bairoch, the Third World's share of global manufacturing production increased by almost half between 1980 and 1990, that is from 11.8 to 16.6% of the global total (Table 3.1). China remains the largest of the Third World countries, with Brazil, Mexico and India being the next largest. Since the early 1970s, the fastest rates of growth outside of the OECD have occurred among the so-called newly industrializing countries (or NICs) which comprise Spain, Portugal, Greece, former Yugoslavia, Mexico, Brazil, Hong Kong, Singapore, Taiwan and South Korea. Among the NICs, the four Asian Tigers have recorded the most impressive growth in manufacturing (Brohman 1996). Between 1974 and 1983, for example, manufacturing employment in Hong Kong increased by 43%, in Singapore by 41% and South Korea by 77% (Watts 1987: 3). Admittedly, by 1990, these three relatively small countries only accounted for 1.5% (and Taiwan for another 1.24%) of world manufacturing production and especially in

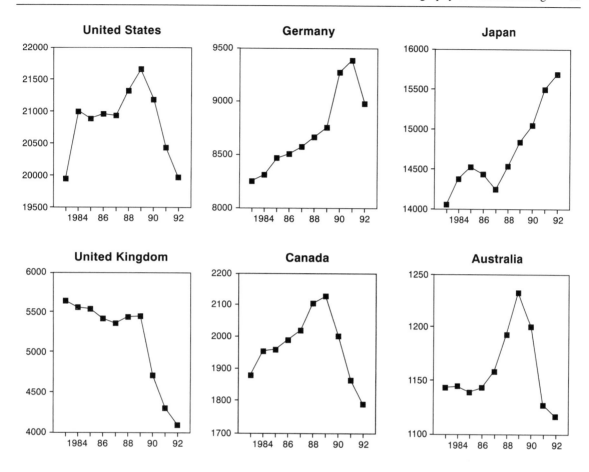

Figure 3.3 Manufacturing Employment for Selected OECD Countries, 1983–1992. All Data are in Millions. Source: *Yearbook of Labour Statistics* (1993)

Singapore and Hong Kong this share is unlikely to increase. Even so, there is a continuing trend in the shift of manufacturing value added, production and employment to new industrial spaces in the Third World, and recent growth of industry in China and Mexico has been especially rapid.

Contemporary global patterns of trade

Historically, there is a strong link between industrialization and trade as export has been a necessary corollary of the development of geographically concentrated, specialized indus-

trial regions. Indeed, trade patterns, especially exports, provide important insights into evolving patterns of industrial competitiveness (Glasmeier et al 1993; Hayter 1986a,b). Trade performance in visible goods in 1989 and 1990, for example, confirms the continued concentration of industrial strength of the OECD while revealing the emergence of the NICs (Table 3.4). Nine of the ten leading exporters in 1990 were long-established industrial powers of the OECD and they accounted for 57.1% of world exports (US$3475 billion) in 1990. In 1990, Germany became the world's largest exporter of visible goods, closely followed by the US, Japan, France and the UK;

Table 3.4 Visible Trade: The World's Leading Exporters and Importers 1989 and 1990

	Exports					Imports			
Rank			Value	Share	Rank			Value	Share
1989	1990	Country	($ billion)	(%)	1989	1990	Country	($ billion)	(%)
2	1	Germany	421	12.1	1	1	United States	517	14.3
1	2	United States	394	11.3	2	2	Germany	359	9.9
3	3	Japan	288	8.3	3	3	Japan	235	6.5
4	4	France	217	6.2	5	4	France	234	6.5
5	5	United Kingdom	185	5.3	4	5	United Kingdom	223	6.2
6	6	Italy	170	4.9	6	6	Italy	182	5.0
9	7	Netherlands	132	3.8	9	7	Netherlands	126	3.5
7	8	Canada	132	3.8	7	8	Canada	124	3.4
10	9	Belgium–Luxembourg	118	3.4	8	9	CIS (previous USSR)	121	3.3
8	10	CIS (previous USSR)	104	3.0	10	10	Belgium–Luxembourg	120	3.3
11	11	Hong Kong	82	2.4	12	11	Spain	88	2.4
12	12	Taiwan	67	1.9	11	12	Hong Kong	82	2.3
13	13	South Korea	65	1.9	15	13	Switzerland	70	1.9
16	14	Switzerland	64	1.8	13	14	South Korea	70	1.9
14	15	China	61	1.8	17	15	Singapore	61	1.7
15	16	Sweden	58	1.7	16	16	Taiwan	55	1.5
18	17	Spain	56	1.6	18	17	Sweden	54	1.5
17	18	Singapore	53	1.5	14	18	China	52	1.5
24	19	Saudi Arabia	44	1.3	20	19	Austria	49	1.4
22	20	Austria	41	1.2	19	20	Australia	42	1.2
20	21	Mexico	41	1.2	21	21	Mexico	42	1.2
19	22	Australia	40	1.1	24	22	Thailand	33	0.9
25	23	Denmark	35	1.0	23	23	Denmark	32	0.9
–	24	Norway	34	1.0	–	24	Malaysia	29	0.8
21	25	Brazil	31	0.9	–	25	Finland	27	0.7
Total Top 25			2 933	84.4	Total Top 25			3027	83.7
Total World			3 475	100.0	Total World			3616	100.0

Source: GATT (1991, 1992) International Trade, Tables 11 and 12.
Note: Visible trade comprises manufactured goods and raw materials.

these five countries supplied over 43% of global visible exports in 1990. Among the next 15 exporters there are several NICs led by Hong Kong, Taiwan and South Korea, although in Hong Kong's case re-exports are included. These Asian NICs, plus Singapore and China, accounted for another 9.5% of world visible exports, while Spain, Mexico and Brazil are other new industrial spaces whose export power is growing. It is also interesting to note that the only former communist country of the Soviet bloc in the top 25 exporters in 1990 is CIS, in tenth place.

The league table of leading importers is close to a mirror image of the list of leading exporters (Table 3.4). The top ten countries are the same, with two differences in order: the US is easily the world's largest importer of manufactured goods and CIS is in ninth place. The nine leading OECD countries purchased 58.6% of global visible imports. Historically, the world trade in visible goods, including manufactured goods, has been overwhelmingly concentrated among established industrial countries. This concentration still exists in 1990 although it is becoming less as increasing flows of manufactured exports from

Table 3.5 Selected OECD Countries Share of Imports from NICs: 1975 and 1992/3

	Share of imports from NICs (%)	
	1975	1993
United States	11.8	20.0
Japan	7.0	12.2
Germany	6.6	9.5[a]
United Kingdom	5.5	9.1[a]
Canada	2.9	7.1

Sources: US Statistical Abstract Yearbook; Japan Statistical Yearbook; UK Annual Abstract of Statistics; Statistics Canada; German Annual Abstract of Statistics.
[a]1992.
Note: NICs refers to Newly Industrializing Countries defined as Spain, Portugal, Greece, former Yugoslavia, Hong Kong, Singapore, Taiwan, South Korea, Mexico and Brazil. Imports refer to visible imports which include raw materials as well as manufactured goods.

developing countries and NICs are destined for established industrial countries (OECD 1979). In the 1960s and 1970s the penetration of OECD markets by NICs was concentrated in a few industries, notably clothing, leather and footwear industries. By 1977, for example, clothing exports from the NICs accounted for almost 40% of the imports of OECD countries (Steed 1978, 1981; Glasmeier et al 1993). With respect to aggregate patterns, NIC exports to the OECD are growing across a range of industries. In the cases of the US, Japan, Germany, the UK and Canada, for example, the NICS (and developing countries in general) have accounted for increasing shares of imported manufactured goods since 1975 (Table 3.5). The US is the dominant export market for the NICs, followed by Japan, and in 1993, 20% of US visible imports were from the NICs.

In general, there is a strong continental dimension to international trade in manufactured goods (Gibb and Michalak 1994). In particular, exports originating in North and South America, Europe and Asia tend to flow within the same continental area. Moreover, within each of these blocs there are long-established industrialized countries and low-wage NICs and the exports of the latter tend to also flow along continental lines. These trading blocs are by no means

fortresses, however, and there are important and growing trade flows between them, especially involving Japan's exports to the rest of the world.

The extent to which manufacturing activities are dependent on exports varies by industry and by country (Table 3.6). As might be expected, given its size and possibilities for interregional domestic trade, the dependence of manufacturing in the US on exports is less than the UK. In 1982, for example, the export sales ratio in the US was 13%, compared to 27% for the UK. In the UK there is also a much bigger variation in export sales ratios between industries. In 1982, for example, while 63% of the shipments from the instrument engineering industry was exported, just 6% of food and drink output was exported. Within countries, regional variations in export sales ratios also exist. In Western Canada, for example, between 1986 and 1989 visible exports to foreign markets from British Columbia and Saskatchewan amounted to 24.5 and 27.4% of gross provincial product respectively, while in Alberta and Manitoba the respective export sales ratios were 19.5 and 13.4% (Chambers and Percy 1992: 16).

If trade and specialization go hand in hand and are a necessary consequence of economic growth, trade also creates tensions among new and old industrial spaces since exports, actually or potentially, threaten and replace domestic

Table 3.6   Export Ratios of Selected Manufacturing Industries in the UK and US

United Kingdom				United States	
Exports as % of manufacturers' sales				Export-related manu-factured product value as a % of all domestic manu-factured product value ($m)	
Sector	1982	1989	Sector	1981	1991
Instrument engineering	63	54	Machinery, not electrical	22	27
Mechanical engineering	48	39	Instruments	19	17
Vehicles	46	33	Primary metals	19	11.5
Electrical engineering	45	–	Tobacco	19	14
Chemicals	41	47	Chemicals	18	14
Leather	35	41	Transportation equipment	18	21
Textiles	32	32	Electrical equipment	17	21
Metal manufacture	27	35	Fabricated metals	12	7.5
Other manufacture	22	30	Paper	11	7
Clothing & footware	19	19	Rubber	11	7
Shipbuilding	18	–	Textiles	10	6
Coal & petroleum products	17	18	Lumber	10	9
Metals n.e.s.	17	13	Stone	9	6
Bricks, pottery, cement	13	–	Miscellaneous manufactures	9	12.5
Paper & printing	11	10	Petroleum & coal	7	4.5
Timber	7	4	Food	5	4.5
Food & drink	6	12	Apparel	3	5.5
			Furniture	3	5
All manufacturing	27	30	Printing	3	2
			All manufacturing	13	12.5

Sources: Central Statistics Office (1985, 1993) Annual Abstract of Statistics; US Bureau of Census (1985, 1993) Statistical Abstract.

production. For individual countries, imports of manufactured goods may or may not be balanced by exports. In terms of visible trade in 1990, for example, the US had a deficit of about $125 billion and the UK had a deficit of almost $40 billion (Table 3.4). In contrast, Japan enjoyed a surplus in excess of $50 billion and Germany a surplus of over $60 billion. To an important degree, the deficits on visible trade (raw materials and manufactured goods) experienced by both the UK and US are offset by surpluses on invisible trade (payments for various kinds of services, including financial, business, entertainment and tourist services). Nevertheless, in the US in particular the massive trade deficits in manufactured goods is a source of friction with several other countries, notably Japan, especially

as jobs are threatened or perceived to be threatened. It should also be recognized that visible and invisible trade are interdependent and that it cannot simply be assumed that invisible exports will inevitably offset visible imports (Singh 1977; Chapter 16). Indeed, manufacturing vitality is a basis for export success in invisibles and Japan, for example, is rapidly developing export surpluses in various kinds of services.

Concerns regarding trade typically occur when imports account for increasing shares of domestic markets and threaten the profitability and employment base of domestic industries. In the case of the UK, import penetration ratios were already high in 1972, but had increased significantly by 1982 when, for example, five major industrial categories recorded import penetration ratios of 40% or more (Watts 1987: 22). In the US, while import penetration ratios are less than those recorded by the UK, they also increased between 1974 and 1981. During the 1980s, import penetration ratios have continued to increase in the US and have remained high in the UK (Table 3.7). In the US, rising imports have frequently been the basis for protectionist action which has typically focused on other advanced countries. When Canadian construction grade lumber reached 25% of the US market in the mid-1980s, for example, protectionist action was mounted against Canada, thus reversing an historic free trade relationship (Hayter 1992). The most significant trade frictions in the 1980s and 1990s, however, involve US–Japan relations, and the auto industry is at the centre of these concerns. Not surprisingly, the import penetration ratio in the transportation equipment industry in the US is one of the highest and it has increased steadily since 1972. Moreover, an increasing share of domestic production in the US is accounted for by foreign auto makers, especially Japanese firms.

Typically, concerns over rising imports and calls for protectionist measures have a strong regional base. This is most obvious in the US, a federal country where political representation is structured on regional lines. Thus, criticism of Canadian lumber exports to the US were led by political leaders and business coalitions from the timber-producing states of the Pacific Northwest (Hayter 1992) while opposition to Japanese auto imports has been mounted by political leaders in the traditional auto-producing states of the Manufacturing Belt and by the 'Big Three' US auto producers. In turn, such regionally based lobbying reflects the regionalized nature of the process of industrialization.

The regional dimension of industrialization

From the beginning of the Industrial Revolution, manufacturing change within countries throughout the globe has been strongly regionalized. As Berg (1985: 110) reminds us, regarding the UK:

"the most striking change in the industrial structure of the country between the end of the seventeenth century and the middle of the nineteenth was its geography. The urbanized industrial zone of the seventeenth century extended along a right angle connecting Bristol, London and Norwich. By the nineteenth century this zone had shifted north and northwest to the coalfields of the West Midlands, the West Riding, Lancashire and South Wales. The old manufacturing centres of southern and eastern England had languished or disappeared."

As the UK industrialized in the latter part of the 18th century, the most significant of the new factories and workshops occurred in northern and central regions of England (the West Midlands, South Yorkshire, the West Riding, Lancashire and nearby areas) on or adjacent to coalfields (Figure 3.4). Thus Abraham Darby's iron-making works at Coalbrookdale, the manufacture of steam engines by Boulton and Watts in Birmingham, Huntsman's crucible steel making in Sheffield, Wedgewood's pottery manufactures in Stoke, Arkwright's textile mills in Manchester and Derby, and new woollen mills near Leeds all occurred within this region during the first long wave or early mechanization techno-economic paradigm. Several of these centres, most importantly Birmingham, Leeds, Manchester and Sheffield, became focal points of major industrial concentrations and in the first part of the 20th century formed the northern spokes of the so-called Axial Belt of industry, first defined in the 1930s (Chisholm 1990: 93–95). As such, the Axial Belt combined the industrial strengths of the northern centres with the market potential of the London region and its attraction to the industries of the electrical and engineering techno-economic paradigm including the newly emerging auto industry.

The second Kondratieff wave strengthened those existing centres accessible to coal resources and created new coalfield-based industrial concentrations in the UK and across Europe, especially the northern part of central Europe (Figure 3.5). Thus, among others, the Sambre–Meuse region of Belgium, the valley of the

Table 3.7 Import Penetration by Manufacturing Industry, United Kingdom (1973, 1982 and 1989) and the United States (1974, 1981 and 1991)

Manufacturing industry	United Kingdom		
	1973	1982	1989
Instrument engineering	46	62	60
Electrical engineering	27	49	52
Leather goods	27	45	52
Vehicles	23	45	51
Textiles	21	40	48
Metal manufacture	21	31	45[a]
Other manufacturing	15	24	45
Chemicals	22	33	42
Mechanical engineering	26	36	40
Clothing	18	35	40
Timber	29	38	30
Paper and printing	19	22	22
Metals n.e.s.	10	16	18
Food and drink	19	15	18
Coal and petroleum	17	17	17
Shipbuilding	56	17	n/a
Bricks, pottery, glass, cement	7	11	n/a

Manufacturing industry	United States		
	1974	1981	1991
Leather and leather products	20	40	58
Miscellaneous manufactures	17	20	38
Apparel and textile products	8	16	29
Electrical, electronic equipment	10	14	27
Machinery, not electrical	6	9	24
Transportation equipment	13	17	23
Primary metals	12	16	16
Instrument engineering	8	13	15
Furniture	3	5	12
Rubber and miscellaneous plastics	5	6	10
Textile mill products	5	6	10
Lumber and wood products	9	10	8
Paper and allied products	6	7	8
Chemicals	4	5	8
Petroleum and coal	16	7	7
Fabricated metals	3	4	7
Food	5	4	4
Tobacco	1	2	1
Printing	1	1	1

Sources: Central Statistical Office (1990) Annual Abstract of Statistics, [a]1985 figure; US Bureau of Census (1992) Statistical Abstract of the United States.

Figure 3.4   United Kingdom: The Axial Belt of Industry. Source: based on Taylor et al (1938)

Scheldt in Belgium and France, and regions along the Rhine such as the Solingen–Remschied area and most notably the Ruhr became important centres of industry. The rise of the Ruhr as Europe's pre-eminent coal, iron and steel and heavy engineering region occurred primarily after 1850. The first coke-fired blast furnace began production in 1849. Between 1850 and 1899 coal production increased from 1.7 million tonnes to 56.0 million tonnes, with major steel works

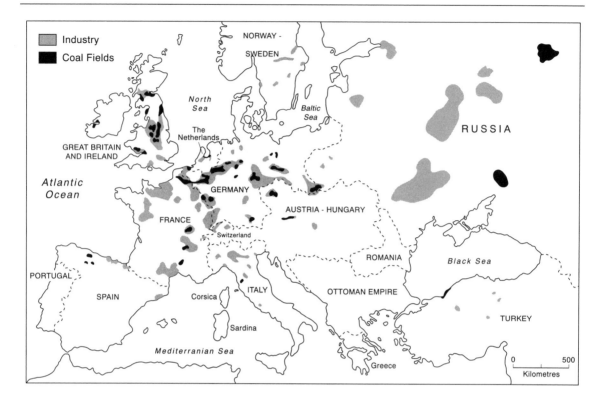

Figure 3.5   Distribution of Industry in Europe *c.* 1875. Source: Pollard (1981: xv)

established in several centres including Bochum, Duisburg, Dortmund, Geisen-Kirchen and Essen, the latter being the centre of the Krupp empire (Alexandersson 1967: 51–53). In the 1870s the coal-tar-based chemical industry, producing dyes and drugs, was also established in the Ruhr. Its resources and extensive transportation networks (including the Rhine) which provided access to European and world markets, combined with its early development, combined to provide the Ruhr with substantial location advantages.

For similar reasons, but on an even vaster scale than the Ruhr, the American Manufacturing Belt, which spills over into southern Ontario and southern Quebec, Canada, rapidly became the industrial heartland for the entire continent (Figure 3.6). Thus, the industrial know-how gained by early industrial developments were

rapidly consolidated in the latter part of the 19th century by well-developed canal and rail transportation networks and by access to coal and iron ore reserves and to rapidly growing national markets. Within the US, the iron and steel industry, and related coal, metalworking and engineering industries, overwhelmingly concentrated in the Manufacturing Belt, especially in the Pittsburgh region but with important centres in Chicago, Detroit, Buffalo, Cleveland, Baltimore, and Hamilton in Ontario, as well as several other smaller centres. In addition, a wide variety of other specialties developed among the communities of the Manufacturing Belt (Pred 1965). Overall, three broad zones evolved within the Manufacturing Belt: a zone biased towards consumer goods in the ports of New York, Boston and Baltimore; a zone between Philadelphia and Cleveland biased towards the produc-

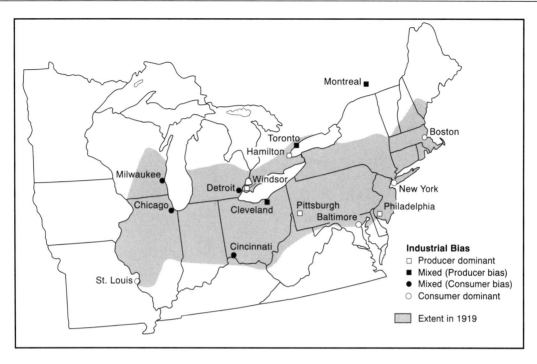

Figure 3.6   The American Manufacturing Belt in 1920. Source: Knox et al (1988)

tion of producer goods (machinery); and a zone further west oriented towards less-specialized consumer goods. The Canadian portion of the Manufacturing Belt similarly developed a wide range of city-based specialisms with a broad distinction between the capital-intensive, high-value products in Ontario and the labour-intensive, lower-value products emphasized in Quebec (Gilmour and Murricane 1973).

Within the Manufacturing Belt, including on the Canadian side, locally based specialisms provided the basis for increasingly complex intraregional linkages, generating important multiplier effects and helping to sustain a variety of related specialisms in wholesaling, finance, warehousing and transportations. Such interdependencies, also evident in the Axial Belt and the Ruhr, served to reinforce the industrial advantages of these agglomerations. Indeed, the northern spoke of the Axial Belt, the Ruhr and the American Manufacturing Belt, until the 1970s and the onset of the ICT techno-economic paradigm, were each considered to represent the

law of circular and cumulative advantage and to be vibrant, diversified regions capable of self-sustaining growth. There were good reasons for thinking so, especially in the case of the Ruhr and the Manufacturing Belt, which were still nationally dominant centres of industry in the late 1960s. By the mid-1960s, for example, the Ruhr, with 9% of Germany's population, still accounted for 60% of Germany's steel production and 80% of its coal, while the Manufacturing Belt, with 16% of the area of the coterminous US and 40% of the population, contained 80% of production and employment in metal manufacturing as well as major concentrations of other industries, including autos, machinery and many consumer goods (Alexandersson 1967: 38). Even the northern spoke of the Axial Belt had for the most part remained industrially healthy in comparison to more peripheral industrial regions such as central Scotland, south Wales and north-east England (Martin 1988).

In the late 1970s and early 1980s, however, each of these regions experienced extensive (and

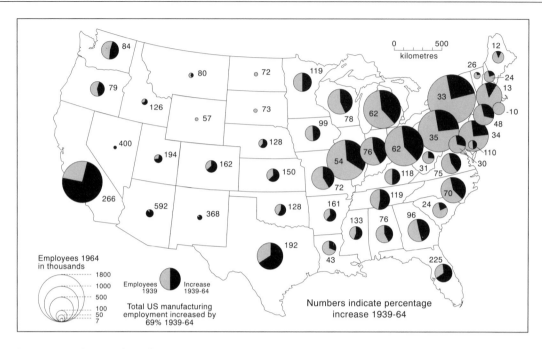

Figure 3.7   US Manufacturing Employment by State, 1939 and 1964. Source: Alexandersson (1967: 24)

unexpected) de-industrialization. Yet, long before this time, manufacturing growth had been faster in other regions. Moreover, spatial shifts in manufacturing at regional scales have typically been complicated by spatial shifts at more localized scales, notably those occurring within metropolitan areas and those between metropolitan and rural (non-metropolitan) areas. The US case provides a good illustration.

Spatial shifts in manufacturing employment: the US case

During most of the period of the fordist techno-economic paradigm, manufacturing employment throughout most of the Manufacturing Belt increased (Figure 3.7). However, between 1939 and 1964 employment in the Manufacturing Belt did not grow as fast as the national average (Alexandersson 1967: 24). The New England and Middle Atlantic States had a particularly low expansion rate. According to Alexandersson

(1967: 25), the relatively slower growth of these parts of the Manufacturing Belt have been evident since the end of the 19th century. In contrast, the fastest rates of growth in manufacturing employment in this period occurred in California and Texas, both of which successfully attracted the new industries of aerospace, electronics and chemicals. Southern states in the south-east, from Virginia to Mississippi, experienced below-average growth as shifts in the clothing industry from New York and the expansion of synthetic fibres were offset by continuing declines in the cotton textile industry.

Over the next decade, much more marked regional differentials in manufacturing employment occurred (Figure 3.8). Thus between 1967 and 1977 the Manufacturing Belt *lost* manufacturing jobs; in the case of the New England and East North Central States manufacturing jobs were reduced by over 15% and the Middle Atlantic States by between 5 and 15% (Watts 1987: 4). The western and southern regions, on

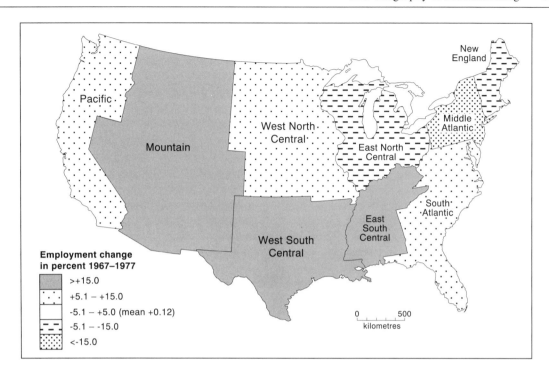

Figure 3.8 US Manufacturing Employment Change by Region, 1967 and 1977. Source: Watts (1987: 4)

the other hand, increased manufacturing employment by at least 5% and often by over 15%. These very substantial regional variations occurred while overall national levels of employment remained more or less the same. Since 1980, regional differences in manufacturing employment change have remained as the nation as a whole shed 11.3% of its jobs (Figure 3.9). Thus between 1980 and 1993 manufacturing employment in the Manufacturing Belt continued to decline and in some cases (e.g. Illinois, Pennsylvania and New York, and in the New England States) job loss was at least 15% and sometimes much greater. New York State, for example, lost about one-third of its manufacturing employment. In addition, significant manufacturing job losses were recorded outside of the Manufacturing Belt for the first time, most notably in California where almost 10% of manufacturing jobs in 1980 were lost by 1993, and in Texas which recorded a decline of 5.9%. However,

most southern states, as well as the mountain states and Washington, experienced manufacturing job growth.

Since the late 1960s, it has become common to discuss regional employment trends in manufacturing in the US as part of a general shift away from northern 'rust' and 'frost' belts in favour of southern (and western) 'sun' (and 'gun') belts (Weinstein et al 1985). In terms of state-level and broad regional-level employment data, this characterization is strongly in evidence from the late 1960s to the late 1970s (Figure 3.8). Since then, again in terms of aggregate state-level manufacturing employment data, this trend is not as clear (Figure 3.9). Certainly, the core states of the Manufacturing Belt had fewer manufacturing employees in 1993 than in 1980. Yet, so too did the sunbelt's powerhouse states of California and Texas. In addition, other sunbelt states did not grow, or grew slowly during this time period. Moreover, since the early 1980s some observers

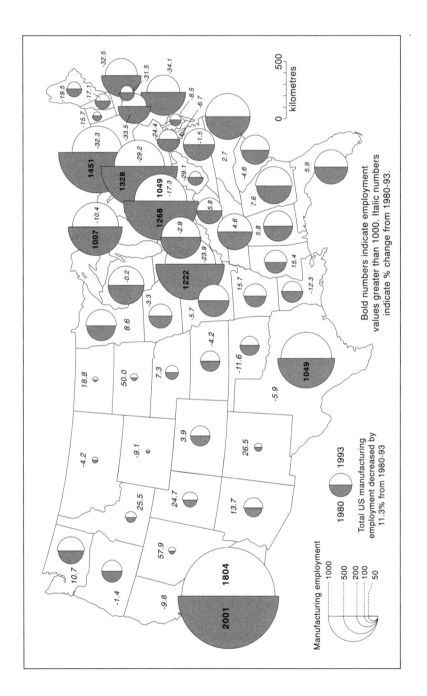

Figure 3.9   US Manufacturing Employment by State, 1980 and 1993. Source: US Bureau of Labor Statistics (1980, 1994)

Table 3.8 The Changing Geographical Distribution of Production Workers: New York Metropolitan Region, 1869–1970

Year	No. of employees (thousands)	No. of employees in Manhattan	Manhattan (%)	Metropolitan region (%)
1869	240.3	131.4	54.7	45.3
1889	683.0	356.5	52.2	47.8
1919	1 158.6	384.7	33.2	66.8
1956	1 483.3	376.8	25.4	74.6
1970	1 889.9	332.6	17.6	82.4

Source: Yeates (1990: 230).

have noted 'economic reversals' in the frostbelt–sunbelt trend. Thus, Weinstein and Gross (1988:10) argue that "The Sunbelt has diminished to only a very few 'Sunspots', while many locations in the Frostbelt have recovered their prosperity". They note that in the 1980s several frostbelt states performed better than many sunbelt states in terms of population migration, employment and total job creation. Florida (1994) has also noted evidence of a revival in the manufacturing fortunes of the Manufacturing Belt (Chapter 16).

These state- and regional-level trends also need to be imprinted on intra-metropolitan trends, metropolitan–non-metropolitan trends and variations among metropolitan areas.

Intra-metropolitan shifts Historically, in the US, as elsewhere, industrialization has been closely associated with urbanization. By 1900, the Manufacturing Belt was anchored by cities that were growing extremely rapidly and it was in these cities that manufacturing investment was strongly concentrated. This close association between industrialization and urbanization continued in the first half of the 20th century, although in an increasing number of metropolitan areas the expansion of manufacturing employment was greater in suburban areas than in the central cities. In the case of New York, for example, the central city area of Manhattan accounted for a consistently declining percentage of manufacturing employees after 1889 (Table 3.8). After the 1920s and until the early 1950s, the

absolute number of manufacturing employees in Manhattan remained more or less the same, while suburban jobs increased substantially. In the 1950s and 1960s, the decline in jobs in Manhattan (11.7% between 1956 and 1970) was more than offset by substantial job growth in suburban manufacturing (see also Meyer et al 1969: 28–30). In fact, this trend was typical for metropolitan areas throughout the US in this period except that for most inner cities the decline was less than that recorded by Manhattan (Table 3.9).

During the 1970s, however, the rate of manufacturing job loss in most inner cities in all parts of the country increased rapidly. In just six years between 1972 and 1977, for example, the inner cities of New York and Miami lost over 19% of manufacturing jobs and several metropolitan areas within the Manufacturing Belt, such as Buffalo, Chicago, Detroit, Milwaukee, Boston, Minneapolis St Paul and Pittsburgh lost at least 10% of manufacturing jobs (Scott 1982: 190). In this time period, some inner cities outside of the Manufacturing Belt continued to attract manufacturing jobs, and metropolitan areas such as Seattle–Everett, Los Angeles–Long Beach and Houston experienced double digit growth, but such occurrences were unusual. Most metropolitan suburban areas, however, continued to increase manufacturing jobs in the 1970s. The few exceptions occurred within the Manufacturing Belt including suburban New York, Buffalo and Pittsburgh which respectively lost 2.4, 4.7 and 8.0% of their manufacturing employment (Scott 1982: 190).

Table 3.9   Changing Manufacturing Employment in 245 SMSAs in the United States, 1947–1967

	Employment		Absolute change	Percentage change
	1947	1967		
Central cities	7 356 733	7 063 426	− 293 307	− 3.9
Suburban rings	4 141 704	8 044 030	+ 3 902 326	+ 94.2
SMSAs	11 498 437	15 107 458	+ 3 609 021	+ 31.4

Source: US Bureau of the Census.

Table 3.10   Changes in Manufacturing Employment in Metropolitan and Non-metropolitan Areas, United States, 1962–1978

	1962–67 (%)	1967–70 (%)	1970–74 (%)	1974–78 (%)
Metropolitan	+ 14.3	+ 0.7	− 2.3	− 1.3
Non-metropolitan	+ 24.1	+ 5.3	+ 10.8	+ 1.2

Source: Haren and Holling (1979: 19).

The non-metropolitan shift   In tandem with the regional shift in manufacturing employment, particularly away from the Manufacturing Belt, and a metropolitan shift, particularly from inner cities to suburbs, with increasing momentum since at least the early 1960s, manufacturing employment in non-metropolitan areas has grown notably faster than in metropolitan areas throughout the US, at least until the early 1980s (Erickson 1976; Lonsdale and Seyler 1979; Norton and Rees 1979; Rees 1979; Cromley and Leinbach 1981; Park and Wheeler 1983). Thus, in the 1960s, manufacturing growth in non-metropolitan areas was significantly greater than in metropolitan areas; and in the 1970s, while the latter lost manufacturing jobs, the former still recorded increases (Table 3.10). Moreover, the growth of non-metropolitan areas at this time occurred in all regions of the country, if more strongly present within the sunbelt.

Over the past decade, there has been growing evidence of more varied employment trajectories in metropolitan and non-metropolitan or rural areas (Mack and Schaeffer 1993). Bechter and Chmura (1990), for example, found that manufacturing job losses in rural counties declined with distance from metropolitan areas, while Bloomquist (1988) found that rural-based manufacturing grew faster in the South and West compared to other regions. At the same time, Bloomquist noted that rural manufacturing jobs in the West were higher quality (waged) than those in the South. More generally, Mack and Schaeffer (1993) claim that while manufacturing employment levels increased modestly more in rural areas than in metropolitan areas during the 1980s, the latter areas did much better in terms of growth in real earnings. There is an argument that rural employment in manufacturing has become increasingly constrained (T. A. Clark 1991) and that an increasing number of metropolitan areas are beginning to regain manufacturing employment.

The unpredictability of locational evolution   Clearly, broad, significant shifts in the manufacturing employment geography of the US have occurred since 1950, even when simply presented in terms of state-level employment levels and shares. Yet, major new trends were not predicted by either conventional or radical theories, or by forecasts. Indeed, the reverse is the case. The

Manufacturing Belt in the 1950s and 1960s was widely portrayed as offering overwhelming locational advantages in terms of infrastructure, employment and market access. Harris' (1954) famous and oft-cited market potential maps, for example, seemed simultaneously to confirm the Manufacturing Belt as the locus of the nation's greatest market potential and most advantageous location for industry (see Chapter 4). Basically, all theories then stressed the permanence of established core areas. Similarly, the shift of manufacturing activity towards non-metropolitan areas in the 1960s and 1970s surprised established theories which emphasized the advantages of agglomeration economies of scale.

Numerous explanations, reflecting alternative theoretical positions, for the changing geography of the manufacturing employment patterns in the US have been offered (Peet 1983; Weinstein et al 1985; Markusen et al 1991). While this chapter is primarily concerned with a geographical overview of industrial change, and without reviewing this literature, three general observations regarding such explanations can be made at this time. First, explanations of employment change will invoke different factors depending upon period and geographic scale. Second, explanations need to recognize variations in global as well as local factors. Industry within the US, for example, has faced a very different competitive challenge from firms in foreign countries since the 1970s and the onset of the ICT paradigm compared to the 1950s when in many respects US industry was in a virtual monopoly position (Weinstein and Gross 1988). Third, employment change in the private sector in the US is primarily implemented by individual firms whose capabilities and judgements, even when operating in similar circumstances, can vary considerably. Thus, firms in the same industry and region can respond to restructuring in different ways: by outright failure, by relocation, by branch plant investment or by *in situ* change, while the characteristics of employment, technological and product changes implemented at new or existing locations can also vary. To the extent that such variations are important, and rein-

forced by variations in the behaviour of other agents such as local and regional governments, geographic trends are themselves likely to be variegated.

Trends elsewhere In other countries there have been significant regional and local differentiation in the performance of the manufacturing sector over time and in given time periods. In the UK, for example, the massive 'northwards' shift of manufacturing employment in the 19th century was arrested in the 20th century. By the 1930s, if not before, the highly specialized coalfield regions of the periphery were already experiencing long-term structural unemployment although the more diversified northern spoke of the Axial Belt, or what Martin (1988) refers to as the industrial heartland, continued to perform well until the 1960s. Since the 1960s, however, while the problems of the northern fringe areas remained, de-industrialization began to take hold of the northern spoke or industrial heartland – a trend that subsequently spread to the southern and eastern regions of the country (Keeble 1976; Fothergill and Gudgin 1982; Martin and Rowthorn 1986; Massey 1988; Healey and Ilberry 1990). Although regional rates of manufacturing employment change have varied, the general theme in the UK is one of de-industrialization; by 1988, only East Anglia had more manufacturing employees than in 1966 (Table 3.11; Figure 3.10).

Throughout the UK, paralleling US experience, a strong trend towards non-metropolitan industrialization occurred in the 1960s and 1970s (Fothergill et al 1985: 149). Indeed, between 1960 and 1981 only rural areas recorded growth in manufacturing employment. Similarly, it is only in what Townsend (1993) calls "remoter, mainly rural" industrial areas that manufacturing jobs have increased in the 1980s; he also argues that in the UK, in contrast to the US and other industrialized countries, the relative shift in manufacturing to rural areas remains strong, with few signs of metropolitan areas arresting manufacturing decline (Table 3.12).

Table 3.11 Regional Manufacturing Employment Change in Great Britain, 1966–1988

| Regions | 1966 (000s) | % change in employment | | | 1988 (000s) |
		1966–1974	1974–1984	1984–1988	
South-East	2 363	− 14.4	− 20.6	− 11.1	1 321
East Anglia	173	+ 18.5	− 6.9	+ 14.7	218
South-West	429	+ 4.4	− 14.9	− 3.2	364
East Midlands	631	− 2.2	− 19.3	+ 1.0	493
West Midlands	1 197	− 9.7	− 28.6	− 1.7	697
Yorkshire and Humberside	860	− 11.2	− 31.7	− 8.7	443
North-West	1 251	− 12.9	− 32.0	− 10.6	600
North	461	+ 1.3	− 33.3	− 5.4	261
Wales	317	+ 5.7	− 31.8	+ 0.5	213
Scotland	726	− 5.1	− 28.7	− 11.3	385
Great Britain	8 408	− 8.4	− 25.5	− 6.2	4 995

Source: Healey and Ilbery (1990: 321).

In most advanced countries tendencies towards some kind of dispersal of industry from established regional and urban concentrations have been identified. This observation applies to Japan although it should be noted that in Japan's case manufacturing industry remains highly concentrated in two major urban regions, the Kanto region anchored around Tokyo and the Kinki region anchored by Osaka. Moreover, in contrast to the de-industrialization that has occurred in London and New York, Tokyo and Osaka remain powerful agglomerations of manufacturing.

In addition to regional and metropolitan trends, unpredicted shifts in the locational evolution of manufacturing industry on an international scale have occurred. In particular, the speed and extent of de-industrialization in the UK, the emergence of the NICs and the rise of Japan as one of the world's two leading industrial powers, were similarly surprising developments from the perspective of existing theories, both conventional and radical.

The manufacturing sector in economic development

Since at least the Industrial Revolution, the richest, most powerful countries in the world have been the most industrialized. For the past 300 years, for example, the countries now comprising the Organization for Economic and Co-operative Development (OECD) dominated the industrialization process while recording consistent and substantial increases in per capita income, as measured by per capita gross domestic product (Table 3.13). With respect to leading countries, the UK (and Australia) enjoyed the highest per capita income in 1820 and 1870 at the time of its industrial (and political) hegemony, while the US (and Canada, arguably its economic colony) achieved the highest per capita incomes by no later than 1950, by which time its industrial output was about four times greater than the UK. In 1989, the US was still the world's biggest industrial producer and the OECD country with the highest income levels. Between 1950 and 1989, however, Japan's per capita incomes increased tenfold and its share of global manufacturing output rose from 1–3% to almost 20%, close to US levels. Germany's income and industrial growth has also grown in tandem. In 1989, the US, Japan and Germany were the three most powerful countries in the world, the richest in terms of absolute measures of wealth and among the richest in per capita income levels. They also accounted for over half of the world's manufacturing output and they were

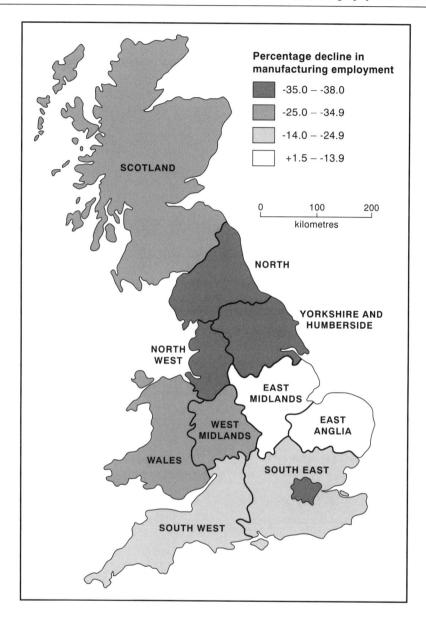

Figure 3.10 United Kingdom: Industrial Employment Decline, 1976–1986. Source: Martin (1988: 397)

easily the three biggest exporters of manufacturing goods.

Historically, as countries have developed economically, the manufacturing sector has grown rapidly and accounted for an increasing share of employment. Among the 16 OECD countries, for example, by 1870 manufacturing

(and construction) industry accounted for almost 27% of employment and this share continued to increase until 1960 when manufacturing accounted for almost 39% of total employment (Table 3.13). Since 1973, the share of manufacturing (and construction) employment has fallen among OECD countries, and is still falling.

Table 3.12 Manufacturing Employment Change: Great Britain 1981–1989, South and North, by type of district (thousands of employees)

	Total		Total change 1981–1989	
	1981	1989	No.	%
South				
Inner London	282.6	180.4	− 102.2	− 36.2
Outer London	401.3	263.3	− 138.0	− 34.4
Non-metropolitan cities	416.0	323.6	− 92.4	− 22.2
Industrial areas	358.0	322.5	− 35.8	− 10.0
Districts with New Towns	200.4	178.1	− 22.3	− 11.1
Resort, port & retirement	169.5	159.9	− 9.6	− 5.7
Urban & mixed urban–rural	695.4	652.4	− 43.0	− 6.2
Remoter, mainly rural	274.0	283.9	+ 9.8	+ 3.6
Total	2 797.2	2 364.1	− 433.5	− 15.5
North				
Principal cities	604.9	413.5	− 191.4	− 31.6
Other metropolitan districts	1 133.6	938.7	− 194.9	− 17.2
Non-metropolitan cities	293.3	251.2	− 42.1	− 14.4
Industrial areas	610.0	568.3	− 41.8	− 6.8
Districts with New Towns	195.5	193.6	− 1.9	− 1.0
Resort, port & retirement	48.9	45.7	− 3.2	− 6.5
Urban & mixed urban–rural	207.6	195.3	− 12.3	− 5.9
Remoter, mainly rural	159.7	169.6	+ 9.9	+ 6.2
Total	3 253.5	2 775.9	− 477.7	− 14.7

Source: Townsend (1993: 211).

Table 3.13 Levels of GDP per Head of Population ($ at 1985 US Prices)

	1820	1870	1950	1973	1989
Australia	1 242	3 123	5 931	10 331	13 584
Austria	1 041	1 433	2 852	8 644	12 584
Belgium	1 024	2 087	4 228	9 416	12 876
Canada	n.a.	1 347	6 113	11 866	17 576
Denmark	988	1 555	5 224	10 527	13 514
Finland	639	933	3 480	9 072	13 934
France	1 052	1 571	4 149	10 323	13 837
Germany	937	1 300	3 339	10 110	13 989
Italy	960	1 210	2 819	8 568	12 955
Japan	588	618	1 563	9 237	15 101
Netherlands	1 307	2 064	4 706	10 267	12 737
Norway	856	1 190	4 541	9 346	16 500
Sweden	947	1 316	5 331	11 292	14 912
Switzerland	n.a.	1 848	6 556	13 167	15 406
UK	1 405	2 610	5 651	10 063	13 468
USA	1 048	2 247	8 611	14 103	18 317
Arithmetic average	1 002	1 653	4 693	10 396	14 456

Source: Maddison (1991: 6–7).

Historically, agriculture's share of employment has declined consistently while service sector employment has increased consistently. According to Fisher (1935) and Clark (1940), a three-sector 'stage' model of economic development is associated with this sequential shift in employment composition, in which the primary, secondary and tertiary sectors all take their turn dominating the economy. The underlying rationale relates to demand and technological changes. Thus, in agriculture, while demand for food has not risen proportionately with income, technological changes created massive productivity improvements, so the need for employment in agriculture drops with economic development. However, manufacturing then 'takes-off' to become the lead sector as technology creates more jobs than it replaces and society's demand for manufactured goods tends to increase with income, i.e. the income elasticity of demand for manufactured goods is high. Finally, services become more important with economic development as society's demands shift towards services and because productivity improvements have traditionally been harder to achieve in this sector.

Does industry matter anymore?

The variation in broad sectoral employment composition with economic growth tends to be more varied in practice than the Fisher–Clark model admits. Nevertheless, this model underlies an increasingly widely held view that in advanced industrialized countries manufacturing does not matter from the point of view of job and even wealth creation. Indeed, Bell (1974) coined the term 'post-industrial' society to suggest that western economies in particular were shifting from a "goods producing society to an information or knowledge society". Certainly, the share of manufacturing to total employment has declined in advanced industrial countries. Thus, for the major OECD countries as a group, manufacturing (combined with the construction and utilities industries) reached its greatest relative importance as an employer in 1960.

Since then, it has gradually declined in relative significance (Table 3.14). Among individual OECD countries, manufacturing employment shares vary but downward trends are clearly evident (Table 3.15). In the UK, the relative decline of manufacturing jobs has been matched by absolute declines in jobs for over two decades. During the 1980s several OECD countries lost manufacturing jobs or have not seen much growth.

There are several problems, however, with a too-literal interpretation of the post-industrial model, which suggests that manufacturing is unimportant. First, as already noted in Chapter 1, the various sectors of the economy are interdependent, and manufacturing industries remain significant sources of dynamism throughout the economy by continually and directly generating new demands in the construction, transportation, business services, utilities, and primary industries, and many government departments are in one way or another designed to serve manufacturing. Defined broadly, the goods-producing sector of modern economies probably still accounts for over half of employment (Britton and Gilmour 1978). Clearly, growth in manufacturing also responds to growth in services but the 'material' basis of modern societies should not be underestimated.

Second, and following on from the first point, the production of goods and the processing of knowledge and information are not two distinct segments of the economy. Rather, manufactured goods are themselves technology, design and knowledge intensive. Thus, manufacturing stimulates a wide range of interesting and intellectually demanding tasks, which translates into high-income, interesting job opportunities. At the same time, sustained economic growth requires that economies innovate and create design and knowledge intensive goods in order to simultaneously maintain market share and high income. Indeed, the manufacturing sector is vital to productivity increases, and while in established industries increased productivity often means job losses, new technology also generates employment through the creation of new demands and

Table 3.14 OECD Countries: Percentage Structure of Employment (Average for 16 Countries)

	Agriculture	Industry	Services
1870	48.8	26.9	24.3
1950	24.7	36.6	38.7
1960	17.5	38.7	43.8
1973	9.3	37.3	53.4
1987	6.0	30.5	63.5

Source: Maddison (1991: 73).
Agriculture = farming, forestry and fisheries; industry = manufacturing, utilities and construction; services = everything else.

Table 3.15 Manufacturing Employment as a Percentage of Total Employment, Selected OECD Countries, 1972, 1983 and 1992

	1972	1983	1992
Australia	25.5	18.1	14.5
Canada	21.8	17.6	14.6
West Germany	36.8	31.3	30.8
Japan	27.0	24.5	24.4
United Kingdom	32.9	23.9	–
United States	24.3	19.8	17.0

Source: International Labour Office (1993).

because technology itself is manufactured in the form of machinery and equipment.

Third, employment is just one measure of the size and importance of manufacturing and other sectors. Value added and income, for example, are other variables which, compared to employment, are likely to assign greater importance to complex, capital-intensive manufacturing. Manufactured goods and raw materials also comprise international trade in 'visible' goods and for individual countries exports in relation to imports provide vital signs of economic health. Invisible trade in services has grown massively in recent years. Again, however, this invisible trade is not independent of visible trade as payments for business services by manufacturing activities are an important component of invisible trade. It is no coincidence that the countries with the largest visible trade are also the largest industrial

powers and the homes of most multinational manufacturing corporations. It is also true that international money flows are now vast and are occurring on a daily basis literally at the push of a button. Yet many of these flows are exchanges of numbers in an account, and have no immediate bearing on the 'real' economy and the actual production of goods and services – except in so far as they are potentially highly destabilizing. A good case can be made to impose restrictions on short-term money transactions whose global circuits are measured in days and have no relationship with actual investments on the ground (including in services!).

Fourth, if direct employment is declining relatively among OECD countries, and even absolutely in some cases, then in many other 'developing' countries it is increasing. For example, in China, the biggest country in the world, manufacturing employment has grown from a very small base in the 1960s to about 16% of total employment in 1990. Moreover, 'developing' countries that have grown rapidly in recent years and have achieved impressive increases in per capita incomes (e.g. Hong Kong, South Korea, Malaysia and Taiwan) are also increasingly important manufacturers and exporters. In this regard, it is important to recognize that while in 1900 a handful of countries served global demands for manufacturing, in the 1990s these demands are met from a variety of sources. Moreover, countries and regions that have lost significant numbers of manufacturing employees (e.g. northern Britain and north-eastern US) have suffered economically and socially. To an important degree, the social problems of inner Detroit, Michigan, and Sheffield, England, are not due to industry being unimportant but because it still is; it has simply declined too fast and too far in these places. Despite its spectacular success, MacDonald's has not been able to produce a hamburger that is as valuable as a Toyota, or incomes that are as good.

Fifth, the massive job creation problems that face the OECD and developing countries, and which are particularly profound for young people, are related to an extraordinarily efficient manufacturing sector whose global supply capabilities have outstripped demand. This near world-wide job problem will not be easy to resolve but the solutions will scarcely be helped by neglecting manufacturing. Certainly, services do not appear as a ready-made alternative solution, and have not resolved the employment problem. It may well be that those economies in control of vibrant innovative manufacturing activities will increase their chances of resolving job creation problems. Moreover, the increasing advocacy of income distribution policies as a way of reducing growing inequalities in countries and offsetting lost job opportunities ultimately depend on wealth being generated in the first place. Those economies which can generate wealth from manufacturing will be in a better position to redistribute income. This may be the worst time to neglect manufacturing.

Finally, the historic relationships between industrial strength, political strength and the standard of living is not coincidental. Recent trade debates between the US, Japan and Europe, for example, over autos and many other goods-based trade matters suggest political leaders of powerful countries still think in these terms and are vitally concerned with the health of their manufacturing industries. Similarly, the continued strong interest among developing countries in stimulating manufacturing structures suggests that their political leaders share similar concerns. The caveat, of course, is that the relationship between industrialization and economic growth is not automatic and attempts to industrialize are not inevitably successful.

National policies towards industrialization: a comment

Once the UK had industrialized and had begun to export its goods, the problem facing other countries was how to develop the industrial skills already developed in the UK. In addition to goods, the UK could provide managers, workers, capital and technology, as well as a working example of industrial operations which could be observed. As other countries industrialized, late developers could then examine alternative models and obtain expertise from a wider range of sources. For national governments, then as now, how best to industrialize constituted significant policy dilemmas concerning such matters as protectionism, industrial priorities, strategy of industrialization, the nature of financing, and the control and organization of industry.

Although national paths to industrialization are complex, a basic distinction, which remains useful, is that between the so-called Anglo-Saxon economies who subscribe to a philosophy of economic liberalism, founded principally on the writings of Adam Smith, and countries favouring a philosophy of economic nationalism, founded principally on writings of Frederick List (1922). According to Smith's (1986) economic liberalism, the wealth of nations is rooted in the division of labour and the pursuit of self-interest which, if regulated by fair competition (the 'invisible hand'), will ultimately serve the needs of the consumer. In this view, governments should generally have a minimal role in economic matters and, within their own countries and among nations, strive to promote 'economic freedoms', in terms of trade, investment and the movement of labour. According to List's (1922) economic nationalism, the wealth of nations depends on the development of 'productive forces', featuring the harmonious interrelationships among (and within) industry, agriculture, commerce and transportation, and equally essentially, the cultural basis for production involving improvements in education, health, religion and family life. In this view, rather than simply leave matters to self-seeking individuals, governments need to comprehensively develop the productive forces necessary to promote an integrated and balanced range of occupations. In this regard, List recognized that by the early 1800s the UK's industrial capacity exceeded that of any other so that industrial diversification in Germany and the US required that their 'infant industries' be protected from British competition. At the same time, List advocated that Germany should promote free trade within the country in order to establish a stronger nation state.

The historical importance of economic nationalism History underwrites List's advocacy of protectionism on the grounds of the infant industry argument; indeed an important exception is hard to identify. Certainly the US, along with Germany, adopted protectionist measures in the 1830s, and in subsequent years, virtually all other industrializing countries did so, including Japan. Even given this historical consensus for protectionism, the different traditions of economic liberalism and economic nationalism underlay significant differences in national trajectories of industrial development. Thus, countries which share the tradition of economic liberalism (e.g. the US, UK, Canada and Australia) tend to have a more deep-rooted scepticism over the role of government involvement in industrial affairs, while according greater stress to the 'independence' of management, 'arms-length' relationships among firms, sectors and between managers and workers, and to financial results in the short term (Dyson 1983). In contrast, countries which share a tradition of economic nationalism, including (in different ways) Germany and Japan, have recognized more active roles for governments, have placed more emphasis on developing interdependencies among firms, sectors and between managers and workers, and have adopted long-term views regarding industrial performance.

Apart from philosophical differences in the role of governments, differences in liberalism and nationalism as a basis for industrial development are more specifically expressed in such matters as attitudes towards technological transfer, the development of a capital goods or machinery sector, skill formation, inter-firm relations, the nature of the links between financial and manufacturing sectors, and, by no means least, the control of industry, including the role to be played by direct foreign investment (DFI). Foreign-owned branch plants, for example, have always been better appreciated in countries dominated by economic liberalism such as the

UK, Canada and the US. On the other hand, those countries favouring some form of economic nationalism have been more cautious towards allowing the entry of foreign-controlled plants. Their underlying concerns are that branch plants do not effectively transfer technology, while at the same time comprising economic and political sovereignty over long-term industrial development. Through formal legislation prohibiting foreign investment, bureaucratic procedures, local custom and planning requirements and as a result of complex interdependencies between manufacturing firms and financial institutions which render acquisition very difficult, countries as different as Japan and Sweden have sought to limit the extent of foreign control. Instead, these countries see value in the domestic control of industry.

In contemporary times, economic nationalism is usually expressed either as statism or as corporatism whereby the government takes leading and participatory roles in economic management respectively. However, defined, it may well be that the lines between nationalism and liberalism are becoming blurred (Katzenstein 1985). Yet, these distinctions continue to underlie important policy approaches and attitudes, including the trade and investment conflicts between the US and Japan, and internal debates in newly industrializing countries such as China. It may well be that distinctions of industrial culture remain vitally important, if more nuanced, at a time when trade and investment liberalization dominates rhetoric.

Canada and Sweden as late developers Canada and Sweden – two late-developing medium-sized industrial powers and resource-rich 'northern' countries – offer an interesting comparison of approaches to industrialization, including with respect to foreign investment. Thus, as exemplars of liberalism and nationalism respectively, Canadian industrialization stressed the independence of the financial and industrial sectors, while Sweden more closely articulated these sectors; within the industrial sector Sweden encouraged important dimensions of co-operation among firms, such as between goods producers and domestic machinery producers, while Canada emphasized competitive relations, encouraging producers to buy machinery from whatever source; and while Sweden strongly favoured the development of domestically controlled industry, Canada equally strongly favoured an open-door policy to foreign investment in the manufacturing sector (Laxer 1989).

Canada adopted protectionism as a key element of its industrialization strategy in the 1870s (as did Sweden) and this policy directly encouraged foreign firms to invest in Canada, whether by investment in new plant or by acquisition. In fact, foreign ownership of manufacturing was already important by the early 1900s and a dominant feature by the 1930s (Marshall et al 1936). In Sweden, on the other hand, a key aspect of its strategy was to maintain control over industry and build up the skills of its own workforce. In the early part of the 19th century, in the critical machine-tool industry, for example, Sweden relied heavily on British exports which, while not the most advanced available at the time, allowed Swedish workers to progressively develop their industrial skills (Gooch and Castensson 1991). Where possible, the secrets of industrial success elsewhere were discovered (e.g. by importing equipment and reverse engineering it), domestic R&D was encouraged and foreigners establishing branch plants were limited and not encouraged. In contrast, Canada relied heavily on foreign investment as a way of acquiring technology that was cheaper and faster than redeveloping the same technology domestically.

Any assessment of the effects of alternative economic philosophies on the industrial structure of Canada and Sweden must recognize that the history, culture and situations of the two countries differ, and Canada's economy, for example, cannot be understood without recognizing its intimate ties with the US. Even so, there is a widespread view that differing philosophies have helped shape distinctive industrial structures in Canada and Sweden (Laxer 1989). In particular, although the smaller country,

Sweden has developed a more innovative and diversified range of globally competitive secondary manufacturing activities than has Canada whose industrial structure is still heavily weighted towards primary manufacturing activities. A particularly noteworthy difference relates to the greater extent to which Swedish industry has invested in research and development (R&D), even in the primary industries, such as the forest industries, which are vital to Canada as well as to Sweden (Hayter 1988). An important factor explaining Canada's lower levels relates to ownership as foreign firms in Canada have primarily concentrated their R&D in their home economies (Chapter 15). A similar point can be made across the industrial structure (Britton 1980). Consequently, while industrial competitiveness in Canada continues to rely on importing technology, there is a stronger indigenous technological capability in Sweden, which increases the chances of employment creation and participation in the global economy in a more diversified way, notably by developing value-added products.

Can the NICs be cloned? The historical relationship between industrialization and economic development is reinforced by the emergence of the NICs whose wealth has increased as they have become important industrial exporters. Indeed, firms based in Singapore and Hong Kong have become important suppliers and organizers of investment and expertise throughout Asia, most notably in helping to fuel China's industrialization plans. Singapore, for example, now constitutes a high-wage economy and it has become less attractive to low-wage, low-skill activities such as textiles and electronics. On the other hand, its industrial base has become more sophisticated and Singaporean organizations are now selling their expertise in industrial park development and management to regions and towns within China (and elsewhere) while also providing crucial and mutually beneficial bargaining services linking MNCs to their Chinese hosts. Meanwhile, Hong Kong based firms have been the leading sources of industrial investment in China ('foreign' investment until 1997!) since the latter's economic reforms of the late 1970s (Sit 1986; Wang and Bradbury 1990; Eng and Lin 1996). Predominantly, these investments are in labour-intensive, low-wage, low-skill activities, while value-added activities are kept in Hong Kong or relocated elsewhere.

Singapore and Hong Kong, as city-states in strategic locations which have long been global cities, have certainly enjoyed circumstances favourable to industrial and economic growth. The larger NICs, however, have become even more important export-oriented manufacturing centres whose growth is continuing. South Korea and Taiwan are particularly noteworthy in the extent to which they have developed; Arrighi and Drangel (1986: 44) suggest they are exceptional in shifting from peripheral to semi-peripheral status. Moreover, as Brohman (1996: 107) points out, despite what is commonly supposed, the NICs' approach to industrial development has involved strongly activist roles by governments along the lines of classic Listian mercantilism, with an emphasis on inward- as well as export-oriented development.

An important question is whether or not more economically disadvantaged countries can 'clone' the NICs, perhaps by pursuing export-based industrialization as a way of increasing standards of living (Arrighi and Drangel 1986; Gereffi 1989; Gereffi and Korzeniewicz 1990). A related question is whether the NICs themselves can sustain their development to become 'core' countries. Gereffi and Korzeniewicz (1990) are pessimistic. They note that since the 1930s few countries have been able fundamentally to change their status within the admittedly crude characterizations of periphery, semi-periphery and core country status. For them, the performance of South Korea and Taiwan (and of Italy and Japan shifting from semi-peripheral to core) are the exceptions that prove the rule of 'immobility'. They note in particular that industrialization in Latin American countries has not sustained wider economic progress. Moreover, they conclude from an analysis of production chains in the footwear industry that core countries are

still in control of the most profitable retail, marketing and higher-value functions, while the early stages of the chain in agriculture and manufacturing, which are in the semi-periphery or periphery, are the least profitable, or at least the lowest waged (see Chapter 11). In their view, attempts to increase value in the semi-periphery by greater innovativeness is offset by continued innovations in the core countries.

Yet if industrialization is a deeply cumulative process it is also highly dynamic. A little over one generation ago there was little discussion at all about export-based manufacturing in developing countries. Even more surprisingly, Japan has been virtually ignored until recently. Industrialization no longer guarantees development in a wider sense but it substantially enhances such possibilities. The very poorest countries in the world remain the least industrial, the very richest the most industrial, and it is the NICs that are most likely to significantly improve standards of living. At the same time, it might be noted that as the global economy becomes more intensively globalized and fragmented, the very notions of core, semi-periphery and periphery, and the facile association between high-value, medium-value and low-value industries, come into question (Glasmeier et al 1993). Indeed, Scott and Storper (1992: 8) suggest that the established core and periphery structure of the global economy is being replaced by a 'global mosaic of regions'. The relationships between industry and local development are increasingly complicated, and they are increasingly demanding that their organization be understood.

Conclusion

The 'geography of industrialization' has only been broadly outlined in this chapter. Nationally, each wave of industrialization is associated with international shifts in technological leadership (Figure 2.3) and this leadership is associated with shifts in industrial production. Within countries, industrialization is also a geographically selective process. Within the UK, for example, Hall (1981, 1985) has linked each long wave of industrialization with particular geographical patterns, while Massey's (1984) geological metaphor of industrialization is based on successive layering of new activities ('sediment') and metamorphosing of existing activities. Yet, this evolving geography is not easy to summarize. On the one hand, several observers (e.g. Berg 1985) are impressed by the concentrating effects of the Industrial Revolution, noting that by the second Kondratieff wave, massive industrial agglomerations were taking shape, the largest being in the UK, Europe and the US. Other observers stress how the technological and institutional innovations sparked off by the Industrial Revolution initiated possibilities for dispersal. Alexandersson (1967: 12–13), for example, emphasizes the growing 'location

freedom' of manufacturing as energy sources, originally tied to water power and then coalfields, have widened to include electricity and oil-pipeline networks which, along with other sources, have essentially rendered power a ubiquitous input. Reference to the progressive expansion of transportation and communication networks and to the emergence of the multinational corporation reinforces this point of view.

If, on the one hand, there are reasons to believe capital is increasingly mobile (Bluestone and Harrison 1982; Harvey 1990), on the other hand, the inertial effects of the 'sunk' costs of capital should not be underestimated either (Clark 1994; Clark and Wrigley 1995). The advantages of relocation are inevitably complicated by the advantages of staying put.

PART II

THE LOCATION OF FACTORIES

The geographically highly uneven distribution of manufacturing activity provides the underlying stimulus for an understanding of industrial location. The role of manufacturing in economic growth and its importance to local, regional and national economic development policies in many parts of the globe gives added relevance to enquiries into the location of industry. The next four chapters briefly address four approaches that have been developed in geography over the last several decades to explain industrial location. In each approach, the principal focus of the discussion is the location of a new factory although other forms of locational adjustment are recognized (Krumme 1969b).

Chapter 4 reviews a well-established typology of location conditions and the changes in location conditions over time, and discusses the nature of locational factors as derived by surveys of manufacturers who have recently established new factories. The next three chapters seek to interpret location factors from the perspective of more formal theory. Chapter 5 reviews neoclassical industrial location theory, especially that theory which interprets the location of factories as a relatively abstract cost-minimizing exercise. Although not incorporated within economic geography textbooks until the 1960s, by 1970 neoclassical industrial location theory, which has its roots in the established and powerful theories

of neoclassical economics, rapidly came to represent, to borrow Galbraith's (1967) phrase, 'conventional wisdom' within industrial geography. However, as conventional wisdom, neoclassical location theory was widely criticized within industrial geography and alternative theoretical approaches were soon mooted. Behavioural location theory, for example, interprets the location of factories as a decision-making (and learning) process and this approach is reviewed in Chapter 6 (Pred 1967). Chapter 7 discusses the 'enterprise' (institutional) approach which interprets location as an exercise in bargaining (Krumme 1969a). In turn, each of these approaches is derived from a different theory of the firm in economics.

Factory location and the theory of the firm

Since location is ultimately a part of an investment decision made by individual firms, any theoretical explanation of location must invoke a theory of the firm. In a general way, firms function to co-ordinate various activities; in particular, firms assemble materials and services in factories in order to add value by the application of factors of production (labour, land, capital and entrepreneurship) and to then ensure the distribution of the resulting output.

Table II.1   Theories of the Firm and Industrial Location

Characteristic	Neoclassical theory	Behavioural theory	Institutional theory
Decision-making type	Economic person	Satisficer person	Technostructure
Decision-making capabilities	Perfect rationality, information	Bounded rationality, information	Strategy and structure, power
Goal(s)	Minimum costs, maximum profit	Aspiration levels or better	Growth, security and profit
Theory of competition	Perfect (and 'fair')	Perfect? (and fair?)	Monopolistic ('unfair')
Nature of economic landscape	Cost and revenue surfaces	Information space, action space	Big business, big labour, big government
Nature of economic relationships	Arms length	Information flows	Bargaining, collusion, persuasion
Location decision-making	Automatic, instantaneous	Learning process	Bargaining process
Location change (long run)	Adapt/adopt to economic forces	Learn, adapt to economic forces	Political economy and technology

While how firms perform these activities has been conceptualized or theorized in many different ways, some time ago Machlup (1967) recognized over 20 different (economic) theories of the firm which he classified into three broad types, notably neoclassical, behavioural and managerial theories of the firm. Within economics, these theories provide alternative ways for analysing prices, profits, production and investment, and of linking changes in markets with changes in supply conditions. Within geography, these theories provide the broader context for distinguishing three approaches to industrial location (Table II-1). In particular, each of these theories is based on different concepts of the firm in terms of abilities and motivations, different concepts of the wider economy or 'environment' in which individual firms function, and different emphases of how firms relate to this wider economy or environment.

The three different theories of the firm and approaches to industrial location, alternatively interpret firms as Economic Persons, Satisficer Persons and Managers or Technostructures. Economic Persons optimize or maximize and are economically rational; they operate within an economic landscape that is interpreted in terms of costs and revenues; they respond automatically or instantaneously to economic forces; profit maximizing is considered socially beneficial

so there is no conflict between firms and the economy. In the long run, neoclassical firms survive by adapting to laws of demand and supply or, unless fortuitously saved by government subsidy or some other form of 'adoption', they fail. Satisficing firms, on the other hand, have limited information and they are boundedly rational; the behavioural economic landscape is interpreted as information flows which firms process through their 'mental maps' in order to make decisions. In the long run, satisficing firms survive by learning (Simon 1957). For their part, managerial firms are dominated by big corporations which plan, develop strategies and structures, seek to grow, make profits and reduce uncertainty, and operate within an economic landscape of institutions interpreted in terms of 'countervailing' powers', and the centre of which is a set of bargaining strategies between business, government and labour (Galbraith 1952). In the long run, industrial evolution depends upon the play of these forces of political economy.

Collectively, these three frameworks serve to remind us that the reasons for industrial locations are complex, perhaps more so than is often realized, and they collectively enrich our understanding of location. In practice, eclectism in contemporary literature on industrial location has blurred the boundaries between the theories.

At the same time, these theories offer a different 'micro' basis for different evaluations of the evolution of industrial location patterns, different justifications of industrial location policy and different interpretations of 'labour' as a location factor. The distinction between neoclassical and managerial theory is of particular importance. In the neoclassical economy firms are abstract agents linking the forces of demand and supply in a manner consistent with the goals of society. In contrast, in the institutional economy it is explicitly recognized that the goals and priorities of firms may differ from that of other actors in the economy as a whole, including local and regional interests. In this latter approach, firms have real, if not autonomous, bargaining power as economic agents able to at least partially influence their environment. Such a review recognizes the legitimacy of conflicting views, the possibilities of alternative courses of action and in general allows for a more critical understanding of the decisions and impacts of firms on local economic development.

4

LOCATION CONDITIONS AND LOCATION FACTORS

Location conditions and location factors define characteristics which vary from place to place and which directly or indirectly affect the viability of factories. There is an extensive literature in industrial geography and related disciplines which has sought to identify location conditions and factors, particularly with respect to why firms choose to locate *new* factories. While there has been a tendency in such literature to interpret location conditions and factors as synonyms, Nishioka and Krumme (1973) prefer to make a distinction. In particular, they define location factors as the specific interpretations made by individual firms of more general location or environmental conditions. According to Nishioka and Krumme (1973), location conditions are the differences among locations that exist for all industries, while location factors refer to the specific importance that is attached to such differences by individual firms when choosing locations for specific factories. In other words, a general bundle of location conditions is 'translated' by firms into a more specific set of location factors relevant to their particular operations. This distinction between location factors and conditions is a potentially useful one for understanding the highly divergent locational reasoning offered by firms in explaining their location choices and in distin-

guishing the perspectives of the community or region from that of firms in surveys of locations and locational characteristics.

This chapter has two main parts. The first discusses a typology of general location conditions. The second part of the chapter examines location factors by looking at selected examples of location surveys which are drawn from different stages of so-called product life cycles. The product life cycle model is an important concept within industrial geography and it is reviewed as part of this discussion. The location surveys, it might be noted, typically involve structured questionnaires administered to individual respondents representing a sample of factories or establishment drawn from a larger population. That is, they are examples of extensive research designs (Chapter 1).

What places offer industry: a typology of location conditions

Places, whether thought of in terms of nations, regions or communities, vary in terms of the bundle of location conditions available to industry. As a way of illustrating these variations, 11 broad categories of location conditions are recognized (Table 4.1). While there is no one

Table 4.1 A Typology of Location Conditions

Location condition	Tangible features	Non-tangible features
1. Transportation facilities	Freight rates	Reliability, frequency, damage, availability
2. Materials	Production costs, transportation costs	Security, quality
3. Markets	Transportation costs, servicing costs	Personal contact, tastes, rivals
4. Labour	Wages, non-wage benefits, hiring costs	Attitude, unionization, skill, type, turnover, availability
5. External economies		
(a) urbanization		Externalities (positive and negative),
(b) localization		labour skills, information sharing, common services, reputation
6. Energy	Costs	Reliability, diversity
7. Community infrastructure		
(a) SOC	Capital cost, taxes	Quality, diversity
(b) EOC	Capital cost, taxes	Quality, diversity
8. Capital		
(a) fixed	Construction costs, rent	Availability, lay-out age
(b) financial	Cost of borrowing	Availability
9. Land/buildings	Costs	Size, shape, access, services, lay-out
10. Environment		
(a) amenity		Worker preferences
(b) policy	Costs, taxes	Local attitudes
11. Government policy	Incentives, penalties, taxes	Attitude, stability, business climate

Note: SOC = social overhead capital; EOC = economic overhead capital.

way of classifying location conditions (and location factors), these 11 categories are similar to others that have been identified. Prior to summarizing the nature of each of these location conditions, extensive discussions of which are available (Estall and Buchanan 1980; Norcliffe 1975; Watts 1987: 75–136), several additional, general introductory points should be noted.

First, location conditions are complex in that each comprises multifaceted characteristics of a tangible and non-tangible nature. This distinction between tangible and non-tangible is based on ease of quantification in terms of cost. In the case of labour, for example, locations vary considerably in terms of the particular bundle of characteristics they supply or potentially supply. To an important degree these variations can be captured by costs based on calculations of wages and benefit packages and even hiring and training costs. At the same time, labour characteristics differ from location to location in

terms of level of unionization, skills, attitudes, willingness to work, gender and ethnicity.

Second, given their complexity, location conditions are hard to measure. Tangible cost-based factors, such as wages, rents, construction costs, taxes and freight rates, can be quantified precisely on a common interval scale. These costs can be aggregated and locations compared in a precise way. On the other hand, non-tangible features, which refer to such characteristics as availability, security, quality, reliability and attitudes, can only be measured along an ordinal scale or more crudely by nominal scale categories. Comparison of locations in terms of government attitudes towards attracting investment as 'aggressive' or 'passive', for example, or in terms of environmental amenity as 'good' or 'bad', represents categorized differences in which the categories themselves are likely to be judgementally derived and the extent of the difference (between being aggressive and passive) is solely subjective. There are non-tangible features whose distribution can be mapped precisely, such as the number or percentage of employees that are unionized, or the number of days workers are absent or on strike. Yet information such as the level of unionization cannot be combined with information on costs along some common scale since such a scale does not exist. Attempts to derive some composite location index are inevitably judgemental.

Third, caution needs to be exercised in the interpretation of individual location conditions. *Ceteris paribus*, low taxes will almost certainly be preferred to high taxes, but without the *ceteris paribus* assumption, low taxes are likely to imply low levels of community services and poor-quality supplies of industrial land. Low wages may imply low skill and low purchasing power.

Fourth, the various categories of location conditions are not mutually exclusive; external economies of scale, for example, includes reference to worker skill, accessibility to markets and suppliers, and access to specialized community infrastructure. The idea of business climate, it might be noted, is particularly ambiguous. While business climate is often interpreted as a non-tangible expression of government attitude towards business (Table 4.1), it is also used as a composite variable to incorporate features of many location conditions including labour, government and taxation levels (Weinstein and Firestine 1978: 137).

Fifth, notwithstanding measurement difficulties, the non-tangible features of location conditions are real and important. Indeed, external (dis)economies of scale, which refer in a general way to the (dis)advantages of factories agglomerating together, is an inherently intangible concept for which there is no commonly accepted quantitative measuring rod. There is widespread agreement, however, that this location condition is of great significance to the geography of manufacturing (see Chapter 13).

Materials, markets and transportation

Factories which utilize and transform materials from a variety of sources to create products for one or more markets incur transportation costs. Simply stated, transportation costs comprise procurement costs, i.e. the cost of transporting materials from sources of supply to the factory, and distribution costs, i.e. the cost of transporting the manufactured product to market(s). Since locations vary in terms of accessibility to materials and markets, which are unevenly distributed across the economic landscape, the locational pull of markets and materials can be measured by transportation costs. In addition, there are intangible features associated with markets, materials and transportation as location conditions.

Transportation costs comprise direct freight charges, while 'transfer costs' refer to both direct costs and indirect costs, notably the cost of insurance on goods and material in transit, interest charges on their capital costs, and losses resulting from damage or deterioration in transit. In general, transportation cost structures are determined by distance, the physical characteristics of goods (weight, bulk and perishability), the value of goods, the mode of transportation (e.g. road, rail, air, pipeline and water carriers), the existence of competition, and back-haul possibilities. An important distinction is between total transportation costs and freight rates per unit (average costs). Thus total costs typically increase with distance and with weight, volume and any special handling requirements of the products and commodities being transported. Freight rates per tonne kilometre, however, typically decrease with distance and according to the quantities of goods transported. In addition, large quantities of standardized materials are typically transported at lower charges per tonne kilometre than higher value goods, partly because the latter can sustain higher charges and partly because the former can be shipped in specialized bulk carriers that can exploit economies of scale. In particular, rail, water and pipeline transportation carriers have high fixed costs which reflect the significant capital investments that are required in port facilities, boats, stations, track and rolling stock. Charges designed to recover (and replace) these capital investments remain the same (are 'fixed') regardless of the distance over which goods are transported. Consequently, average transport costs decline significantly with distance. Fixed costs, on the other hand, are much lower for road transportation, which is most cost-effective, but by no means limited to, short hauls of relatively small quantities. It is also reasonable to expect that competition among and within transportation modes places downward pressures on freight rates. In the case of back-haul transportation, especially low rates occur to attract freight in carriers that would otherwise be empty on 'return' journeys.

More generally, the actual prices charged for transportation services may or may not strongly reflect the real costs of transportation. Free-on-board (f.o.b.) pricing is sensitive to actual transportation costs since the buyer pays for the good (plus the cost of loading) at the point of production and then arranges for transportation. In c.i.f. (cost, insurance and freight) and c.f. (cost and freight), sellers pay the cost, insurance and freight (or cost and freight) pricing of delivering goods to the consumer. Uniform-delivered pricing systems, however, which are sometimes a special case of c.i.f., charge all consumers the same price regardless of location. In basing-point pricing systems, transportation rates are charged from the designated basing point even if the goods are manufactured somewhere else. Both uniform-delivered and basing-point pricing, which are not based on actual cost structures, are 'discriminatory' in the sense that some consumers pay more than costs warrant and others pay less. Under uniform-delivered pricing, for example, consumers nearer to the sources of supply are typically discriminated against and consumers further away are favoured. In addition, while manufacturing firms are motivated to minimize transportation costs, independent transportation carriers seek to increase their revenue and will be encouraged to charge 'what the market will bear'.

As a location condition, transportation has important, if sometimes underestimated, non-tangible features which relate to the frequency, speed and reliability of distribution (and supply) systems, and the need to provide fast and reliable servicing requirements (Wallace 1974). Just-in-time (JIT) delivery systems, such as Toyota's kanban system, reflect these types of advantages; they place a premium on proximity and can be more readily obtained in some places than others (Linge 1991).

All manufacturing activities use transport services to access inputs and distribute outputs which are highly unevenly distributed. Locations near inputs lower procurement costs, and locations near markets lower distribution costs. For primary manufacturing activities that utilize large quantities of raw materials, especially those that lose weight, bulk and perishability during the manufacturing process, transportation costs can be significantly reduced by close access to the required raw materials. Similarly, manufacturing activities that utilize inputs which increase in bulk, perishability and weight can reduce transportation costs by locating near markets. Over the past 100 years, however, greater efficiencies in the processing and shipping of raw materials and the increased tendencies towards the recycling of materials, as well as shifts in the composition of the manufacturing sector from heavy industry to high value-added industry, have reduced the importance of the distribution of raw materials as a location condition (Norcliffe 1975). On the other hand, market access as a location condition reflects distribution costs and significant non-tangible features related to the need for close personal contact with consumers to provide accurate and timely information on consumer preferences, servicing requirements and rival behaviour.

For many observers, since the 19th century and at least until the fordist paradigm, markets as a location condition have become increasingly important for individual industries and manufacturing as whole (Harris 1954; Ray 1965; Clark 1966). Indeed, during the 1950s and 1960s, maps of market potential, or the closely related idea of economic potential, for the US, Europe and UK, tended to overlap closely with centres of industrial concentration (Figure 4.1). Essentially, market (economic) potential shows the relative demand of places within a given market area, e.g. the US, based on the distribution of people (or regional income in the case of economic

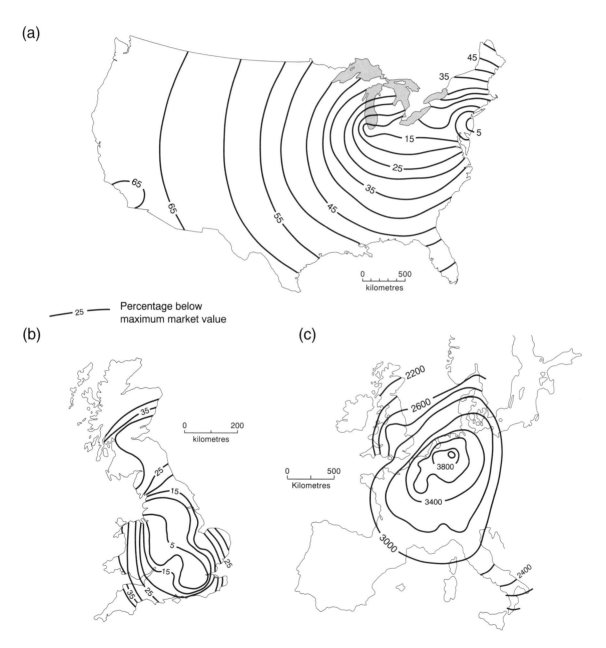

Figure 4.1 Market Potential: United States, United Kingdom and Europe. Sources: Harris (1954: 346); Gudgin (1978: 22); Clark (1966: 11)

potential, or industry sales in the case of industry potential). The basic formula for market potential is

$$MP_i = \frac{\sum M_j}{\sum T_{ij}}$$

where MP_i is market potential at i, M_j is the population (or income) of j, T is the distance (or transport cost) between i and j, and i and j are the places in the system.

In the case of the US (Figure 4.1(a)), market potential centred on New York and the Manufacturing Belt; in the case of the UK, London and the Axial Belt (Figure 4.1(b)) occupy the centre; while in Europe (Figure 4.1(c)), the Ruhr and nearby areas constitute the highest potentials. Indeed, high levels of statistical association were found between the distribution of industrial activity and market potentials in cross-sectional studies (Beattie and Watts 1983). As such, the potential concept strongly reinforced belief in the accumulating attractions of already developed areas and for the idea of circular and cumulative causation. Studies have also shown a correlation between increases in employment and potential (e.g. Keeble et al 1982). However, the market potential model is a crude measure of market access: the distribution of population, for example, also roughly measures the distribution of other location conditions such as labour and infrastructure. In addition, industrial growth (and decline) inevitably implies population growth (decline) so that measures of market potential are not independent of the processes of industrialization. In this regard, it is interesting to note that in the 1950s and 1960s manufacturing employment growth in the US (Figure 3.8) was relatively stronger outside the areas of highest market potential (Figure 4.1(a)). Moreover, for the US, UK and Europe as a whole, the centres of highest market potential in the 1960s have subsequently experienced some of the highest levels of manufacturing job loss.

Labour

Labour costs comprise wages and non-wage benefits such as contributions to medical plans, unemployment insurance, vacation time and pay, and pension schemes. Labour costs vary by industry, union and non-union sector, gender and location. Thus, within British Columbia, as of mid-1994, wages for 'blue collar' work vary from the minimum wage of $6.50 an hour (no benefits) in the non-union sector to over $30.00 per hour (plus benefits) in the unionized forest industry. In contrast to the high-income region of British Columbia, in Mexico the average wage in manufacturing was estimated at US$2.35 per hour in 1992, about one-third the minimum wage in British Columbia. Perhaps the lowest wages in the world are paid to young women in various Asian countries where hourly rates of less than US$0.20 were still being paid in the early 1990s. Even among advanced countries, wages and benefits can vary. For example, a recent report revealed that Germany pays the highest manufacturing wages on average in Europe, while wages in the UK are relatively low (Table 4.2).

Within countries, substantial variations in wage rates also exist. In the US in 1993, for example, the average hourly earnings of production workers in South Dakota were 60% of the earnings in Michigan (Figure 4.2). Although there are exceptions, such as South Dakota, in general, the earnings of production workers are lowest in the sunbelt states and highest in the frostbelt states (Peet 1983).

Table 4.2 Hourly Wage Costs in Manufacturing ($US), 1992

Germany	22.17
Canada	17.31
US	15.45
France	15.26
Japan	14.41
UK	13.42
Taiwan	4.42
Brazil	2.55

Source: Reported in *Vancouver Sun* (and based on data provided by World Wide Watch Institute, The Bureau of National Affairs Inc., Forbes Magazine, Statistics, Canada).

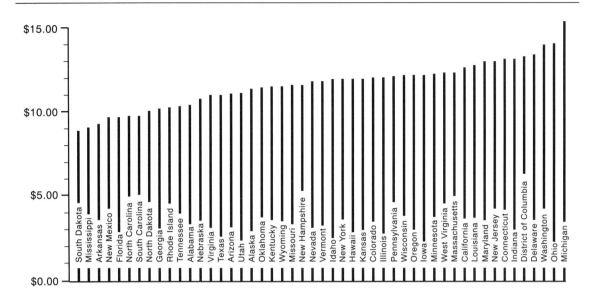

Figure 4.2 US Average Hourly Earnings of Production Workers in Manufacturing Industries by State, 1993. Source: US Bureau of Labor Statistics, Employment and Earnings (monthly 1993)

Hiring and recruitment costs for labour can vary significantly by location, principally depending on availability. Generally speaking, hiring costs are lower in metropolitan labour markets where there are large pools of labour, and higher in isolated communities which may have to recruit over considerable areas to find appropriately skilled workers (Hayter 1979). For these reasons, voluntary labour turnover costs for firms in isolated areas are also higher than for firms in metropolitan areas.

With respect to non-tangible features, the degree of unionization is an important characteristic which varies considerably among and within countries. Among advanced countries, the US has relatively low (and declining) levels of unionization and by the late 1980s unions accounted for just 12% of the workforce, compared to 36% in Canada. Within the US, however, the degree of unionization varies considerably and is generally highest among the core frostbelt states of the Manufacturing Belt and lowest in the so-called 'right to work' states

such as South Carolina where workers are not obligated to join a union, and in the rural states of the West (Figure 4.3). Apart from potential implications for costs and productivity, unions generally enhance labour power, although the nature and extent of this power does vary geographically. On an international scale, German unions are considered to be particularly strong and, by law, have the right to sit on the boards of large corporations and to participate in investment decisions. In North America, on the other hand, neither management nor workers have wanted such arrangements and prefer a more adversarial form of collective bargaining. While unions have won significant concessions in these agreements, control over the direction of companies, and levels of employment, has remained with management.

Another important intangible characteristic of labour is skill. Defining 'skill' is difficult since skills can be specific to industries, firms and even machines, and can involve the use of the senses (touch, smell and observation), physical strength

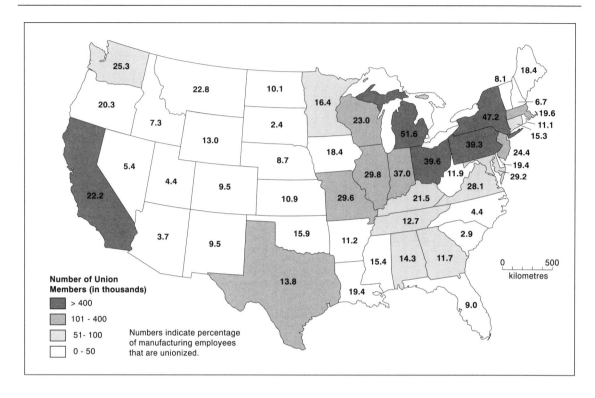

Figure 4.3   US Union Membership in Manufacturing By State, 1989. Source: US Bureau of Labor Statistics (1993)

and endurance and the ability to execute tasks in a precise way. Some labour skills, notably trades skills such as electrical, welding and pipe-fitting skills, involve written examination and certification. Other skills in the manufacturing process have been developed through practical experience featuring on-the-job training under the supervision of already skilled workers, sometimes supplemented by seminars, workshops and trade school courses. Some production skills can be learnt in hours, while others take years. In general, skilled workers engage in tasks which require a 'long' time to learn and involve problem-solving capabilities which have important implications for the quality of output (Kioke and Inoke 1990). Unskilled workers typically engage in highly repetitive tasks that require little

time to learn and which do not involve problem-solving.

Skills also change radically (and ambiguously) over time, principally because of technological change. For Braverman (1974), an enduring feature of technological change since the Industrial Revolution is the deskilling of the workforce, a process significantly enhanced by Taylorization. Other observers have noted that technology can enskill; for example, the application of computerization to a wide variety of industrial processes has introduced a vast array of new problem-solving challenges (Zuboff 1988). In this debate over technology and skill, it is important to recognize that choices can be made. Zuboff (1988), in the context of the North American pulp and paper industry, notes that in some mills computerization means workers are 'screen watchers' whose main task is to communicate the existence of problems to superiors. In other mills, workers are more actively engaged in writing programs and in analysing the vast quantities of information generated. Thus, in the same industry and country, skill levels can vary considerably between locations. Internationally, the same point can be made. A comparison of robot development in the British and Japanese construction industry, for example, found that the British were seeking a highly sophisticated technology which required limited labour skills, while the Japanese were developing more specialized robots that were seen as extending the skills of workers (Gann and Senker 1993). This study also concluded that the Japanese were far more likely to be successful.

From a dynamic perspective, Kioke and Inoke (1990: 7) define skill formation "as whatever contributes to an increase in [worker] productivity while the machinery (K) and quantity of labour (L) inputs are held constant." By observing Japanese operations, Kioke and Inoke argue that the main ways in which labour efficiency is enhanced is through the intellectual ability of workers to deal with routine problems and more significant changes on the production line, e.g. with respect to new methods and products (Patchell and Hayter 1995: 345). In turn, the intellectual ability of workers is gained by the width and depth of career experience and on-the-job training. In this way, worker skill is directly related to flexibility and to particular choices as to how skills will be developed, which can vary from place to place. An alternative (or complement) to increasing labour productivity (defined simply as the ratio of total output to total labour inputs) is to provide workers with more capital.

Clearly, labour as a location condition is particularly complex and multifaceted. Labour costs need to be related to skill and productivity, while unionization simultaneously raises issues of high wages, rigid work practices and limits to managerial control with high levels of skill, a structured work environment and a willingness to negotiate change. Moreover, there are ethnic and gender variations among places, and variations in population growth, which in turn has implications for labour supply. While it is often assumed that capital is more mobile than labour, the labour force is not immobile. Indeed, the migration of people (and labour) continually adds to local labour market variation and creates options for firms to attract the labour characteristics they desire to a particular place rather than search for a location with an existing supply of desired labour characteristics.

Among economic geographers, labour has emerged as a central consideration in understanding industrial locational dynamics (Peck 1996; Sayer and Walker 1992). With deepening globalization, as more and more firms are contemplating locations on a continental and even global basis, variations in labour conditions have become even more marked and have reintroduced debates over labour exploitation.

External economies of scale

As a location condition, external or agglomeration (dis)economies of scale are conventionally dichotomized as urbanization and localization (dis)economies of scale. The former refer to the (dis)advantages of locating in a larger city rather than a smaller one, while the latter refer to the (dis)advantages of locating among a concentration of firms in the same and related industries. Although external economies have long been thought of as a critical location condition, they essentially comprise a set of non-tangible characteristics which are themselves summarized as 'positive and negative externalities'. Externalities are generated over time as the collective result of the decisions of many firms rather than the deliberate actions of any one firm, so that an externality can be simply defined as an unintended consequence of decisions by firms on other firms. These consequences may be good (positive) or bad (negative).

Thus urbanization economies (positive externalities) refers to the benefits firms derive from locating in larger cities rather than smaller ones in terms of access to more diversified, reliable and cheaper transportation services, access to larger labour pools and access to a wider range of business services and amenities in general. Large cities can also impose diseconomies of scale (negative externalities) in the form of congestion, pollution and crime.

Localization economies of scale in the manufacturing sector evolve over time and are most obviously revealed in so-called 'industrial districts' which may offer positive externalities to firms in the form of such tangible advantages as lower costs of production by reducing transportation and processing costs of closely related functions (Chapter 13). In addition, localization economies are associated with a variety of non-tangible characteristics, including the following:

- an established reputation for goods;
- a highly skilled labour pool (and low recruitment, turnover and training costs);
- specialized services in marketing, research and development, and worker training, while industry associations may be present to lobby on behalf of industry (which can be of benefit even if a firm does not pay membership fees);
- patterns of local interdependencies and an industrial atmosphere in which a variety of small specialized firms are available to supply highly specialized services (specialized subcontractors) and additional productive capacity (capacity subcontractors) to other firms.

In addition, while firms may compete they also co-operate in sharing information on markets and technology, in helping one another out, and in determining acceptable forms of business behaviour. On the other hand, negative externalities can develop in such industrial districts, especially if competition among firms becomes cut-throat and private pecuniary gain overwhelms community spirit, i.e. trust breaks down, or if the firms in the district collectively become lethargic or conservative.

Energy

Energy sources were a significant locative condition for traditional industries prior to the Industrial Revolution, but have declined in importance since. Thus, for many traditional industries, access to water power provided narrowly defined limits to industrial sites. During the first two techno-economic paradigms the emergence of steam power and the steam engine widened these constraints, although for the critically important iron and steel industry access to vast quantities of coal was important. Subsequently, the development of electrical power and progressive improvements in the transmission of electricity have allowed electrical energy to be transmitted over very long distances cheaply, while oil, a major source of industrial energy during the 20th century is not particularly cheaper nearer to its source (Chapman 1985). Similarly, nuclear power can be transmitted over long distances. Hydropower sites, which offer huge quantities of power relatively cheaply, remain an important locational consideration for the aluminium industry. Otherwise, the general tendency is for energy to decline as a location condition.

Community infrastructure and amenity

All manufacturing activities require access to community infrastructure, most notably economic overhead capital (EOC), such as roads, railways, port facilities, power lines and service facilities, and social overhead capital (SOC), such as schools, universities, hospitals and libraries. Infrastructure is enormously expensive to build and for most manufacturing activities the existing stock of infrastructure provides physical restrictions on location possibilities (Norcliffe 1975; Peck 1996). Once established, community infrastructure constitutes a significant source of inertia, i.e. the tendency of existing locations to attract new investment in order to save on fixed costs. New capital has a supply price but old capital does not. In some instances, particularly when involving resource development, manufacturing investment occurs in remote areas where entirely new infrastructure has to be provided. Such provision inevitably adds significantly to the costs and uncertainties of projects, and typically requires government financial support (Bradbury 1979). In many cases, industry does not pay the full or even any of the construction costs of new SOC or EOC. In these cases, existing centres still have the advantage of facilitating faster development.

Since infrastructure is a *sine qua non* for manufacturing investment, with gathering momentum an increasing number of national, regional and local governments located beyond existing centres of industry have sought to create and designate 'industry parks' or 'industrial estates' which provide ready-made infrastructure in the form of serviced industrial land and buildings with bulk transportation access and access to existing or new population centres (Peck 1996). Among already industrialized countries, industrial parks have expanded rapidly since the 1950s (Bale 1977; Barr 1983; Barr et al 1984). In recent years, a special category of industrial park – innovation or science and technology parks, designed to encourage research-intensive activities – have become widespread (Ewers and Wettman 1980; Steed and DeGenova 1983; Gibb 1985; Schamp 1987). Since the mid-1960s and early 1970s, developing countries have increasingly sought to provide the essential infrastructure for industry in the form of special economic zones (SEZs) in the case of China, or export processing zones elsewhere (Figure 4.4). Before the mid-1960s there were two export processing zones, in India and Puerto Rico. By 1986 there were 116, principally located in Asia and Central America. Not all of these zones have been successful and it is worth repeating that infrastructure does not 'cause' industry to occur. Clearly, however, the dispersal of manufacturing within advanced countries and around the globe beyond the traditional concentrations of manufacturing, where infrastructure has often become dilapidated and a negative factor for new industrial development, has been facilitated by the spread of industrial parks, innovation centres, SEZs and export processing zones.

In addition to infrastructure classified as EOC or SOC, communities vary considerably in terms of what they provide in terms of cultural, social, sports, life-style and prestige amenities. Among advanced countries at least, such amenities are thought to have become more important as a location condition in relation to abilities to hire skilled professionals and workers. While the more sophisticated community amenities are found in established metropolitan centres, industrial parks increasingly cater for contemporary life styles regarding health, recreation and aesthetics.

Capital

As a location condition, it is important to distinguish physical ('fixed') capital in buildings and equipment from financial capital. Fixed capital costs are measured by costs of construction, and related design costs, which are strongly affected by particular site characteristics. There is also evidence (e.g. from the United States) that these costs vary from region to region (Smith 1971: 283). On the other hand, buildings can be rented, purchased and/or converted, or existing plant expanded. The use of existing buildings pre-empts the need for the fixed costs of construction and saves time. Traditionally, older industrial areas have provided a large supply of 'ready-made' buildings although in recent decades industrial parks have often featured a supply of empty new buildings ready for immediate use. In general, industrial capital, like community infrastructure, is powerfully affected by the force of inertia.

Financial capital is highly mobile and while interest rates vary by location, systematic geographical variations are not evident. Some time ago, Smith (1971: 38) summarized the widely held view that "The cost of financial capital is thus not very influential in location choice in the modern industrial state." Smith emphasized this point especially in the context of large MNCs who can both draw upon substantial internal resources and readily access international equity and portfolio capital on the world's major financial markets. However, in the last two decades or so, there has been substantial growth in venture

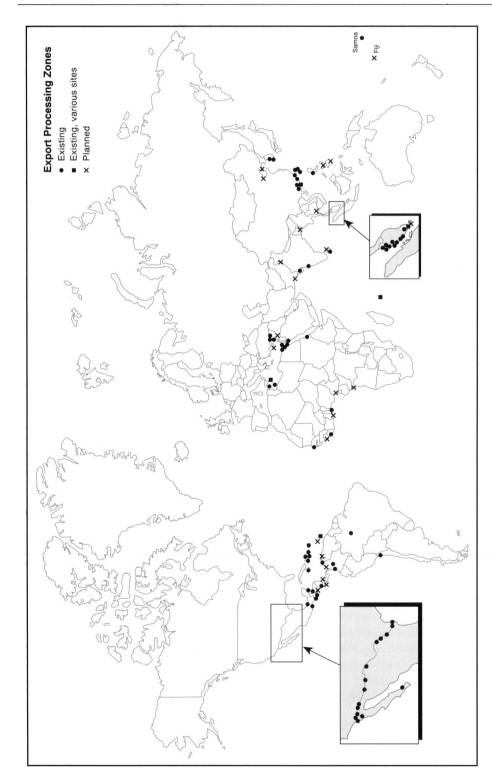

Figure 4.4 Export Processing Zones in Developing Countries, 1980. Source: UNIDO (1980) and Dicken (1992: 182)

capital for high-risk developments, often by small firms, in various high-tech fields. Moreover, venture capital has a distinct geography in terms of both demand and supply. In particular, on an international scale the US is the pre-eminent supplier of venture capital for high-tech companies. Within the US, venture capital firms, which function as financial intermediaries providing capital to new firms from such sources as pension funds, corporations and even families (Florida and Kenney 1988b), are themselves concentrated in a few places, notably New York, San Franciso–San Jose (Silicon Valley) and the Boston–Hartford region (Figure 4.5(a)). These same places also dominate the distribution of venture capital (Figure 4.5(b)). In high-tech activities, capital is an important location condition (Florida and Smith 1993).

Land

The cost of land varies considerably within as well as among countries, with the highest costs being recorded in the larger metropolitan areas. During Japan's bubble economy of the late 1980s, land values in Tokyo reached an astronomical US$35 000 per square foot in 1990, or about US$15.2 billion per acre or US$37.7 billion per hectare (Edgington 1995: 381). In fast-growing, large metropolitan areas in advanced countries, land costs are typically significantly higher than in outer suburbs, smaller cities and peripheral areas. Industrially zoned land is also lower priced than if zoned for commercial or residential uses. Nevertheless, especially bearing in mind the space required by many industrial users, land prices (and rents) constitute an important location condition.

In addition, there are significant non-tangible features associated with land as a location condition. Thus, land units vary in terms of size, shape, accessibility, and whether or not it is serviced. For example, some municipalities have a reputation for offering land that is rocky, infertile, swampy and unserviced. While such land may be cheap, services have to be installed and construction costs may be unusually high.

Land can also be provided with or without buildings, which vary in terms of design, size and layout. Land availability is another feature. According to the 'constrained location' hypothesis, for example, the lack of space for physical expansion constituted a major problem for many British firms located in large urban areas in the 1970s (Fothergill and Gudgin 1982: 99). Locations can of course become unconstrained through the widespread failure of plants.

Environment

The environment as a location condition raises two distinct sets of issues: spatial variations in environmental amenities and spatial variations in environmental policies and regulations. Within industrial geography, Ullman's (1954) pioneering paper explicitly recognized the emerging role of environmental amenity, as reflected in aesthetic considerations, as well as accessibility to various outdoor recreational pursuits (e.g. hunting, fishing, hiking, golfing, skiing, sailing, diving and mountaineering) and the reliability of warm, sunny weather, as a location condition within the US. Thus, Ullman argued that manufacturing activities (and other activities) were growing fastest in those regions of the country, such as the southern and south-western parts of the US, that offered superior lifestyles attractive to footloose entrepreneurs, who could use access to such lifestyles as incentives to attract desired scientists, engineers and workers (see also Alexandersson 1967: 24). Environmental amenity is nevertheless an inherently intangible location condition and it might be added that while there are obvious, regionally based climatic differences in a country as large as the US, which form the basis of the sunbelt–frostbelt dichotomy, all regions offer some form of environmental amenity.

In recent years there has been growing interest in the role environment regulations, especially with respect to air and water pollution, play in industrial location (Chapman 1980, 1982; Thornton and Koepke 1981; Stafford 1977, 1985; Robinson 1995). Manufacturing industries,

Figure 4.5 (a) Geographic Distribution of Venture Capital Offices in the US by Metropolitan Statistical Area, 1987. (b) Geographic Distribution of Venture Capital Investments in the US by Metropolitan Statistical Area 1982–1987. Source: Florida and Smith (1993: 439, 441)

especially heavy industries such as chemicals, pulp and paper and iron and steel, have traditionally been massively polluting and a source of health problems, landscape deterioration and long-term environmental damage. In most advanced countries, 'clean air' and 'clean water' policies of one kind or another were introduced after World War II and legislation, strongly provoked by emerging environmental lobbies, has gathered momentum since 1970. In a general sense, legislation has required existing factories to reduce pollution and new factories to

incorporate technology to restrict pollution (to certain specified levels) or to minimize its effects on local populations and local environments.

Environmental regulations are a location condition because they vary in content and commitment across political jurisdictions, because of the uneven geographic distribution of polluting industries, and because costs of meeting regulations vary from one location to another. Environmental regulations are generally the most restrictive, and most effectively implemented, among the OECD countries, while the industrialized economies of the former Soviet Bloc and developing industrial countries have done little until recently to curb environmental damage. Within OECD countries environmental legislation also varies regionally; for example, in the US, individual states have developed specific policies within overall federal guidelines (Robinson 1995). Recycling legislation, for example with respect to paper, varies among states. Regulations can also be localized and even site-specific. In the Canadian pulp and paper industry, for example, water effluent regulations have long been stronger in relation to salmon bearing rivers than elsewhere (Hayter 1978). More recently, Canada's most stringent regulations specifically targeted the Port Alberni pulp and paper mill in British Columbia.

While environmental regulations do pose tangible, locationally variable costs, which vary significantly among industries, the incorporation of state-of-the-art technology in a new factory typically represents a relatively small, if not insignificant, share of overall capital costs. In the pulp and paper industry, for example, environmental technology is often judged to be about 5–10% of the cost of a new mill. However, once again with variations according to jurisdiction, new mills are frequently required to undertake environmental assessments and reviews which are expensive and time-consuming (Chapman 1981). On the other hand, resolving environmental problems at existing locations is normally extremely expensive. Environmental concerns, and sometimes laws, also involve intangible considerations, particularly in attempts to maintain aesthetic, quality of life and even spiritual values. In this regard, there are variations in the extent to which communities welcome or resist major new industry which pollutes or is perceived to pollute.

Government

In a wide variety of ways, different levels of government affect the nature of location conditions. To understand this influence, Watts (1987) distinguishes explicit, implicit and derived spatial policies. Explicit spatial policies are 'top down' policies implemented by national or supranational governments which seek to stimulate industrialization in 'designated' regions, but not others, by some form of incentive or subsidy. Incentives may take the form of a grant to help build a new plant or modernize an old one, a low-interest loan for the same purpose, a payroll subsidy, or some form of income tax break. In the UK, the carrots of industrial incentives for designated regions were combined with the stick of restrictions in the established growth areas. During the 1950s, 1960s and 1970s, many national governments provided such incentives; these were particularly oriented towards attracting branch plants of multinational firms to the designated regions, which were invariably regions with higher levels of unemployment, higher rates of out-migration and lower per capita incomes than the national average (Todd 1977). Explicit spatial policy is also known as regional policy.

By the early 1980s, in countries such as the US, the UK, Australia and Canada, explicit spatial policies were discarded or at least reduced in scale and profile (Rees 1989; McLoughlin and Cannon 1990). The election of conservative governments ideologically opposed to public subsidies, growing balance of payments problems and the increasingly national nature of unemployment problems are some of the key reasons why governments have withdrawn or reduced investment in industrial location policy (see also Friedmann and Weaver 1979). Yet these policies, regardless of their effectiveness in meeting long-term goals, did significantly change location conditions. Moreover, if explicit spatial policy is less important than it once was, it does still exist. The EC, for example, formally designates problem regions and provides funds to assist industrial development, while in Japan an important thrust of contemporary government policy is to promote development in the peripheral regions of the country (Edgington 1994, 1995). Meanwhile in Germany, in the last few years, massive flows of funds have been used to help reconstruct the dilapidated industrial structures in the east of the country.

National government policies can also generate important implicit spatial effects (Armstrong and Taylor 1978). Thus, policies that are conceived of in primarily national terms, such as trade and tariff policies and defence policies, may have regional effects, which may or may not be intended. In any event, these effects are implicit to the main goals of policy. In Canada, for example, the national tariff policy was designed to protect secondary manufacturing industries, overwhelmingly located in Ontario and Quebec. Defence policies also generate regionally discriminatory effects. In the US, for example, a large share of military spending has traditionally gone to prime contractors located in California and a few states in the Manufacturing Belt (Rees et al 1988), while Barff and Knight (1988) argue that New England's economic turnaround resulted from the expansion of firms supplying the US military. While the implications of implicit spatial policies for location conditions are not easy to identify, since the policies potentially apply nationwide, there are some trends. While prime contracts by the US Defense department, for example, are awarded throughout the country, subcontracting awards typically show some form of distance decay around the dominant recipients. In the UK, there is some evidence that military spending has to some extent targeted particular regions (Law 1983).

Derived spatial policies occur whenever regional and local governments pursue different industrial polices. There is considerable variation among local governments in their enthusiasm and willingness to seek manufacturing investment. This variation, which defines derived spatial policy, is often related to the (intangible) idea of local 'business climate' (Weinstein and Firestine 1978). From this perspective, communities with a favourable business climate actively seek to present a pro-business attitude by one form of 'boosterism' or another, including by providing entrepreneurs with information on a variety of matters, facilitating access to government programmes, facilitating co-operation within business and among government, labour

and business, creating industrial parks and investing in local development offices. Other communities are unable or unwilling to pursue investment aggressively and offer a less favourable business climate. Apart from encouraging feelings of confidence and friendliness, aggressive local communities are a potentially important supply of information, thus helping to save the time and costs involved in location search. Similar remarks can be made regarding national governments and whether they have pro-business attitudes or are antagonistic towards free enterprise.

As a final point on the impact of government economic policies for industrial location, it should be emphasized that among many OECD countries, regional and local governments are playing more pro-active roles, regardless of whether spatial policies exist or not. In the US, where governments often profess sharp criticism of subsidies to industry, branch plants of MNCs continue to be courted by virtually all states offering wide-ranging forms of support. Every foreign auto manufacturer that has located in the US, for example, has received substantial inducements from individual states and sometimes from local governments (Perrucci 1994). Such support, even if it has become more *ad hoc*, is by no means unusual. Indeed, it may be that in a period when explicit and formal regional policies are in decline, local support for industry has never been greater throughout the OECD.

What industry seeks in places: the nature of location factors

Location factors express how firms assess places. Location conditions have different implications for individual firms contemplating investment in new facilities. Firms interpret, or translate, location conditions as location factors which reflect their specific requirements for specific investment decisions. Firms may assess the same location condition in different ways depending on which tangible or non-tangible features are considered important. Nishioka and Krumme

(1973: 203) note that access to markets as a location condition has several interpretations as a location factor:

"(1) advantages in the selling price...and accordingly in total profits...resulting from the size of the market and/or the spatial incidence of a monopolistic range; (2) market advantages resulting from being able to keep close contact with the market (e.g., low cost of information gathering activities and...savings   in...activities   related   to ...customer   services,   including   repairs);   (3) transport cost advantages."

Similarly, other location conditions can imply a range of location factors to individual firms. Access to labour, for example, may mean advantages in terms of the cost of production, productivity, desired forms of labour relations, desired gender characteristics, availability of pools of labour, or combinations of these considerations. It is reasonable to expect that the location factors emphasized by firms in a particular period of time will in a general sense vary by product (and technology) and geographic scale. Thus different types of manufacturers make products using technologies which require different mixes of inputs and serve different markets, which in turn imply different rankings of location factors. For any particular firm, decisions to locate in a particular country or region are at least potentially distinct from decisions to select particular sites within neighbourhoods. Moreover, location factors are influenced by the organizational constraints, value systems and preferences of individual decision-makers or decision-making groups. Unionized labour, for example, can be a desirable, undesirable or inconsequential characteristic to firms. In addition, there are personal values and organization-specific constraints which shape location factors but have no clear link with location conditions. For example, location factors such as entrepreneurial preferences to live in the places of their birth and the desires of branch plants to maintain 'close' communication with the head office have to be related to specific firms.

In order to identify, and if possible rank, location factors, a plethora of questionnaire surveys have been conducted in numerous regions and time periods to ask decision-makers why firms chose the locations they did for a factory or establishment (Katona and Morgan (1952) in the US and Luttrell (1962) in the UK provide well-known early examples). The specific nature of the research designs employed has varied in several ways (Table 4.3). A basic distinction is between those 'partial' surveys designed to assess the importance of a particular location factor and those 'general' surveys which seek to assess the relative importance of all location factors. Partial surveys have given considerable emphasis to the effects of taxes on industrial   location   (e.g.   Erickson   and Wasylensko 1980), while in recent years more

Table 4.3 Location Surveys: Sources of Variability

Component of research design	Options
1. Survey objective	Partial surveys assessing one location factor; general surveys assessing all location factors
2. Industry/activity focus	Industry of cross-sectional surveys
3. Geographic focus	Locality, region, nation, comparisons
4. Sample characteristics	Case study, a 'few' firms, a 'large' survey; sample based on distinctiveness or representativeness
5. Timing of survey	Interviews immediately or sometime after the decision
6. Questionnaire design	Number of questions; the phrasing of questions; open or closed; forms of ranking
7. Classification of location factors	Established before or after research
8. Data analysis	Qualitative listings and rankings; quantitative techniques

attention has been paid to environmental policy and regulations (see Robinson (1995) for a review). Interestingly, in more widespread general surveys, these particular factors are rarely given special importance, although Robinson (1995) notes some alternative views and the crude ways in which the importance of location factors are conventionally measured.

There are other ways in which the research designs of location surveys vary (Table 4.3). For the most part, these so-called 'location surveys' focus on plants operating in a wide cross-section of industries (e.g. Moriarty 1983), although case studies of individual location choices (Whitman and Schmidt 1966) and of a few firms in one or a few industries (Stafford 1974; Hayter 1978) have been conducted. Geographically, surveys have varied in terms of scale (locales, regions or countries) or national context and a few have utilized a comparative dimension. Surveys have varied in terms of the firms interviewed (from one case study to several hundreds), the respondents interviewed, the length of time between the survey and a decision being made on the location, the nature, scope and structure of the questions asked, how the location factors are actually classified, and the manner in which the data are summarized and analysed. In addition, it might be noted that since these surveys have been conducted for over 50 years, changes in the underlying conditions must be recognized.

It is not easy to summarize or integrate the results generated from the vast number of location surveys conducted around the world. The reasons for this difficulty relate to the inherently judgemental nature of location factors and the wide variations in research designs employed, while some observers have noted the lack of clear conceptual yardsticks by which to evaluate observed behaviour (Nishioka and Krumme 1973). Nevertheless, these surveys do serve to reveal both the complexity and the subjectivity associated with the idea of location factors. As one way of recognizing the 'wide range of factors' influencing location, the product cycle model is reviewed and used as a basis for selecting examples of location surveys for different types of activity.

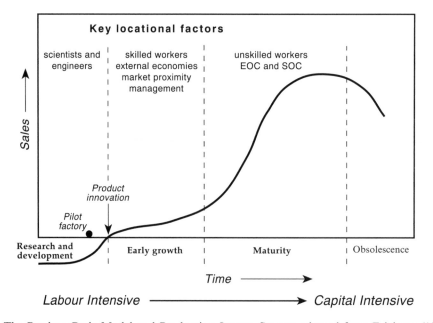

Figure 4.6   The Product Cycle Model and Production Inputs. Source: adapted from Erickson (1976); Erickson and Leinbach (1979)

The product cycle model

The product cycle model, as developed by Vernon (1966) and associates (Hirsch 1967), was developed in the 1960s to explain international investments by US-based multinationals that were manufacturing consumer durables such as electronic products. Since then the product cycle model has been widely adapted to link product cycle and location dynamics with respect to *in situ* employment changes in advanced countries (Krumme and Hayter 1975; Healey and Clark 1985), and non-metropolitan industrialization, especially within the US (Erickson 1976; Erickson and Leinbach 1979; Norton and Rees 1979; Rees 1979; Markusen 1985). At the same time, it has continued to be used to model branch plant investments in poor countries (Suarez-Villa 1984).

In brief, the central (locational) thesis of the product cycle model is that as products are researched, developed and standardized (i.e. as they 'mature'), the optimal bundle of location conditions shifts from high-wage regions (in the original model, specifically in the US) to low-wage countries (Figure 4.6). Thus, the product cycle theory is a form of 'filtering down' theory. According to this model, products evolve through various 'life-cycle' stages analogous to human beings. Vernon (1966) suggested products evolve through stages of invention, research and development (R&D), innovation, rapid growth, maturity, obsolescence and maybe death. During this evolution, technology shifts from laboratory equipment and pilot plants to relatively small-scale operations to large-scale plants utilizing standardized machinery designed for mass production. That is, over time, products shift from (skilled) labour-intensive activities to capital-intensive activities employing unskilled labour. From a geographical perspective, as products evolve through the various stages, the underlying input conditions change, which in turn may lead to related shifts in location conditions and factors.

The product cycle model argues that during the R&D and innovation stages the optimal location conditions are found within the US, particularly in locations providing access to scientists, engineers, technicians, skilled labour and various kinds of external economies of scale in the form of universities and their expertise and skills, specialized engineering and machine tool firms, the machinery manufacturers themselves and, if required, supplies of venture capital. In addition, advanced countries, and notably the US in the 1960s, provide access to large, high-income markets containing consumers able and willing to try new products.

In the late stages of the product cycle, when the product is standardized and mass produced, increasingly strong competition encourages firms to seek lower costs; for example, by building large-scale factories which fully exploit any available economies of scale. In this mature stage, firms can replace skilled labour with relatively unskilled labour operating sophisticated but proven technology. If labour is the cost component that geographically varies the most, the optimal bundle of location conditions shifts to low-wage regions at this stage. Indeed, Vernon (1966) argued that many consumer products that were initially produced in the US were relocated in low-wage countries in their mature (and capital-intensive) stage, from where they were imported into the US. In the context of non-metropolitan manufacturing, the proponents of the product cycle model essentially 'complicate' Vernon's view by noting that products filter down from major metropolitan areas to rural regions instead of, or perhaps prior to, international relocation. In either case, filtering down to lower-wage locations can occur even though technology is capital-intensive. The argument is that if the cost of capital does not vary much spatially, wages most certainly do, so that labour becomes the decisive locational influence.

The product cycle model has been criticized for various reasons. One important concern is that the model lends itself to explanations based on technological determinism, when in reality product cycles take on many different forms (Storper 1985). Vernon (1979) himself suggested that the model was less relevant to explaining the location of foreign branch plants by US-based MNCs, in part because increasingly rapid product cycles and ever increasing automation of late-stage processes were reducing the attractiveness of dispersal to other poor countries. Indeed, several studies have argued that plants may agglomerate in late stages of the life cycle rather than disperse (Schoenberger 1986; Storper and Christopherson 1987). A second important concern is that the product cycle oversimplifies the range of location factors affecting branch plants (Cromley and Leinbach 1981; Norcliffe 1984; Taylor 1986). Taylor (1986) particularly noted the neglect of market factors (see also Kolde 1972), while Clark (1981) sought to widen the model's emphasis on labour costs to include the nature of labour relations (see Chapter 7). However, the model, or at least its terminology, continues to be widely discussed, and it reinforces the dynamic nature of industry in a way that at least raises questions about the relationships between location and technological change. If the criticisms are kept in mind, for the purposes of this chapter the product cycle model also offers a useful organizing device and template with which to selectively review a few of the countless location surveys that have been conducted over the years.

Control functions

For single-plant firms, production and related head-office and R&D activities are combined in one location, and location decisions are typically the prerogative of individual owner-managers. On the other hand, for large, multi-plant firms, head-office and R&D activities, are typically geographically separate from production activities, and locations are typically chosen by professional managers who may or may not work in the same location. The geographical separation of head-office and R&D activities also permits firms to locate specialized functions according to more specific location factors.

Head-office location factors   Corporate head-offices are overwhelmingly located in major metropolitan centres and typically in downtown cores or central business districts (Tornqvist 1968). The key locational attraction for the geographical concentration of head-offices is to facilitate person to person contact in the exchange of information. If the underlying functions of head-offices are to 'control' operations and formulate long-term strategies, their activities focus on the collection, interpretation and dissemination of information, and related negotiation and bargaining activities, on personal contact bases. Technostructures assess trends in business environments, plot strategies towards uncertain futures and make bargains ('deals'). These functions typically require face to face communication with representatives of other organizations and personal contact is facilitated by geographic proximity. Hutton and Ley (1987) investigated head-office activities in Vancouver, a metropolitan city with a population of almost two million and the control centre for British Columbia's resource hinterland and an increasingly important gateway to the Pacific Rim. They noted that the principal advantage of a downtown location is to facilitate personal contact with related businesses, including representatives from a variety of producers services (Table 4.4). A location in Vancouver's downtown offered other advantages of a largely intangible nature such as the 'availability and quality of office space', the 'prestige' of a downtown address, the variety of readily accessible 'amenities', and

Table 4.4 Head-Office in Vancouver, British Columbia: Advantages of Existing Locations

Location factor	Response (%)
Business contacts	71
Labour force	33
Availability/quality of space	67
Amenities	38
Prestige	40
Contact with other operations of firm	28
Rental costs	35
Other costs	3
Other	10

Source: Hutton and Ley (1987: 133).

'labour', specifically the ability to readily recruit labour with the necessary skills. In this survey, only one location factor involved costs and this was the rental costs of a downtown location. Over a quarter of the respondents also mentioned the ability to 'contact other operations of the firm' as an important locational advantage. While other location factors affect the location of head-offices, closeness to producer services and other head-offices and related services clearly predominates. The locational advantages reported by Hutton and Ley (1987; Ley and

Hutton 1987) are in fact consistent with similar studies in other metropolitan areas, including London and New York. A factor mentioned in other studies, it might be noted, is access to international airports to facilitate personal contact elsewhere.

R&D location factors With respect to R&D location, an innovative discussion is provided by Malecki and Bradbury's (1992) survey of the locational assessments and preferences of managers and employees employed in 13 R&D

Table 4.5   Ratings of Location Factors by Management and Employees: Research and Development Facilities of Large US Corporations

Location factor	Management rankings			Employee rankings	
	Present location	Ideal location	Employee assessment	Present location	Ideal location
Recreational opportunities	1	13	4	1	3
Proximity to major airport	2	5	7	2	14
Community business atttitudes	3	3	16	11	6
Restaurants and shopping	4	17	11	7	13
Environmental quality	5	1	1	5	2
Overall business climate	6	8	14	—	—
Accessibility to headquarters	7	10	24	—	—
Economic growth potential	7	10	12	8	11
Climate	7	19	8	3	5
Proximity to university	7	6	5	10	11
Cultural amenities	22	13	9	9	9
Cost of housing	12	6	3	4	1
Quality of private schools	12	21	19	14	17
Cost of living	14	10	6	—	—
Availability of professionals	15	3	16	—	—
Quality of public education	15	1	1	13	7
Accessibility to suppliers	15	13	25	—	—
Accessibility to market	18	8	19	—	—
Alternative employers for spouse	18	20	12	17	16
Traffic congestion	20	16	9	12	4
Entrepreneurial opportunities	21	22	23	15	18
Proximity to other research facilities	22	17	16	—	—
Alternative employers for employee	22	23	19	16	9
Proximity to similar firms	24	25	22	—	—
Nearness to family	25	24	14	—	—

Source: Malecki and Bradbury (1992: 128–9).
Note: Management rankings are based on 13 responses, and employee rankings on 700 responses. Tied ranks occur when average scores are tied.

operations of large manufacturing companies (Table 4.5). The design of this location survey is particularly noteworthy because of its focus on one clearly defined type of activity (large R&D laboratories of large US manufacturing corporations), by distinguishing between the rankings of management from employees and by distinguishing between locational assessments of existing locations, established at varying times in the past, from the preferences of 'ideal' locations. The survey also identifies a relatively detailed listing of 25 location factors.

In general terms, Malecki and Bradbury's (1992) survey confirms the assertions of the product cycle model. Thus, in R&D management's ideal location, 'the availability of professionals' ranks third, while the joint first ranking factors – 'environmental quality' and 'quality of public education' – may be thought of as factors needed to attract employees, a point which is further suggested by the R&D employees' ranks for an ideal location (especially with respect to environmental quality). Similarly, 'cost of housing' is an important location factor in management's ideal location and the most important factor in the ideal location of employees. However, interesting discrepancies arise in the ideal rankings of 'recreational opportunities', 'climate' and to some degree 'cultural amenities', which might have been expected to correlate closely to the priority attached to 'environmental quality'. In fact, employees rank these factors highly but, unexpectedly, much more so than management. Yet management has an accurate perception of what they believe employees' assessments of these three factors would be.

Similarly, R&D management and employees share the view that in an ideal location 'community business attitudes', 'economic growth potential' and 'proximity to university' are important. In addition, management ranks 'access to markets' and 'proximity to a major airport' relatively highly as desired location factors. Malecki and Bradbury's (1992) survey also refers to a number of factors not often mentioned in previous studies. Some of these factors, such as

'traffic congestion', 'entrepreneurial opportunities' and 'alternative employers for spouses', are not ranked highly. One factor, however, namely 'accessibility to headquarters', is relatively important and reflects the need for personal communication between corporate and R&D management. In this context, a suburban location for R&D operations can balance the need for scientists and engineers 'to be left alone' to concentrate on long-term projects while permitting regular personal contact with downtown head-offices, libraries, universities and airports.

Malecki and Bradbury's (1992) survey reveals that the location of R&D laboratories is influenced by a wide range of factors which, for the most part, are difficult to cost precisely for individual firms. One factor not mentioned in this survey, however, is government incentives, perhaps in part because they were not involved in the establishment of existing laboratories. Yet the differences that exist between the rankings of factors for existing and ideal locations suggest that relocation is a possibility and government support of one kind or another is frequently available for high-tech activities (Table 4.6). Indeed, according to the US Congress (1984) and Christy and Ironside (1987) surveys, such incentives are important location factors for high-tech activities and there is no reason to believe R&D operations would be ineligible.

Innovative manufacturing

The kind of location factors underlying high-tech manufacturing are reflected in the summary rankings of four American and Canadian studies that are summarized here (Table 4.6). Generally speaking, high-tech activities are associated with the early stages of the product life cycle, which emphasizes the importance of access to skilled scientific, engineering and factory workers and to external economies (Figure 4.6). The surveys, in an albeit subjective way, confirm these expectations, especially with respect to skilled labour which is ranked first in two surveys (one American and one Canadian), and third and

Table 4.6 The Rankings of Location Factors by High-Technology Companies in Four North American Surveys

Joint Economic Congress (US) Survey (1982)		Office of Technology Assessment (US) Survey (1984)		Christy and Ironside's Alberta Survey (1987)		Bathelt and Hecht's Ontario Survey (1990)	
Rank	Factor	Rank	Factor	Rank	Factor	Rank	Factor
1	Labour skills/availability	1	Founding entrepreneurs lived there	1	Overall business climate	1	Availability of skilled labour
2	Labour costs	2	Close to existing operations	2	Founding entrepreneur lived there	2	Proximity to the place of education/birth/residence of the founder
3	Tax climate	3	Labour skills/availability	3	Access to markets		
4	Academic institutions	4	State government support	4	Labour skills/availability	3	Proximity to universities
5	Cost of living	5	Local transportation	5	Political stability	4	Proximity to customers
6	Transportation	6	Quality of life	6	Proximity to university	5	Access to transportation networks
7	Access to markets	7	High-technology business climate	7	Local government incentives		
8	Regional regulatory practices	7	Universities	8	Proximity to international airport	6	Land availability
9	Energy cost/availability	9	Availability of suitable sites	9	Proximity to domestic airport	7	Wage levels
10	Cultural amenities	10	Overall business climate	10	Proximity to university	8	Proximity of suppliers
11	Climate	11	Financial incentives	11	Provincial government support programme	8	Land costs
12	Access to raw materials	12	Venture capital availability	12	Availability of venture capital	8	Socio/cultural quality
				13	Recreational opportunities		
				14	Local transportation		
				15	Access to raw materials		
				16	Energy costs/availability		
				17	Cost of living		
				18	Cultural amenities		
				19	Labour costs		
				20	Proximity to government departments		
				21	Climate		

Source: Christy and Ironside (1987: 235, 246); Bathelt and Hecht (1990: 228); Joint Economic Congress (1982); Congress of the United States (1984).

fourth in the other two. While two surveys identify 'labour costs' and 'wage levels' as location factors, they are ranked less important than labour skills, and the US Congress (1984) and Christy and Ironside (1987) surveys make no mention of labour cost. It is also possible that such factors as 'cultural amenities', 'climate', 'quality of life', 'recreational opportunities' and 'socio/cultural quality' are important to the attraction and maintenance of professional and skilled (and mobile) labour.

Some of these same factors may also point to the effects of external economies of scale, as might such factors as access to markets, suppliers, venture capital, airports and academic institutions, including universities. In this respect the two Canadian studies offer an interesting comparison. Thus Christy and Ironside (1987: 248) note that external economy effects among the fledgling and small-scale high-tech companies in the sparsely populated resource periphery of Alberta, particularly with respect to networking among themselves, are not as evident as in regions where a greater concentration ('critical mass') of high-tech companies exist, including the Waterloo region of Ontario where Bathelt and Hecht (1990) conducted their study. This comparison is reinforced by reference to the importance of academic institutions. Thus, as did the Joint Economic Congress Survey (1982) in the US, Bathelt and Hecht's (1990) Ontario survey identified 'proximity to universities' as an important (third ranked) location factor, providing high-tech companies with the advantages of personal contact with graduate labour supplies, research expertise, sources of innovations and entrepreneurs, as well as being convenient for professionals wishing to update their skills (see also Oakey 1984). The Alberta survey, however, found proximity to universities to be of some but clearly lesser importance: only 21% of the sample assessed such proximity as 'important' or 'very important' while 40% considered proximity to universities to be 'unimportant' (Christy and Ironside 1987: 245). Indeed, the influence of universities on the location of high-tech companies, whether considered as an external economy

effect or not, is controversial although the balance of opinion is emphasizing their significance (Malecki and Bradbury 1992: 132).

The US Congress (1984), Christy and Ironside (1987) and Bathelt and Hecht (1990) surveys all find the birthplace or established place of residence to be an extremely significant influence on the location of high-tech companies. Indeed, in these three surveys, this factor was ranked first or second. While this finding seems surprising, since birthplace appears to be a somewhat idiosyncratic location factor which is not readily related to an underlying location condition, similar results have been found in many locational surveys since the 1950s. In practice, this finding reflects the importance of owner-managed, generally small companies within the chosen samples and their decisions to build their companies in their home environment. This tendency is an important one and, as will be noted in subsequent chapters, has an economic as well as a personal foundation. In surveys focusing on branch plants, such as the Joint Economic Congress Survey (1982), this factor is not present (Table 4.6; Moriarty 1983).

Branch plants

Most location surveys have sought to understand the location factors underlying new secondary manufacturing plants in particular regions, including those of branch plants controlled by head-offices based outside of the region. Moriarty's (1983) study of 530 manufacturing plants established in North Carolina between 1969 and 1974, including 60 branches of firms headquartered elsewhere in the US, is a good example. In this study, firms were asked to assess whether 58 location factors were essential, very important, important, unimportant or not considered. For the 60 non-locally owned branch plants, Watts (1987: 170) listed the 49 most important (Table 4.7). It might also be noted that during the time period of the survey North Carolina experienced the lowest manufacturing wages in the US and so is the kind of region where mature products would be expected to locate (see Johnson 1985).

Table 4.7 Location Factors for Branch Plants in North Carolina

Location factors	Essential or very important (%)	Unimportant or not considered (%)
Labour factors		
Labour costs[a]	70	8
Availability of semi and unskilled labour	75	10
Availability of skilled labour	52	17
Labour productivity	70	8
Extent of labour unionization	72	11
Labour climate	80	5
Right-to-work law	63	22
Accessibility factors		
Suitability of rail service	43	43
Suitability of motor freight service	78	3
Suitability of air service	35	32
Suitability of access roads and highways	72	10
Proximity to production material sources	43	28
Proximity to national markets	45	38
Proximity to regional markets	52	35
Proximity to large city (> 50 000)	18	57
Community factors		
Physical attractiveness	22	37
Community attitude towards industry	80	7
Social make-up of inhabitants	22	35
Suitability of housing	22	25
Community race relations	43	27
Fire protection and insurance	47	13
Police protection	32	20
Adequacy of local schools	37	32
Suitability of medical facilities	30	28
Suitability of shopping facilities	22	48
Business climate factors		
Suitability of repair and maintenance services	20	43
Suitability of business, financial and legal services	22	43
Compatibility of other industry	37	40
Suitability of building codes	52	22
Suitability of zoning restrictions	55	22
Suitability of environmental regulations	58	22
Availability of public technical training	35	30
State of local planning assistance	23	52
Utility factors		
Suitability of electrical service	78	7
Suitability of telephone service	57	15
Availability of natural gas	42	42
Adequacy and cost of water supply	57	22
Suitability of waste disposal service	35	35
Plant site factors		
Suitability of site parking facilities	27	38
Site development and construction costs	47	17
Room for expansion	77	5
Plant site topographic features	45	18
Plant site flood risk	50	27
Plant site adequacy and costs	62	10
Plant buildings for sale or lease	10	77
Financial and special factors		
Hometown of company official	2	90
State taxes	50	13
Local taxes	52	10
Availability of direct loans	12	73

Source: Watts (1987: 170) (adapted from Moriarty 1983: 70–71).
[a]Italicized factors cited as essential or very important by 50% or more of all respondents.

This survey further confirms the wide range of factors relevant to the location of branch plants. At the same time, labour factors are the most important. Moreover, in accordance with product cycle model expectations, the surveyed branch plants give particular emphasis to locations with low labour costs – but also to low levels of unionization, right to work laws, the availability of labour, especially unskilled and semi-skilled labour, and high levels of labour productivity. In this study, the labour climate reflects a preference for a malleable and co-operative workforce. In fact, all seven labour factors were cited as essential or as very important by at least 50% of the respondents. In addition, three accessibility factors, one community factor, three business climate factors, three utility factors, three plant site factors and two financial factors were judged essential or very important by at least half of the respondents.

Factory closure

Product life cycles imply that eventually old products will be replaced by new ones. Indeed, it is often supposed that product life cycles have speeded up over long periods of time. The obsolescence and death of a product does not necessarily equate with factory closure. Old products can be replaced *in situ* with new ones

and factories can manufacture more than one product. However, the de-industrialization of many cities and regions in the US and Europe has meant widespread factory closures of unprecedented proportions. A growing number of surveys have sought to elicit reasons for closure (Healey 1981, 1982; Watts and Stafford 1986; Stafford and Watts 1990; Stafford 1991; Watts 1991a). Watts (1991a), with specific reference to multi-plant firms, distinguishes between two types of closure, cessation and selective, and for selective closures, area and plant level factors (Figure 4.7). Cessation closures occur when multi-plant firms cease to manufacture a product altogether and consequently close all factories specializing in that product. Selective closures occur when multi-plant firms selectively close one factory while maintaining the product line in others. Area and plant level factors respectively relate to the characteristics of plants (e.g. age of buildings and machines, size, state of repair, labour productivity) and characteristics of local area (e.g. wage rates, labour relations, government subsidies). While these two dichotomies are blurred in practice, they provide a useful framework for assessing reasons for plant closure which, for the obvious reason that appropriate respondents are hard to find, are more difficult to elicit than for new plant locations.

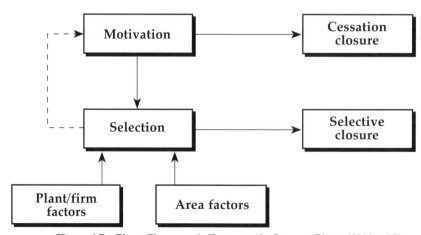

Figure 4.7   Plant Closures: A Framework. Source: Watts (1991a: 38)

Watts' (1991a) survey of plant closures in Sheffield, UK, identified 77 factories employing more than 100 workers which were closed down completely between the late 1970s and late 1980s. Most were branch plants closed by multi-plant firms; 20% were cessation closures and the remainder were selective closures. The managers representing 18 selective closures were interviewed in two stages: in the first stage managers were encouraged to discuss the context of the closure decisions, and in the second stage managers were asked to rate the Sheffield plant in relation to similar plants kept open on the basis of 27 location factors.

The closure factors most frequently cited by the respondents relate to site features, number of production activities at the plant, the age of building and plant, labour relations and size of plant (Table 4.8). That is, several of the most important location factors underlying closure emphasized fixed capital (buildings) and land, factors that are not noticeably in evidence in the opening of new branch plant (Table 4.7). The most frequently cited factors also tend to be the dominant closure factors (Table 4.8). Although unanticipated, one important closure factor relates to the number of production activities on one site. Thus a number of firms emphasized that the Sheffield plant was closed because it undertook just one activity on site while

competing sites could perform a wider range of ancillary activities. Another important closure factor was the small size of the Sheffield plants that were closed in relation to the ones kept open. In addition, the sample firms complained about restricted sites which were too small to move materials in or the product out, were too small to expand or modify or were inconveniently shaped. Several firms also reported labour relations' difficulties related to manning levels for new equipment, the introduction of more flexible labour practices and declining wage levels. Management–labour problems also have to be seen in the context of the relatively old age of plant and buildings. Indeed, generally speaking, the Sheffield factories that were closed had old-fashioned (multistorey) building structures that were in a dilapidated state.

In a second part to this study, for 20 selection closures in Sheffield by multi-plant firms which retained plants elsewhere in the UK to manufacture the product, Watts (1991b) is able to distinguish between plant and area level factors (Table 4.9). He indicates that plant level factors are the more important (Watts 1991b: 814). As expected, smaller plants with older machinery and buildings, employing old, simple technology in association with unproductive labour, were more likely to close. In addition, the highly specialized plants were more likely to close. The

Table 4.8   Factors Affecting the Closure of Factories in Sheffield, UK

Closure factors	Number of mentions	'Dominant' closure factors
Site features	9	Number of activities on site
Number of activities on site	8	Size of plant
Age of building/plant	7	Site features
Labour relations	7	Age of building/plant
Size of plant	6	Labour relations
Reputation	2	Reputation
Local authority rates	1	Distance from market
Distance from head-office	1	
Distance from market	1	
Total mentions	42	

Source: Watts (1991a: 49, 51). The number of respondents (all multi-plant firms) is 18.

Table 4.9 Selective Closure Factors in Sheffield

Category	Plant	Area
Access to markets	—	—
Access to supplies	—	Accessibility to head-office
Land	—	Space for expansion
Capital	Age of machinery	—
	Age of buildings	
	Size of plant	
Labour	Labour productivity	Labour relations
Organization	—	—
Technology	Appropriateness of technology	—
	Activities on site	
Policy environment	—	Local taxes
Personal	—	—

Source: Watts (1991b: 814).

most significant plant level characteristics, however, are low levels of labour productivity and technical sophistication. Given their lesser importance, several area level factors (notably the greater the distance between a factory and head-office, constricted sites, poorer labour relations and higher local rates) are deemed relevant to closure decisions.

Conclusion

The results from location surveys are admittedly not easy to interpret in precise ways, while the non-standardized nature of research designs also creates difficulties for comparisions of surveys. Yet, time and again, location surveys reveal, as Watts (1987: 169) summarizes, that there are many factors which influence factory locations and, it might be added, many of these factors are of a subjective nature. Thus, taken literally, the product cycle model which predicts branch plant dispersal to seek cheap labour does over-simplify, whether discussed in the context of foreign investment in low-wage countries or low-wage rural areas within advanced countries. In these contexts, labour costs have frequently been revealed to be important location factors, but other factors invariably have to be taken into account.

Location conditions and location factors are deceptively simple ideas. Location conditions and especially location factors convey subjective meanings which are difficult to measure and interpret. These meanings also vary with geographic scale. Pulp mills in Scandinavia or northern British Columbia which are attracted by timber resources also need access to markets. Indeed, Lindberg (1953) argued that market access was the most important factor for pulp mills. He reached this conclusion, however, by focusing on regions within Sweden and noting that plants were not located in the middle of the timber supply area but were often pulled to peripheral sites which provided good transportation access to foreign markets. Globally, a firm may prefer Mexico to China because of better market access to the US; while within North America, Mexico is preferred because of lower wages. Whichever site is chosen, basic infrastructure has to be provided. In some way or another, numerous factors influence location choice (Hoare 1973).

5

FACTORY LOCATION AS A COST-MINIMIZING EXERCISE

Conventional neoclassical theory explains the location of factories principally in terms of spatial variations in cost structures. The rationale for this view ultimately rests on principles of competition. In neoclassical economic landscapes, competition is the great regulator of economic behaviour, including locational behaviour. If competition is strong and fair, according to neoclassical theory, rational patterns of economic behaviour, whether with respect to price, production, technology or location, will inevitably result as the forces of competition eliminate the weak or incompetent. From this perspective, there is no (or limited) need to examine decision-making processes or the internal structures of firms in order to understand general patterns of economic activity. So long as competition exists, the conditions of production and markets impose an economic reality on firms that they cannot ignore if they wish to survive. In the context of location, this economic reality takes the form of spatial cost and revenue structures or surfaces: to survive, firms need to locate where revenues cover or exceed costs.

Put crudely, neoclassical location theory is a form of economic determinism in which economic forces 'dictate' the location of factories (see Table II.1). In essence, neoclassical location theory interprets the firm as an Economic Man or Person (*Homo Economicus*) who has the perfect information and perfect rationality necessary to compute an 'optimal' location in the sense of minimizing costs or maximizing profits. In this formulation, the firm is a 'mental construct' or 'black box' which, acting as if it were a super-fast computer, analytically and instantaneously links location conditions, as represented by cost and revenue surfaces, to locational outcomes. The presence of competition, preferably 'perfect competition' featuring large numbers of independent suppliers and consumers linked by 'arms length transactions', ensures that only economically rational outcomes survive. Over time, factories may need to adapt to changing location conditions, or risk failure.

Within geography, neoclassical location theory has been developed along two main lines, namely the 'profit-maximizing' (or 'locational interdependence') and 'cost-minimizing' schools. Both approaches are based on *Homo Economicus* and both derive ideal patterns of location fundamentally from a consideration of transportation costs and scale economies. Profit-maximizing models focus more on transportation costs to spatially distributed markets (distribution costs) and incorporate the effects of rival behaviour on location. Cost-minimizing models focus more on the transportation costs of (spatially distributed) inputs and incorporate the effects of location conditions on spatial variations in cost structures. Traditionally, but by no means exclusively, profit-maximizing models have been applied in the context of personal and retail services while cost-minimizing models have evolved primarily with respect to manufacturing. However, profit-maximizing models can be applied to manufacturing and attempts to synthesize the two approaches have been made (Smith 1971).

Following tradition in the context of the location of manufacturing factories, this chapter emphasizes the cost-minimizing version of neoclassical thinking. In particular, the chapter follows the tradition originally rooted in the work of Alfred Weber (1929), a German economist who sought to determine general principles of location. In broad outline, the first part of the chapter outlines the key elements of his cost-minimizing or 'least cost' industrial location theory, even more briefly comments on profit-maximizing approaches, and then notes how locational evolution is treated in neoclassical thinking. The second part of the chapter examines the nature of cost structures in selected primary and secondary manufacturing industries, and a concluding comment refers to the nature of industrial location policy in a neoclassical landscape. Fuller discussions are provided by Smith (1966, 1971) who also extensively reviews profit-maximizing models. As a final point of introduction, it might be noted that 'neoclassical' location theory is derived from neoclassical economics, which developed in the late 19th century as an extension of the pioneering 'classical' economists, such as Adam Smith (1986). Thus neoclassical economics emphasized the importance of competition, especially perfect competition and related principles such as free trade, and refined the abstract, deductive approach to theorizing which remains the basis of conventional economics. As neoclassical thinking evolved, however, it tended to become increasingly abstract and increasingly less inclined to recognize political, social and historical context. Thus neoclassical thinking is

'abstract' in at least two ways: first, it emphasizes deductive reasoning and, second, it focusses narrowly on 'purely' economic considerations.

Firms as black boxes: landscape as space cost (revenue) surfaces

Alfred Weber (1929) wrote his pioneering least-cost approach to industrial location at the beginning of the 20th century, when manufacturing was dominated by heavy industry in which transportation costs were a significant consideration and the Ruhr was one of the world's great industrial districts. Weber considered transportation, classified as procurement and distribution costs, to be the most important general principle of location. In the Weberian approach, the first step is to (deductively) assess the effect of transportation costs on location and then, once a minimum transportation cost location is derived, successively and deductively assess the effect of other ('economic') location conditions. For Weber, given the primacy of the general principle of transportation, the two most important regional principles of location were labour and external economies of scale.

The effects of transportation costs on industrial location

According to Weber, if freight rates are assumed to vary linearly with distance, *ceteris paribus,* the effects of transportation costs on industrial location depend upon the nature of the physical characteristics of inputs or raw materials. Weber classified inputs as ubiquitous, pure (which experience no change in physical characteristics during processing) and impure (which experience change in physical characteristics during processing). In this regard, it might be noted that even under highly simplified transportation cost conditions, location matters in terms of minimizing costs (Figure 5.1). Thus, *ceteris paribus,* in a region in which a factory serves one market and which utilizes ubiquitous inputs, the optimum (cost-minimizing) location for the factory is at

(a) Region with one market (M), ubiquitous inputs

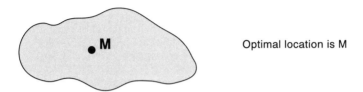

Optimal location is M

(b) Linear region with several markets (A–F), ubiquitous inputs

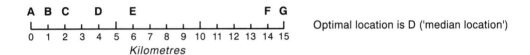

Optimal location is D ('median location')

(c) Region with one market (M), one pure input (I)

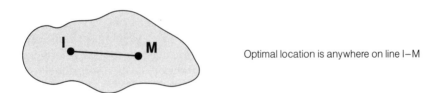

Optimal location is anywhere on line I–M

(d) Region with one market (M), one impure input (I)

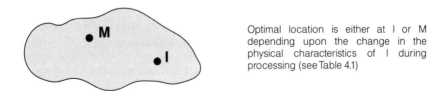

Optimal location is either at I or M depending upon the change in the physical characteristics of I during processing (see Table 4.1)

Note: Factory location is based on minimizing transportation costs.

Figure 5.1   Simple Weberian Transportation Cost Surfaces. (b) is based on Alonso (1964)

Table 5.1   Types of Input-oriented and Output-oriented Activities

Process characteristics	Orientation	Examples
Physical weight loss	Input	Smelters; sawmills
Physical weight gain	Output	Soft-drink bottling; manufacture of cement blocks
Bulk loss	Input	Compressing cotton into high-density bales
Bulk gain	Output	Manufacturing containers; sheet-metal work
Perishability loss	Input	Fish processing
Perishability gain	Output	Newspaper (and job) printing; baking bread
Fragility loss	Input	Packing goods for shipment
Fragility gain	Output	Coking of coal
Hazard loss	Input	Deodorizing captured skunks
Hazard gain	Output	Manufacturing explosives; distilling moonshine whisky

Source: Based on Hoover (1971: 47).
Note: In some of these cases, the actual orientation reflects a combination of two or more of the listed process characteristics. Thus, some kinds of canning and preserving involve important weight and bulk loss as well as reduction of perishability.

the market, which is the only location which minimizes distribution costs while procurement costs are zero everywhere (Figure 5.1(a)). On the other hand, in the case of a region in which a factory serves one market and utilizes one pure input, the optimal location is anywhere between these two locations (Figure 5.1(c)). Alonso's (1964: 60) linear market location problem is slightly more complicated (Figure 5.l(b)). In this example, seven consumers (A to G) are distributed at locations along a linear highway which is 15 km long. The factory (e.g. a bakery) wishes to locate along the highway so as to minimize distribution costs, whether measured by consumer trips or by a salesperson making deliveries to each consumer separately. A measure of centrality is the mean centre found, for example, by summing the distances each consumer is from either end, say A, and dividing this total (42 units) by the number of consumers (7) to give the mean centre in terms of number of units from A, which is 6 units or E. In fact, the best ('least cost') location is D, which is the median centre, i.e. the location which bifurcates a distribution in half. Indeed, for Alonso (1964), this simple example reinforces the advantages of large cities, which in many regional contexts constitute the median location, for market-oriented activities.

In the case of impure inputs or raw materials, the changes in physical characteristics that occur during processing affect the structure of procurement and distribution costs, and just how the physical characteristics of materials change will determine the optimal location (Figure 5.1(d)). Thus, in activities that utilize impure inputs which lose weight or bulk or perishability, procurement costs are more significant than distribution costs. Accordingly, the sources of these inputs exert a stronger 'pull' on location than the location of markets. Conversely, in activities that utilize inputs which gain weight or bulk or perishability, distribution costs are more significant than procurement costs. Consequently, the location of markets exerts a stronger pull on location than the sources of inputs. Indeed, Weber (1929) developed a simple index, the Material Index (MI), to predict input or output orientation:

$$MI = \frac{\text{weight of localized raw materials}}{\text{weight of final product}}$$

If MI > 1 then activity is input-oriented; if MI < 1 then activity is output-oriented.

As Hoover (1971: 47) has noted, there are a number of manufacturing activities which utilize one principal input, normally a raw material, whose physical characteristics change considerably during processing so that their location remains strongly constrained by transportation costs (Table 5.1). Thus, activities which process one principal raw material which loses weight, bulk, perishability, fragility and/or even hazardousness, are input-oriented in the neoclassical view since procurement costs are greater than distribution costs. Activities which illustrate input-orientation for one or more of these reasons include sawmills, smelters, cotton compression, and fish processing. Conversely, activities which process one principal raw material which gains weight, bulk, perishability, fragility and/or even hazardousness are 'output-oriented' because distribution costs are more significant than procurement costs. Examples of output-oriented activities include paper-box manufacture, soft-drink bottling, bakeries and the coking of coal.

In the slightly more complicated case in which an activity utilizes more than one impure input from different sources (and possibly serves more than one market), then the minimum transportation cost location will be 'intermediate' between input sources (and markets). One method of calculating the least-cost location in such a situation is provided by so-called isodapane analysis (Figure 5.2). The first step in the analysis is to calculate isotims, which are lines of equal transportation cost around each location factor and which are derived from the procurement costs for each input at all locations (or some subset of locations such as communities) in the region under consideration and the distribution costs to the market(s) from all possible locations (or some subset of these locations). In other words, a transportation cost surface is calculated for each location factor (Figure 5.2(a) and (b)). Next, for all (or some) locations in the region, the transportation costs relevant to each location factor are summed to create a total transportation costs surface and the least total transportation cost location (P) is identified. Around P, lines ('cost contours') linking locations with the same total transportation costs define isodapanes which show the transportation cost penalty of

not locating at the least transportation cost point (Figure 5.2(c)). In this particular case, input C has a relatively stronger pull than input O or the market M, reflecting the fact that its procurement cost is relatively more important than the procurement cost for O or the distribution cost to M.

The simplified assumptions employed in this example can be readily extended to incorporate more realistic notions of transportation costs. Thus, because of fixed or terminal costs, the average cost of transportation tends to decrease with distance and this is particularly true for rail and sea transportation. Consequently, transportation costs in such situations 'taper' with distance, the effect of which is to reduce the pull of intermediate locations such as P where the benefit of the 'taper' might be lost or reduced. Indeed, it was to offset this problem that some 'intermediate communities' between agricultural supply regions and markets in the US offered in-transit privileges to flour millers, which allowed the latter to get the benefit of the taper. More generally, from the perspective of individual factories contemplating new locations, the form of the transportation cost surface is determined by the rates charged.

Labour, external economies and other costs   Given transportation cost surfaces, similar cost surfaces for other location factors can be derived and combined ('overlaid') to form a total cost surface. For Weber (1929), after transportation, the two key location factors to be considered are labour and external (or agglomeration) economies of scale. As Weber's analysis formally recognizes, for any particular activity, the location which minimizes labour costs (L) or maximizes external economies (E) may *not* be the same as the minimum transportation cost location (P). If so, the question arises as to whether a factory should locate at P or L (or E) or somewhere between. Analytically, this question can be pursued by reference to isodapane analysis.

Assume an already derived transportation cost surface and the question is whether or not our

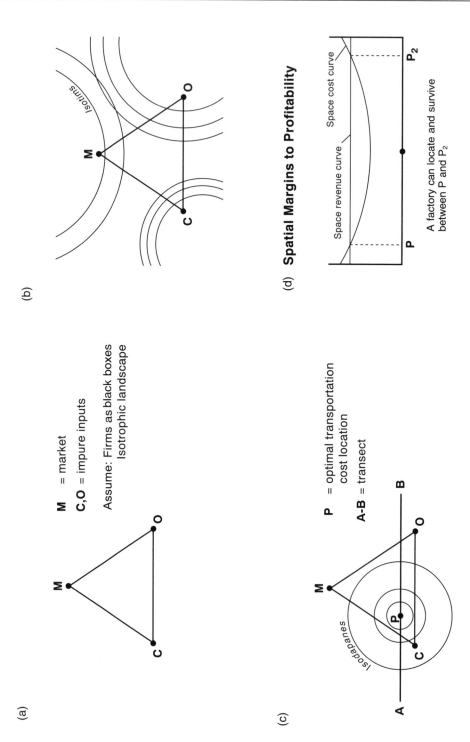

Figure 5.2   The Weberian Triangle

(a)

(b)

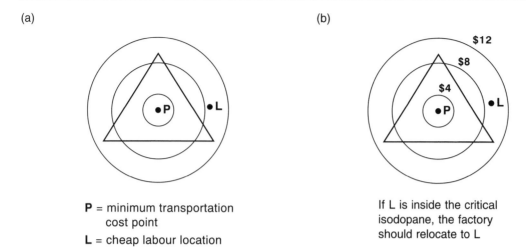

P = minimum transportation
cost point

L = cheap labour location

If L is inside the critical
isodopane, the factory
should relocate to L

Figure 5.3   The Incorporation of a Cheap Labour Cost Location Within the Weberian Triangle

factory should locate at P or move to a cheap labour cost location, L (Figure 5.3(a)). For Weber, the answer to the question depends upon the location of the critical isodapane which is defined as that isodapane which equals the labour cost savings. If L is outside the critical isodapane the neoclassical factory should relocate there since the savings in labour costs more than offset the increases in transportation costs. If L is inside the critical isodapane then the factory should choose P. Similarly, if a factory is faced with a choice between P and a location that maximizes external economies of scale, E, if the cost savings at the latter more than offset the transportation cost penalty then the optimal location is at E (Figure 5.3(b)). In this iterative fashion, other location factors, which minimize energy costs, taxes, environmental costs or infrastructure costs, can be assessed. Alternatively put, in this model, each location factor can be mapped as a cost surface and each cost map overlaid on one another and aggregated to derive a total cost surface.

Spatial margins to profitability   For a plant to survive, a factory need not necessarily find the optimum location; so long as revenues exceed costs, plant locations are viable. Plant locations

which are viable are defined by 'the spatial margins to profitability' (Rawstrom 1958; Smith 1966, 1971). Spatial variations in costs may be graphically summarized by a space–cost transect, A–B, derived from a cost surface (Figure 5.2(c) and (d)). Whether or not any of these locations are viable depends upon revenues exceeding costs. In the hypothetical example shown, revenues are assumed to be the same, regardless of location, and the point at which revenues exceed costs defines the spatial margin to profitability. The spatial margins to profitability also provide one answer to the question of the validity of explanations based on costs when actual location choice is strongly influenced by intangible and non-cost considerations. Quite simply, personal locational preferences can be exercised so long as they remain within the constraints of profitability.

The principle of substitution

Isodapane analysis reveals a fundamental principle of neoclassical location theory, i.e. the principle of substitution. Indeed, it may be argued that in a neoclassical landscape, the location of factories is an exercise in substitution as 'trade-offs' are made among the various location factors. Thus procurement and distribution costs are substituted for one another; a location at the market means that procurement costs substitute for distribution costs, while a factory location near raw material sources substitutes (or trades off) distribution costs for procurement costs. In the case of a cheap labour location (Figure 5.3), the factory trades off the effect of lower labour costs against higher transportation costs (or lower transportation costs against higher labour costs). Indeed, the thesis of the product cycle model (Chapter 4) is that firms will be encouraged to disperse production to low-wage regions to make exactly this form of substitution as products mature; the low-wage region is not expected to be either the source of inputs or the market for mature products and so is not the 'least transportation cost' solution. In this context, there is an argument that the multinational corporation (MNC) provides a close representation of *Homo Economicus* in the real world (Dicken 1977). In this view, MNCs have the ability to collect and rationally assess data on literally a global basis, and in effect to conduct globalized isodapane (and substitution) analysis.

Similarly, it might be noted that neoclassical factories can substitute the cost savings of locations offering cheap energy, lower taxes, lower building costs, lower financing costs and government subsidies with increased costs in other location factors. The landscape is a cost surface and movement to the least (total) cost location is compromised only by the forces of capital inertia. Thus, once paid-off and still operational, old plant and equipment does not have a supply price, while new plant and equipment does. While operating revenues cover operating costs, factories can rationally remain in present locations even when superior locations are known (Auty 1975).

As an exercise in substitution, neoclassical location theory is readily incorporated within neoclassical production theory as a whole, thus establishing the interdependence of location, scale (or size) and technology (Moses 1958). Thus, for a factory of a given size, it may be possible to combine (i.e. to substitute among) the factors of production, in different ways. These factors of production are usually generalized in terms of land, labour and capital (and sometimes entrepreneurship); for example, many manufacturing operations, can within limits, substitute capital for labour, and vice versa. According to production theory, whether or not a factory should be more or less capital (labour) intensive depends upon the relative cost (price) of capital and labour within the realm of substitution possibilites. In turn, choices over particular combinations of inputs affect location to the extent that input costs vary by location. If labour is substituted for capital (or for more expensive labour), cheap labour cost locations, become more attractive, *ceteris paribus*. It may also be expected that with increasing size of factory, new combinations (as well as increased amounts) of inputs become optimal. For both these reasons, decisions regarding the size and location of factories are related.

This argument is graphically illustrated in production theory terms (Figure 5.4(a)). In this particular (hypothetical) situation, an isoquant, or line of equal output or scale, shows how a factory can utilize different combinations of two inputs, X and Y, within certain limits of substitution. The factory, in other words, can substitute X for Y, and vice versa. According to production theory, the choice of the optimal combination of inputs depends on their relative cost or price as revealed by isocost lines, which are lines of equal cost for different combinations of inputs. The slopes of the isocost lines vary depending on whether X is relatively cheaper than Y (Figure 5.4(b)) or Y is relatively cheaper than X (Figure 5.4(c)). The best combination of inputs for a given size of factory is defined as that point on the isoquant where costs are lowest (Figure 5.4(d)). In the case illustrated, the optimal combination is provided at the intersection of X^* and Y^* since any other combination will be higher cost. The particular combination of inputs chosen in this way in turn influences the optimal location in the neoclassical landscape. We might further note that at higher scales of output, new technologies

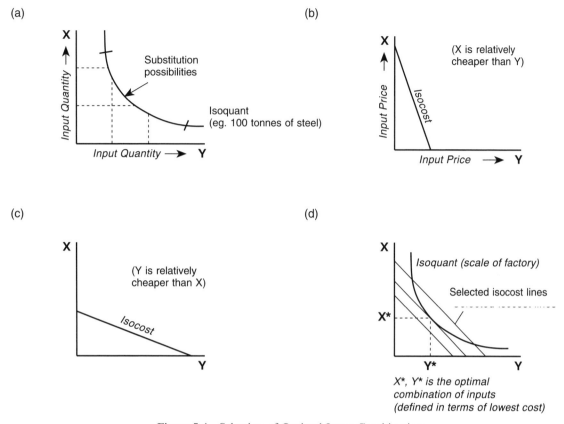

Figure 5.4   Selection of Optimal Input Combinations

may define different input substitution possibilities (as expressed by differently sloped isoquants) so that, even with the same relative costs, the optimal location may change.

In a neoclassical landscape, especially if location, scale and technology are allowed to vary, there is potentially an endless variety of substitution possibilities. In other words, the neoclassical landscape is infinitely divisible as economically rational firms trade off the pros and cons, or more strictly the opportunity costs, of every possible location.

In this context, a comment might be inserted regarding labour. From the neoclassical perspective, labour is considered as an input to the production process just like any other input and, as costs dictate, is to be substituted for, or by, other inputs. The critical isodapane precisely defines this interpretation; the advantages of a

low labour cost location are to be substituted with the disadvantages of higher transportation costs (Figure 5.3). Indeed, more generally, neoclassical production and price theories interpret labour as a commodity to be traded, like other commodities, according to the laws of demand and supply. If price and other characteristics are appropriate, firms hire labour, while the 'free' movement of labour among firms ensures that demand and supply is properly regulated by the principles of competition. In this view, unions are often interpreted negatively because they are seen as a cause of rigidity in wage levels (at least in terms of downward movements) and as constraints on managerial discretion in a range of matters pertaining to employment conditions. In other words, in this view unions increase costs and restrict mobility and substitution possibilities. Fundamentally, from the neoclassical

(a) Single good: single seller

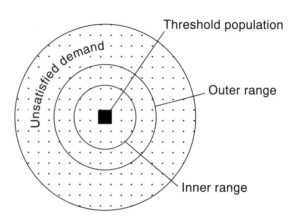

(b) Single good: several sellers

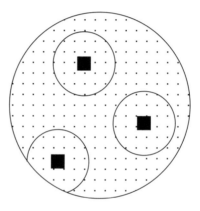

(c) Single good: optimal location pattern minimizing distribution costs

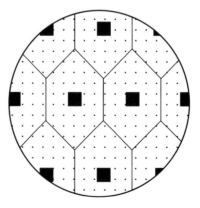

Figure 5.5   Location of a Market-Oriented Activity

perspective, labour is an instrument of production, no more and no less.

Locational interdependence and revenue surfaces

In the locational interdependence school of neoclassical location theory, explanatory emphasis is on how competition among rival firms affects the location and market areas of factories. The traditional theories illustrating this approach recognize spatial variations in demand and typically ignore spatial variations in costs. The best known theory of this type is Christaller's central place theory (1966). While conceived in the context of service activities, the focus of central place theory on distribution costs and inter-firm rivalry introduces general principles of location which are relevant to manufacturing activities. Indeed, in neoclassical terms, there are numerous manufacturing activities whose locational orientation, like that of many services, is dominated by the need to access spatially distributed demand. The location problem of the factory utilizing ubiquitous inputs and serving one market is the simplest hypothetical example in this respect (Figure 5.1(a)).

Central place theory   The point of departure for central place theory is the location of a factory producing a single good (e.g. bread or beer) and serving a spatially distributed population (Figure 5.5(a)). Assuming *Homo Economicus* and a homogenous plain in which inputs are ubiquitous and transportation (distribution) costs are the same in all directions, the factory can locate anywhere the *range* of a good incorporates the *threshold* population. In central place theory the range of a good defines the maximum distance consumers are willing to travel to purchase the good or, alternatively stated, the maximum distance the good can be economically transported to consumers. Since transportation costs add to the price of the good, according to the law of demand, demand will decrease with distance from the factory. The threshold population defines the minimum level of demand necessary to sustain factories of at least the minimum

economically viable size, which may be considered as the size where economies of scale are first fully realized (the 'minimum optimum size' or MOS; see Figure 2.5). The inner range of a good defines the distance within which the threshold population exists.

If one factory cannot satisfy all demands within a region, or if surplus profits are made, under competitive conditions, additional factories are possible until all demands are met, and surplus profits eliminated. In a Christallerian landscape in which demand is distributed evenly over a homogenous plain and there are numerous rival factories each producing an identical good, additional factories disperse to ensure that the threshold population exists within the (inner) range of the good (Figure 5.5(b)). This 'marketing' model predicts an optimal location pattern in the sense of minimizing distribution costs from a set of factories which serve market areas that are hexagonal in shape, thus ensuring all possible demands are met within a spatially distributed market (Figure 5.5(c)). In more complicated situations where there are factories producing different goods, each with distinctive thresholds and ranges, Christaller's marketing principle predicts a hierarchical structure of production. Thus, the most accessible locations will attract factories which produce goods with the largest threshold/range requirements, as well as goods with lower threshold/range requirements. Less accessible locations only attract the latter.

In central place theory, including the marketing model of this theory, products of a given category are differentiated only by distance and consumers are assumed to prefer nearer (and cheaper) sources of supply than more distant (and expensive) sources of supply. Under these conditions, dispersed patterns of product are to be expected. Product differentiation in terms of quality, style, taste or function, however, may encourage forces of concentration as markets are less restricted by distance but more uncertain. Each factory, with a slightly different product, may need access to a larger threshold population and there may be uncertainty over the extent to which one product can be substituted for

(a) Location of two competing sellers

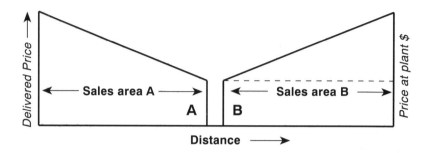

(b) Location of two branch plants owned by one seller

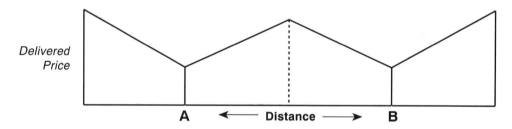

Figure 5.6 Locational Interdependence and Uncertainty: Hotelling's Duopoly Model

another. Similarly, uncertainty over rival behaviour in a neoclassical landscape may encourage concentration. In the well-known Hotelling duopoly model (Smith 1971: 138–140), for example, two sellers supply a homogeneous product to a spatially distributed but linear market (Figure 5.6(a)). Assuming *Homo Economicus,* Hotelling deduced that both sellers would locate at the centre of the market. Any other location, Hotelling claimed, would be unstable because of uncertainty over rival behaviour. If one seller located 'off-centre', for example, the other seller could locate 'next door' but with closer access to a larger share of the market. Distribution costs would of course be minimized if the sellers dispersed and located at the quartiles (Figure 5.6(b)). On the other hand, it is not clear why the bakers do not talk to each other – after all, *Homo Economicus* has perfect information and rationality.

In neoclassical theories of location the presence of uncertainty, including that regarding rival behaviour, has elicited different responses. In one argument, uncertainty denies the possibility of single optimal outcomes and allows that locations which are individually rational (e.g. Figure 5.6(a)) may not be socially optimal (Webber 1972). Another argument is that while uncertainty is relevant at the level of the individual firm and factory, economic costs and revenues nevertheless limit the range of permissible outcomes in the long run. This argument is traditionally expressed in the form of the adapt–adopt dichotomy.

Uncertainty and the principles of adaptation and adoption

With few exceptions, the issue of uncertainty is not a central concern of neoclassical industrial location theory, especially least-cost theory, even when dealing with the question of factory location (see Webber 1972). Yet location itself is one dimension of a broader investment decision which involves a significant commitment of resources, including fixed capital, over long-term time horizons. Over time, possibly even before an investment has generated sufficient returns to recoup fixed expenditures, the assumptions underlying the investment can change as a result of, for example, technological innovation, rival behaviour, market dynamics and resource depletion and discovery. Firms also miscalculate, even if for no other reason than in reality they do not have perfect information.

Least-cost industrial location theory argues that such uncertainty can be ignored because in the long run only those factories that behave in an economically rational manner will survive. In a neoclassical landscape, the location (and closure) of factories is therefore ultimately to be found in underlying spatial variations in cost (and revenue structures). Firms may be ignorant about such cost structures and they may choose in practice to locate in an area for apparently non-economic motives related to place of birth and recreational opportunity. While there have been some attempts to interpret such preferences as maximizing 'psychic income', thereby broadening the motivations of *Homo Economicus* without undermining rationality, non-economic considerations are of little importance in a neoclassical landscape. Whether for reasons of good luck or judgement, to survive in the long term, neoclassical factories need to be viably located within the spatial margins of profitability. Even if once rationally located, as environmental conditions change, factories will have to adapt, be adopted or risk failure. Thus, this adapt–adopt dichotomy provides a rationalization for theories which do not incorporate uncertainty or consider the 'details' of decision-making processes (Alchian 1950; Tiebout 1957).

While adaptation implies some deliberate response on the part of the firm, adoption implies the factory has been saved as a result of the actions of others, notably governments. Thus, uneconomic factories being maintained, for social and political reasons, by government subsidies is a form of (environmental) adoption. In this regard, the chances of a government 'bailout' are doubtless much greater in the case of large factories and large corporations which operate numerous factories than for small plants and firms. The adoption of geographically marginal plants may also occur because of entirely unforeseen changes and as an unintended consequence of government action. Thus an unexpected devaluation of a currency may boost the chances of survival of a marginally located export-oriented plant.

Location adjustment possibilities

The adaptation or adjustment possibilities available to firms in response to changes in their economic environment vary. Firms can relocate, open branch plants, adjust their existing operations in some way, rely on contracting out, or merge or acquire existing firms. According to Krumme (1969b), adjustment possibilities can be classified along the three dimensions of *space*, *organization* and *time*. In terms of the space dimension, firms can adjust their operations on-site, change them between sites ('inter-site adjustments') or develop new sites. On-site *in situ* changes to expand, contract, modernize or change a factory's operations in terms of product-mix, technology, markets and/or input patterns, represent 'location' decisions in the sense of reinforcing or qualifying past location decisions and because of the opportunity costs involved. Inter-site, *in situ* changes occur when firms reorganize their production among a set of existing locations. Although data are not recorded systematically, investment at existing locations is often greater than at new locations (Stafford 1969). In terms of new site adjustments, firms can either relocate or establish branch plants.

As economic circumstances change, factories are continually modified, sometimes radically. Well-known examples of comprehensive *in situ* shifts in technology and products are provided by the cotton textile mills in Lancashire, England. As cotton manufacture became increasingly uneconomic in this region in the 1950s, many factories were converted to entirely new uses in a wide variety of activities ranging from electronic products, toy manufacture, rope and twine manufacture, to poultry processing (Bale 1976: 24–25). With respect to inter-site changes, many examples are provided by multinational corporations that have reorganized their existing factories in response to changes in tariff levels. The 1965 Auto Pact between Canada and the US, for example, eliminated tariffs between the two countries in the industry and permitted the auto companies to specialize their factories on products serving continental rather than national markets (Holmes 1987, 1992). Progressive decreases in the tariffs between EC countries have likewise provoked MNCs to re-integrate their factories in different, usually more specialized ways, including in the auto industry (Dicken 1991). At the same time, it is investment in new locations that gives the clearest indications of geographical shifts in cost structures.

With respect to the organizational dimension, the main distinction is between small single-plant firms and large multi-plant firms. In terms of on-site change, the small firm may have greater discretion in responding to change in the sense that it will possibly be able to make faster decisions in comparison to a branch-plant manager whose autonomy is tightly circumscribed. On the other hand, the multi-plant firm has inter-site adjustment possibilities unavailable to the small firm. Moreover, it has been argued that the decision-making characteristics of multinational corporations resemble the relentlessly rational capabilites of *Homo Economicus* more closely than is the case with small firms (Dicken 1977). In this view, for the multinational firm, cost and revenue surfaces are global in scope while the effect of continuing improvements in transportation costs has been to widen ('push out') isodapanes. Consequently, other locational considerations, such as those pertaining to labour, become more important and cheap labour cost locations within poor countries become more viable.

With respect to the time dimension, Krumme (1969b) distinguishes short-, medium- and long-term planning horizons. Thus, in the short term, for the single-plant firm, spatial adjustments are limited, for example, to using plant more intensively by hiring more labour or offering overtime, using more inputs and running down inventories (or vice versa). A multi-plant firm may make some adjustments among its plants in the short term by, for example, shifting an order from one plant to another. In the medium term, spatial adjustments are again limited for the single-plant firm but for the multi-plant firm some expansion and/or consolidation is possible, as it may be possible to lease production facilities, and develop subcontracting. Long-term plans for locational adjustment are dominated by new site location options and by mergers and acquisitions strategies which also provide firms, especially large firms, with opportunities to geographically expand their operation (Fleming and Krumme 1968; Leigh and North 1978; Green and Cromley 1984).

Locational evolution in neoclassical perspective A new optimal location pattern does not instantaneously replace an old, 'obsolete' one. In a neoclassical landscape such shifts are at least slowed by investments in fixed capital, both in the form of infrastructure and capital equipment. The capital costs of building new infrastructure and factories are significant. An old factory operating in an increasingly marginal location, may survive for a considerable period because its capital costs are zero. In effect, old factories substitute low capital costs for high operating costs (Auty 1975). Thus patterns of locational adjustment may be expected to be complex, with surviving factories adjusting in various ways over a period of many decades. In this regard, examples of case studies of adaptation, adoption and failure are provided by Auty's (1975) and Watt's (1971, 1974) analyses of the sugar beet industry in the Caribbean and England respectively, and Hunter's (1955) examination of the North American pulp and paper industry. In the latter case, the development of new wood-pulping technologies in the latter part of the 19th century radically altered location conditions and new rounds of investment overwhelmingly shifted the pulp and paper industry to northern locations accessible to softwood forests. Yet, while many plants failed in the established centres of the paper industry, such as New York, other plants continued to survive half a century later, typically on a small scale, utilizing old infrastructure and by serving, higher-value market niches. More recently, as recycled paper has become an increasingly important input to paper-making, locations within major urban agglomerations are again viable for major new plant in paper-making.

For factories that are unwilling or unable to adapt, and are not adopted by changing environmental conditions, failure constitutes a third option. Despite adjustment possibilities, entire industries die out while others shrink in size and plants do fail. More generally, models of industrial evolution typically recognize that over time there will be a shift from locational and organizational patterns dominated by large numbers of small firms and plants, often in agglomerations, to a consolidated pattern in which many fewer, large plants dominate the industry from a more dispersed set of locations (Markusen 1985, 1987). In Markusen's (1985) version of the profit cycle model, for example, these changes are driven by underlying changes in economic conditions, particularly in the nature of competition, market conditions and technology, which shifts from an emphasis on

product innovation to process innovation (Utterbach and Abernathy 1975). Such thinking is consistent with the neoclassical emphasis on economic laws governing behaviour. Ultimately, in a neoclassical landscape, the opening and closing of factories, the de-industrialization of regions and the opening up of new industrial spaces reflects the inexorable power of economic reality.

Industrial location policy in a neoclassical landscape

In a neoclassical landscape, capital is viewed as roaming about the landscape looking for the 'best', usually expressed as 'least-cost', location. Neoclassical theory typically prefers minimal policy interference by government in this locational selection process. Policies that are advocated tend to favour increasing the mobility of capital and workers, less by subsidies than by the loosening of restrictions for capital and providing education and training for workers, so as to increase their scope as economically rational agents. It might be noted that at the beginning of the Industrial Revolution classical economic thinkers strongly advised increasing the freedom of workers, then constrained by guild and other types of restrictions.

According to some neoclassical theorists, freely operating market forces, unhindered by government interference, will resolve regional problems (Figure 5.7). In this view, in the core regions, the concentration of investment in factories puts pressure on labour supply and as the demand for labour increases, wage levels increase. As a result, firms are encouraged to seek locations in peripheral regions where there are available labour supplies and where wage rates are lower. Over time, however, as more factories locate in peripheral regions, the demand for labour increases, which in turn puts pressure on wage levels. In addition, movements of labour from peripheral to core regions reduces pressure on wage levels in the former and increases pressures on wage levels in the latter. Thus, in a neoclassical landscape, freely mobile capital and

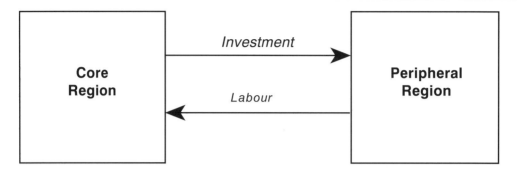

Figure 5.7   Towards Regional Equilibrium in a Neoclassical Landscape

labour define tendencies towards equilibrium wage and unemployment rates among regions.

Neoclassical justification for industrial location policy emphasizes those approaches which enhance the mobility of workers and capital or investment. In the context of investment, for example, and to anticipate a behavioural argument in the next chapter, subsidies to factories located in designated regions can be justified as encouragement to firms to investigate cost and revenue structures as fully as possible, including designated regions they would not have otherwise considered. On the other hand, subsidies, especially in the form of one-shot capital grants, may only establish locational viability in the short term. Indeed, an underlying concern of neoclassical theory with industrial location policy is that subsidies based on social and political rather than economic grounds will lead to factory locations outside the spatial margins of profitability (Figure 5.8). In the 1950s and 1960s, when it was thought that these policies could be afforded, neoclassical economists and economic geographers interpreted such policies as involving a trade-off between national efficiency and social and political equity. Even in this context, the location of factories is seen as a substitution problem – more politics and less economics. Such policies are seen as distorting the economically rational pattern because, it was claimed, they promoted inefficiency and an unneccesary duplication of facilities, and even undermined the viability of the rationally located. Evidence in

support of these charges is available. In Atlantic Canada, for example, industrial subsidies were frequently, indeed mainly, awarded to firms that were planning on investing in the region anyway and in some cases competing firms were all subsidized thus encouraging excess capacity (Springate 1978; Hayter and Storey 1979). In Cannon's (1975) terms, these studies did not generate incremental effects. On the other hand, there are designated regions, in Canada (Cannon 1980; Hayter and Ofori-Amoah 1992) and elsewhere (Keeble 1980), where industrial incentives did generate substantial incremental effects.

From the 1950s to the 1970s, many national governments rejected market solutions to regional problems. Critiques paid particular attention to the unrealistic assumptions of equilibrium tendencies. Two points are relevant in this context. First, the mobility of workers, including unemployed workers, is inevitably constrained by social and family commitments and a highly localized knowledge of opportunities. Migration to other regions involves considerable costs and uncertainties, which are further compounded if workers have to change occupation. Second, while continued investments in core regions are likely to increase pressure on wages, they are also likely to reinforce agglomeration economies and the effects of inertia. In other words, the regional pattern of growth may be disequilibrating (Myrdal 1957; Krebs 1982).

In the highly competitive times of the 1980s and 1990s, however, many national governments

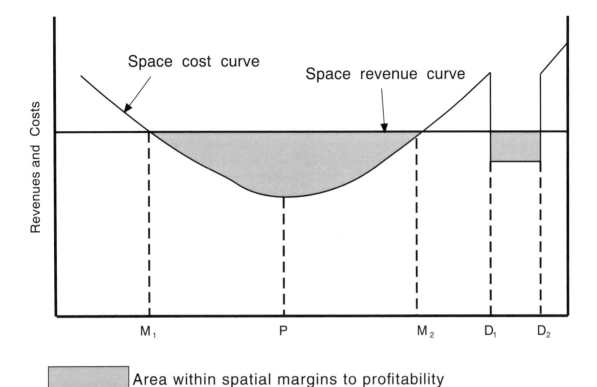

Figure 5.8   Location Subsidy and the Spatial Margins to Profitability. Source: based on Smith (1966: 106)

have reduced or eliminated commitments to industrial location policy (McLoughlin and Cannon 1990). Indeed, especially in countries such as the UK, the US, Australia and Canada, the demise of these policies has been associated with the adoption of neoclassical thinking in which market solutions to socio-economic problems are once again in the ascendancy. In addition, unemployment problems have become national in scope, and governments have become more concerned with balance of payments and national debt problems. Moreover, policies that

offer subsidies to some regions (and firms) but not to others have been politically contentious, one result of which was the tendency to continually broaden the geographical and industrial terms of reference of the policies (Alden and Morgan 1974). In the UK and Canada, for example, the areas represented by designated regions progressively expanded to incorporate the larger part of national territory. The market solution, in theory at least, leaves regions and communities to their own abilities and interests in promoting industrial development.

Cost structures and locational orientation

In a neoclassical landscape, location matters because costs and revenues vary over space. Since different industries have different cost (and revenue) structures the locational patterns of industries also vary. The well-known tradition of classifying manufacturing activities by locational orientation, for example according to 'transport orientation', 'labour orientation' and 'agglomeration orientation', is directly derived from Weberian cost considerations (Hurst 1972; Morrill 1974; Berry et al 1993). At the same time, such typologies are inevitably relatively crude representations of actual cost structures which are continually subject to forces of change. Indeed, the 'common sense' argument of the neoclassical approach is that revenues must cover costs for factory operations to be viable; that explanations of factory location need to be rooted in an understanding of underlying cost (and revenue) structures, and that over time locational evolution requires adaptation to changes in economic circumstances.

In practice, the cost structures of factories and industries are complicated and detailed, comparative information is often not readily available. In advanced countries, the census typically provides data on costs of production for individual industries. However, apart from information on labour, the census breakdown of production costs is typically crude and, for example, does not include reference to procurement costs, selling costs or the various items that comprise material costs. Moreover, classification procedures vary among countries. Within censuses or statistical abstracts there is often information on some individual cost components which can be readily mapped and tabulated (Smith 1971), although such generalized cost components cannot simply be added together to form meaningful cost structures. There are also consulting agencies which have databases on the locational economics of individual industries, such as Jaakko Pöyry, a Finnish-based firm which has accumulated considerable information on the cost of production on individual pulp and paper mills throughout the world. However, such information is typically confidential and available only for a fee.

Calculations of the cost structures of individual industries operating in different regions are either based on actual costs of production or on estimated (hypothetical) costs of production for brand new state-of-the-art mills. In the former approach, most calculations are based on the average costs of all mills operating in a region, which is essentially what happens when data are drawn from the census or from a sample of mills operating in a region. Of course, average cost structures may not reflect actual conditions at any one mill, and even within a region, variations between mills can be substantial. Alternatively, individual factories can provide case studies illustrating the 'best' case, the 'worst' case or a 'representative' case, assuming such cases can be identified. In the second approach, engineering estimates of the cost structures of state-of-the-art mills may or may not accurately represent inherited locational economics, but they are particularly useful in terms of assessing the location of new investments.

Within the same industry and region, cost structures can vary considerably. However, a basic distinction can be made between primary and secondary manufacturing. In the former, material costs and transportation typically remain important, while in secondary manufacturing labour inputs are relatively more important, even if the nature of these labour inputs vary.

Primary manufacturing

Sawmill cost structures As an example of a primary manufacturing activity that utilizes one principal raw material, specifically timber, which loses bulk and weight during processing, sawmilling is conventionally classified as input-oriented (Table 5.1). State-of-the-art engineering estimates of the cost structures for two sizes of sawmills in the Pacific North West (PNW) region of the US and the US South confirm this

Table 5.2   Annual Operating Costs: Lumber Mills in the US *c.* 1984

Cost	Southeast Region		Pacific Northwest Region	
	Large	Medium size	Large	Medium size
Wood	67.7%	59.2%	85.0%	78.0%
Labour	3.4%	7.0%	2.1%	4.6%
Maintenance	9.1%	10.6%	5.6%	6.9%
Overhead	2.2%	7.8%	1.4%	5.1%
Energy	17.6%	15.4%	5.9%	5.4%
Total ($000)	11 135	3 195	18 014	4 916
Cost/MBF	$111	$128	$180	$196

Source: Tillman (1985: 196).
Note: MBF = 1000 board feet measure.

orientation (Table 5.2). In this particular example, the cost structures are hypothetical in that they represent engineering estimates for new, state-of-the-art mills, as of the mid-1980s. As expected, wood costs dominate the cost structures in both regions and although transportation costs are not distinguished as a separate category, it is reasonable to assume that procurement costs are a significant element of wood costs. It might also be noted that while both regions are within the spatial margins to profitability for sawmilling, overall costs are noticeably lower in the South than in the PNW primarily because of lower wood costs. In the PNW, wood costs are higher as a result of higher tree growing and harvesting costs, while in recent years the decreasing availability of wood in the region (due to environmental legislation) has been a further cost-enhancing factor. Indeed, these data indicate that the US South is the least-cost location for primary forest product manufacture in North America, if not the world.

On the other hand, while wood costs are relatively lower in the South, energy costs are relatively lower in the PNW. In effect, to some extent, the PNW is able to substitute for its relatively high wood costs with relatively cheaper energy costs to maintain its competitiveness. Lumber firms in the PNW have also sought to maintain competitiveness by producing higher value products, although this strategy is not

evident from the data (and no information on revenues and profitability is presented). In both regions, labour in new mills is anticipated to be a relatively insignificant cost component. At the same time, the average cost of production is less in the larger mills in both regions and an important source of economies of scale relates to the use of labour. Thus, the relative importance of labour costs is reduced by half in the bigger mills as bigger machinery does not imply more workers. In addition, overheads are also much reduced in relative importance. In contrast, wood costs become more important in the bigger mills since as the size of mill increases, the consumption of wood increases more or less proportionately, while economies are achieved in the use of other inputs. In sawmilling, scale economies have been important for some time in pushing down the average costs of production and threatening the survival of smaller mills in these regions. In fact, as sawmills have become larger, there has been a parallel tendency to concentrate in the communities that are most accessible to the timber supply. That is, with increasing scale, technology has become more capital-intensive and procurement costs have imposed increasingly stringent locational requirements within each region.

It needs to be emphasized that these estimates are for state-of-the-art new mills, and that existing mills typically exhibit higher costs,

Table 5.3   Newsprint Product Costs ($CAN): Principal Producing Regions

Costs	1990 Delivered cost per finished tonne ($CAN)				
	US South	US West	Canada	Sweden	Finland
Other mill costs	111	149	149	219	298
Chemicals	29	36	17	7	5
Energy	98	85	92	114	108
Labour	88	71	124	87	72
Other mill costs	109	80	82	104	75
Corp. & selling	20	15	33	14	36
Delivery	32	52	86	83	96
Total	487	488	538	628	690

Source: Forest Sector Advisory Council (1991).

including with respect to labour costs which may also be of greater relative importance. Indeed, new mills provide a significant competitive threat to existing operations, encouraging the adaptation of more efficient processes.

Newsprint industry cost structures   As another example of primary manufacturing, the cost structures that have been outlined for the newsprint industry in the major producing regions of the world in 1990 are based on the average costs for a sample of existing mills in each region (Table 5.3). The Forest Sector Advisory Council (1991) has collected data on these same mills several times now, so that trends over time can be made, and the use of common procedures facilitates comparison (Holmes and Hayter 1994: 19–21). In 1990, according to these data, the US South and West are the low-cost newsprint-producing regions, enjoying significant advantages over Canada and even more so over Sweden and Finland, the traditional optimal locations when softwoods were the desired raw material. The US South now enjoys the crucial advantage in wood costs: almost CAN$40 per finished tonne over Canadian producers and 100% lower than Scandinavian rivals. The US regions also have substantial advantages in terms of distribution costs to markets. To some extent, Canada and the Scandinavian countries offset

their higher wood and delivery costs with lower chemical costs. However, whatever energy cost advantages these regions did enjoy in relation to the US have been eroded by investment in more energy-intensive pulping methods.

Labour costs are approximately similar, apart from Canada which has relatively high labour costs. Indeed, in Canada, labour costs account for 27% of total mill level costs, compared with 17–20% in the US, 15% in Sweden, and less than 13% in Finland. Yet wage rates per hour in Canada are slightly lower than in the US, and the difference in labour costs per finished tonne is almost entirely due to differences in labour productivity. In Canada, a finished tonne of newsprint required 3.6 operating hours to produce, but only 2.5 hours in the US South and 1.8 hours in the US West. Canada's lower labour productivity in turn reflects use of smaller (and older) paper-making machines and traditional labour-intensive pulping technologies.

However, cost structures can change, sometimes rapidly. From 1986 to 1990, Canada's competitive gap in newsprint in relation to the US widened considerably. The increasing value of the Canadian dollar was a major factor in this decline in this period; another factor was increasing energy costs. In response to much higher costs, Canadian paper producers (and even more so Scandinavian producers) have shifted away from standard newsprint to the manufacture of higher value papers, and from fordist to more flexible methods of mass production (Hayter and Holmes 1993). In Canada, efforts to further reduce labour costs by replacing labour with technology and by the adoption of more flexible employment practices is also well under way (Hayter and Holmes 1994; Chapter 12).

Other raw material using activities   There are manufacturing activities which utilize one principal raw material that cannot be simply classified as either input- or output-oriented. In petroleum refining there is little change in the physical characteristics of the raw material (i.e. petroleum is a 'pure' input in Weberian terms), so that a location either at the input source or at or close to markets is feasible without undue transportation cost penalties. In Canada, for example, the traditional concentration of refineries near central Canadian markets has been complemented in recent decades by the establishment of refineries on the Alberta oil patch (Chapman 1985). Aluminium refining is another example of an activity that utilizes one principal raw material (alumina) whose location is neither input- nor output-oriented even though the procurement costs of alumina and the distribution costs of ingots are important. In this case, aluminium smelting requires vast quantities of electric power and prefers locations which can guarantee large supplies of relatively cheap hydro-electric power, such as Dienes, Iceland, and Kitimat, British Columbia. Both these locations are a considerable distance away from the least transportation cost point; in the case of Iceland's smelter, alumina is imported from the Caribbean and the ingots are shipped to Europe. In Weberian terms, the low cost of power in Iceland more than compensates for the resulting higher costs of transportation: Iceland is within the critical isodapane as an energy supplier (see Skulason 1994).

The modern iron and steel industry provides examples of transportation orientation at locations intermediate to a range of input sources and market destinations. Thus, Fleming (1967) noted that the major new integrated iron and steel works that had been built during the 1950s and 1960s in the US and Europe frequently favoured tidewater locations, allowing economical access to alternative supplies of coal and iron ore, including those on different continents such as Africa, South America and Australia, and economical access to major markets. For Fleming, the transportation cost advantages of such locations were reinforced by their 'locational flexibility' to alternative input sources and markets, which in turn provided insurance against short-term disruptions or permanent loss in any one location factor.

Secondary manufacturing

In secondary manufacturing, labour and external economies of scale are widely thought to be the most significant principles of location. Indeed, labour costs are often the major component of the cost structures of individual factories in secondary manufacturing industries. There are also secondary manufacturing activities in which labour, although not the most important cost component, is still the most important locational influence. With respect to industry averages in the US in 1982, wages varied from 11.1% (tobacco products) to 40.0% (leather and leather goods) of value added (Watts 1987: 86). Given that these figures are averages, within each industry there are considerable variations in labour's share of value added among individual factories and products, even within particular countries.

Cost structures in the auto industry   The automobile industry is an example of an industry manufacturing a complex product which requires a high level of value added by labour. The auto industry has traditionally been concentrated in particular regions in industrialized countries in order to take advantage of large pools of higher skilled, productive labour as well as scientific and engineering expertise and other forms of external economies. Even so, the importance of labour in

Table 5.4 Labour's Share of Value Added in 12 Major Auto Companies, Average 1980–1991

Company	Labour's share of value added (%)	Company	Labour's share of value added (%)
Toyota	42.2	Nissan	68.5
Honda	50.9	Mazda[a]	64.9
GM	73.8	Ford	71.5
Chrysler	76.6	VAG	72.2
BMW	63.1	Fiat[a]	64.3
PSA[a]	71.9	Ford UK[a]	66.6

Source: Williams et al (1994: 18).
Value added is defined as labour costs (including social benefits), depreciation and net income or profits pre-tax.
[a] Calculations based on periods other than 1980–1991, namely 1987–1991 (Fiat), 1984–1990 (PSA), 1982–1991 (Ford UK), and 1982–1991 (Mazda).

Table 5.5 Employer Labour Cost per Employee Hour for Major National Auto Industries 1980–1991 (US$)

Country	1980	1985	1990	1991
Japan	7.40	11.15	18.03	20.52
USA	12.67	22.65	20.24	21.24
France	10.03	10.20	15.57	16.42
Germany	13.70	13.72	26.03	26.95
Italy	9.05	10.40	16.94	19.13
UK	7.63	8.70	15.48	16.15
Spain	6.95	6.76	16.31	17.91
Sweden	15.70	12.33	26.40	27.52

Source: Williams et al (1994: 58).

auto production has always encouraged firms to seek ways to reduce labour costs, and over the past decade or so firms have increasingly invested in selected low-wage countries, such as Mexico and Brazil, while South Korea and Spain have also rapidly developed their auto industries.

In the auto industry, labour's share of overall costs is relatively high. Indeed, among 12 of the most important auto companies in the world, labour's share of value added in the 1980s and early 1990s (apart from Toyota and Honda) is around two-thirds (Table 5.4). Even in Toyota, the most successful major company in recent years at reducing labour inputs, labour's share of value added, while very much lower, is still 42%. Moreover, since Japanese companies (other than

Toyota and Honda) are similar to American and European companies, Toyota and Honda's distinctiveness in this regard, indicates the importance of company-level factors rather than simple regional differences. At the same time, national variations in labour costs per employee hour do exist among the major national auto industries (Table 5.5). These differences were strongly apparent in 1980 when the low-cost producer (Spain) had hourly labour costs less than half those of the high-cost producer (Sweden). In 1980, it might be noted Japanese hourly labour costs were just 60% of those in the US. By the early 1990s, the wage differentials among countries had clearly narrowed. In 1991, for example, the low-wage producer (the UK) had hourly labour costs that were 59% of those of the high-cost producer (still Sweden) – a significant differential but much less than previously. It might also be noted that in 1991 Japan's hourly labour costs were almost the same as the US and were higher than several European countries.

In addition to labour costs, national variations in labour productivity in the auto industry have been calculated. In this context, Williams et al (1994: 61) offer 'build hours', (i.e. the number of hours required to build a vehicle) as a significant measure of competitiveness (Table 5.6). Their calculations, based on both auto and truck manufacture, include work performed by suppliers as well as the major companies, so the

Table 5.6   Build Hours per Vehicle: International Comparisons 1970–1989

Year	USA	Germany	France	Japan	Korea
1970	189	278	267	254	n.a.
1975	174	279	292	176	1475
1980	202	318	252	139	1255
1985	155	258	220	139	572
1989	170	286	162[a]	132	352[a]

Source: Williams et al (1994: 61–2).
[a] Data are for 1988.

Table 5.7   Auto Assembly Plant Productivity: International Comparisons Productivity (hours/vehicle) for 1989

Region	Origins of firms	Best practice	Weighted average	Worst practice
Volume producers				
Japan	Japan	13.2	16.8	25.9
North America	Japan	18.8	20.9	25.5
US	US	18.6	24.9	30.7
Europe	US/Japan	22.8	35.3	57.6
Europe	Europe	22.8	35.5	55.7
NICs	Various	25.7	41.0	78.7
Luxury producers				
Japan	Japan		16.9	
US	US	33.3	35.7	37.6
Europe	Europe	37.3	57.0	110.7

Source: Womack et al (1990: 85, 89).
Note that the volume producers refer to the Big Three in North America, Fiat, PSA, Renault and Volkswagen in Europe and all companies in Japan. Luxury autos from Europe include Daimler Benz, BMW, Volvo, Saab, Rover, Jaguar, Audi and Alfa Romeo; from the US, Cadillac and Lincoln; and from Japan the Honda Legend, Toyota Cressida and the Mazda 929.

statistics are comparable. The most significant trend in these data is the rapid decrease in build hours in Japan, especially during the 1970s. In the US, build hours declined noticeably in the 1980s, and according to Williams et al (1994: 62) the US–Japan gap is 'always relatively narrow'. Yet, bearing in mind that in 1970, US build hours were fewer than in Japan, in 1980 Japan required 31% fewer build hours than the US to build a car, and in 1991, 23% fewer hours. These differences could be interpreted as a significant gap. The unusually large number of build hours in the German auto industry might also be noted; indeed Germany combines high wages (Table 5.5) with low productivity – clearly a problematical situation.

The assembly plant productivity data obtained by Womack et al (1990: 84–89) for 1989, based on the hours required to assemble autos in various countries, also suggest that, on average, Japanese producers in Japan enjoy a productivity advantage (Table 5.7). However, their data show that within each region, including Japan, assembly plant productivity does vary between best and worst plant practice. In general, assembly plant productivity of auto plants in Japan are better than in the US, although best plant practice of firms in the US are better than the worst plants in Japan. According to these data, plant productivity in Japan and the US is notably higher than for European-based plants owned by European countries. In addition, the variation between best and worst is much greater in Europe and the NICs than in either Japan or the US.

Even among advanced countries, labour costs and productivities underlying auto production vary from location to location. Moreover, over time, costs and productivity differentials between locations have changed. In 1970, for example, labour costs in the Japanese auto industry were among the lowest in the world; by the 1980s they were among the highest. Labour productivity in Japan, however, rapidly shifted in the other direction. Simultaneously, following tendencies in other high-cost regions, such as Sweden, Japanese firms have continued to shift to higher value production. Indeed, although hourly labour costs have risen rapidly, Japan has become the most competitive auto-producing region in the world, forcing other countries to adapt in some way to the Japanese challenge.

Indeed, using data for the early 1980s, Bloomfield (1991) reports that Japanese firms could deliver a typical compact car in the US that was over $1700 cheaper than rival products made in the US (Figure 5.9). If transportation costs are ignored, the manufacturing cost advantage is even greater. It is also important to recognize that these estimates identify the major source of Japan's cost advantage to be 'management systems and techniques' (organization), while wages and fringe benefits are the second but much less important source of cost advantage. Shop-floor labour relations are other sources of cost advantage. But why should Japanese firms have such an advantage in organization and what is its nature? Suffice to say that recognition of the organizational basis of cost advantages implies that cost structures are not simply a 'given' datum but that they can be created.

In an increasingly competitive global economy, companies in many industries, including the auto, garment and micro-electronic industries, are looking to reduce costs, including by locating branch plants in low-wage regions. The ability to do so, however, is limited by skill and produc-

tivity constraints (Figure 5.10). In general, it is the labour-intensive components of these industries which locate in low-wage locations such as export-processing zones. In the case of micro-electronics, a particularly important distinction is between the design and wafer fabrication stages on the one hand, and the assembly stage on the other. Each stage has different production characteristics and there is no particular advantage for geographical proximity. Traditionally, it has been the assembly stage that has been located in low-wage regions. However, it might be noted that in this industry the relative importance of labour costs has been substantially reduced from 47% of the industry's total costs in the early 1960s to 30% in 1976. Labour costs are much less important in the production of complicated and high-value integrated circuits than in simple devices. One US estimate, for example, suggests that labour costs for assembly operations were 33% for simple integrated circuits and 4% for complex integrated circuits (UNIDO 1981). In general, the product cycle dynamics of semiconductors are becoming more complicated, with each 'stage' showing diverse locational trends. In contrast, the production of televisions is following a more conventional life cycle experience as automation is vastly reducing the number of components in a TV set, so that production is much simpler and attracted to low-wage regions (Scott 1987).

Conclusion

In essence, the neoclassical view is that factory location is driven by powerful forces of economics, and industry can only ignore economic reality at its own peril. This view has had a pervasive influence on economic geography. Practically, the neoclassical approach has a strong 'common sense' appeal in that it stresses the axiom that for factories to be viable revenues must exceed costs and that explanations for factory location should accordingly emphasize costs and revenue considerations. Moreover,

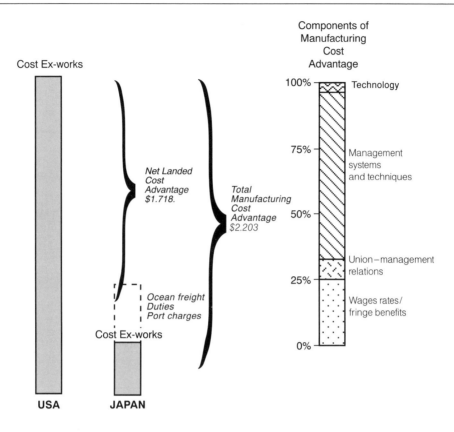

Figure 5.9   Japanese Manufacturing Cost Advantage for a Typical Compact Car, Early 1980s. Source: Bloomfield (1991: 28)

at a time of rapidly globalizing competition, the neoclassical emphasis on the relentless and rational pursuit of lower costs and more profits captures an essential dimension of contemporary economic dynamism.

Yet the neoclassical approach towards industrial location dynamics is controversial and it has received extensive criticism (Krumme 1969b; Massey 1979; Storper 1981). In a recent paper, for example, Barnes (1987) has noted that the assumption of *Homo Economicus* means that economic agents do no more than respond to universal laws of rationality. There are no other motives. Indeed, the neoclassical approach

assumes that the competitive process, if left to itself, allows for the most socially beneficial allocation of resources. In this model, the interests of firms and communities are the same. But as Barnes (1987) notes, in a neoclassical landscape places are no more than spaces where capital may or may not be deposited. As such, the neoclassical model, by denying or simplifying roles for local agency and local social and political context, demeans the richness of economic geography, an enduring theme of which is its variability.

There are also many interesting questions concerning industrial location dynamics that are

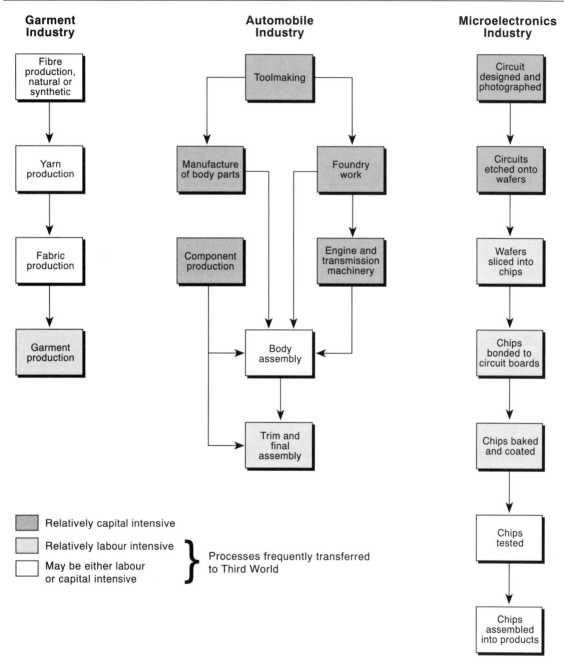

Figure 5.10   The Labour Process in Three Manufacturing Industries. Source: Crow and Thomas (1985)

unaddressed by neoclassical frameworks with their underlying assumptions of universality and economic rationality. In the next two chapters, two alternative approaches to industrial location are explicitly concerned with the processes underlying the location of factories.

6

THE LOCATION OF FACTORIES AS A DECISION-MAKING PROCESS

In the behavioural approach, factory location is expressed in terms of the preferences of decision-makers. Explanations of these preferences focus on decision-makers' search for, and evaluation of, information. From this perspective, behavioural theory is more 'realistic' than neoclassical theory – it emphasizes that in the real world decision-makers make choices based on limited and uncertain information. Thus, in the best-known behavioural theory of economic behaviour, decision-makers are characterized as 'satisficers' to reflect the reality of decision-makers who have limited information and limited ('bounded') rationality. Satisficers seek to make choices, including locational choices, which at least meet aspiration levels (and which are hopefully economically viable). In order to make these choices satisficers collect, code and evaluate information, and learn. In behavioural theory, satisficer firms are information processors; the environments of firms (or the wider economy) comprise information spaces or beds; and the key interactions between firms and environments are in the form of information flows (Table II.1). Moreover, since in the real world learning cannot be perfect and because decisions commit firms to the future which cannot be known precisely, behavioural theory stresses how decision-makers cope with uncertainty.

Within economic geography, Pred (1967, 1969), Townroe (1969, 1971) and Stafford (1969, 1972) pioneered the behavioural understanding of industrial locational choice processes from the perspective of the individual firm. They argue that in practice locational choice is part of a strategic or long-term investment decision that is complex, uncertain, inherently subjective and conducted by individuals or groups of decision-makers who do not have the capabilities of *Homo Economicus*. In this view, factory location reflects locational preferences, which shape, and are shaped by, decision-making processes.

The behavioural rationale for detailed analysis of industrial location choice processes is at least twofold. First, behaviouralists point to evidence which suggests that for an increasing number of activities the spatial margins of profitability are extremely wide so that factories can potentially invest in a large number of viable locations. The basis for such a view rests on a decline in the importance of transportation and communication costs, the widespread investment in basic infrastructure and wide-ranging possibilities for locational adjustment. One New Zealand study (McDermott 1973), for example, showed that while some industries had restricted locational choices, others, such as leather and stationery, could profitably locate anywhere in the country!

In the UK, Taylor (1970) made similar comments regarding small firms in the iron foundry industry, while Taylor (1975a) found on the basis of isodapane analysis that a fireworks factory could locate virtually anywhere throughout the Axial Belt and parts of Wales without incurring any penalty in relation to its existing location. Even in the iron and steel industry, spatial limits to survival are broad (Warren 1970: vii). Thus, an important rationale for the behavioural approach towards factory location is rooted in the premise that decision-makers have legitimate choices that are not economically preordained and only crudely limited by economic constraints. Second, the behaviouralists suggest that, regardless of whether spatial margins to profitability are sharply defined, analyses of locational decision-making can be useful to regional planning agencies seeking to attract industrial investment and to firms seeking guidelines for their own decisions (Townroe 1976; Stafford 1979; Schmenner 1982).

The chapter is in three main parts. First, the central concepts of behavioural location theory are outlined, specifically with respect to the characteristics of the environment, decision-makers and firm–environment relations. Second, the literature that has analysed locational choice in terms of the decision-making process is reviewed. Third, the behavioural underpinnings of the locational preferences of small and large firms are noted and, especially in relation to multinational corporations (MNCs), related to the so-called 'hyper-mobility of capital' thesis which envisages such firms scanning the globe at will, looking for new locational opportunities. Reference is also made to the behavioural interpretation of industrial location policy.

Firms as information processors: landscapes as information beds

The behavioural theory of the firm in economics stemmed primarily from the work of Simon (1955, 1957) and Cyert and March (1963). Thus

Simon sought to replace the idea of the firm as a Economic Person or Black Box with another picture of the firm: "that of a learning, estimating, searching, information processing organism" (Simon 1959: 269). In the real world, firms neither enjoy perfect information nor act in a perfectly rational way with the information they have. For Simon, the idea of 'optimal' decisions, and of minimizing and maximizing, is a theoretical abstraction. Rather, in practice, firms accumulate information in a variety of ways over time, often make decisions in response to crises within limited time-frames, and solve problems sequentially and in a manner that continually invokes subjective judgements. In turn, judgements are shaped by the personal abilities and perceptions of decision-makers, whether acting alone or as part of a group.

To provide an alternative to *Homo Economicus*, Simon offers Satisficer (and Administrative) Person as a characterization of 'real world' decision-makers. Satisficers have limited information horizons and bounded rationality and so cannot possibly evaluate all relevant information, tangible and intangible, on all possible alternatives. Bounded rationality does not imply irrational behaviour but recognizes limitations to the abilities of decision-makers in evaluating information. In addition, decision-makers typically pursue a decision-making process, which at least informally involves goal setting, identification of criteria, time-frames and decision-making structures, which makes sense in their circumstances ('procedural rationality'). In general, according to behavioural theory, firms (a) consider a limited number of choices; (b) search and evaluate alternatives in a strongly sequential way; and (c) choose the first solution that is 'satisfactory'. For Satisficers, a satisfactory solution is one that meets their 'aspiration levels'. While many factors potentially influence aspiration levels, the experience of decision-makers is particularly important.

A complementary if more general idea of satisficing is that decisions reflect individual perceptions, rather than 'objective' definitions, of reality, and that these perceptions incorporate

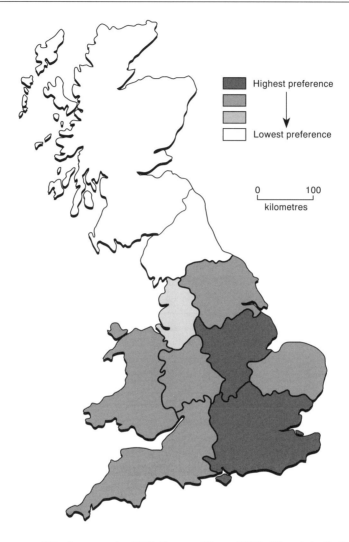

Figure 6.1   Space Preferences of Businesspeople, 1974. Source: Watts (1987: 69); originally based on Pocock and Hudson (1978: 118)

spatially biased information horizons or 'mental maps' (Abler et al 1971: 519–530; Gould and White 1974; Pocock and Hudson 1978). Typically, the perceptual maps individuals have about a country are strongly influenced by where they live and, while there may be other effects, there is noticeable distance-decay effect in terms of the 'desirability' of places (e.g. in terms of where to live and work or to invest). From this perspective, the mental maps of corporate decision-makers frequently reinforce established core–periphery contrasts. The space preferences of business people in the UK in 1974, for example, revealed a strong overlap between the core regions of the South-east and Midlands and the regions of highest preference (Figure 6.1). With distance from this core, space preferences decline. Similarly, when high-technology firms in Ottawa were asked about the suitability of Canadian cities for their operations, Ottawa itself and

Table 6.1 Ottawa's TOC: Executives' Ratings of the Ability of 11 Cities to Satisfy Their Firms' Locations Requirements, 1981

City	Total number of responses	Very satisfactory 1		2		Acceptable 3		4		Very unsatisfactory 5		Less than acceptable (columns 4 and 5) %
		No.	%	No.	%	No.	%	No.	%	No.	%	
Halifax	38	0	0	3	7.9	5	13.2	8	21.1	22	57.9	79
Montreal	39	5	12.8	7	17.9	8	20.5	11	28.2	8	20.5	49
Ottawa	44	28	63.6	12	27.3	3	6.8	1	2.3	0	0	2
Toronto	40	15	37.5	16	40.0	6	15.0	2	5.0	1	2.5	7
Hamilton	37	3	8.1	5	13.5	10	27.0	11	29.7	8	21.6	51
Kitchener–Waterloo	37	4	10.8	7	18.9	11	29.7	9	24.3	6	16.2	41
London	38	1	2.6	7	18.4	10	26.3	10	26.3	10	26.3	52
Winnipeg	38	1	2.6	5	13.2	7	18.4	13	34.2	12	31.6	65
Edmonton	38	2	5.3	6	15.8	7	18.4	12	31.6	11	28.9	61
Calgary	38	3	7.9	11	28.9	8	21.1	6	15.8	10	26.3	42
Vancouver	36	4	11.1	7	19.4	11	30.6	8	22.2	6	16.7	39

Source: Steed and DeGenova (1983: 275).
TOC: Technology oriented centre.

Toronto, Canada's most powerful economic centre and also located in Ontario, easily received the highest rankings in terms of the 'very satisfactory' criterion (Table 6.1). These ratings are significantly higher than those for the next three cities, nearby Kitchener–Waterloo and Montreal, and the distant, but fast-growing Vancouver. On the other hand, no respondent rated Halifax, Nova Scotia, within Canada's 'poor' periphery, as very satisfactory.

Such mental maps are not the same as location decisions. But they do indicate that how decision-makers view the world is spatially limited and biased, and they reinforce the argument that location decisions ought to be understood in terms of information search and learning strategies.

The behavioural matrix

The key to behavioural explanations of industrial location is how firms perceive, code and evaluate information and the factors which influence cognitive and choice processes. In this view, firms are information processors, the environ-

ment is an 'information bed' and the links between firms and environment occur as information flows. Thus, Dicken's (1971) discussion of locational decision-making in (large) manufacturing organizations defines the behavioural environment as that part of the 'objective environment' which represents the total sum of information in the economy (whether considered globally or regionally), with which the firm receives and sends information flows and signals of one kind or another (Figure 6.2). In practice, the geographic core of the behavioural environments of individual firms comprise their operating locations (and associated communities) where firms have intimate and to some extent firm-specific knowledge. The behavioural environment is also defined by the various business connections firms develop with suppliers, consumers and governments. The actual knowledge firms have about these connections varies according to length, frequency and type of contact.

Information exchange occurs in a variety of ways and an important distinction is whether information is exchanged by direct personal

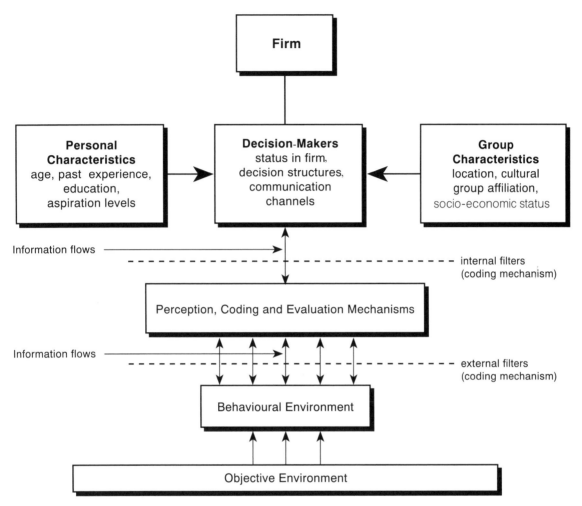

Figure 6.2   Firm–Environment Relations in a Behavioural Landscape. Source: Adapted from Dicken (1971) and Huff (1960)

contact or in some other way (e.g. written correspondence, fax, e-mail or telephone). Personal contact is a unique form of information exchange vital to decision-making, including most notably investment decision-making, and the geographic constraints on personal contact are more stringent than for other forms of information exchange. Regardless of communication channels, the perception and evaluation of information is affected by numerous personal and group characteristics of individual decision-makers, related, for example, to their location,

cultural group, socio-economic status, personality, age and past experience, education and aspiration levels.

While behavioural environments vary considerably among firms, a generalized, 'polar' distinction is between small, single-plant firms and giant multinational firms. First, the behavioural environments of the latter, by definition, are geographically more extensive than the former (Taylor 1970). Second, in small firms, information channels are typically more highly personalized around one or two key decision-makers and rely

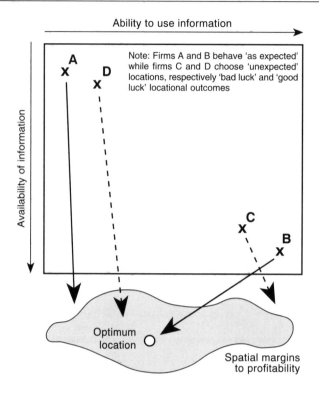

Figure 6.3   The Behavioural Matrix. Source: based on Pred (1967: 92)

more on informal contact than is the case in MNCs. The latter are more bureaucratic, which in behavioural terms implies that the coding, filtering and deliberation of information is conducted within the framework of group decision-making and subject to corporate rules, procedures and culture (Schoenberger 1994; Edgington 1995). Indeed, in MNCs, communication channels within the firm are problematical which are resolved by explicit corporate policies.

In Pred's (1967) classic study, the complex range of factors affecting how firms process information and make locational choices are summarized in terms of a 'behavioural matrix' (Figure 6.3). Pred argues that the particular locational choices made by firms reflect an interplay between factors influencing 'the availability of information' to firms and another group of factors influencing 'the ability to use information' by firms. The information available

to firms is approximated by the behavioural environment, as just defined, and is therefore geographically structured. The ability to use information by firms relates to the competence of decision-makers in dealing with specific decision situations and is affected by various personal and group characteristics whose geography is difficult, perhaps impossible, to map. In summary, a firm with a high ability and a high level of information would typically be in a better position to make a 'good' locational choice than a firm with low ability and a low level of information. In this context, it might be noted that Pred's idea of 'good' locations are those that fall within the spatial margins of profitability and the best ones are close to the neoclassical optimum (Figure 5.3).

Within the context of the behavioural matrix, there are two main polar types of behaviour. First, firms with high abilities and information levels can be expected to locate closer to the optimum and, second, firms with poor abilities and information levels may be expected to lie near the spatial margin of profitability, or even beyond it and fail (Figure 6.3). Clearly, a continuum of behaviour exists between these two polar situations. In addition, unexpected outcomes can occur. Thus, a firm with high ability and information can choose an unprofitable location (the 'bad luck' outcome) while a firm with low ability and information levels can choose a profitable location (the 'good luck' outcome). Firms may be able to adjust to unexpected circumstances. The reality of unexpected outcomes, however, reflects the fact that strategic decisions face uncertainty.

Uncertainty

From a behavioural perspective, uncertainty is inherent to strategic decision-making and has an important impact on behaviour and the outcomes of behaviour. Uncertainty raises the possibility of unexpected outcomes and while decision outcomes can be better than expected, they may be less than expected and projects can fail. In broad terms, behavioural interpretations of location decisions, or indeed any kind of investment decision, recognize that uncertainty results from both imperfect information and because the future cannot be predicted. As Simon (1957) recognized, decision situations vary considerably in terms of uncertainty. On the one hand, programmed decisions are characterized by a high degree of frequency, short time horizons, negligible or limited investments of new resources by firms and by little or no uncertainty. In contrast, non-programmed or strategic decisions are rarely made, one decision is never exactly the same as another, the resource investment is considerable, time horizons are long and levels of uncertainty are significant.

Mack (1971) makes a useful distinction between the uncertainty associated with 'knowledge gaps' and 'true uncertainty'. Knowledge gaps define the difference between what a decision-maker knows at the beginning of the decision-making process compared to what the decision-maker needs to know and can be learnt during the decision-making process in order to make an informed decision. In general, the extent of knowledge gaps depends on how different a new decision problem is from previous decisions encountered by the firm. Thus, firms building a new factory who wish to diversify their product-mix or the geographic base of operations face bigger knowledge gaps than firms who locate a new factory close to existing locations and duplicate an existing product line. Uncertainty in this case arises due to a failure to close knowledge gaps. Thus, decision-makers may not be able to define the knowledge gap precisely in the first place, while a lack of resources, a willingness to accept uncertainty and incompetence may also leave knowledge gaps unclosed. True uncertainty is temporal and results from the fact that decisions to commit resources at one point in time must make assumptions about economic conditions at a future point in time. Bearing in mind that several years may elapse between the time when a decision to go ahead with an investment is made and the time when the plant is finally constructed and operating, changes in technology, government policy and regional stability may all threaten the viability of new projects as they start up or shortly after. As a result, firms making investment decisions incur significant costs which are certain, while potential revenues await an uncertain future.

Uncertainty means that investments can fail before they generate a return or earn a profit or pay back the investment. Thus, projects may be unlucky in that some unforeseen change (true uncertainty) in the business environment undermines its economics. In addition, projects may fail because of a failure to properly identify and close the knowledge gap, i.e. to obtain all the information necessary to ensure the project's success. The knowledge gaps are frequently of a technical or marketing nature. The Ford 'Edsel', for example, is an oft-cited symbol of those new products which fail soon after production begins because technical problems or deficiencies were undetected during the research and development

Table 6.2 Reasons for On-site Expansion, Reasons for New Plants: A Fortune 500 Perspective

On-site expansion		New plant openings	
Reason	Counts	Reason	Counts
1. Keeps management together	103	1. No space at existing location	74
2. Achieving economies of scale	92	2. Insurance against labour strife or other problems in one location	52
3. Quicker form of expansion	87	3. Desire to escape unproductive labour	41
4. More effective spread of overhead	85	4. Too many workers at existing plants	39
5. Cheaper construction costs	66	5. Improve market access, lower transportation costs	36
6. Difficult to separate out a product for manufacture elsewhere	49	6. New technology required a new facility	32
7. Difficult to separate out a part of the production process	48	7. Urgency for space need argued for buying building	20
8. Good existing workforce, technical support, wage rates	16	8. Improved access to suppliers	8
9. Raw material availability	3	9. Better labour rates	6
10. Other	8	10. Environmental considerations	3
		11. Other	11

Source: Schmenner (1982: 89, 94).
Note: The number of plants citing at least one reason favouring on-site expansion is 170 and the number of plants citing at least one reason favouring new plant openings is 158. The count column sums the responses for each reason by all plants. The plants belong to firms 'drawn almost exclusively from the Fortune 500' list (Schmenner, p. xv).

(R&D) process and marketing surveys misunderstood consumer preferences or were incorrectly communicated to the R&D department. Knowledge gaps are also geographical in nature. Firms that choose to build branch plants for the first time in other regions and countries, for example, typically have much to learn about local conditions, and failures to collect or properly interpret such information has caused many problems and even doomed projects.

Learning and decision-making experience

A summary interpretation of Pred's behavioural matrix is that smart firms make smart location choices. While appearing tautological, this prediction does raise questions about what is meant by 'smart', especially for firms (assuming that smart locations are within the spatial margins of profitability). This issue is complex and, as noted, is potentially affected by a wide range of group and personal characteristics of decision-makers and forms of decision-making

(Figure 6.2). However, Pred's prediction does draw attention to the importance of learning and to the past experience of decision-makers in understanding location choices. In fact, there is evidence that more experienced, better informed decision-makers make more economically viable location choices and are less likely to make mistakes. Watts (1971, 1987: 71–72), for example, compares the choices made by two large firms, the Anglo-Scottish group and the Anglo-Dutch group, in locating new sugar plants in the UK in the 1920s. Anglo-Scottish, an engineering firm with experience in sugar manufacture, chose to build small plants dispersed throughout England. In contrast, Anglo-Dutch, an established sugar manufacturer, concentrated fewer, larger plants in the main sugar beet growing region of East Anglia. As the better informed, experienced decision-maker, Anglo-Dutch located within the spatial margins to profitability while the sugar business of Anglo-Scottish recorded losses.

In contemporary times, there are a plethora of 'industrial albatrosses' which stem, in part at least, from poor location choices and, more generally, from managerial inexperience. Two examples are provided by John Sheenan's oil refinery built at Come-by-Chance, Newfoundland, and Daniel Ludwig's pulp and paper mill built in Amazonia. Sheenan's large-scale oil refinery at Come-By-Chance, Newfoundland, for example, failed to make a profit and failed within two years of start-up in the mid-1970s; despite numerous attempts by the Newfoundland provincial government to resurrect the facility, it remains mothballed. Despite hiring management and engineers who had been involved in oil refinery construction and start-ups around the world, there were technological flaws in the refinery's design that simply could not be corrected. Arguably the best-known industrial albatross of recent decades is the Jari Project in Amazonia, Brazil (Hecht and Cockburn 1989: 129–132). This project, financed by Ludwig, began in 1967 as a wood pulp mill based on imported Japanese technology (the mill was actually imported in the mid-1980s on two barges from Japan across the Indian and Atlantic Oceans and up the Amazon) and plantations of Gmelina, a fast-growing East India tree. Unfortunately, although 3 million acres of plantation had been planned, less than one-tenth had been created 10 years after clearing had started because of inappropriate construction technology and a very high labour turnover. In addition, most of the seedlings of Gmelina failed and yields were less than expected primarily because the specie was not suitable to Amazon conditions. In 1982, after investing $750 000 million, by which time the project had been extended to include (capital-intensive) rice cultivation (which was also uneconomic), cattle ranching and a kaolin mine, Ludwig sold the operations to a consortium of companies backed by the Brazilian government for $280 000 million. The project survives apparently as an example of environmental adoption; it does not pay income tax and enjoys other subsidies.

At both Come-by-Chance and Jari, the decision-makers were investing in a geographic region with which they were unfamiliar and where there were no existing, similar factories in operation from which to learn or copy. For this reason alone, the knowledge gaps that needed to be reduced by the decision-making process were relatively wide. Moreover, the owners and parent companies involved in the two projects were not established oil-refining or pulp-making companies. For these decision-makers, the uncertainties of entering a new and remote geographical region were compounded by the uncertainties of operating new and complex industrial technology. In these cases, these uncertainties proved too great for the levels of competence of the decision-makers.

Not all post-start-up problems threaten a plant's survival and in many cases, firms are able to adjust to the problems they experience. Hayter (1978), for example, discusses problems of varying magnitude faced by pulp mills shortly after starting up facilities in remote regions of British Columbia, principally because decision-makers lacked previous experience with the decision situations they faced. Two projects did not properly assess their timber supply base and introduced inappropriate technology; another failed to conduct an analysis of its river water supply, which contained a high (and unexpectedly abrasive) silicon content; while still another only recognized an important climatic implication of a northern, interior location when its huge (outdoor) store of wood chips froze. While one of these projects failed, the others successfully, if expensively, adjusted their operations and subsequent pulp mill projects in the region learnt from these experiences and evaded similar problems.

In practice, in cases in which there is failure to close knowledge gaps, whether by oversight or misinterpretation, it is difficult to distinguish the role of incompetence of the decision-makers from a lack of experience with decision situations. Moreover, problems of inexperience are compounded for pioneering firms; for example, in firms that are the first to build a particular type of manufacturing plant in an isolated region, decision-makers have no exemplars from which to observe, copy and learn. Once established, such factories then become a source of information for other mills. In most industries, in fact, there is a considerable diffusion of information about new developments, through industry magazines and industrial association meetings, particularly with respect to large-scale investments, about such matters as technology, marketing, supply systems, work organization and any problems a new factory might be experiencing in its new location. Such problems can be readily appreciated by other firms. Even so, individual firms only rarely make decisions to locate new factories, and new locations are often chosen by a few senior individuals with limited experience in actually making locational choices and in the places that they are considering for investment. Factories are highly sophisticated and complex operating systems and the plans of even the largest projects can go wrong. Inevitably, uncertainty is a significant dimension in questions of location choice.

Stages in the locational decision-making process

Within the manufacturing sector, even among large firms, new site location decisions are rare. There are certainly instances when firms expand rapidly through the establishment of branch plants. Monsanto, for example, established an average of two branch plants per year between 1958 and 1969 (Rees 1974). Such behaviour would seem to be well above average, even for the period of the 'long boom'. Thus MacMillan Bloedel, Canada's largest forest product corporation, substantially expanded its corporate system in British Columbia after 1950 mainly by investing at existing sites (Hayter 1976). In addition, its interregional and international expansions, beginning in the early 1960s, were largely accomplished by acquisition of existing facilities. Similarly, a 'longitudinal' survey of three Scandinavian firms found that they established an average of one new branch plant every 5.3 years, with a tendency for each branch plant to be a feature of the tenure of each chief executive officer (Laulajainen 1982).

For the individual firm, the new site location decision is one form of locational adjustment that is of a long-term nature (Krumme 1969b). Indeed, it is important to recognize that for firms the question of 'locational' choice is in practice an integrated part of an *investment* decision-making process which also involves choices in many other respects, including regarding plant size and technology, employment (numbers, type and labour relations), financing, management, marketing and distribution, engineering and construction. The actual time lines involved in this process can be extremely long. For example, large-scale pulp and paper mills built in British Columbia during the 1960s and 1970s were planned over periods as long as 15 years, including a two- to three-year period necessary for construction alone (Hayter 1978). Even for less capital-intensive and smaller projects than pulp and paper mills, investment decision-making for new mills can be a relatively lengthy, complicated process.

For firms, locational assessments are one particular perspective on wider investment planning processes. Moreover, to a significant degree, locational assessments in practice emphasize intangible features or what German firms and professionals label 'soft factors' (*weiche standortfaktoren*) more so than tangible features or 'hard factors' (*haute standortfaktoren*). (see Table 6.2). Indeed, Townroe (1969), Stafford (1969), Rees (1972) and Nishioka and Krumme (1973), argue that a proper understanding of the relative importance of location factors needs to recognize the integration of location factors within the overall investment decision-making process. To provide an analytical basis for this argument they suggest the disaggregation of the decision-making process into particular 'stages'. In practice, two types of approaches to the disaggregation of decision-making stages have been adopted. The first is based on the identification of distinct processes and the second is based on geographic scale. In the first approach, decision-making is analysed in terms of decision stimuli, search, evaluation and choice processes, investment go or no-go decisions, and post-locational

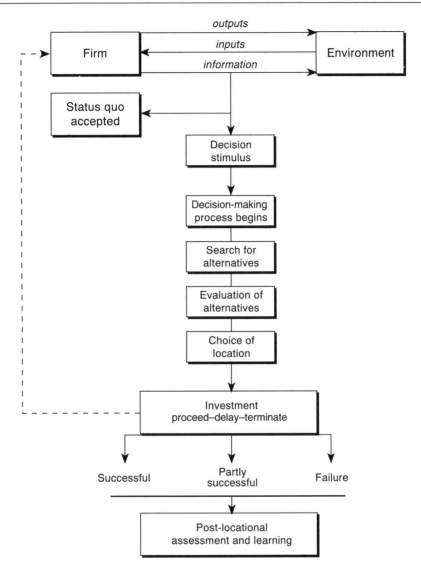

Figure 6.4 Stages in Locational Decision-Making Based on the Decision Process. Source: adapted from Dicken and Lloyd (1977: 322); Hayter (1978)

assessments or learning (Figure 6.4). In the second approach, decision processes are related to the selection of country, region, community, sites and even buildings (Tables 6.3 and 6.4). These frameworks are not mutually exclusive and discussions of location decision-making necessarily involve both. In the following discussion, the

second approach is integrated within that of the first.

Decision-making processes

The investment decision-making process comprises sets of interwoven search and

Table 6.3 Number of Regions and Sites Considered by Firms in Two Surveys

Alternatives	British survey		Canadian survey	
	Region	Sites	Region	Sites
1	17	5	3	
2–4	13	13	4	2
5–7	13	10		2
>7	16	31		3

Sources: Townroe (1971: 56–7); Hayter (1978: 244).
Note: Townroe's British survey principally comprises secondary manufacturing branch plants and Hayter's Canadian survey comprises case studies of seven pulp and paper mills, one of which was not built.

Table 6.4 Stages in Locational Decision-Making Based on Geographic Scale: Townroe's (1976) Generalization of British Experience

Geographic scale	Location factors
Region	Government regional policy, strategic communications, labour relations, markets
Area (town)	Transport and communications, ties with parent plant, labour supply, cost and training, supplies of materials and components, access to services, local and central government services, amenities
Site	Intra-urban location, physical characteristics, tenure (leasehold or freehold), the availability of buildings, access to services and utilities, price
Buildings	Structure, decoration, heat and light, office requirement, ancillary space, tenure

Source: based on Townroe (1976: 54–71).
Note that Townroe's disaggregation of location choice is based on several surveys of about 150 companies making location decisions *within* Britain.

evaluation processes occurring at different scales and with different objectives in mind pertaining, for example, to location, scale and technology. In other words, modelling the decision-making process in terms of stimuli–search–evaluation–choice is a simplification which is not meant to imply that these processes are so discrete and straightforwardly linear (Figure 6.3). For satisficing firms, however, these processes capture the nature and essence of decision-making.

The decision stimulus or trigger

Satisficer firms exist as open learning systems and their market relations with other firms (input and output flows) are governed by various forms of information exchange (Figure 6.4). If there are

no reasons to change established practices, firms are 'satisfied' with their relations with the environment. However, any disturbance in either the operations of a firm or in established patterns of exchange between firms and environments creates new 'decision situations' or problems to be solved. For Dicken, such disturbances are labelled 'stresses', which he defines as "any influence whether it arises from the internal environment [that is, within the firm] or external environment which interferes with the satisfaction of basic needs" (Dicken 1971: 432). Examples of internal sources of stress include changes in corporate philosophy following the arrival of a new manager, acquisition by a new owner, growing dissatisfaction with internal plant configuration or labour relations, while externally generated stresses might come from actions taken or threatened by customers, rivals or governments, the termination of a building lease, urban renewal schemes or from unexpected occurrences which range from the loss of factories to hazards such as fires and earthquakes, or from shifts in government monetary policy such as a currency devaluation.

Whatever their source, stresses, or what others call stimuli or triggers to decision-making, may occur suddenly or evolve more slowly until a point in time is reached when decision-makers within the firm decide to consider ways of resolving the problem or responding to the opportunity in question. Some stimuli to decision-making may be locationally specific. Thus, the stimuli to several firms to consider building pulp and paper mills in British Columbia were provincial government announcements that particular sources of timber were available on long-term timber leases on the condition that lessees would build facilities within or adjacent to the timber leases (Hayter 1978). In many other cases, firms have been encouraged to contemplate the establishment of branch plants to serve local markets in distant regions as a result of the imposition of tariff barriers, threats from competitors or because of transportation difficulties in serving vital markets through exports.

The location decision as an alternative

New site location decisions are one alternative among a range of possibilities and in many cases alternative strategies are available to accomplish similar market goals. Often firms compare new site investments with the option of expansion at existing sites. In practice, there are different sets of reasons favouring the expansion of existing sites compared to opening up new plants (Table 6.2). In Schmenner's (1982) survey of plants belonging to large American corporations, for example, the four most important reasons given in support of on-site expansion were the ability to keep management together; the ability to achieve economies of scale by expanding an existing plant; the fact that capacity could be more

quickly added at existing plants; and because overhead costs could be more effectively spread by expanding existing plants. Expansion at existing sites, however, can also aggravate existing problems, including those related to materials handling, congestion, and complexity of production control (Schmenner 1982: 93). Indeed, among the reasons cited in Schmenner's survey for opening up new plants, problems of existing locations such as lack of space and different kinds of labour problems at existing plants were often emphasized. Other surveys have made similar comments (Townroe 1971).

On-site expansion and new plant openings are not mutually exclusive forms of locational adjustment possibilities, and firms, especially large firms, may wish to consider and implement particular combinations of on-site and new-site investments. Moreover, the combinations considered may evolve as the decision-making process progresses. As Whitman and Schmidt (1966) show in their pioneering case study, General Foods was concerned that its major factories in the north-east region of the US were obsolescent and its initial plan involved their modernization. Subsequently, it entertained various plans involving combinations of plant closure and modernization in association with the establishment of new plants in entirely different communities and sub-regions (although still within the US north-east). Thus new-site and in-site investments can substitute for one another or occur together.

Locational search processes

The spatial biases in the existing behavioural environment of firms, or the mental maps of decision-makers, are important influences on locational search processes. Thus, small firms, whose behavioural environments are relatively restricted geographically, choose locations which are typically (but not always) within close proximity to existing plants or where the owners live (see Chapter 8). That is, new and small firms locate new factories – whether involving first plants, relocations or the first time establishment of branch plants – within the existing mental maps of the decision-makers. Establishment of factories in a new region, however, requires more explicit consideration of location choice and has more immediate implications for regional development, especially as longer distance moves are more likely to involve larger firms and larger branch plants. In this respect, the locational preferences of foreign firms are especially interesting and problematical and will be discussed shortly.

Given the importance of the distinction between large and small firms, the search (and evaluation) processes underlying new-site locational decisions vary considerably. Several broad generalizations are possible, however. First, there is consensus that locational decision-making involves considerable elements of judgement and that processes of search and evaluation are conducted, or closely supervised, by owner-managers or senior executives and managers. Large firms are more likely to delegate some functions of locational search and evaluation, and to conduct analyses which include hard data on the hard factors (*haute standortfaktoren*) or tangible features (Rees 1974: 197; Townroe 1969: 166). Even so, senior management often directly participates in these processes and invariably makes the 'final' judgements regarding location.

Second, locational search and evaluation involves time and cost which can be significant. On this issue, precise evidence has proven difficult to find, in part because search and evaluation processes are intermittent, often secretive and one part of the responsibilities of owners or senior managers. In addition, firms themselves often only have a partial idea of what the costs and time commitments of locational search and evaluation processes are, partly because it is not always clear what constitutes the full extent of these processes which are not conventionally classified as a distinct budgetary item. Rees (1974) and Townroe (1971), based on the experience of branch plants in the US and UK respectively during the 1960s, indicate site search and evaluation times of between six months and two years. These studies, however, concerned moves within a country (and even regions) which principally comprised secondary manufacturing activities. International investments and those involving resources may be expected to be more time-consuming (and expensive). Thus investment in a foreign country potentially requires an assessment of its distinctive social, economic and political conditions, i.e. learning costs, while additional costs may be incurred in getting access to information. In the resource sector, regional evaluations of the physical quantitative and qualitative characteristics of resources, such as those involved in oil drilling, mining explorations and timber cruises, can take years and involve several millions of dollars and require highly specialized work teams. Such evaluations would normally not include a national level assessment or the identification of suitable communities and sites for manufacturing investment.

Third, there is evidence that when firms, especially medium-size and large firms, expand interregionally they frequently contemplate more than one region (Table 6.3). In Townroe's (1971) survey of 59 firms establishing branch plants in the UK, for example, just 17 (28.8%) focused exclusively on one region, and virtually as many considered more than seven regional alternatives. In a detailed case study, Krumme (1981) notes that Volkswagen considered at least a dozen States before opting for Pennsylvania to locate its first manufacturing plant in the US. Bearing in mind the relative fixity of resources, Hayter (1978) also found that four of seven pulp and paper firms contemplated at least two regions for their pulp mill projects (Table 6.3). Yet, Nishioka and Krumme (1973) argue that investment proposals typically are narrowed down to a particular regional context early in the decision process. Such a view is readily explained by the costs and time involved in interregional evaluations and the need to establish at least the regional parameters of an investment in order for the 'non-locational' aspects of investment decisions to be pursued in any detailed way. These constraints would appear to be particularly tight for small firms and the evidence does indicate that the spatial planning horizons of such firms when establishing another plant, or relocating, is more limited than for larger firms. In the case of Townroe's branch plants and Volkswagen, it may be that interregional comparisons were not detailed. Moreover, in Volkswagen's case (and probably for some of Townroe's branch plants), State (regional) governments underwrote at least some of Volkswagen's expenses of locational search and evaluation.

Fourth, in general, firms identify and assess more sites than regions (Table 6.3). Over half of Townroe's sample branch plants considered more than seven sites, for example, and only five considered just one, while all seven of Hayter's pulp and paper firms considered at least two sites and in one case over 30 were identified. Both studies report, however, that typically only three to five sites were evaluated in any detail so that many site options were eliminated by an initial screening process. At this geographic scale at least, most firms in these two studies did not strictly conform with satisficing behaviour in the sense of identifying sites in a sequential manner and choosing the first, 'satisfactory' alternative as a limited amount of comparative analysis was performed. Other surveys, it might be noted, have concluded firms behaved in choosing sites more along the predicted satisficing lines (Cooper 1975). At the regional level, probably the same mixed range of behaviour occurs, while in some cases opportunities are regionally specific and the choice facing the firm is either to go ahead or not.

Locational evaluation

For firms, 'locational choice' potentially implies several decisions made at different geographic scales. Implicitly at least, all location decisions involve choices about countries, regions, towns, communities and sites. In practice, some of these choices may simply be taken for granted and not given any formal consideration by firms. Moreover, what decision-makers mean by 'community', 'metropolitan', 'regional' and 'national' is likely to vary and should not be assumed to coincide with conventional geographic definitions. Decision-makers among firms in the same

locality may associate different boundaries (and mental maps) with the idea of any geographically defined region. In practice, locational surveys themselves vary in how geographic scale is interpreted, particularly with respect to region. Given this caveat, several studies have nevertheless recognized that the geographic scale of thinking by decision-makers about location factors varies during the decision-making process, and as scale varies so does the relative importance of location factors. One such study is provided by Townroe's (1976) survey of British firms (Table 6.4).

While firms in this study apparently did not contemplate any location outside Britain, or at least were not asked if they did so, they were asked to identify location factors at four other scales, namely region, town, site and building (Table 6.4). In this study, the principal factors governing the selection of a region relate to (national) government regional policy, labour relations, markets and strategic communications, which refers to the general transportation and communication requirements of a new plant. Once the regional context is established, labour, transport and communications remain important. The characteristics of labour that are thought important at the regional scale (cost, supply and training), however, are different than those considered at the national scale (labour relations). In addition, at the regional scale there are other important location factors, notably ties with parent plant, access to services, local government services and amenities that are not important at other scales. Similarly, there are distinctive groupings of location factors associated with the choice of site and buildings. With respect to site, for example, the location within a city, the physical characteristics of land, tenure conditions, the availability of services and land prices are relevant.

Surveys of location decision-making in other countries have similarly found that the relative importance of location factors varies depending on scale (Table 6.5). Stafford's (1974) detailed 'content analysis' of six firms locating in Ohio, for example, found location factors varying in

Table 6.5 Stages in Locational Decision-Making Based on Geographic Scale: Stafford's (1974) Analysis of Six Firms Locating in South-east Ohio

Location factors	Total no. of counts	Geographic scale			
		National	Sub-national	Regional	Local
Personal contacts	117	1	23	27	66
Labour productivity	108	18	25	27	38
Labour availability	50	2	8	19	21
Local amenities	49	0	2	15	32
Transport facilities	42	4	14	18	6
Labour rates	39	4	6	18	11
Dispersion tendencies	28	2	14	4	8
Executive convenience	27	1	4	12	10
Facilities and utilities	25	5	2	8	10
Corporate communications	20	0	3	9	8
Supplies accessibility	17	1	5	1	3
Induced amenities	13	0	0	7	6
Market accessibility	12	3	5	1	3
Taxes	1	0	1	0	0

Source: Stafford (1974: 177).

Note that the locational factors cited and the total number of counts were derived by 'content analysis' of open-ended questionnaires with respondents; each time a locational factor is mentioned it is 'counted'.

overall importance in according to national, subnational and regional scales. While regional policy was not a factor in this study, in contrast to the results found by Townroe, labour was found to be significant and, similar to Townroe's findings, Stafford (1974) found labour to imply different concerns at different scales. Thus, in Stafford's survey, labour productivity is important at all scales while labour availability and wage rates become more influential at regional and local scales. Another important finding of Stafford's survey is the importance attached to personal contacts at regional and local scales.

Neither Townroe's nor Stafford's analyses address locational factors at a global scale and they are not particularly concerned with foreign firms. It might be noted, however, that a distinctive factor that firms mention at this scale is political stability. In practice, MNCs not only want stable governments but also pro-development governments, and in these respects whether governments are democratic or not seems to be a secondary issue.

From the point of view of individual firms, the nature of locational choice for new plants and the

closure of existing ones can vary considerably. Moreover, the narrowing down of countries, regions, communities and sites need not be in a specific order; investment opportunities can be regionally and even community specific, inviting an initially focused geographic search and evaluation, while later in the decision-making processes an assessment of wider regional or national circumstances can occur. For some firms and some situations 'location' is a 'given' at least at particular scales. New firms, for example, may simply locate in the neighbourhood in which the founders live (see Chapter 9), while even large MNCs contemplate opportunities in which 'location' is only a matter of choice at community and site levels.

Methods of locational evaluation   The methods employed by firms to compare locations vary considerably. There are at least two broad tendencies (but not rules) that can be drawn from existing evidence (Townroe 1969; Rees 1974; Stafford 1974; Hayter 1978) First, in general, large firms are more likely to engage in some kind of formal, systematic analysis of locational alternatives than are small firms. Second, formal, systematic analyses of locational alternatives are more likely to be conducted at community/site scales of analysis than at regional or international scales. At the same time, it should be recognized that small firms do conduct systematic locational evaluations and some MNCs have adopted frameworks for international comparisons. With more spatially disaggregated data sets accessible by microcomputer, trends in these directions are likely to increase.

On the one hand, some firms have always used 'muddling through' *ad hoc* procedures and 'back of the envelope' assessments to evaluate locations. On the other hand, an increasing number of firms conduct comparative cost analyses of tangible location factors and 'weight ranking' schemes which allow the incorporation of tangible as well as intangible factors. In Schmenner's (1982) survey of new plants located by some of the largest US corporations, for example, some form of comparative cost analysis was conducted in over half the cases. Smith (1971) also offers a few examples of firms and locational consultants calculating cost structures (and even surfaces) of alternative locations within the US for various activities.

Nevertheless, the intangible nature of location factors necessarily implies that locational evaluations cannot be simply and solely reduced to a mechanistic calculation of costs. In this regard, an increasing number of firms systematically establish locational requirements prior to the search and evaluation of locations, e.g. sites. Indeed, there are weight-ranking schemes that have been developed which allow firms to systematically measure intangible or soft locational factors at any one particular scale (Kepner and Tregoe 1965; Townroe 1976). According to Kepner and Tregoe's method of site evaluation (1965), for example, decision-makers first identify a set of 'musts' or 'minimum requirements' that are essential features of a site if the factory is to be viable and these 'musts' are used to screen or narrow down sites. Once alternatives meeting these musts have been identified, Kepner and Tregoe suggest firms draw up a list of 'wants',

which are desirable locational features. In turn, these wants are assigned a weight and each locational option is assigned a score, and the score and weight are multiplied for each locational factor. By summing the results, Kepner and Tregoe suggest that a 'preliminary' decision may be obtained.

In practice, this scheme has been used, for example, by firms in British Columbia's forest sector to locate various kinds of facilities. In one case, the firm assigned three engineers to conduct a survey of pulp mill sites (in the southern Okanagan region of British Columbia). They collectively agreed on lists of 'musts' and 'wants', identified alternatives meeting the former, and for these alternatives each engineer assigned their own weight and score to each locational factor. Finally, the three results were averaged out. This engineers' report was then considered by senior executives as an input to a more informed decision about location which they still ultimately regarded as a judgemental matter. At that time (Hayter 1978), executives in this particular firm also noted that Kepner–Tregoe's method of locational analysis was used only at community and site scales in regions where the firm was already established; evaluations at broader geographical scales were regarded as more uncertain and judgemental and they clearly had less confidence in 'formal' analysis in these contexts.

Investment decisions and post-locational assessments

Whether or not a project will be given the go-ahead, and the precise timing of any go-ahead decision, ultimately depends upon an investment analysis. Whatever the qualities of the location, if investment prospects are not considered good enough, or for other reasons such as a change in ownership, projects may be delayed and possibly abandoned. In this regard, many North American and British firms have a reputation for short-term conservatism which often means that, in practice, major investment decisions are ratified during market upswings when future scenarios look good. Of course, this behaviour can also mean that the start-up of new facilities, following the time taken for construction, can occur during market downturns!

The start-up of an expensive new facility is a critical point in time for decision-makers since how the mill performs at this time, or soon after start-up, reflects on the adequacy of the decision-making process and the viability of the mill. Whatever their cause, the problems that firms experience after the start-up of brand new factories reinforces the idea of locational choice as a learning process. The smoothness of start-ups reflects directly on the effectiveness of the planning process. Smooth start-ups confirm procedures and the factors considered, while difficult start-ups point to mistakes of omission or commission during the decision-making process. The performance of new factories, good and bad, ultimately becomes incorporated in corporate experience and provides exemplars for others to observe and from which to learn.

The locational preferences of foreign firms

The behavioural underpinnings of locational preference is readily evident in the context of new and small firms (see Chapter 9). In these firms, behavioural environments are strongly localized and decision-makers have typically limited capacity for geographic search, evaluation and for accepting the uncertainty of moving beyond regions of familiarity. In contrast to small single-plant firms, MNCs enjoy geographically extensive behavioural environments and participate in a labyrinth of formal and informal information networks, based around existing manufacturing plants and related linkages, which continuously document a wealth of information on operational matters and serve to monitor and assess relevant information on industry matters on virtually a global basis (Aguillar 1967). In many instances, strategic planning departments within MNCs have responsibility for information monitoring and assessment. In addition, in relation to small firms, decision-makers within

MNCs typically have prior experience in choosing new locations for manufacturing plants; more resources to plan for new locations; and the ability to afford specialized, locational consultants.

Given the decision-making discretion and power that MNCs undoubtedly have, it is perhaps not surprising that the locational behaviour of MNCs is difficult to generalize. In fact, such behaviour is controversial. The first and most general controversy centres on the extent of the mobility of capital (for productive investment), i.e. on the geographical limits to the discretion and power of MNCs. Second, particularly with respect to branch plant locations in foreign countries, there are alternative views as to the nature of locational preferences.

With respect to capital mobility, what might be called 'the world is our oyster hypothesis' essentially dismisses behavioural arguments as, at best, an anachronism (Frôbel et al 1980; Bluestone and Harrison 1982; Christopherson 1989; Harvey 1990). According to this hypothesis, the largest MNCs are already global and familiar with all cultures and territories. Moreover, this argument stresses that since the Industrial Revolution, an enduring theme of technological change in transportation and communication has been to reduce the friction of distance, and that space–time compression, in the present ICT techno-economic paradigm, has reached the point where communication is virtually spatially costless. In addition, MNCs are recognized as relentlessly promoting the homogenization of tastes and the standardization of production; as the power of MNCs increases, the culture differences (and information barriers) between places decline. Trade liberalization and deregulation further promote capital mobility and the freedom of MNCs. Rather, the 'world is our oyster' hypothesis states, or at least implies, that MNCs are free to locate and organize production and related functions regardless of national boundaries. The globe is essentially a space over which capital roams, touching down only when profitable opportunities present themselves.

If the 'world is our oyster' thesis recognizes powerful tendencies enhancing the mobility of capital, the 'power of geography' thesis defines limits and opposition to these tendencies (Häger-strand 1970; Johnston 1984; Wolch and Dear 1989; Hirst and Thompson 1992; Daly and Stimson 1994; Edgington 1995). This power is reflected in the geographical structures of location conditions, by the resistance of local culture to universalizing tendencies and by profound behavioural forces that continue to narrowly circumscribe the geographical extent of daily lives. In this view, the most mobile of productive capital ultimately requires people whose lives are place- or even neighbourhood-specific, and nations are still influential forms of organizing territory (Hägerstrand 1970). Moreover, there is an ongoing literature which emphasizes the continuing importance of advanced industrial countries for direct foreign investment (O'hUal-lacháin 1985; Schoenberger 1985; Dicken 1992), and that even for established MNCs, locating in foreign countries, remains a problematical exercise (Michalak 1993; Hayter and Edgington in press).

The behavioural underpinnings of foreign branch plant location

From a behavioural perspective, the factory locations of MNCs are determined by decision-making processes. Moreover, the decision-makers of MNCs operate within geographically structured behavioural environments, have their own mental maps and face limits to locational search and evaluation processes. Within industrial geography, considerable attention has been given to the locational preferences of foreign firms inside national territories. These studies reach different, often contradictory conclusions, in large part because they have drawn evidence from different regions, time periods and groups of firms. Clearly, care has to be taken in generalizing about location behaviour even with regard to MNCs, the ultimate global actor. The following discussion summarizes, in the form of hypotheses, the main behavioural arguments

offered for the locational preferences of foreign firms.

1. A well-established hypothesis is that foreign firms favour established core regions of 'host' countries (Hymer 1972; Hamilton 1976). In turn, such behaviour reinforces core–periphery region contrasts. The rationale for the hypothesis is that core regions are typically centres of communication and transportation, which facilitates travel to foreign countries and personal contacts with host-country decision-makers. Core regions also typically define areas of highest market potential. In addition, core regions, and the largest metropolitan centres of these regions, tend to be well known, part of the mental maps of foreign decision-makers and are associated with lower uncertainty. Edgington (1995), for example, notes that Japanese investors in North American real-estate markets strongly focus on a relatively few centres, notably New York, Los Angeles and Hawaii. For the larger firms, the spatial biases of senior managers in Japan in favour of these centres tends to be reinforced by the information channels established by the branch offices in the US – in part because the branch managers find it easier to get investment proposals in known places ratified by parents if the proposals are in known places. Within the manufacturing sector, Blackbourn (1974) found that in six of seven industrial countries reviewed, foreign-owned plants were concentrated in core regions (although foreign ownership by acquisition as well as by branch plant establishment is included). In Australia, Edgington (1990: 225) found that between 1971 and 1983, Japanese branch plants were strongly concentrated in core metropolitan areas, especially Melbourne and Sydney, and offset marked declines in overall manufacturing employment in these centres.

2. Contradicting the above-mentioned arguments, a second hypothesis suggests that foreign firms are more likely to invest in peripheral areas than core areas, especially when incentives are offered in the former. The behavioural rationale is that since foreign firms have no particular commitment to one region, they can be readily

attracted by the incentives and information provided by local development agencies. In turn, as noted in the next section, industrial policy can itself be justified in terms of offering locational signals to foreign firms and compensation for locating in areas once outside or marginal to their mental maps (Cannon 1975). In the UK and Eire, especially during the 1970s, several studies have concluded that locational incentives and information packages did successfully deflect branch plants of foreign firms to designated regions (Watts 1979; Dicken and Lloyd 1980; Law 1980; O'Farrell 1980; Dicken 1983). Kemper and de Smidt (1980), on the other hand, suggest that during the 1970s locational incentives were not particularly important in the Dutch, German and Danish experience.

3. A third hypothesis suggests that foreign firms prefer to concentrate in particular regions of the country. Thus, by following the experience of successful pioneering firms, latecomers can reduce search costs and perceived risks of locating in an unfamiliar place (McConnell 1980). In this context, particular reference has been made to the behaviour of Japanese firms with their localized concentrations of activities in South Wales (Trevor 1983) and Dusseldorf, West Germany (Dunning and Norman 1983). Japanese firms do not inevitably herd closely together, however. In the late 1980s and early 1990s, for example, Japanese auto assemblers and parts manufacturers located within a 'central corridor', broadly overlapping with but also extending existing auto production concentrations, in the US and Canada (Rubenstein 1991; Holmes 1996; Table 14.6). Such a regional orientation provides access to national markets and suppliers. Yet, within this geographically extensive corridor, Perrucci (1994) and Rubenstein (1991: 130) note that Japanese assemblers have deliberately chosen to avoid locating in a state which already has a rival Japanese auto assembler, in part 'to establish unique relationships with state and local governments' (Figure 6.5). In addition, Rubenstein further notes the Japanese auto makers' preferences for small towns, which, at start-up, had median population sizes of 10 000. Such

preferences, he argues, reflect a desire to avoid existing concentrations of unionized auto workers in order to faciltiate the introduction of more flexible work practices and to block organizing attempts by the UAW, motives which also reveal location choice to be a bargaining process (see Chapter 7).

4. A fourth hypothesis argues that national culture is important in understanding the location preferences of foreign firms. The above-noted, albeit ambiguous behaviour of Japanese firms to concentrate branch plants, at least in some cases, is thought to reflect the 'grouping' tendency of Japanese society (Edgington 1990: 21). There are other cases in which the location preferences of one national group of foreign firms differ from another. Watts (1980c, 1982), for example, found that the spatial distribution of EEC-based manufacturing plants in the UK differs from the distribution of US-owned branch plants, and Edgington (1990) finds the spatial distribution of new Japanese branch plants in Australia to be significantly different from the established British and American ones. While these differences may reflect differences in mental maps, there may also be differences in underlying location conditions and corporate motivations. Thus, Edgington (1995: 220–221) notes that in the manufacturing sector the Japanese corporate preference for Melbourne and Adelaide reflected specific industry needs and local market access.

5. A fifth behavioural hypothesis recognizes that foreign firms, as with other firms, may exercise the equivalent of personal preference in choosing locations. Corporations do develop distinct cultures which help shape corporate strategies (Schoenberger 1994), and in turn corporate cultures may incorporate particular locational preferences; for example, those derived from founding entrepreneurs. One example is provided by MacMillan Bloedel (MB), whose entrepreneurial founder (whose first company was a trading company) insisted that MB's manufacturing locations in British Columbia be on tidewater. This preference became part of corporate thinking long after the founder retired, including during MB's internationalization

(a) 1979

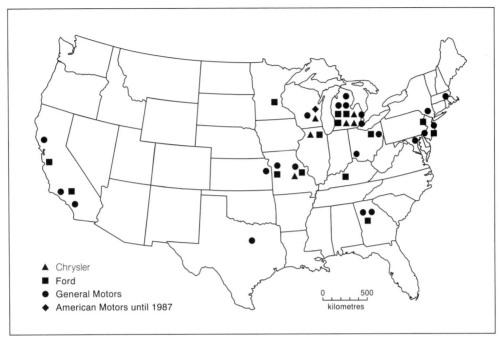

(b) 1989

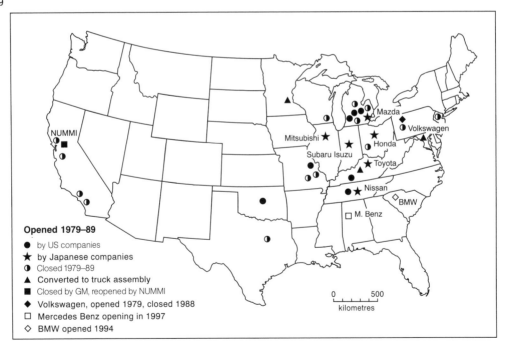

Figure 6.5 The Location of Auto Assembly Plants in the United States, 1979 and 1989. Source: Rubenstein (1991: 130, 132)

strategy (Hayter 1976). It also helps to account for MB's non-participation in the forest boom in British Columbia's interior in the 1960s and 1970s. Another example is provided by Michelin which developed a reputation for corporate secrecy; this had a locational expression in a preference for small, isolated communities. In Nova Scotia, Michelin insisted on this preference despite attempts by federal and provincial governments to encourage it to locate in a large, designated 'growth centre'.

6. A sixth hypothesis suggests that foreign firms change locational preferences following initial entry into a country. The rationale for such changes relates to a growing awareness of locational possibilities in association with expanding behavioural environments as branch plants seek out local suppliers and markets. McConnell (1980), for example, noted that foreign firms in the US initially reinforced existing location patterns and then subsequently became part of the trend away from the Industrial Belt and towards non-metropolitan areas. Indeed, O'hUllacháin (1985) suggests that by 1981 foreign firms were expanding more rapidly in the sunbelt than domestic manufacturers.

Clearly, the location preferences of new branch plants established by MNCs are subject to various influences. These influences can change over time. In many countries, one shift is the decline of traditional forms of regional policy and attempts to deflect foreign firms from investing in core regions to designated ones. Yet, subsidies continue to be available to the factories of foreign firms. From this perspective, the traditional behavioural rationale for regional policy still has relevance.

Industrial location policy in a behavioural landscape

As noted in the last chapter, the neoclassical position is inherently suspicious of government interference in private sector decision-making and its justification for industrial location policy relies primarily on the promotion of mobility arguments with respect to both capital and labour. Behaviouralists are more particularly concerned with the effects of industrial incentives on capital mobility and, more specifically, whether or not industrial location incentives change the location preferences of entrepreneurs in favour of designated regions (Cannon 1975). If policy is not meant to change such preferences, why have it? As noted in the previous chapter, there is evidence that industrial incentives do generate incremental investments in targeted regions and in this chapter particular reference has been made to the location of foreign-owned branch plants in this regard. It is also true that incentives have often been awarded to firms without influencing location choice.

In the behavioural landscape, industrial location incentives can serve to change locational preferences in two main ways. First, incentives serve as signals to firms to encourage them to at least consider designated regions. As studies of location choice reveal, while firms often compare regions and communities, search processes are geographically limited, so that such signals may be vital in advertising designated regions. Second, given that the decision-making process itself is costly and time-consuming to firms, behaviouralists further justify industrial location policies as offering compensation for any additional learning costs or uncertainties the firm might incur by contemplating investment in an hitherto unfamiliar, peripheral region. Indeed, incentives in the form of subsidies and tax breaks are often supplemented by information packages on designated regions. In fact, the information made available by development agencies can be extensive (Lloyd and Dicken 1972) and such agencies can be proactive in getting this information to potential investors and in providing various kinds of follow-up services in terms of facilitating business contacts, answering questions and showing potential sites. More generally, it can be argued that policies which increase the information available to firms also increase the chances of economically rational behaviour. On the other hand, it can also be suggested that offering incentives and support in economically marginal areas increases the chances of attracting short-term opportunists. Questions of this latter sort often plagued industrial location policy in Atlantic Canada, for example, and similar criticisms have been made in other jurisdictions (George 1970, 1974).

If national governments have tended to play down, even terminate, centralized and explicit industrial location policies, regional and local development policies of one kind and another remain important in promoting industrial development. For regional and local governments, economic development policy continues to be important in terms of providing signals, information and inducements to mobile capital.

Conclusion

It is ironic that there has been little direct assessment of satisficing theory in studies of industrial location choice, although comments are offered from time to time (Townroe 1971). Similarly, industrial geography has made little effort to operationalize the behavioural matrix. Yet, it can equally be argued that behaviour is a pervasive theme in industrial geography which is so grounded in observing and explaining actual locational patterns and dynamics. But in addressing this reality, a weakness of the behavioural approach is that it provides no sense of conflict or the controversies that often surround matters of location. In behavioural location theory, imperfect information and bounded rationality modify the decision-making capabilities of *Homo Economicus*, while neoclassical cost and revenue surfaces are similarly modified by information surfaces or mental maps. Thus, in both theories the environment or economy is seen in passive terms and the implication of both theories is that the interests of the economy are reflected in the goals and behaviour of individual firms. The next chapter directly questions this assumption.

7

THE LOCATION OF FACTORIES
AS A STRATEGIC PROCESS

Enterprise geography (and institutional) approaches interpret factory location as an expression of the investment strategies of firms, especially large multi-plant corporations. Strategies define the competitive positions of firms in the long term and new factories are critical to this definition. Consequently, in this approach, factory location is explained in terms of the factors that influence strategy formulation. In particular, emphasis is placed on how the geography of corporate strategies is guided by 'internal' long-term motivations, accumulated expertise and established corporate structures, and by the 'external' strategies and structures of other business organizations, especially rivals (but also consumers and suppliers) and by other institutional forms and interest groups, notably labour organizations and governments.

Large corporations, particularly multinational corporations, are at the core of the enterprise approach (McNee 1958, 1960, 1963; Table II.1). As the dominating agent in modern industrial societies, MNCs are governed by 'technostructures', i.e. professional, specialized management bureaucracies (Galbraith 1967), who have power to influence the behaviour and performance of other agents. This power is exercised in various ways and, for example, is evident in bargaining processes with other firms, labour and govern-

ments. Indeed, for MNCs in particular, location conditions and factors are not simply given datum, in the form of cost and revenue surfaces (or information spaces). Rather, wage levels and other labour characteristics, transportation rates, prices, taxation levels, the supply of infrastructure, and a host of regulations influencing spatial profitability, are subject to negotiation, persuasion and bargains of one kind or another. Technostructures are by no means omnipotent, however. Rather, corporate strategies are constrained in turn by the 'countervailing powers' of equally large rivals, labour, governments and increasingly environmental groups (Galbraith 1952).

The assertion that economic agents have power provides a different, more radical concept of the firm, the environment and firm–environment relations than that assumed by neoclassical (and behavioural) location theory (Storper 1981; see Table II.1). For Galbraith (1967), neoclassical theory represents 'conventional wisdom', but not contemporary wisdom. Thus, in Galbraith's modern industrial economy, black boxes and perfectly competitive forces exercised by 'invisible hands' are replaced by technostructures and the very visible strategies and structures of large corporations which create monopolistic or oligopolistic forms of competition (Veblen

1932; Heilbroner 1985, 1992). Moreover, in landscapes of countervailing powers, conflicts exist among economic agents who pursue goals related to security or control, market share, size and growth as well as goals of efficiency and profitability. Indeed, in the institutional landscape, differences among the priorities and preferences of business, labour and government help shape locational outcomes.

This chapter is in three main parts. First, the key concepts of corporate strategy and structure and countervailing power are defined and ways in which rival strategies shape locational outcomes are identified. The second part interprets factory location as a strategic bargaining process, specifically with respect to the negotiations between MNCs and host governments regarding new investments and the implications of labour relations for location. The final part briefly comments on industrial policy as a social bargain from national and local perspectives.

Firms as institutions: landscapes as countervailing powers

Organizations, as institutions, are formal structures embedded in society whose behaviour reflects particular rules, tradition and values (Oinas 1995b). Business organizations exhibit a wide variety of forms and pursue a complex range of goals (Starbuck 1971; Heilbroner 1992; Ahern 1993a). As Ahern (1993a) notes, there is one view which emphasizes that organizations seek to use resources efficiently and another which emphasizes that organizations seek to access and control resources. In the context of large firms, growth is usually considered an important goal in itself and one that relates to others, such as executive income, prestige and power (Starbuck 1971). From this perspective, the so-called managerial models of the firm argue that large firms seek a rate of growth that is not too slow, thereby causing financial problems and threats from stockholders and rivals, and not too fast, thereby generating internal problems of co-ordination, efficiency and planning (Baumol

1959; Penrose 1959; Marris 1968). In practice, firms can pursue a range of goals. Galbraith (1967), for example, suggests that large corporations seek growth, subject to security and profitability constraints. Of course, plans do go wrong, the economy can change in unexpected ways and firms do restructure and 'downsize' in an attempt to survive. Yet, among capitalist societies, the desire for growth ('accumulation') is deep-seated, and the underlying motivations relate to the control, power and prestige that growth and increasing size brings, as well as to profits (Veblen 1932; Heilbroner 1985). In large firms, strategic goals are the context in which the location of specific factories occurs.

Corporate strategy and structure

Firms grow by implementing strategies (Hayter 1976; Harrington 1985). According to Ansoff (1965), strategies are ultimately driven by the investment decisions of firms and Ansoff refers to the allocation of resources and selection of policies designed to achieve long-term objectives and associated market roles. Thus, the formulation of strategies, whether implicit or explicit, inherently incorporates notions about the actual or anticipated behaviour of other organizations in the environment or economy. In theory, the range of strategic options for firms is extremely wide. In practice, although by no means invariably so, the formulation of strategy, especially when emphasizing investments in new facilities, tends to reveal 'common threads' with corporate history and established corporate structures.

In theory, individual firms have wide strategic choices. In practice, these choices are constrained by accumulated know-how, assets, expertise, and, more generally, by competitive or entry advantages with respect to production, marketing, technology, access to raw materials and/or financing (Marris 1968: 113; Caves 1971; Kolde 1972: 178; Langlois and Robertson 1995). Indeed, according to Langlois and Robertson (1995), at the core of each firm's competitive advantage is some (precise) idiosyncratic or firm-specific knowledge that is not shared with other firms. Specific strategies seek to extend the firm's accumulated advantages of know-how, resources and skills, as well as sources of power and size. Occasionally, firms may wish to diversify away from established markets if those markets are in decline. Such diversification, however, is best achieved by acquisition and joint venture.

Types of corporate strategy   There are several classification schemes of corporate strategy. The best-known scheme emphasizes the industrial direction of growth in relation to existing activities of firms (Figure 7.1). According to this scheme, corporate strategies refer to (backward and forward) *vertical integration* (expansion to internalize inputs and markets respectively), *horizontal integration* (expansion of existing products to increase market penetration) and *horizontal diversification* (entry into new products for the same markets). Ansoff (1965: 115–118) also distinguishes *concentric growth* (diversification of the product mix to serve new markets) and *conglomerate growth* (simultaneous diversification of products, markets and technology). Each of these strategies is further distinguished in terms of *internal* and *external* growth. The former refers to investment in new plant and equipment, and the latter to the acquisition of existing plant and equipment. Traditionally, strategies of horizontal and vertical integration have been, and remain, particularly important and have featured both internal and external growth. Conglomerate growth strategies, which became important in the 1960s, usually involve external growth. Subsidiary companies within conglomerates, however, typically pursue horizontal and vertical integration strategies. Moreover, since the early 1990s, the rationale of the conglomerate – the ability to achieve a superior financial performance by moving funds from declining to growing businesses rapidly – has been questioned even by the conglomerates themselves and several have chosen to spin-off their empires into more specialized parts.

There are other ways of classifying corporate strategy (Rumelt 1974; Freeman 1982). Freeman (1982), for example, distinguishes strategy in terms of technological innovativeness, specifically offensive, defensive, imitative, dependent and traditional strategies. Thus technologically offensive firms are on the leading edge of industrial R&D in order to remain the leading product innovators, while defensive firms also invest in very large-scale R&D in an effort to 'catch up' with the leaders. Strategies of technological imitation involve firms in copying the successful technology developed by others, while dependent firms simply purchase 'off the shelf' or lease mature technology with perhaps some modest adaptation for local circumstances. Traditional firms rely on technology that has not changed for at least one Kondratieff cycle. Clearly, the classifications of corporate strategies by the nature of industrial integration and technological innovativeness, as well as in other respects, are not mutually exclusive.

The corporate structure follows strategy thesis
In general terms, the deployment of corporate physical and human assets among its various manufacturing plants and offices and the manner in which these operations are integrated and governed defines corporate structure (Caves 1980). In practice, corporate strategies are closely intertwined with corporate structures (Chandler 1963; Caves 1980). Strategies emerge from structures and in turn modify them. Chandler (1963) argued, on the basis of American experience, that corporate structures are adapted following the implementation of corporate strategies (Figure 7.2). According to Chandler's 'structure follows strategy' thesis, US-based corporations such as Ford, implemented strategies that created larger-scale and more complex operations, which in turn required changes in the structures of command and communication.

In an idealized form, as (some) firms grow, entrepreneurial functions are decentralized and bureaucratized as individual entrepreneurs are progressively replaced by groups of specialized managers; groups of specialized managers with supporting departments; divisions based on functions (such as accounting, production, marketing and personnel); divisions based on

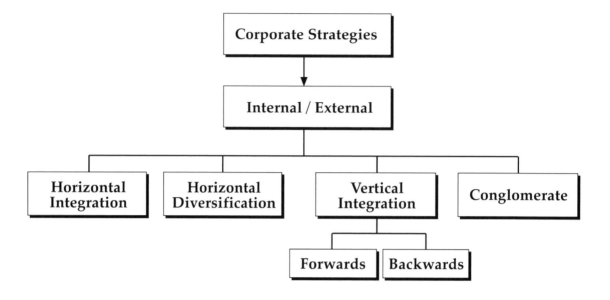

Figure 7.1   Types of Corporate Strategy

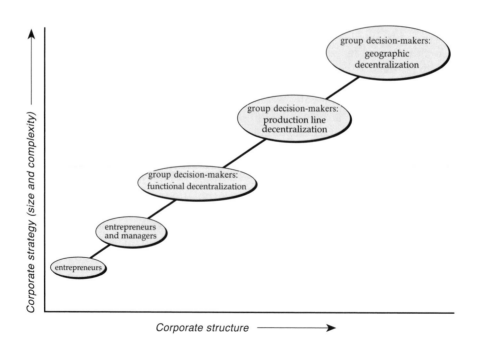

Figure 7.2   The Corporate Structure Follows Strategy Thesis

functions and product lines; and divisions based on functions, product lines and geographic spheres of operation. Indeed, in the US, particular emphasis has been placed on a hierarchical, 'multi-divisional' or 'M' form of organization. In tandem with the decentralization of decision-making functions, firms develop new ways of integrating and co-ordinating increasingly dispersed and diversified operations. From this perspective, integration relates to the organization of information flows within the firm, the insertion of checks and balances on local autonomy, and the maintenance of some form of centralized responsibilities. Even in highly decentralized companies, budget control and investment planning, long-range R&D and an information flow system that has its ultimate apex in the parent's head-office, provide ways to co-ordinate operations.

In large firms such as MNCs, decisions as to strategy and structure are the responsibility of technostructures. While typically characterized by strong senses of hierarchy and loyalty, technostructures are bureaucracies in which 'coalitions of interest groups' holding differing views on strategy and structure can occur, even if such differences are rarely reported (Krumme 1981; Soyez 1988a, 1991). The most significant division of interests within the firm is that between management and labour, whether the latter is unionized or not. In a union firm, these differences are formalized and subject to explicit negotiation under sets of rules to which both parties may contribute. Moreover, unions exist within the firm and are part of the firm's environment to the extent that a particular union or union local is affiliated to a bigger union or simply loosely related to a wider union movement.

Economic landscapes as countervailing powers

In a neoclassical landscape populated by numerous small firms, economic behaviour is regulated by competitive forces comprising arms length and independent relationships which can be readily substituted for one or another. Indeed, in a perfectly competitive world, the 'invisible hand' is a metaphor for the (apparently free) self-regulation of markets. While some markets work in this way, in landscapes of countervailing powers big business dominates and their behaviour is interdependent rather than independent (Galbraith 1952, 1967). Indeed, reaction to, and anticipation of, rival behaviour constitutes one form of 'countervailing power' which constrains the autonomy of any one giant firm (Fellner 1949). In this landscape, the critical features of corporate interdependence from a long-term perspective are investment decisions, particularly those incorporating product and process innovation, rather than simply price (Schumpeter 1942). Thus, in countervailing landscapes dominated by a few big firms concerned with size, market share, security and growth, strategies implemented by one firm, which feature product innovations that undermine existing products and investments designed to access new markets or resources, inevitably draw reactions from rivals.

Oligopolistic rivalry: locational hypotheses Within oligopolies, interdependent corporate investment strategies can range from competitive to collusive, with different types of strategy creating different types of countervailing landscape morphologies (Watts 1978, 1980a). In a competitive situation, for example, equally large rivals may be encouraged to pursue locationally overlapping branch plant investments in order to gain a share of each market or a share of a particular natural resource (Figure 7.3(a)). Vernon (1971, 1985: 67–70) interprets such behaviour as a response to risk and uncertainty as firms who do not match the locational initiatives of rivals potentially forfeit sales and profits in new markets (or potentially forfeit access to low-cost resources in new supply areas). A number of studies in the resource and especially the secondary manufacturing industries have demonstrated that corporate investment strategies move in a geographically parallel manner (Vernon 1971; Knickerbocker 1973; Rees 1978; Gwynne 1979; Laulajainen 1982). In the petroleum industry, for example, the leading UK and US

(a) Competition: the locational overlap model

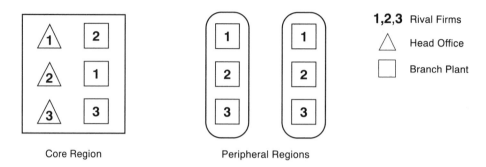

Core Region   Peripheral Regions

(b) Competition: the exchange of threats model

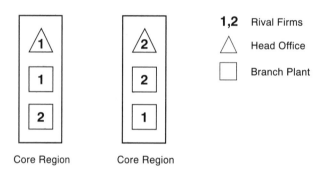

Core Region   Core Region

(c) Collusion: the spatial monopoly model

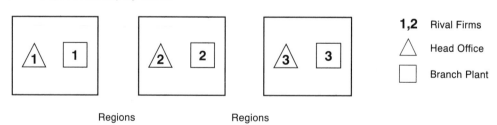

Regions   Regions

Figure 7.3   Landscapes of Countervailing Power

oil companies collectively invested in the Gulf of Mexico region in the 1930s, and in the Persian Gulf region in the 1950s and 1960s (Vernon 1985: 69). In secondary manufacturing, investments by large corporate tyre manufacturers (Gwynne 1979) and by leading auto manufacturers (Baranson 1969) in various developing countries also contributed to overlapping facilities, in some cases establishing capacity far in excess of local demand.

In countries with relatively small markets, including advanced countries, this form of competition can encourage an inefficient industrial structure if no individual plant can reach minimum optimal scale. In Canada, particularly before trade barriers started to decline in the mid-1960s, US corporations in several industries established branch plants which served the Canadian market with a wide range of the parent companies' product mix. As a result of low-volume product volumes, however, product-level economies of scale were frequently not realized, contributing to what is known as the 'miniature replica' effect (Britton and Gilmour 1978: 93).

A modification of the 'locational overlap' hypothesis is provided by the so-called 'exchange of threats' thesis (Vernon 1985: 70). This idea is based on the observation that firms in the same industries but based in different core countries establish branch plants in the other country (Figure 7.3(b)). For example, US-based firms that established branch plants in Europe were often in similar industries to the European-based firms in the US (Hymer and Rowthorn 1970: 80–82). According to the exchange of threats thesis, this behaviour occurs because the leading firms in both countries, threatened by the establishment of foreign-owned branch plants in their home market, respond by establishing branch plants in the invading firms' markets. This cross-investment constitutes a 'warning' to invading firms that strong competition in foreign markets will be countered by similar efforts in the home markets of invaders. In still other cases, for example in the American sugar beet industry at the turn of the century (Eichner 1969) and the British brewing industry in the 1960s (Watts 1980b), powerful firms have been able to 'block' rivals from entering a region and competing for sources of supply or markets.

Spatial forms of collusion Collusion creates different landscapes. Although many forms of collusion are illegal and not easy to identify, collusive behaviour has exerted important impacts on industry evolution and probably continues to be important (Watts 1987: 123). Collusion may take the form of shared participation in which supposed rivals agree to invest in a region only if they do so together. According to the 'red-line' agreement of 1928, for example, the world's leading oil companies agreed (for over a decade) not to undertake any new developments in the indicated territories, principally in the Middle East, except in partnership with the others (Vernon 1985: 69). The most obvious examples of spatial collusion, however, occur when firms agree to carve up markets (or resources) among themselves to create market cartels and regional monopolies (Figure 7.3(c)). The evolution of the electrical industry, following the discovery of the light bulb in 1879 and until the 1940s, is an example of how the dominant firms in the US (notably General Electric) and in Europe (notably AEG of Germany) literally carved up the global market, on the basis of cross-licensing agreements, patent pooling and joint ventures, into "exclusive territories, non-exclusive territories and territories excluded from the agreements" (Newfarmer and Topic 1982: 38). These agreements not only stipulated where products could be sold but also at what price and where they were to be manufactured. Watts (1980a) cites another example from the UK sugar beet industry in the 1920s of a collusive agreement which divided up supply areas and which directly influenced the location of factories. Thus, the two rivals which had been located close to each other and involved in disputes over stealing each other's supplies, agreed to build new plants no closer than 30 miles (48 km) apart (Watts 1980a: 301).

In landscapes of countervailing power, the price (as well as investment) behaviour of firms may be interdependent and organized by or within dominant companies in various ways (Figure 7.4). Under basing-point pricing, for example, the price charged to a customer is the price of the good at the designated basing point plus transportation charges from the basing point, regardless of where the good is actually produced (Figure 7.4(a)). The result is that markets distant from the basing point and served by local producers are discriminated against to the extent that 'phantom freight charges' are incurred. The advantages of such schemes for producers are that they are easy to administer, provide protection to basing points and potentially create windfall profits. In the US, basing-point pricing, such as the Pittsburgh Plus system in the iron and steel industry, was widespread at the beginning of the 20th century although it has since been judged an illegal practice which unfairly discriminates against consumers (Machlup 1949). In Europe, however, basing-point pricing is still prevalent and officially sanctioned in several industries throughout the European Community. In terms of the location of new factories, basing-point pricing systems typically reinforce existing locations for both producers and consumers. Basing-point producers are never at a price disadvantage throughout the market area while consumers can only guarantee not paying phantom freight by locating at the basing point.

Even if illegal, there is the possibility that producers will secretly collude to fix prices in particular markets (Figure 7.4(b)). It is difficult to generalize the locational effects of such price fixing although a tendency towards encouraging concentration is likely.

Significant levels of transactions in contemporary economies, involving services and goods, occur within large corporations (Figure 7.4(c)). These transactions are clearly not voluntary, independent or determined by the activities of many competing buyers and sellers, the central characteristics of perfectly competitive markets. Rather, the prices charged by affiliated companies are 'administered' and subject to the policies of particular corporations. Within the internal flow of goods and services, large firms have some discretion as to pricing and this discretion can have important implications for location. For example, oil-rich resource peripheries have often complained that the international oil firms deliberately kept the price of crude oil low to reduce *ad valorem* taxes at the source of supply. In turn, low prices (and *ad valorem* taxes) shift

the location advantage for new refineries towards markets. More generally, large firms enjoy similar discretion over the prices charged to subsidiaries for head-office, R&D and related services.

Governments, labour and other sources of countervailing power Corporate power is not only constrained by rivals but by the strategies and structures of governments, labour, other organizations such as consumer groups and, most notably in recent decades, by environmental groups, as well as by society's values and customs. There is a Marxist view which emphasizes the subjugation of government to the interests of business. In Bradbury's (1979) theory of the state, for example, the function of government is to legitimate capital and provide the necessary infrastructure for capital to exploit labour effectively. From this perspective, the notion of countervailing power is irrelevant. Of course, it is to be expected that governments in capitalist societies would promote capitalism and for corporations to seek to further their own interests, including by influencing government policy. Given these expectations, an alternative 'pluralistic' view of the state permits governments to both support and regulate business and to mediate among a wider range of social interests. Thus, in addition to providing services, incentives and infrastructure for business, government legislation also regulates business in terms of employment, safety, pension and environmental standards, as well as the conditions of competition, and through a plethora of taxes to provide public goods and services.

The most general point to be made about governments in relation to business is that policies, attitudes and forms of mediation vary from one jurisdiction to another. Such variations exist at international scales, for example related to different polices towards foreign investment (Laxer 1989), and at regional scales as reflected in, for example, different business climate measures and commitment to boosterism (Weinstein and Firestine 1978; Cox and Mair 1988, 1991). A similar general point can be made about

(a) The basing-point pricing system

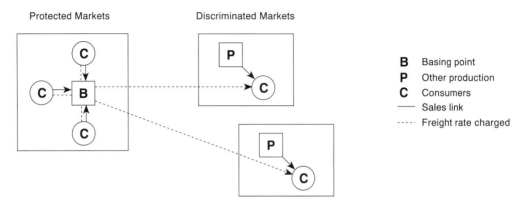

(b) Price fixing in a regional market

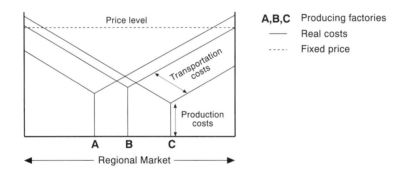

(c) Administered prices

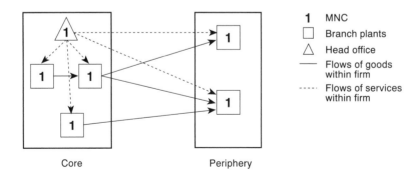

Figure 7.4   Interdependent Pricing: Examples of Collusion in Landscapes of Countervailing Powers

labour. Thus, at international and national scales there are significant geographies of labour power. Management–labour relations, for example, have evolved on different lines in the UK, Japan and Germany and, even if union power in the UK (and the US) has declined and become less adversarial, unions in these countries constitute different kinds of countervailing power to business. Within countries, variations in labour power are reflected in the regional distribution of unions (Figure 4.3).

If in the 1960s, countervailing powers principally comprised business, government and labour, in the 1990s numerous 'environmental' groups have emerged to mount increasingly effective campaigns against business behaviour that destroys environmental values. Environmental groups, of varying philosophy, have particularly targeted primary activities such as fishing and mining, and the primary manufacturing industries, most notably the forest industries, which are involved in the exploitation of scarce natural resources or are involved in unusually extensive environmental destruction. In recent years, environmentalist opposition to industrialization has occurred for several reasons (Soyez 1996). First, environmental groups have successfully mobilized local coalitions of quite distinct interest groups, such as aboriginal tribes, local tourist operators, fishermen and nature lovers, to oppose industrial development. Second, environmental groups have been successful in 'internationalizing' environmental conflicts over industrial projects, especially by targeting potential opposition among consumers of export-oriented projects. Third, some environmental groups have been more willing to take businesses to court if environmental laws are broken. Fourth, other groups have been willing to break the law themselves to protect environmental values. Finally, and ultimately of most importance, in advanced countries at least, there has been growing social concern over environmental values. In fact, environmental regulations in advanced countries are much stronger and more effective than in poorer countries where it is more difficult, but not impossible, to mount environmental opposition. In any event, geographical variation in environmental regulation of, and opposition to, industrial development exists.

Locational evolution   In landscapes of countervailing powers, evolving patterns of industry are explained in terms of the interaction of long-term corporate strategies, government policy and technological change (Britton and Gilmour 1978; Markusen 1987). From this perspective, the spatial margins to profitability are shaped by public and private sector policies, rather than simply by 'given' underlying distributions of resources, markets and other location conditions. The particular mix of policies, as well as of location conditions, vary among nations and regions. In the case of Canada, for example, the roots of its particular manufacturing structure originate with the national policy of the 1870s which established tariffs to protect secondary manufacture and an open-door policy to forest investment (Britton and Gilmour 1978). As a result, the tariff encouraged foreign firms to establish branch plants to serve local markets, while exports were dominated by the resource industries. Indeed, one of the historically distinctive features of Canada's manufacturing is the high level of foreign ownership (see Chapter 15).

Corporate strategy: bargaining over location

Locational choice occurs as part of investment decisions that have strategic purposes, i.e. they are made to meet the basic motivations of corporate strategy (growth, profits and security) by gaining access to new markets or sources of supply in a way that makes sense to individual corporations. Moreover, locational choice is not simply a matter of identifying alternatives, evaluating them and choosing the best, as if the alternatives were all given datum. Rather, technostructures consider alternative bargains, contracts or deals with labour, suppliers and consumers, and discuss options with different and varying levels of government on such matters as infrastructure provision, tariff levels, profit repatriation, taxes, subsidies, zoning, energy supply and environmental impact analyses and regulations. In this regard, the spatial mobility of 'new' capital provides technostructures with a fundamental bargaining advantage in relation to 'local' labour and governments. While labour and governments have geographically fixed planning horizons, technostructures do not; new capital has wider location options.

Theoretical frameworks which have explicitly interpreted location as a bargaining process have emphasized either bargains between MNCs and host-country governments (Gregerson and Contreras 1975; Goodman 1987; Kobrin 1987) or the nature of labour relations (Clark 1981, 1989). There are also case studies of locational bargaining processes that have examined the roles of firms, governments and labour, and which provide the basis for this part of the chapter (Krumme 1981; Soyez 1988a; Alvstam and Ellegard 1990).

Multinationals and nation states

It is generally argued that during the 20th century, MNCs have becoming increasingly powerful in relation to nations and in their ability to influence nation states (Harvey 1982; Galbraith 1983; Peterson 1988). At the regional scale, this observation becomes more emphatic (Krumme 1970). There is debate, however, as to the extent to which this balance in bargaining strength and ability has shifted. Peterson (1988: 159) argues that in the face of a rapidly globalizing world economy, in which productive and especially financial capital have become extremely mobile, nation states have lost their power to manage. A greater weight of opinion suggests that interest groups that are antagonistic towards, or do not entirely share, business values also influence governments who retain considerable powers to regulate their economies (Hirst and Thompson 1992). Hirst and Thompson (1992) also suggest that the majority of large firms that are international in scope remain MNCs, which have distinct national (and regional) 'homes', and many fewer that consti-

tute true transnational companies (TNCs) that are state-less with no particular national loyalty. In any event, "Bargaining between firms and nations goes on, spawning business arrangements of varying complexity" (Goodman 1987: 133). One important context in which this bargaining occurs is over direct foreign investment (DFI) when a MNC wishes to build a large new factory in a foreign ('host') country.

According to Kobrin (1987), the bargaining power of MNCs and host-country governments (HCGs) comprises three dimensions: the relative demand by each of the two organizations for resources which the other controls; the constraints on each organization that affect the translation of bargaining power into control over outcomes; and bargaining ability (Figure 7.5). Briefly, the power of MNCs is rooted in their technological and managerial expertise and complexity, their financial resources, international marketing channels and differentiated products, reinforced by powerful advertising campaigns. Their ability to exercise this power is constrained by competitors who wish to negotiate with HCGs and by the relative importance of the HC as a market (or resource supplier). The power of HCGs, on the other hand, is related to the size of domestic markets, resources and skilled labour pools, the availability of infrastructure and the political situation in the country, especially with respect to stability. This power, in turn, is constrained by high levels of global corporate concentration, competition from rival countries, existing high levels of dependence on MNCs and by balance of payments problems. In addition to their respective 'power resources' and 'constraints', the bargaining abilities of MNCs and HCGs are affected by other factors, including past experience in similar projects: bargaining processes are also learning processes. However, MNCs typically have clearer and more narrowly focused objectives than HCGs; they typically have much better knowledge of the nature of their activities and can better anticipate the impact of investments on the local economy (Zurawicki 1979). To an important degree, the expertise of the

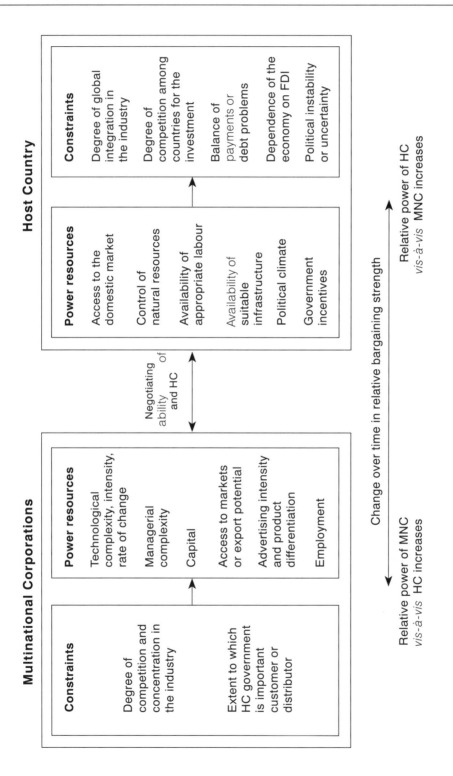

Figure 7.5 Components of the Bargaining Relationship between MNC and HC. Source: Skulason (1994) and adapted from Kobrin (1987: 628)

MNC is embodied within its international operations, and its investment planning horizons are inherently more global than those of HCGs. Once MNCs have established factories in HCs, however, bargaining power between MNCs and HCGs can change (Vernon 1970; Korbrin 1987; Auty 1993). The obsolescing bargain hypothesis, for example, predicts that over time bargaining advantage shifts towards the HCG as MNCs can no longer move fixed investments and HCs learn more about the MNC operations (see Chapter 11 for a fuller discussion of this hypothesis).

Bargaining between HCGs and MNCs typically focuses on specific investment proposals. Gregersen and Contreras (1975) and Goodman (1987), for example, analyse the investments of US-based MNCs in the forest product industries of Latin America (Figure 7.6). In their model, bargaining power depends upon the interactions of motivations and 'minimum requirements' as well as resources and constraints. For HCGs, minimum requirements may be expressed as laws or guidelines, and for MNCs in terms of internally defined rules. Three types of bargaining situation then arise depending on whether motives are complementary, competitive and conflicting. Conflicts result when minimum requirements are not met over a particular matter. In this regard, key issues often relate to levels of profit repatriation allowed, taxation levels, the location of factories, and the degree to which the proposed factory will export, buy local components, train local workers and hire local management. With respect to location, for example, HCGs frequently want (for various reasons) foreign investment to be located in 'designated' regions or 'special economic zones'. In many, perhaps most instances, this demand results in a complementary situation since foreign-based MNCs may have no particular commitments to any one region and designated areas typically offer the minimum requirements firms need in a location, particularly in relation to infrastructure, labour supply and incentive programmes. In some instances, however, conflict arises and locations have to be negotiated.

The Icelandic aluminium industry In the planning of foreign investment, MNCs and HCGs frequently negotiate locations and specific location factors. Thus, the MNC may seek to 'play-off' one nation, region and community against another to gain a better deal. MNCs can also negotiate location factors in specific places in the absence of any formal comparisons and explicit threats that it will locate elsewhere. Until an investment is made in a particular place, MNCs always have the power of not doing so. In addition, MNCs can argue to HCGs that they have superior knowledge about locational requirements, especially when they represent a new industry to a country. In this context, Skulason (1994) offers an interesting case study of the bargaining processes between MNCs and governments in the establishment of the aluminium industry in Iceland. Globally, the aluminium is highly concentrated and the leading firms are horizontally and vertically integrated. Iceland offered to the aluminium MNCs a low-cost power base and tidewater access to affiliated sources of bauxite and refining operations in Europe. Alusuisse, a MNC based in Switzerland, built the first aluminium smelter in Iceland in 1969, and the Atlantal-group (a joint venture of three – originally four – MNCs) have negotiated the details of a second smelter, the go-ahead decision for which is still pending. The first set of negotiations lasted almost 10 years if completion of the smelter is defined as the end point, or about six years if 'final' approval of the Master Agreement by the Icelandic Parliament and Alusuisse is used. The second set of negotiations began in 1987 and construction is yet to start, although tentative agreement has been reached (Table 7.1(a) and (b)).

Alusuisse's original proposal touched off a contentious debate in Iceland about allowing foreign investment of any kind, but particularly that of a large MNC, into such a small country (of about half a million people). This debate, which constituted part of the context of the entire set of negotiations, was only marginally resolved in Alusuisse's favour and underlines the point that the positions taken by HCGs are themselves

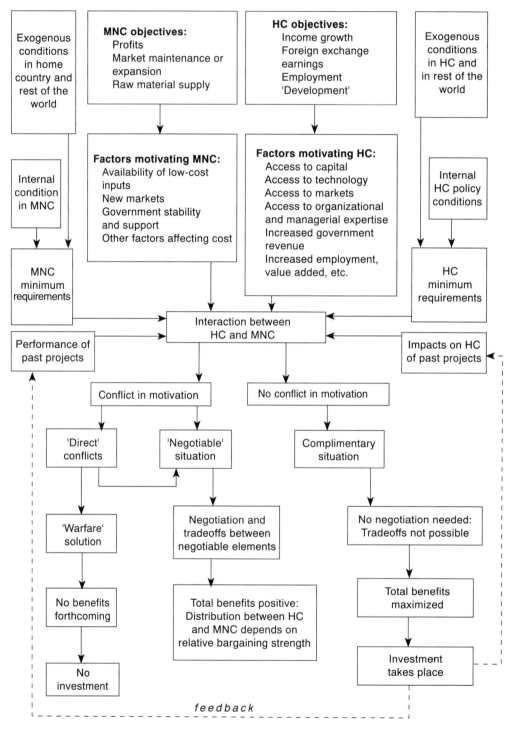

Figure 7.6 Interactions Between MNC and HC for a Given Investment Proposal. Source: adapted from Gregersen and Contreras (1975: 49, 50)

compromises and bargains of potentially highly variant views. Nevertheless, negotiations continued because Alusuisse and the Iceland government (if not the opposition parties) had overlapping motives: Alusuisse saw Iceland as a large source of cheap hydropower accessible to its European manufacturing base, and the government wanted to diversify Iceland's economy by utilizing its water resources.

The negotiations themselves focused on the interrelated questions of location, power supply and cost, scale of plant, taxes, import duties (on the alumina), length of contract and the legal status of the plant. The power supply issue, for example, raised questions as to the number, size and location of new power plant, which in turn depended on the size and location of the smelter, and the price of this power to Alusuisse. At the time, the government entertained some hopes to use the smelter to meet regional development goals, particularly to spread growth away from the capital region of Reykjavik. Of five alternatives considered, however, Alusuisse strongly preferred the Reykjanes (Straumsvik) site, near Reykjavik, which was determined as the low-cost option. Since the government's desire to attract the aluminium industry to Iceland was far greater than its concern over location within the country, this location was the one chosen (Figure 7.7). On a legal matter, however, the government was forced into a far more difficult compromise, one which almost scuttled the project. While the government wanted Icelandic law to govern the operations of the smelter, Alusuisse insisted on the primacy of international courts of arbitration. The final compromise allows Alusuisse to use international courts, if necessary, but these courts should take into account (in some unspecified way) Icelandic law.

The Atlantal-group has also negotiated an agreement with the Icelandic government. Much the same sets of issues were involved in this negotiation although in this case location within Iceland involved more controversy. In particular, while the government felt more strongly about promoting regional development within Iceland, Atlantal's smelter will be built adjacent to the existing one if the project goes ahead. Yet Atlantal is not affiliated with Alusuisse and its choice of location seems to have been based on a rather odd cost comparison in which the alternatives considered (Figure 7.9) were required to include expenditures for environmental control while the chosen site was not! The Atlantal group, however, strongly preferred Straumsvik which is close to both Reyjkavik and the international airport, thus facilitating access for parent company executives to government decision-makers and the proposed smelter. The existing smelter is also proof of the viability of this location. As it happened, the signing of a Master Agreement was delayed in the early 1990s by the Gulf War and because Russia has been trying to obtain hard currency by selling off its aluminium stocks which in turn is driving down prices. If the project once more becomes economically feasible there may of course be some reconsideration of location conditions

Location as a bargaining strategy with labour

Do firms use location as a bargaining ploy with labour? New capital investment, particularly that by large corporations, is relatively more mobile than labour and by investing in new locations firms can find and develop new kinds of labour bargain while the threat of investment elsewhere can potentially be used to extract concessions from an existing union in an existing factory. Within the geographical literature, early textooks (Estall and Buchanan 1980), and numerous locational surveys since the early 1950s, frequently noted concerns by firms regarding labour costs, attitude and unionization (as well as for skill and availability). For many firms contemplating investment, unions are perceived negatively for their ability to negotiate high wages, non-wage benefits and limits to the scope of managerial autonomy and to organize strikes. On the other hand, firms build factories that are planned in co-operation with unions.

The significance of new locations It is widely recognized that factories in new locations provide

Table 7.1a   Aluminium Refining in Iceland: The Bargain Between the Government and Alusuisse

Key elements	Comments
Location	ISAL is located in Straumsvík on Reykjanes.
Power supply	ISAL is provided with electricity from Búrfell power plant. According to the Agreement, the price was supposed to be 3.0 mills per KWh for the first six years of operation and then drop to 2.5 mills. Actually the price never dropped; instead it kept rising.
Taxation	ISAL pays a fixed amount per produced ton of metal. According to the agreement, the smelter will pay US$12.5 per ton the first six years of operation, then US$20 for another nine years and the amount would rise to US$35.
Import duties	ISAL is exempt from import duties on machinery and raw materials.
Legal issues	ISAL is an Icelandic firm, but 100% owned by Alusuisse. Disagreements regarding the Master Agreement will be solved using international court of arbitration, which, however, will use Icelandic laws as guidelines.
Duration	The duration of the Agreement is 25 years, but the scope of the Agreement is 45 years as each party has the option to extend the Agreement for 10 years.

Source: adapted from Skúlason (1994).
Note: ISAL is the Iceland Aluminium Co.

Table 7.1b   Aluminium Refining in Iceland: The Proposed Bargain Between the Government and Atlantal

Key elements	Comments
Location	Keilisnes on Reykjanes.
Power supply	The proposed smelter will be provided with electricity from Blanda power plant and other yet to be constructed power plants. Price will be a certain percentage of the price of aluminium.
Taxation	The proposed smelter will be taxed pursuant to Icelandic law with some exceptions because of the size and specialization of an aluminium smelter.
Import duties	The proposed smelter is exempt from import duties.
Legal issues	Disputes are supposed to be settled by Icelandic courts pursuing Icelandic laws.
Duration	The agreement is supposed to be 25 years, with the firm having the option to extend the contract by two additional five-year periods.

Source: adapted from Skúlason (1994).

the most favourable opportunities with respect to hiring new labour forces and creating new labour bargains (Massey 1984; Barnes et al 1990). Once factories are established, existing work practices, attitudes and agreements are difficult to change. Indeed, whether or not labour characteristics are important location factors, the hiring of labour at the start-up of plants has long-term impacts. Thus in many primary manufacturing activities, labour costs are a small component of overall costs and labour factors are of little importance in the location of new factories. Even so, new factories provide firms with unique flexibility to hire workers with what they consider to be desirable characteristics. This flexibility is particularly apparent in strongly unionized industries where once a workforce has been hired, job mobility along particular lines of progression is constrained by seniority or the hiring date of workers.

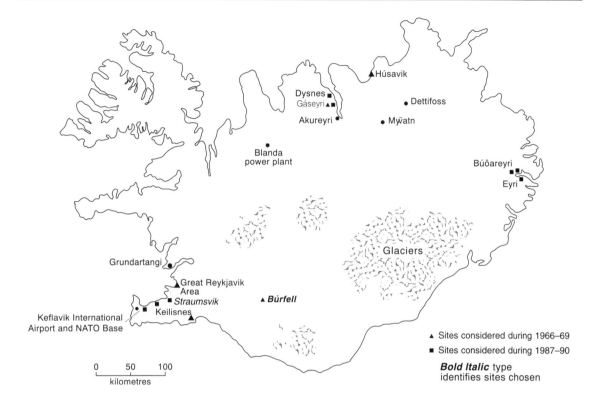

Figure 7.7   Iceland: Sites Considered for Aluminium Smelting. Source: based on Skúlason (1994)

New pulp mills built in Canada in the 1970s, including one built in the 'new town' of Mackenzie, British Columbia (Figure 7.8), recruited extensively across Canada, particularly to obtain experienced, married pulp and paper workers from other small, isolated towns. The firm deemed that workers with these kinds of characteristics would be more productive and more stable and so invested considerably in its recruitment drive to increase its choice of workers. Once in operation, seniority became a guiding principle of internal employment change so that the start-up of the mill offered a choice of employees for positions other than starting positions. Similarly, from the perspective of the workers recruited from established pulp mills, the new mill at Mackenzie offered workers the chance to 'jump' their position in the line of progression or job ladder. Once hired, workers also became 'locked-in' within the vertical line of progression of the Mackenzie mill. It might be noted that another firm which built a pulp mill in a remote British Columbia location at the same time, but which did not invest seriously in worker recruitment, experienced much greater problems of labour turnover and productivity (Hayter 1979; Ofori-Amoah and Hayter 1989).

In secondary manufacturing activities, the desire for a new labour bargain may itself be an important location factor. The relationship between labour bargains and location is demonstrated by Clark's (1981) re-interpretation of the product cycle model.

Re-interpretation of the product cycle model

Clark's (1981) re-interpretation of the product cycle model accepts its premise that as products mature, firms wish to relocate from urban agglomerations to more rural locations. However, Clark suggests that such filtering down can be interpreted as a labour control strategy as well as, or even rather than, a desire to reduce labour costs. The large state variations in levels of unionization provide a plausible underpinning to this argument (Figure 4.3). Clark begins by recognizing that labour as a location condition is not just a cost component. Rather labour enters into complex contractual negotiations with management which in several countries occur within the framework of collective bargaining. Within such a framework, wages, social benefits and how work is to be organized is negotiated (Figure 7.9). These negotiations form what Clark calls 'the employment relation', which comprises two main conflicting impulses. First, the relation is one of mutual antagonism as management and union each seek to control the production process (in terms of tasks, speed and organiza-tion) and the conditions of employment (in terms of hours worked, wages and social benefits). Second, the employment relation is one of interdependence because workers need jobs and firms need labour.

There are laws, such as minimum wage laws and working age, which are nation or region wide, but conditions of employment and control over the production process is to some extent negotiable. In these matters management and labour are trying to control each other. More-over, in Clark's argument, management wants to develop different kinds of bargains with the scientists, engineers and skilled labour associated with research and development (R&D) and product innovation in the early stages of the product life cycle than with the relatively unskilled workers employed in the mature stages of the product cycle. From this perspective, the maintenance of R&D personnel (and highly skilled workers) in established urban core regions and the dispersal of mature products and unskilled work to rural locations represents a bargaining strategy with labour.

Basically, Clark argues that firms spatially separate these labour groups in order to treat them differently. On the one hand, the firm has professional scientists and engineers who have considerable formal education and firm-specific training, are 'mentally and physically agile' and who are 'problem solvers' responsible for inter-esting, complex and uncertain tasks which take considerable time to complete. For the firm, such scientists and engineers are relatively expensive to recruit and expensive to train in the sense of learning how to apply their education and experience to the specific needs of the firm and in relation to other team members. The loss of such employees is costly to the firm because replacement costs are high. Consequently, firms may want to provide extra incentives to these employees to stay with the firm in the form of high wages, good non-wage benefits, attractive work conditions, and high levels of stability, even during recessions. Indeed, the preferences of these workers is important to the location of R&D facilities (Table 4.5). Some idea of the importance firms can attach to their in-house R&D groups is revealed by IBM's decision in the 1970s to take to court some of its scientists who were trying to leave the firm. For IBM these employees were clearly critical to its competitive advantage, and to its competitive disadvantage if they should move (Krumme and Hayter 1975: 330).

In contrast, in the mature stages of the product cycle model, firms employ low-skilled workers with little formal education and who require limited training to perform boring, repetitive jobs. These workers are relatively easy to recruit and turnover costs to firms are relatively low. Consequently, as Clark notes, firms are more interested in limiting the wage and non-wage benefits of unskilled workers and in treating them as true variable factors of production, hiring and firing them according to demand conditions. However, in Clark's (1981) view, if different groups of workers are in the same location then the firm's bargaining powers are reduced with respect to the unionized employees. Thus, if in the same location, the union may gain bargaining power by knowledge of benefits received by other groups and, especially through 'strategies of

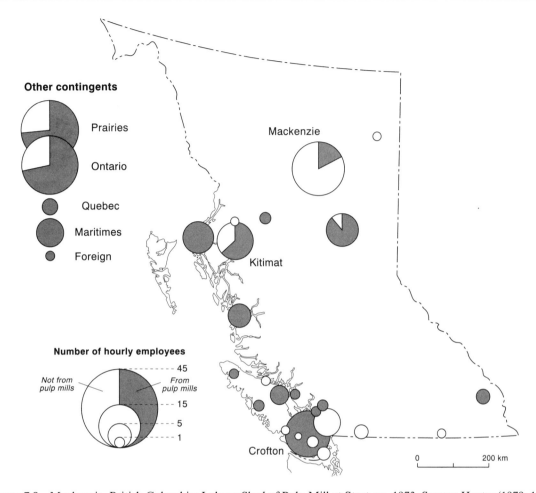

Figure 7.8   Mackenzie, British Columbia: Labour Shed of Pulp Mill at Start-up, 1973. Source: Hayter (1979: 169)

continuous negotiation' in which unions place firms under constant pressure to grant concessions, unions may be able to obtain 'cascading' benefits within the context of a series of collective bargaining agreements. Clark also suggests that in urban agglomerations unions can focus on the weakest firms to gain concessions and force other firms to follow suit.

Accordingly, in this view, firms derive bargaining advantages beyond that of hiring lower-waged labour, by decentralizing low-skilled work to peripheral locations. In particular, the decentralization of unskilled work to rural areas can break the 'demonstration effect'

in which unions derive goals from benefits achieved by other groups; break the cycle of continuous negotiation and cascading benefits; and re-assert managerial authority over the production process and conditions of employment. Firms may also find that in new locations it is easier to innovate technology as there is no need to negotiate with unions on how such technology will be introduced, which employees will have rights to be retrained and how the new process will be run.

In a large scale survey of almost 400 branch plants manufacturing mature products, located in non-metropolitan areas of the US South,

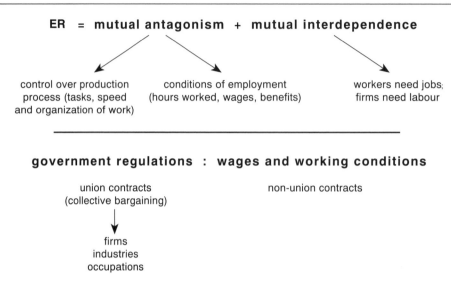

Figure 7.9   The Employment Relation (ER). Source: based on Clark (1981)

Johnson (1991: 402) did find labour costs and availability to be the two most important local location factors. He agrees with Clark, however, that "many companies . . . are less concerned about low wages [or labour availability] specifically than their ability to control the relationship between management and labor". This same study further emphasized the importance of "nonlabor factors in understanding the locational behavior of late-stage activities". Johnson (1991: 406) particularly noted the importance of good trucking connections, a favourable tax environment and an abundance of land. Thus, this study corroborates Moriarty's (1983) earlier survey of plants locating in North Carolina, which found that while a range of labour factors were important, a wide range of other location-specific factors were also relevant (Table 4.7).

More complex bargains

In the context of branch plants manufacturing mature products, bargaining processes are often more complex than anticipated by Clark's (1981) revised product cycle model, and related literature (Sayer and Walker 1992). This model interprets location as a labour-control strategy.

In practice, the motivations and forms of labour bargains are more varied. For example, firms may wish to establish labour bargains which enhance worker responsibility and participation in work organization and which reflects the needs of the workforce rather than established management traditions. Moreover, new locations can involve firms in bargains with governments, suppliers, consumers and other organizations. In addition, for multi-plant firms, the bargains struck at new locations may still be in some way affected by its bargains at existing locations. An example of a relatively complex bargain, which featured arguably the most radical labour relations experiment of recent decades, is provided by Volvo's new, albeit short-lived plant at Uddevalla.

Volvo's Uddevalla experiment   In the mid-1980s, Volvo decided to embark on a horizontal integration strategy in Sweden and build new, integrated auto assembly capacity, and to restructure its labour relations. In Sweden, Volvo's main existing facilities are at Torlsanda, west of Gothenburg, where an assembly plant is integrated with a body shop and a paint shop

(Figure 7.10). This plant was opened in 1964 and another assembly plant was opened in Kalmar in south-eastern Sweden in the early 1970s. Both these locations are supplied from other Volvo plants, including Skövde (engines) and Köping (transmissions), as well as from a large number of subcontractors (Fredriksson and Lindmark 1979). As Alvstam and Ellegard (1990: 188) note, a new body shop, paint shop and assembly operation could have logically been added to the Kalmar site, which at that time was only an assembly operation, or at Olofström which had a press shop. In either case, new integrated facilities would have increased the efficiency of existing flows of components and processes within the Volvo system.

In practice, Volvo eschewed its Kalmar and Olofström options in favour of a new location, specifically Uddevalla to the north of Gothenburg (Figure 7.10). Several key factors underlay this choice, notably the availability of skilled labour, government incentives and reasonable accessibility to Volvo's existing operations. In addition, Volvo desired "to develop new ideas of work organization" which it perceived could be most effectively accomplished at a new site (Alvstam and Ellegard 1990: 189). There were several critical sets of negotiations involved in the establishment of the Uddevalla factory, which opened in 1988. First, Volvo sought the union's co-operation to develop a radical, more humanistic form of work organization. Volvo's motives behind the creation of more interesting, varied and responsible work experiences was to offset the problems of increasing rates of turnover and accident rates associated with short cycle repetitive work, and to offer jobs acceptable to better educated Swedish youth, a declining cohort in the Swedish population (Berggren and Rehder 1993: 195). While more details of the new work practices are provided in Chapter 12, it might be noted that the new forms of work organization, based around the idea of entire cars being assembled by small teams of around 10 workers, took management and labour several years to develop and several plans were discarded before the final one was accepted (Alvstam and Ellegard

1990). This form of labour bargain is entirely different from that anticipated by labour control models.

A second set of negotiations occurred between Volvo and the Swedish (and local) government (Alvstam and Ellegard 1990: 190–192). After the mid-1970s, the Swedish shipbuilding industry declined rapidly and the government took over the shipyards and the responsibility of finding jobs for former shipbuilding workers. While Gothenburg's economy grew sufficiently to compensate for lost shipbuilding jobs, at Uddevalla closure of the shipyard threatened significant problems for the small community of 45 000 people. Consequently, the government offered substantial incentives to firms locating in Uddevalla and improved highway connections to Gothenburg. Volvo was able to take advantage of these incentives and even negotiate additional ones related to financing, cash grants and tax deductions based on the accumulated losses of the former shipyard which Volvo had acquired. (Interestingly, on the basis of this bargain, SAAB, Volvo's Swedish rival, convinced the Swedish government to grant it similar concessions; Alvstam and Ellegard 1990: 206). However, and bearing in mind the region has been hit by acid rain problems, Volvo was unable to convince provincial authorities to grant it a license to emit 1700 tonnes of solvents annually into the air. Volvo then decided against a paint shop and Uddevalla was established only as an assembly operation.

A third set of negotiations involved Volvo's internal politics. As noted, Uddevalla was chosen from among alternatives based at Volvo's existing operations and there are suggestions that the Uddevalla experiment in labour relations was not entirely accepted throughout Volvo management and that as Volvo's markets became increasingly sluggish, vested union interests were not interested in preserving Uddevalla at the expense of Gothenburg. Moreover, the Uddevalla plant was allocated the responsibilities to produce older Volvo models. Even so, according to Berggren and Rehder (1993), in terms of productivity, quality, cost-effective technical

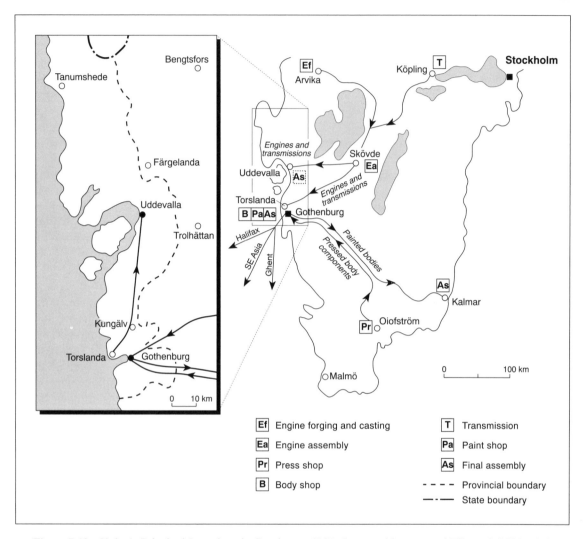

Figure 7.10   Volvo's Principal Locations in Sweden, *c.* 1990. Source: Alvstam and Ellegard (1990: 187)

solutions, cost savings related to health, lower tool and training costs and in reducing delivery times to customers, the Uddevalla plant recorded rapid improvement after 1990 which either met or surpassed the performance in the Torslanda plant. In their view, when Volvo rationalized capacity, it was easier to close the newer (least senior) and small Uddevalla plant than the bigger and older Torlsanda plant which was supported by the power of vested management and union interests.

Contesting closure   As de-industrialization gathered momentum, first in the UK then elsewhere, the closure of factories in old, specialized industrial regions in the 1970s and 1980s was often contested by workers and communities striving to protect local employment opportunities (Hudson and Sadler 1986). As Hudson and Sadler (1986) note in a European context, anticlosure campaigns were mounted by workers in several countries in several industries, but most notably in steel (Hudson and Sadler 1983, 1989;

Hudson 1984; Sadler 1984). While in some places these campaigns were fought by workers in the specific places threatened by closure, in other places workers formed alliances with local community business interests and were supported by broader, regional worker solidarity. The anti-closure campaigns were especially strong in nationalized industries and where it could be demonstrated that threatened plants were viable so that proposed closures reflected particular political choices (Hudson and Sadler 1986: 182).

Closure, or threat of closure, is a potentially, significant bargaining chip for firms trying to convince workers to change the existing employment relation. The problem for unions is that the granting of concessions to firms in order to keep a factory open may enhance their vulnerability to further concessions without necessarily eliminating the problem of closure. Such bargaining at existing locations is particularly difficult (see Chapter 12). Whether or not unions and communities are asked for concessions, factory closures, especially large-scale closures in specialized communities, raise questions about the rights of workers and the communities that are directly affected (Clark 1991; Ettlinger 1990; Watts 1991b). As a minimum, it is suggested that advance notice should be given about closures. Others have suggested that unions and local communities should be formally involved in negotiations over closures to assess underlying motivations and to propose alternatives where appropriate. However, jurisdictions vary in laws governing plant closure. In an increasing number of cases, firms, governments and labour co-operate to provide counsel to those laid-off, including psychological help, family assistance, and advice on training programmes. In some cases financial assistance is provided. There is no easy way to bargain over closures, however, especially for older workers specialized in skills in declining demand and where employment opportunities are difficult to generate. Moreover, given the widespread occurrence of unemployment, 'migration' as a solution is less easy to justify than in previous times.

The internationalization of bargaining processes

For MNCs, bargaining processes are inherently world-wide to the extent that negotiations at one location are compared to those at others. Increasingly, however, the establishment of foreign factories can lead to bargaining processes involving, management, labour, governments and other organizations, such as environmental groups, on an international scale. Krumme's (1981) study of Volkswagen's (VW) decision to establish a factory in the US provides an example of how bargaining between VW management and various levels of governments, labour unions and car suppliers was on a trans-Atlantic scale.

Volkswagen's first North American strategy
VW's plant was eventually located in 1976 in Scranton, Pennsylvania, and VW's decision to invest in the US is conventionally explained by market access factors (in terms of location) and the value of the Deutschmark (in terms of timing). But this decision was by no means inevitable and there was considerable pressure on VW to make other decisions. As Krumme (1981) points out, the bargaining process involved lengthy and contentious negotiations among two federal governments, several state governments and labour unions in Germany and the US, while VW also became entangled in discussions and disagreements with potential suppliers, as well as established rivals in the US.

In location terms, VW assessed investment alternatives in different regions of Germany and the US. For VW, pressures within Germany stemmed from the fact that VW is partly owned by private shareholders and partly by the State, notably Lower Saxony (20%) and the Federal Government (20%). In addition, worker representatives comprised fully one-third of the membership on VW's 'Executive Board' and actively participated in the decision as to whether or not VW should invest in the US. Indeed, VW's executive board is overtly pluralistic and political, which ensures that internal debates within VW are controversial, especially bearing in mind that VW's plants in Germany were located in 'development areas'. As Krumme (1981: 346) notes:

> "The fact that both governments and unions were already integral parts of the Board led . . . to further attempts to expand this foothold by getting 'their men' into management – attempts which . . . resulted in almost constant personality and ideological clashes with and among members of management. Many concessions had to be made to avoid or postpone further direct participation of political figures in management affairs. . . . During the 1974–75 VW crisis several issues, including the US project, were prime election issues at state and federal levels: indeed, it was the VW recession – unlike any other previous corporate event – that brought many of the conflicts and controversies to the surface. There was blame to be distributed and decisions about capacity reductions to be made – unique opportunities for political intrigue and power play."

The upshot of the exercise of the "political intrigue and power play" was that VW decided not to close any of its German plants, while its decision to enter the US was significantly delayed until the cost advantages of operating in the US became obvious and VW was threatened with a declining market share. In the US, VW also became embroiled in further "political intrigue and power play" games. Responses to VW among American firms varied. Thus, VW complained that US auto suppliers initially offered components at relatively low prices but that as negotiations proceeded, prices were raised. In addition, rival US auto manufacturers periodically criticized VW as a government-owned company which gave it unfair competitive advantages. Indeed, the US auto manufacturers encouraged a Federal investigation into foreign car-makers regarding 'suspected dumping', a non-tariff barrier frequently employed by American industry. In this context, Krumme emphasizes the 'intriguing' role played by Congressman John Dent, who was a key backer of the Federal investigation into dumping by foreign car companies but at the same time actively soliciting VW to invest in Pennsylvania.

In looking at locations in the US, VW enjoyed a favourable bargaining position as a result of the intense competition among states and cities for its investment. One report suggested that all 48 continental states made overtures to VW, while Krumme (1981: 350–351) lists the efforts of Tennessee, Mississippi, Arkansas, Ohio and Pennsylvania and those of the cities of Baltimore and Cleveland. In the event, Pennsylvania, which provided a lucrative incentive package to VW, 'won' the competition (Table 7.2). Such was the enthusiasm among governments in the US to get the VW investment, that the situation was less of a MNC 'playing off' regions as regions 'playing off' each other. VW was apparently able to keep negotiations open with Ohio until the Pennsylvania deal went through. At the same time, in Germany, VW made a deal with Lower Saxony regarding parts supply to the Scranton plant. (This plant has since closed and VW has recently built a plant in Mexico.)

Industrial location policy as a bargain

At a national level, institutionalists argue that industrial location policy may be a good social bargain for the country as a whole by promoting regional economic equality and political and social stability. Moreover, in contrast to the neoclassical position, social equality is not necessarily achieved at the expense of economic efficiency (Kuttner 1984). Rather, the use of public funds to promote social well-being potentially not only produces a more egalitarian society but a more skilled, informed, healthy and productive one. In addition, from this perspective, so-called externality effects are potentially important considerations.

Thus, regional policy which offers incentives to firms to locate in designated regions may well be a 'good deal' for society if such intervention can generate positive externalities and/or if it can reduce negative externalities (Figure 7.11). In brief, this argument is based on the idea that the deflection of investments in new factories from 'core' to 'peripheral' regions will reduce negative externalities in the core and increase positive

Table 7.2   The Major Items in Pennsylvania's Incentive Package to Volkswagen

1. A $US40 million loan by the Pennsylvania Industrial Development Authority to the (non-profit) Greater Greensburg Industrial Development Corporation to be used for land and plant purchase and renovation, to be repaid by Volkswagen over 30 years at an interest rate of 1.75% over the first 20 years and 8.5% over the last 10 years. Purchase price paid to Chrysler was reported to be $US28 million. (Chrysler estimated the cost of completing the plant to be about $US100 million.)
2. Highway improvements (through a $US26.9 million bond issue) and a rail spur into the plant (through a $US6.7 million bond issue); both received legislative approval.
3. Originally, two large state pension funds for public employees had offered to lend VW $US135 million over 15 years at 9% interest. The interest rate, however, was slightly higher than had been promised by Pennsylvania as part of its original financing package. Volkswagen eventually accepted only a loan of $US6 million on these conditions (and financed the remainder through the private capital market).
4. Tax concessions were offered. Under a revised plan: 5% for two years after production begins; 50% for another three years; 100% thereafter. According to a county official, the revised plan would give the VW corporation a $6 million tax break over $6\frac{1}{2}$ years.
5. A very 'intense' programme, using federal and other funds, *to train workers for employment at the Volkswagen plant*. The *Wall Street Journal* reported (1 June 1976) that "critics of New Stanton location have asserted that the immediate area lacks the pool of skilled labor offered by other sites, such as the Cleveland area".

Source: Krumme (1981: 352).

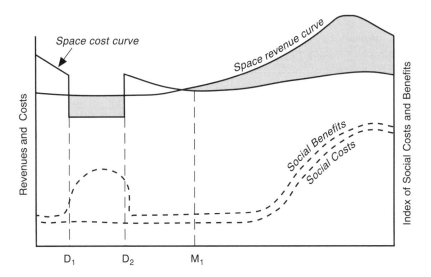

Figure 7.11   Location Subsidy and Social Benefits and Costs. Source: based on Abler et al (1971: 302)

externalities in the periphery. The negative externalities, or social costs, in the core relate to inflation, pollution and congestion. The positive externalities, or social benefits, in the periphery relate to the absorption of unemployed labour within the workforce and fuller use of existing social and economic infrastructure which is available at little or no extra cost – the same infrastructure would have to be built in the fast-growing regions. For individual firms, the pattern of social costs and benefits may be of little consequence in why a particular location is chosen. A subsidy provides a meaningful incentive for firms to take externalities into account. In addition, regional policies need not necessarily lead to the support of otherwise uneconomical locations since spatial margins of profitability are already broad for many activities.

Moreover, to the extent that regional policy enhances economic equality, which in turn leads to greater levels of national and social cohesiveness, national economic efficiency may be further promoted. The alternative – more regional inequality and less political and social cohesiveness – not only reduces demands for goods and services but also potentially creates political fragmentation and conflict. Thus, the economy as a whole may be better off as a result of regional policy.

It needs to be emphasized that the circumstances of the social externality / regional equity arguments vary among countries and over time. In UK, for example, in the 1960s the social externality / regional equity arguments reinforced each other to help support a substantial regional policy programme and the delivery of substantial industrial incentives to designated regions. The same can be said for Canada in the 1960s, whereas in contemporary Japan the social externality rationale for regional policy encouraging the decentralization of industry away from the massive concentrations of people in the Tokyo and Osaka metropolitan areas is a powerful one. In the 1990s in the UK, however, the economic problems throughout the country have substantially reduced the appeal of the externality argument even if regional political

and social equity arguments still exist. In Canada, where the social externality arguments could never be as strong in such a huge country, it is now clear that industrial location policy and other forms of regional development have not achieved political stability and social cohesiveness. Indeed, Quebec, the most important provincial target for regional policy initiatives of all kinds, has never been more likely to separate and to formally create a francophone state in North America. Ironically, as more firms leave Quebec for political reasons, the resulting economic problems continue to justify federal subsidies to Quebec. Clearly, whether or not regional policy is nationally a good bargain is complex and needs to be understood in specific national circumstances.

Industrial incentives as local bargains As previous chapters have noted, while traditional forms of regional policy in many countries have been eliminated or watered down in recent years, regions and communities have themselves remained committed to economic development and in many cases industrial incentives of one sort or another are still offered to firms. The incentives, in the form of grants, tax relief and financing, offered by communities and regions to firms in return for investment, jobs, income and exports may be considered good or bad bargains. From the latter perspective, for example, there is an argument that community-based industrial incentives operate as a zero sum game in which individual regions and communities compete with one another for a limited supply of mobile investment, which in turn can 'play off' one region against another to bargain for more incentives (Cox and Mair 1988, 1991). In this scenario, one community's gain is another's loss and the former is always susceptible to a better offer from yet another community. The previously mentioned Volkswagen story demonstrates that firms are able to play off one region against another simply as a result of regional initiatives and a desire to compete.

Another view argues that from a local perspective, industrial incentives can be a good bargain. In this view, there are limits to the mobility of firms and in any case there is no reason why principles of competition should not apply to communities as well as firms. Cannon (1975, 1980) suggests that from a local perspective, the crucial yardsticks by which to assess the local efficacy of industrial incentives are incrementality and significant benefit, the latter including the survivability of assisted plants (see Hayter and Ofori-Amoah 1992). As previously noted, incrementality in this context refers to the effectiveness of incentives in changing the investment preferences of entrepreneurs, most notably by encouraging them to locate in a designated region rather than in some non-designated region (locational incrementality). Cannon further notes that even if locational behaviour is not changed, incentives can affect the timing, scale and financing of investments (non-locational incrementality). Of course, incentives may have no effect on entrepreneurial incentives. It might be noted that incrementality effects are hard to assess both *a priori* and *ex post*. The *a priori* problem faces local planners who try to determine whether or not a subsidy is actually needed in order to convince a firm to locate in their community, and if it is, what form this should take. In the previously discussed Volkswagen story many states and cities within the US reached the conclusion that incentives were necessary to attract investment but offered differing incentive packages, each of which was subject to negotiation. The *ex post* problem faces regional planners who wish to evaluate the extent of the incremental effects generated by past policies. This exercise is difficult since it raises 'counter-factual' arguments concerning what would have happened in the absence of the policy.

The significant benefit criterion recognizes that firms vary significantly in what they offer to communities in terms of levels of employment, the quality of jobs, employment equity, skill development, stability, longevity, local linkages and environmental impacts. VW, for example, offered Scranton substantial significant benefits in terms of the number of jobs created, job

quality (e.g. in terms of wages and training programmes) and through purchases of local components and supplies. Given that significant benefits are subject to bargaining, their determination is judgemental. What is an appropriate target for the extent of local purchases, for example? The EC demands that foreign auto makers purchase 80% of their supplies from within the EC, while the US has less stringent local rules. Similarly, what is a reasonable requirement for the longevity of jobs? There are examples of firms who have received incentives and never invested, and others who have stayed for maybe a year or two before relocating elsewhere; presumably such behaviour does not constitute a benefit. VW, on the other hand, lasted about a decade in Scranton – a period that may be long enough to generate long-term benefits within the community but also short enough to raise questions about the wisdom of the incentive package. Local development bargains can be extremely difficult to judge.

Conclusion

In different parts of the global economy firms engage in different kinds of bargains with labour and governments (and other organizations) regarding employment, work and other matters. In low-wage, low-skill locations, basic training and skill formation as well as jobs typically are high priorities. In advanced countries, in addition to jobs, advanced skills and interesting work are important. In some cases, firms bargain directly with unions in accepting responsibility for providing work that is appropriate to local values, income levels and education, i.e. for local development. In other cases, firms have little interest in local development *per se* and prefer to scan communities in search of labour, inducements and other deals which they can exploit and adapt to their needs.

In practice, firms have varying commitments and obligations to local development. As Oinas (1995a: 177) states:

"In their operations, decision-makers in some enterprises tend to be willing to take questions of long-term local development into account in their decision making and in their local political action. In other enterprises, on the other hand, decisions are made in a manner that is favorable to their own particular (short-term) interests only, even when those decisions have consequences in their local environments."

Similarly, communities are not passive actors in shaping the commitments of firms. At one extreme, for example, communities can emphasize keeping costs low as a way of attracting industry. On the other hand, communities can strive for more complex bargains in which firms fully participate in the development of local economies. The relationships between local community and firms are complex and need to be understood in terms of the geographical dynamics of firms – the theme of the next part of the book.

PART III

THE MANUFACTURING FIRM AND ITS GEOGRAPHY

The geography of enterprise established that the organization of industrial activities is a significant dimension of location, location change and local development (McNee 1958; Krumme 1969a; Dicken and Thrift 1992). Indeed, since the late 1960s a massive literature has emerged within geography which has focused on 'the firm' as the unit of analysis. Much of this literature, following the initial rationale for the geography of enterprise, focused on large firms, but interest in small firms also grew rapidly. While a much fuller appreciation for the nature of inter-firm relationships has been gained in recent years (see Part IV), there are now substantial literatures dealing with large and small firms. Part III reviews these literatures and adds a further dimension to the discussion by explicitly recognizing a category of medium-size firms that differs from what is normally implied by this term.

By way of introduction, Chapter 8 comments, in a highly generalized way, on the size distribution of firms and on the internal spatial distribution of employment within large corporations. Chapter 9 discusses the formation and function of new firms, regional variations in new firm formation rates and some of the characteristics of small firms of relevance to location and local development. Chapter 10 addresses the above-mentioned and important gap in the industrial geography literature, namely medium-sized firms which represent a group of firms that have characteristics different from the large majority of small firms and the relatively few giant firms. Chapter 11 examines the growth of the inter-regional and international firm (the international firm can be seen as a special case of the interregional firm). Chapter 12 then explores aspects of the restructuring of large firms, especially from an employment perspective.

8

THE SIZE DISTRIBUTION OF FIRMS: GEOGRAPHICAL PERSPECTIVES

To provide a broader context for the examination of the strategies and location behaviour of individual firms that is considered in the remainder of Part III, this chapter begins by briefly examining the size distribution of firms, which raises the question of why small and large (and medium-sized) firms exist in modern economies. Geographically, many manufacturing firms are small and exist at a single site. Others are huge and occupy multiple locations. The second part of the chapter outlines the geographic scope of large firms, particularly with respect to their (internal) employment geographies. The final part of the chapter discusses the 'spatial division of labour' (SDL), particularly with respect to the implications arising from the employment distributions of large firms. Although problematical, the SDL (and related idea of the new international division of labour) does give pre-eminence to the power of corporations in shaping the spatial distribution of employment opportunity.

The size distribution of firms

In capitalist economies there is a tremendous range in the sizes of firms. Although the great majority of firms have sales of less than US$10 million, there are a few multinational corpora-

tions (MNCs) that have sales bigger than the gross national product (GNP) of some medium-sized countries. For over 25 years, General Motors (GM) has been the biggest industrial corporation in the world. Its sales of over US$120 billion per annum in the early 1990s (and US$168 billion in 1995) were significantly larger than the GNP of countries such as Norway and South Africa, and about twice the GNP of countries such as Greece or Portugal. Even after over a decade of downsizing, GM still has more employees (around 709 000 in 1995) than Iceland has people (around 500 000). It has more research and development (R&D) employees than most universities have faculty. GM is not simply big; it is a colossus.

The existence of corporate giants and impressive variations in the size distribution of firms have been important characteristics of industrial economies throughout this century, even if not popularly discussed until the 1960s (Galbraith 1967; Servan-Schreiber 1968; Barber 1970). In the early 1900s in the US, for example, the 100 largest manufacturing firms accounted for over 40% of industrial output (Wilkins 1970). Similarly, by then, high levels of economic concentration existed in the leading industrial sectors in which a relatively few, large firms ('oligopolists') accounted for the lion's share of

The Positively Skewed Distribution

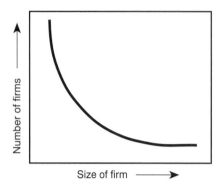

The Population Pyramid

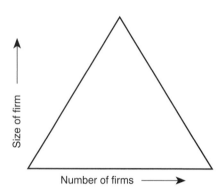

Figure 8.1   Summary Representations of the Typical Size Distribution of Firms

production and markets. Indeed, the rapid emergence of oligopolistic and even monopolistic market structures led many, including Marx and others of a more conventional view, to anticipate the demise of the small firm.

Yet a size distribution still exists. More specifically, in capitalist economies, and for most industrial sectors, the largest firms (and factories) are typically very few while the smallest firms are typically very many. Indeed, in many economies and sectors it has long been suggested that as firms increase in size, the number of firms declines in a more or less steady way (Hart 1962). That is, the relationship between the number and size of firms can be summarily represented by a positively skewed distribution or as a population pyramid (Figure 8.1). In this context, a well-known hypothesis is Gibrat's law, which states that among a population of firms, the *rate* of change of (average) firm size (in any particular size class) is independent of size. From this perspective, while individual firms may experience different growth and decline experiences, the net effects of these variations among a population of firms offset each other to produce a stable size distribution over time (Hart 1962).

Regardless of the particular form of statistical relationships, the traditional explanation for the size distributions of firms emphasizes the effects of internal economies (and diseconomies) of scale. In this model, firms only differ in the

extent to which they take advantage of economies of scale. An alternative 'dual economy' model (Figure 8.3), while recognizing the importance of technological forces or 'imperatives' for the growth of firms, emphasizes a basic distinction ('segmentation') between large and small firms in terms of motivation, structure and behaviour. Moreover, the nature and extent of business segmentation can vary among societies, even at similar levels of development.

The economies of scale argument

Simply put, economies of scale occur when the average cost of production declines with increasing size of activity. Similarly, diseconomies of scale exist when average costs of production increase with increasing size of activity. In theory, for firms (or factories) the relationships between output levels and efficiency, as measured by the average costs of production, can take on various shapes (Figure 8.2; see also Figure 2.5). From this perspective, within an economy or industry, the size distribution of firms reflects the differential effects of (dis)economies of scale on factories and firms. Thus, big factories which are able to realize economies of scale (plant- or factory-level economies of scale), imply firms of at least equal size to finance, control and operate them. In addition, the potential for reductions in the average cost of

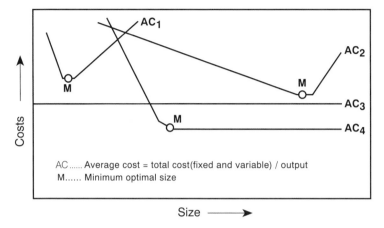

AC...... Average cost = total cost(fixed and variable) / output
M...... Minimum optimal size

Figure 8.2 Economies of Scale: Alternative Configurations

production with increasing size of firm (firm-level economies of scale) implies that large multiple-plant firms are more efficient than small single-plant firms. Diseconomies of scale recognize that at some point increasing size may become 'inefficient' with respect to either factories or firms. In this regard, it might be noted that in addition to factors operating within factories and firms, market size may also place limits on the realization of economies of scale.

In summary, the size distribution of firms exists in an economy or industrial sector because of highly variable opportunities to take advantage of economies of scale. Small firms exist either because factory- and firm-level economies of scale are extremely limited or because, even if they are technically possible, they cannot be exploited because the market is not big enough. On the other hand, if economies of scale are substantial and the market is big enough, at least a few large factories and firms can exist. Medium-sized firms exist because there are possibilities between these two polar situations.

The bases for (dis)economies of scale are well known (Robinson 1931; Bain 1954, 1959; Scherer 1980). Thus, at the factory level, it is argued that big factories are better able than small factories to specialize their workforces according to specific tasks, thus allowing workers to increase productivity by improving their dexterity, eliminating time losses involved in moving between different tasks, and creating possibilities for developing specialized machinery. In addition, fixed costs, which by definition do not vary with output levels, will necessarily contribute towards economies of scale as they are spread over a larger output. Such fixed costs refer, for example, to payments for plant, machinery, municipal services and infrastructure while labour recruitment and training costs are also a form of fixed cost. If bigger machines can be designed which generate a bigger increase in output than in cost (of purchase) then economies of scale can also be realized. Factories may be able to further exploit 'technological' economies of scale if, for example, integrated processes permit savings in energy. If linked machines produce different-sized outputs, average costs will be reduced if all machines are kept running all the time, which means the firm will require more of the smaller machines.

At the level of the firm, large multi-plant firms exploit economies of scale by spreading the fixed costs of expensive managerial functions such as marketing, distribution and research and development (R&D) over the larger output generated by multiple factories. For example, as firms build or acquire more factories, each factory can use the firm's existing marketing and distribution channels, which brings down the average costs of production. Related economies of scope, which define the ability to utilize existing resources for alternative purposes without significant transition costs, may also exist. In addition, big firms enjoy bargaining advantages, for example, with suppliers, consumers and financial institutions which can translate into lower costs. Furthermore, large firms can withstand high levels of uncertainty, especially in relation to large, complex projects involving substantial investment. Finally, once a few firms have become large in particular markets, high levels of fixed costs in plant, equipment, marketing and distribution channels, and R&D constitute important 'barriers to entry' to new firms or to medium-sized firms wishing to participate in the same markets.

As operations increase in size, possibilities for economies of scale can be exhausted and average costs level off or even increase to create diseconomies of scale. In factories, possible causes of such diseconomies include technological limits on machine size, problems of supervision and communication, and labour alienation over excessive specialization. Transportation costs may also impose a constraint on size as big factories require longer supply lines and bigger market areas. In firms, diseconomies of scale relate to the costs of administration, or 'governance', and the issue of 'bureaucratization'. Thus, decision-making in large firms inevitably becomes more complex, time-consuming and subject to formal rules of procedure as committee structures comprising many individuals with differing mandates feed into decision-making structures. In addition, within bureaucracies, combinations of opportunism, personality differences and obsequiousness can undermine trust and effective communication, thereby having an adverse effect on productivity. Similarly, lack of effective communication within the firm may lead to misinterpretation and even contradictory policies.

The dual economy argument

According to the dual economy argument, economies comprise, on the one hand, giant corporations which represent 'planning system firms' and, on the other hand, small firms which represent 'market system firms' (Figure 8.3). As Taylor and Thrift (1983) note, planning and market system firms are qualitatively different: they perform different functions, they have different strategies and structures, and their social rationale is different. They are literally different institutions (Galbraith 1967). In Robertson's (1928: 85; Malmgren 1961: 399) metaphorical terms, corporations exist as "islands of conscious power in [an] ocean of unconscious cooperation like lumps of butter coagulating in a pail of butter-milk".

In the dual economy model, the size distribution of firms is not simply a sliding scale of technological opportunity, constrained only by market size, over which firms increase in size by progressively moving down an average cost curve as markets permit, or by stepping to a new, slightly lower average cost curve as technology permits. Rather, planning system firms exercise greater control over their own destiny. Certainly, among planning system firms economies of scale are important and the need to realize productivity gains by increasing the scale of operations is a significant impulse underlying their growth. Indeed, according to Galbraith (1967), large-scale business organizations are an inevitable consequence of the 'imperatives of [modern] technology' which is extraordinarily expensive, highly uncertain and requires conscious long-term planning by large groups of highly trained specialists in very large-scale organizations. Yet, it cannot be assumed that the largest firms are the most efficient firms. In many economies and industries, there is evidence that firms (and factories) grow to sizes that are far in excess of the notion of 'minimum efficient size' and even to sizes where diseconomies of scale may be present (Blair 1972; Scherer 1980). A recent reminder that the biggest does not equate with the most efficient is the debate within GM about whether or not to break up the company in order to increase overall profitability and efficiency (Taylor 1996). Indeed, there appear to be no obvious limits to corporate size and no mechanical rules for determining an optimal size distribution for firms. In the dual economy

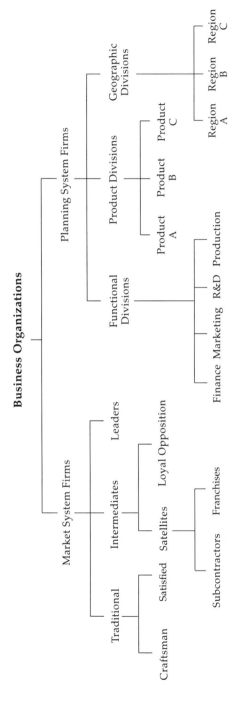

Figure 8.3 Dual Model of Business Organization. Source: based in part on Taylor and Thrift (1982: 1605)

model, planning system firms seek to grow not only to pursue efficiency but for other reasons, including to gain access to resources, to reduce uncertainty (over, for example, access to markets, rival behaviour, supplies), to gain bargaining power and for growth itself (Baumol 1959; Penrose 1959; Marris 1968; Starbuck 1971). From this perspective, efficiency and size are not related in a simple linear fashion.

In the dual economy model, planning and market system firms are different and, to some extent, separate. Thus, some types of small firms are 'laggard' and 'craft' firms which provide traditional products in small markets and essentially exist outside the ambit and concern of planning system or core firms. At the same time, as Edwards (1979: 73) notes, "core firms cast a giant shadow over small firms". The nature of the shadow cast by large firms over small firms does vary depending on whether the latter are 'satellites' which, for example, have direct business links with core firms, 'loyal opposition firms' which directly compete with large firms, or dynamic small firms which manufacture new products and which are potential acquisition targets for the core firms. For satellites, planning system firms control their markets, while for the other small firms core firms potentially dominate, whether intentionally (e.g. through acquisition) or unintentionally (e.g. following the closure of a large plant which reduces local demands).

The issue of subcontracting explicitly raises the question of the boundaries of firms and whether or not core firms should manufacture a product themselves or subcontract. This issue is at the heart of the nature of production systems and will be examined in Part IV of this book. In the meantime, the fact that the size distribution of firms varies among societies at least raises the suggestion that this question generates different answers.

International variations in firm size distributions If economies of scale are the sole or predominant force affecting the size distribution of firms then this size distribution should be more or less the same among industrialized countries.

Instead, we see substantial variations. Thus, small firms are considerably more important in the German and Japanese economies than they are in the US or UK (Table 8.1). While the statistical definition of the small firm sector, or more precisely in this case the small and medium-sized firm (SMEs) sector, does vary, numerous other studies have consistently reached the same conclusion, including in comparisons of the same industry. There seem to be some national or cultural factors, as well as economies of scale, affecting the size distribution of firms. That is, to some extent the size distribution of firms involves choice. From this perspective, Japan and Germany are societies who have 'chosen' to give greater priority to the role of SMEs in the manufacturing process than either the US or the UK. Indeed, historically, core firms in Japan have relied more on subcontracting than core firms in the US.

There is little evidence to support the idea that national variations in the size distribution of firms relate to national variations in competition or 'anti-monopoly' or anti-trust policies which are designed to confront problems arising from high levels of corporate concentration (and too little competition). In most industrial countries, including the US, the UK, Germany and Japan, various forms of these policies have been established. For the most part, however, these policies are practised on an *ad hoc* basis. Occasionally, mergers between already giant firms are prevented and in the US a few mammoth corporations have been forced into 'smaller', if still huge, companies. After 1945, the US used its anti-trust laws to break up the Japanese Zaibatsu 'groupings', while the Japanese government also stimulated competition among its firms. Whether the threat of anti-trust action results in implicit, self-imposed constraints on the growth of firms and so exercises a broader influence on the size distribution of firms is more debatable. In the US and the UK, for example, merger and acquisition activity has been a major cause of corporate concentration. In addition, governments have supported the growth of large firms, both domestic 'champions' and by

Table 8.1   International Variations in the Importance of Small Firms in the Manufacturing Sector, 1986

	Employment		No. of firms	
	% SMEs	Total	% SMEs	Total
Japan	74.4	13.3 million	99.5	874 471
US	46.9	17.8 million	96.2	348 385
UK	39.2	4.9 million	97.0	141 617
West Germany	60.0	10.5 million	99.2	515 917

Source: based on Patchell (1991: 52–54).
Note that small and medium-sized enterprises are defined somewhat differently: Japan: < 300 employees; US: < 250 employees; UK: < 200 employees; West Germany: < 300 employees.

encouraging foreign investments by MNCs. Whether any of these polices have exerted a systematic effect on national variations in the size distribution of firms, however, is questionable.

Rather, national variations in firm size distributions have deep historical roots. In the case of Japan, for example, Fruin (1992) argues that since the beginning of the country's modern drive to industrialization in the Meiji Period, manufacturing firms have favoured highly specialized strategies; with firms relying on each others' specialisms. Such strategies, in this view, were strongly encouraged by the energies required to learn technologies already developed in the West (see also Patchell 1992). It is now widely recognized that the small business sector in the UK has long been of relatively less importance than in the same sector in comparable countries (Bolton 1971; Dewhurst 1989; Storey 1994: 21). Moreover, Dewhurst (1989) rejects the view that small firms in the UK have been discriminated against by government policy. Instead, he suggests that the manufacturing industry as a whole, and small manufacturing businesses in particular, have a negative social perception, particularly in the area of production, and have attracted less capable individuals than more desirable occupations (see also Weiner 1982). While social attitudes to entrepreneurship are changing, the UK is still "trying to correct a huge negative bias against such organizations" (Dewhurst 1989: 78).

In most industrialized countries, there is evidence that the small firm sector is playing a greater role; for example, in relation to total employment (Loveman and Sengenberger 1991). In the US, sales of *Fortune*'s largest 500 firms declined from 55% to 42% of total sales in the US, and that their share of national employment declined from 17% to 11% in the same period. At the same time, the number of new firms in the US is on the rise (Birch 1979). In the UK, in the early 1980s the birth rate of new firms rose considerably, while the death rate remained the same (Burns 1989: 32). However, the basic pyramidal structure of the size distribution of firms remains, and the biggest firms remain very big. In this regard, it might be noted that management discussions to possibly break up GM, whose 1995 sales were US$168 billion, would still leave four very large companies, probably all within the largest 100 industrial companies in the US (Taylor 1996).

The geographic distribution of employment within big firms

The size distribution of firms is mirrored by variations in the geographic scope of firms. Company regions, to use McNee's (1958) term, vary greatly in size including the extent to which they straddle local, regional and national boundaries. While many firms are single-plant operations, others control from two to many hundreds of plants which vary in size and are geographically dispersed in a variety of patterns. Several models have been outlined which attempt to

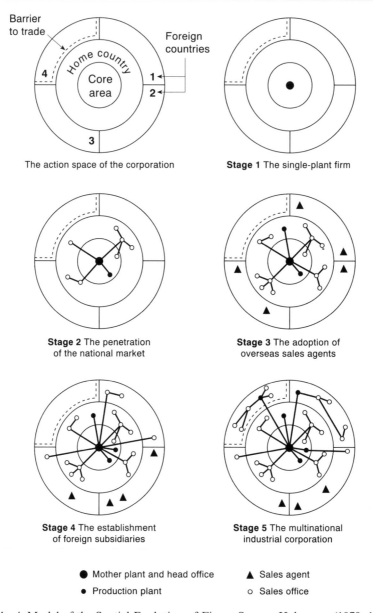

Figure 8.4   A Model of the Spatial Evolution of Firms. Source: Hakanson (1979: 131–135)

generalize the spatial form of long-term corporate expansion paths, usually by the identification of geographically defined stages of growth (Taylor 1975b; Watts 1977; Hakanson 1979). Hakanson's model, for example, antici-pates that firms first establish a strong local or regional base, then grow nationally and finally internationally (Figure 8.4). In this model, sales linkages through sales offices and sales agents extend the firm's knowledge of more distant places and give direction to investment in new factories.

In practice, many firms have created significant regional cores to their activities and have subsequently become major national companies before investing in foreign countries. There are also alternative expansion paths and wide variations in the extent to which equally large firms are geographically diffuse. The geography of the behavioural environments of firms are not only shaped by distance-decay effects around existing centres. As firms become larger, growth by acquisition and merger typically becomes more important so that the geographic scope of operations can occur in 'big leaps'. In addition, locational adjustment at existing sites continually occurs as firms expand, modernize and change some sites while closing others. Watts' (1980b) detailed analysis of planning system firms within the UK brewing industry, for example, reveals some interesting differences in spatial trajectories and morphologies of national firms. Perhaps the most significant criticism of models of the evolving spatial morphology of firms is that they do not adequately address underlying causal processes and isolate firms from their environmental context, focusing solely on 'spatial' distribution (Hayter and Watts 1983: 169). At the same time, actual (changing) maps of corporate activities provide useful points of departure for analysing processes of location change within the firm and raise interesting questions, not least with respect to the spatial division of labour.

National employment maps of corporations

From the beginning of the fordist techno-economic paradigm until the early 1970s, many corporations enjoyed more or less continuous growth. Expanding corporations did close down factories but these closures were typically small plants and part of modernization plans or the rationalizations that followed acquisitions and mergers. From the early 1950s to the mid-1970s, for example, the core firms of the UK brewing industry all expanded their corporate systems, especially by acquisition (Watts 1980b). In the

case of Whitbread, a national system of breweries was developed mainly by acquiring competitors, and while numerous sites were closed two new ones were built and others expanded (Figure 8.5). By 1976, however, in comparison to the other core firms, Whitbread's plants were relatively smaller and more dispersed, providing potential for more sweeping rationalization (Watts 1980b: 236–239).

Since the 1970s, the employment geographies of large corporations, especially in Europe and North America, have been volatile. While it is often difficult to collect appropriate data, a number of studies have documented employment changes throughout corporate systems within particular nations (Peck and Townsend 1984; Healey and Watts 1987; de Smidt and Wever 1990) and even internationally (Clarke 1982). The biggest changes, at least in terms of job-shedding, have occurred within corporations in mature industries such as iron and steel, ship-building and autos (Townsend 1983; Peck and Townsend 1984). In Europe and North America, employment volatility has also been a feature of corporations manufacturing technologically sophisticated products and products for which consumer demand remains high. Three examples are provided by Cadbury Schweppes in the UK, Standard Electric in Germany, and Philips in the Netherlands (Figures 8.6–8.8). The latter two are both in the high-tech vanguard of companies while Cadbury Schweppes serves stable consumer demands.

Cadbury Schweppes is a diversified food, soft drink and household hygiene MNC whose 22 900 workers in the UK in 1984 represented 60% of its world-wide workforce (Healey and Watts 1987). Cadbury Schweppes has four divisions and its plants are scattered throughout the UK, although there are regional concentrations, within the Axial Belt (Figure 3.4), around London, the West Midlands and the Liverpool area (Figure 8.6). This is a changing geography, however (Watts 1987: 12). Thus between 1972 and 1984 the firm shed 4000 workers from this system, almost 60% of which were concentrated in its main production region of the West

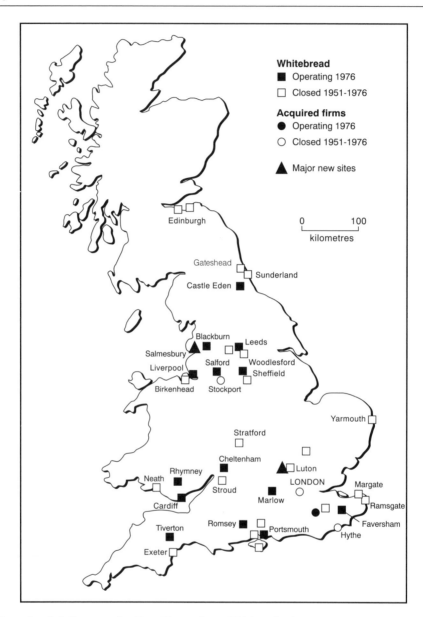

Figure 8.5   Locational Adjustment by Core Enterprises: Whitbread, 1951–1976. Source: Watts (1980b: 237)

Midlands. Given that employment gains occurred in several other locations, the West Midlands' share of the firm's employment dropped from 43% to one-third in this time period.

In the case of Philips, the Dutch-based electronics MNC, internationalization began in the 1930s, and by 1939 the firm employed more

people abroad than in Holland (de Smidt 1990: 56). After 1950, Philips expanded impressively, increasing its employment from 90 000 to 359 000 world-wide by 1970, when one-quarter of its workers were in Holland (Figure 8.7). Within Holland, its core region around Eind-hoven in the south-east of the country accounted

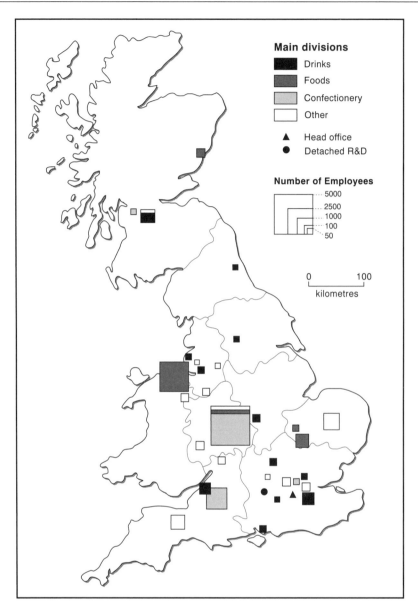

Figure 8.6 Cadbury Schweppes Corporate System in the UK, 1984. Source: based on Healey and Watts (1987: 152)

for 45% of Philips' jobs in 1945 and 41% in 1973 (de Smidt 1990: 65) as the firm aggressively built, acquired and expanded plants in all regions of Holland, including the development areas. From the late 1960s smaller plants have been closed; since the late 1970s, Philips' employment map within Holland has changed more dramatically.

Between 1970 and 1988, for example, Philips reduced employment in all regions of Holland (and in contrast to Cadbury Schweppes) fairly evenly between its core region and the peripheral areas (Figure 8.7). As de Smidt (1990) observes, these losses involved both blue and white collar workers and have been combined with a general

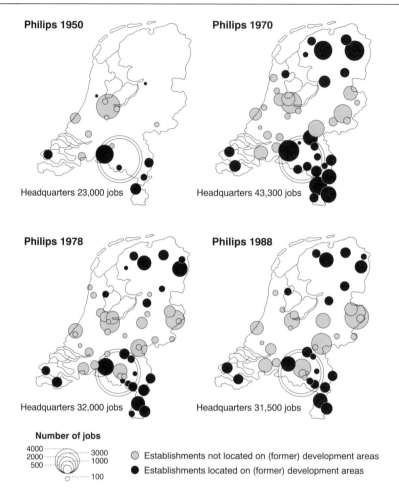

Figure 8.7 Locational Patterns of Philips in the Netherlands. Source: de Smidt (1990: 67)

(but not uniform) shift towards the hiring of more highly qualified employees within Holland as Philips globally integrates its world-wide operations in each of its major divisions.

Standard-Electrik Lorenz (SEL) is an example of a subsidiary company whose operations are within Germany (Figure 8.8). SEL was part of the international corporate empire of ITT of New York from 1958 until 1986, when it was acquired by the French-based Acatel SA (Fuchs and Schamp 1990: 82). The acquisition signalled the onset of a major rationalization of SEL. In particular, the entertainment electronics division was sold off to the Finnish corporation Nokia. Some 2000 jobs were shed in the remaining plants and SEL became more tightly integrated within its new parent company. Geographically, SEL has become more regionally focused on its core region in the Stuttgart area, where half of SEL's 22 800 employees work. Most of the plant closures and divestments have occurred elsewhere, including, all its operations in the Cologne–Dortmund area of the Rhine and Ruhr. As Fuchs and Schamp (1990: 84) note, SEL has become a regional multi-plant corporation whose performance is closely monitored by its French parent.

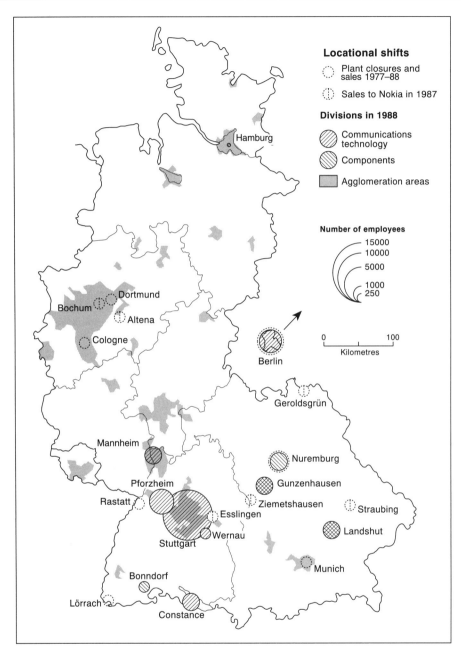

Figure 8.8 Development of the Location Pattern of SEL in West Germany, 1977–1988. Source: Fuchs and Schamp (1990: 83)

In contrast to Europe and North America, the leading corporations of Japan have largely been able to maintain employment growth trajectories within Japan, at least until relatively recently.

Nissan's closure of an auto manufacturing plant in Tokyo in 1994, for example, was its first. Similarly, in South Korea, recent decades have witnessed unprecedented growth among its

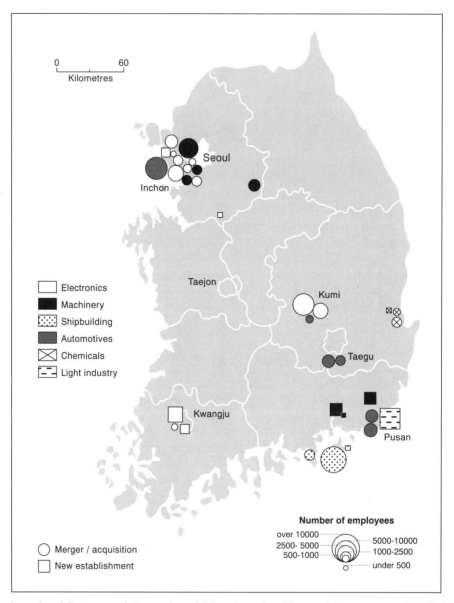

Figure 8.9 Locational Pattern and Dynamics of Manufacturing Plants of Daewoo: total 1967–1988. Source: Park (1990: 225)

leading corporations. A good illustration is provided by Daewoo (Figure 8.9).

Daewoo began as a small (textile) trading company in 1967 and experienced particularly rapid growth in the 1970s, principally by acquisition and merger (Park 1990). Following a brief restructuring phase in the early 1980s, Daewoo embarked on further rapid growth, this time with a greater emphasis on internal growth, and by 1988 had sales of over US$15 billion. By this time Daewoo had become a highly diversified conglomerate involved in construction, trade and a wide range of manufacturing activities, notably heavy industrial and chemical products, steel and metal products, electric and electronic products, machinery, chemicals, vehicles and ships, garments and textiles, and others. Daewoo also had 15 R&D establishments located in Korea, and international trading, construction and manufacturing operations in several countries in Asia, Europe, Africa and North America (Park 1990: 219–220). Within Korea, Daewoo has concentrated its growth in the Pusan region, its original location, and the Seoul–Inchon region (Figure 8.9). In addition, a few branch plants were located on industrial estates near regional cities such as Kwangju and Taegu to take advantage of labour availability and government support for regional development. So far, Daewoo's national corporate system has expanded in tandem with its growing international operations.

The significance of foreign operations to giant firms

Not all foreign manufacturing firms are big. Many small and medium-sized firms have established foreign operations. However, historically, overall levels of foreign investment in the manufacturing sector have been dominated by large firms such as GM, Philips, Cadbury Schweppes and Daewoo. Indeed, if for no other reason than the limitations of domestic markets (or resources), as manufacturing firms grow there is typically pressure to internationalize operations by exporting and eventually by investing in foreign-based facilities.

In practice, considerable variations exist in the extent to which large corporations have internationalized their operations (Table 8.2). Thus, in the late 1980s, Japanese corporations such as Mitsubishi Electric, NEC and Matsushita and the German BASF had less than 10% of their (1985) workforces in foreign operations. Even so, the absolute levels are not minor: NEC, for example, employed about 10 000 workers in foreign plants in 1985. At the other extreme, Nestlé employed almost all of its workforce outside Switzerland, its home country, while BAT Industries of the UK had 86% of its

workers in foreign plants. The other examples cited had between 18 and 53% of their workers in foreign plants, a range which probably captures the behaviour of most giant MNCs with sales of several billion dollars. Simply by virtue of its absolute size, special mention might also be made of GM, whose foreign workers totalled over 200 000 in 1985.

The data shown in Table 8.2 are of course a snapshot of a point in time. Changes constantly occur. In GM's case, while its sales rose to over US$168 billion in 1995, its employment declined to 709 000. In some cases, firms are becoming 'more foreign' but for different reasons. During the 1970s and early 1980s, ICI, for example, radically restructured its global operations but its employment downsizing emphasized its British-based plants, especially the older ones (Clarke 1982). In contrast, Japanese MNCs, whose interest in foreign investment was stimulated by the rising value of the Yen, have increased levels of foreign investment while striving to maintain domestic workforces, although in recent years this has become more difficult (Edgington 1994).

McNee (1958) recognized some time ago that the foreign operations of MNCs were integrated with domestic operations. Recently, rapid improvements in communications and production technology and declining tariff barriers, have constantly changed the operating environments of MNCs, which in turn has encouraged rethinking about integration. The integration of MNCs has thus become more intimate and varied. Whereas MNC branch plants in peripheral countries once either provided resources for export or relatively small assembly plants to serve local markets, they are now integrated in more complex systems. Moreover, MNCs originate from a much wider variety of countries than used to be the case in the 1960s, and include countries such as South Korea which not so long ago was classified as part of the periphery.

The significance of acquisitions and mergers

Corporate systems evolve from a combination of internal and external forms of growth, and from

Table 8.2 'Vital Statistics' of Some of the World's Leading Transnational Corporations in Manufacturing, 1989 and 1995

Corporation	Country of origin	Sales ($million)		Total employment		Foreign employment % of total 1985
		1989	1995	1989	1995	
General Motors	US	126 974	168 829	775 100	709 000	31
Ford	US	96 933	137 137	366 641	346 990	53
IBM	US	63 438	71 940	383 220	252 215	40
Toyota	Japan	60 444	111 052	91 790	146 855	20
General Electric	US	55 264	70 028	292 000	222 000	21
Hitachi	Japan	50 894	84 167	274 508	331 852	18
Matsushita	Japan	43 086	70 398	193 088	265 538	3
Daimler-Benz	Germany	40 616	72 256	368 226	310 993	19
Philip Morris	US	39 069	53 139	157 000	151 000	28
Fiat	Italy	36 740	46 468	286 294	237 426	18
Nissan	Japan	36 078	62 569	117 330	139 856	–
Unilever	UK/Netherlands	35 284	49 738	300 000	308 000	–
Du Pont	US	35 209	37 607	145 787	105 000	23
Samsung	South Korea		51 215		196 000	–
Volkswagen	Germany	34 746	61 489	250 616	242 420	34
Siemens	Germany	32 660	60 674	365 000	373 000	31
Toshiba	Japan	29 469	53 047	125 000	186 000	–
Nestlé	Switzerland	29 365	47 780	196 940	220 172	96
Renault	France	27 457	36 895	174 573	139 950	29
Philips	Netherlands	26 993	40 148	304 800	265 113	–
Honda	Japan	26 484	44 056	71 200	96 800	30
BASF	Germany	25 317	32 259	136 900	106 565	32
NEC	Japan	24 594	45 557	104 022	152 719	9
Hoechst	Germany	24 403	36 409	169 295	161 618	43
Peugeot-Citroën	France	24 091	33 074	159 100	139 300	–
BAT Industries	UK	23 529	24 033	311 917	170 412	86
Bayer	Germany	23 021	31 108	170 200	142 900	–
ICI	UK	21 889	16 204	133 800	63 800	52
Procter & Gamble	US	21 689	33 434	79 300	99 200	39
Mitsubishi Electric	Japan	21 213	36 380	85 723	111 585	6

Source: based on *Fortune*, 30 July 1990 and 1996, p. 47; UNCTC (1988), Annex Table B.1.
–, Data unavailable.

rationalizations and closures. From the firm's point of view, investments in new plant and equipment at new and existing locations, and acquisitions and mergers both serve to meet long-term, strategic motivations and provide closely interrelated ways in which corporate systems develop (see Chapter 7). Thus, both investments in new facilities in new locations and acquisitions and mergers then become focus for subsequent *in situ* investments. As alternatives to new site location decisions, the potential advantages of acquisitions and mergers to corporations lie in

providing for faster, larger-scale and less-risk growth. Moreover, although an alternative to choosing new locations, acquisition behaviour as a whole has a distinct geography or to use the terms of Green and Cromley (1984: 299) cities (and firms within them) have 'acquisition fields' (see also Leigh and North 1978). In practice, acquisitions and mergers are extremely influential to the growth of firms and are an important process underlying the reduction of competition and increasing economic and geographic concentration of power. In North America and Europe

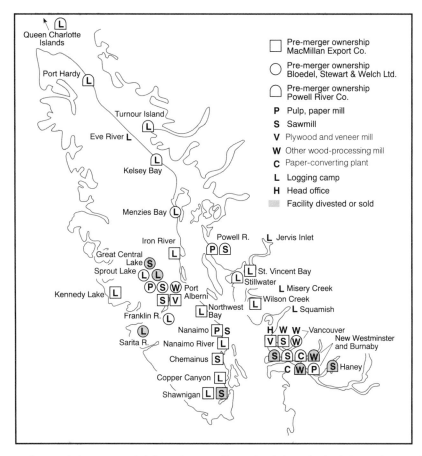

Figure 8.10   Location and Corporate Origins of MacMillan Bloedel's Principal Operating Facilities in British Columbia. Source: Hayter (1976: 218)

at least, it is probably difficult to find a major corporation that has not relied on acquisition to grow.

For the individual firm, spatial as well as aspatial considerations are often important criteria by which acquisitions and mergers are assessed. In the case of MacMillan Bloedel (MB), a Canadian forest product giant, the core of its contemporary corporate system in the coastal region of British Columbia (BC) is founded on two major mergers among then-leading, rival companies in the 1950s (Figure 8.10). The mergers restricted competition and increased levels of concentration in wood supply rights.

The mergers were also perceived in terms of pooling complementary organizational and locational resources (Hayter 1976: 219–220). Thus research and development programmes and global marketing networks, to which each firm brought different strengths, were consolidated and more effectively used. The spatial rationality underlying the mergers principally derived from the proximity of logging camps, manufacturing facilities and the optimality of individual manufacturing sites, all of which were on tidewater. Indeed, even by the 1950s, mergers provided the only way of obtaining such advantageous sites. Immediately after the mergers, MB was able to rationalize logging systems and flows of materials between production nodes, and close down smaller facilities.

Prior to the mergers, each of the three founding companies had built new facilities and acquired others. Since the mergers of the 1950s, investments to expand and modernize, as well as decisions to rationalize, have been substantial but virtually all have been *in situ*. Apart from a new engineered wood mill built in Vancouver in the late 1980s, the principal locations of MB in BC are the same in 1996 as they were in 1950, one year before the first merger.

The rapid evolution of MB's corporate system in BC and elsewhere during the 1960s and early 1970s reflects broadly based trends in the corporate sector (Hayter 1976). Thus, as with other corporations throughout North America, and indeed in other countries, as testified by the experiences of Daewoo (Park 1990) and Philips (de Smidt 1990), acquisitions and mergers have been vital to the creation of its 'core' region, in MB's case in BC. Indeed, as the Hakanson (1979) model predicts, MB developed by first creating a strong regional core (Figure 8.4), the general structure of which has remained intact for a long time. Changes within the individual nodes of the core system, enlargement and diversification from the 1950s to the 1970s, restructuring and rationalization in the 1980s and 1990s, have been driven by *in situ* investments. In contrast to the Hakanson model, MB's national and international expansions occurred more or less at the same time (Table 8.3). However, consistent with broad trends in the pattern of direct foreign investment, MB has pursued strategies of horizontal and vertical integration and in some cases these expansions have been directly linked to its BC base. Its acquisition of paper-box companies in the UK, and to a lesser extent in the US, for example, was to provide a captive market for the paperboard it was (then) manufacturing in BC. Moreover, in expanding outside of BC, the acquisition of existing companies has been important. Where new investments have been made, joint ventures have been the preferred method of entry. Typical of many North Amer-

ican and European corporations, recessionary crises in the 1980s and 1990s have provided an immediate stimulus and context for extensive restructuring of MB's international corporate system, including by the divestment of several facilities in foreign countries (Barnes et al 1990).

Capital, specifically physical capital, it may be argued, is both mobile and immobile. In the case of MB, capital's mobility is revealed by the global reach of MB's investments and divestments. At the same time, capital's immobility is revealed by the selective nature of MB's investments (and divestments) and, most of all, by MB's constant retrenchment of core facilities within BC for at least the past 50 years.

The spatial division of labour

Market and planning system firms collectively organize the (private sector) spatial division of labour. Two basic types of spatial division of labour (SDL) are distinguished (Massey 1984). First, the sectoral division of labour occurs when regions specialize in particular industries and all the related skills. Second, the intra-sectoral division of labour occurs when within individual industries firms choose to specialize tasks and occupations by location. It might be noted that the term 'spatial division of labour' is used in both senses so that its meaning has to be surmised from the context in which it is expressed.

Historically, the underlying division of labour is primarily a sectoral one as during the 19th century the surge of industrialization sweeping through Europe and North America was largely organized by regionally focused firms creating regions specialized on particular industries, or related groupings of particular industries. Thus cotton textile, iron and steel, shipbuilding and engineering, chemicals and lumber regions each developed distinctive ('sectorally specific') labour skills which were organized by distinctive regional populations of firms. Within each region, there was a vertical concentration of all the necessary occupations, ownership and

Table 8.3 Interregional Growth of MacMillan Bloedel's Manufacturing Facilities, to 1973

Product and location	Method of growth
Vertically integrated expansion	
Western Canada	
Corrugated container plants in Winnipeg, Regina, Edmonton, Calgary (and Vancouver).	Acquisition in 1954 of the Martin Paper Box Co. by the former Powell River Co.
Europe	
Corrugated container plants in Hatfield, Nelson and Southall (UK). Subsequent construction of new plants at West Auckland and Weston-Super-Mare.	Acquisition in 1963 of Cook Containers Ltd. and Ily-grade Containers Ltd.
70 000-ton-per-year capacity fine paper plant, Lanaken (Belgium).	Acquisition of a 36% (now 46%) interest in Royal Dutch Paper Mills (KNP), a Dutch fine paper producer, and subsequent $13 million investment in the Lanaken plant completed in 1968.
25 000-ton-per-year capacity fine paper plants, Algeciras (Spain).	Joint venture with KNP and Celupal SA (30% interest acquired by MacMillan Bloedel) in a $25 million investment completed in 1969.
Geographical diversification	
United States	
Corrugated container plants in New Jersey and in Baltimore.	Acquisition in 1966 of the two plants from the St Regis Paper Co. and the Mead Corporation.
Integrated forest product complex at Pine Hills, Alabama, with capacity for 270 000 tons of linerboard, 50 MFBM of lumber, and 100 million square feet of plywood.	Project completed in 1968 for $70 million as a joint venture with the United Fruit Box Co. (which had a 40% interest in the linerboard mill). Became sole owner in 1970 and announced a $10.5 million particleboard plant in 1973.
Corrugated container plant, Odenton, Maryland.	Acquisition in 1971 from the Hoerner Waldorf Co. (Baltimore plant subsequently phased out).
Ten corrugated container plants in New York, New Jersey, Ohio (two), Illinois, Indiana, Arkansas, Texas, Mississippi and California.	Acquisition in 1971 of the Flintkote Corporation.
Canada	
180 000-ton-per-year capacity newsprint mill in St John (New Brunswick).	Acquisition in 1969 of a 54% interest in partnership with Feldmüble AG of Germany. The mill (constructed in 1964) subsequently expanded by 180 000 tons per year.
Aspenite mill, Hudson's Bay (Saskatchewan)	Acquisition from the provincial government of Saskatchewan in 1965.
Plywood mill, Nipigon (Ontario).	Acquisition in 1973 of Multiple Plywood Ltd.
Waferboard plant, Thunder Bay (Ontario).	$9.4 million investment announced in 1973.
South-east Asia	
Plywood and blockboard plant, Pekan (Malaysia).	Completed in 1973 as 30% partner of Mentegon Forest Products Sdn. Bhd.
Europe	
Hardwood pulp plants in France (three) and Belgium (one), with a 600 000-ton annual capacity.	Acquisition late in 1973, reportedly securing a 40% interest in La Cellulose d'Acquitaine.

Source: Hayter (1976: 227).

control, decision-making and the production skills associated with particular industries.

The dynamic underlying the intra-sectoral spatial of labour is the growth of interregional and international corporations in the 20th century (Hymer 1960; Simon 1960). According to this new SDL thesis, big corporations, as they increase in size, horizontally distribute different functions and the associated occupations among different regions which provide 'appropriate' location factors. In the most general situation, within each industry, 'control' functions and occupations, notably decision-making and research and development (R&D), are spatially separated from 'basic work processes', to use Simon's (1960) terminology. More specifically, in this thesis the control functions are concentrated in a relatively few metropolitan ('core') centres while basic work processes are dispersed among a wide variety of ('peripheral') locations. While the SDL thesis was primarily presented within the context of advanced industrial countries (Simon 1960; Pred 1974), the establishment of basic work processes, particularly in relation to resource industries and low-skill manufacturing activities, in developing, 'peripheral' countries was recognized (Hymer 1960). More recently, a *new international division of labour* (NIDL) thesis has been proposed which is defined by geographical shifts of basic work processes from core to peripheral and semi-peripheral countries, thus 'hollowing out' the former (Fröbel et al 1980; Bluestone and Harrison 1982; see Chapter 16). The revision of the new SDL thesis by the NIDL thesis, itself underlain by the emergence of the ICT techno-economic paradigm, implies that capital has become significantly more mobile and significantly less committed to particular places.

Locational hierarchies

Within the context of the division of labour *within* corporations, both the new SDL and NIDL models equate functional occupational hierarchies with locational occupational hierarchies (Massey and Meegan 1982: 207). While the models differ in their implications for the dispersal of basic work processes, both anticipate the agglomeration of control functions at the top of the urban hierarchy. The corporations cited in this chapter provide examples of locational hierarchies. Cadbury Schweppes, for example, has placed its head-office (and technostructure) in central London and its main R&D laboratory in nearby Reading, the latter being neither adjacent to the head-office nor to the operating locations which are dispersed throughout the UK (and elsewhere). Similarly, Daewoo has located its major control functions in the Seoul and Pusan regions, and Philips' head-office and most important R&D facilities are in the Eindhoven agglomeration. Indeed, the strategies and structures of these firms have contributed towards broader spatial divisions of labour within their respective countries. At a regional scale, MB, with its head-office in central Vancouver, its R&D group in a nearby suburb and its operating facilities spread throughout coastal BC, provides a locational hierarchy which contributes to established core–periphery contrasts within the province.

There is an extensive literature which provides the rationale for the agglomeration of the control functions of large corporations, especially head-offices and R&D laboratories or centres (Hayter and Watts 1983; Malecki 1993). In some countries, such as the UK, Japan and Canada, the geographic concentration of head-offices is unusually marked: in the UK, 74% of the head-offices of the 100 largest (by sales) manufacturing firms are in south-east England, principally London (Healey and Watts 1987); in Japan, head-offices are overwhelmingly concentrated in Tokyo and Osaka (Edgington 1994); and, in Canada, Toronto and Montreal have traditionally been the location for over 75% of the head-offices of the country's largest manufacturing firms (Semple and Phipps 1982; Ley and Hutton 1987). In the case of the United States, while head-offices reveal a more dispersed pattern than in the UK or Canada, 'concentration' is still the predominant feature (Ullman 1958; Pred 1974; Borchert 1978). New York, for example,

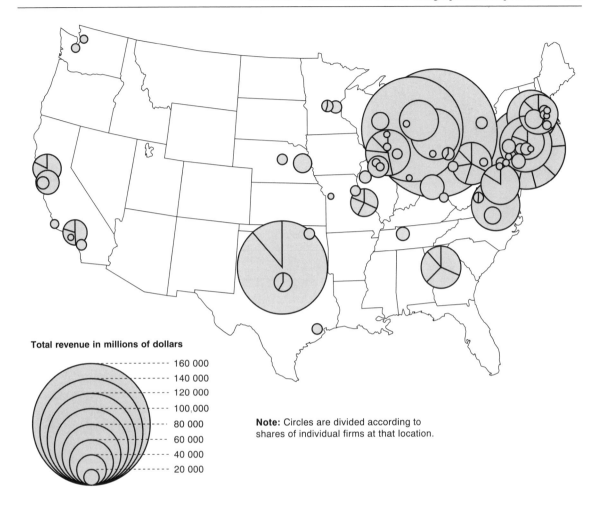

Total revenue in millions of dollars

160 000
140 000
120 000
100,000
80 000
60 000
40 000
20 000

Note: Circles are divided according to shares of individual firms at that location.

Figure 8.11 Headquarter Locations of the 100 Largest US Manufacturing Companies Ranked by Revenues ($m), 1995. Source: *Fortune* July 1996

accounted for about 30% of the head-offices of the 500 largest manufacturing corporations in the US in 1972 (Pred 1974: 113), while in terms of value of assets, New York's importance increased to 41% (Borchert 1978: 219).

The geographic distribution of the head-offices of the 100 largest manufacturing companies in 1995 in the US is generally consistent with previous studies (Figure 8.11). In terms of the number of head-offices and the size of revenues controlled, the most important agglomerations of control centres lie within the old Manufacturing Belt (Figure 3.6). The extraordinary size of the largest, led by General Motors, Ford and Exxon, automatically imparts significance to their headquarter locations (respectively, Detroit, Dearborn in Michigan, and Irvine, Texas). In terms of the 100 largest manufacturing firms, New York State, including New York City, comprises the most important concentrations of control centres. The regional agglomeration of control centres of the largest corporations has traditionally been paralleled by highly localized concentrations within the central business districts

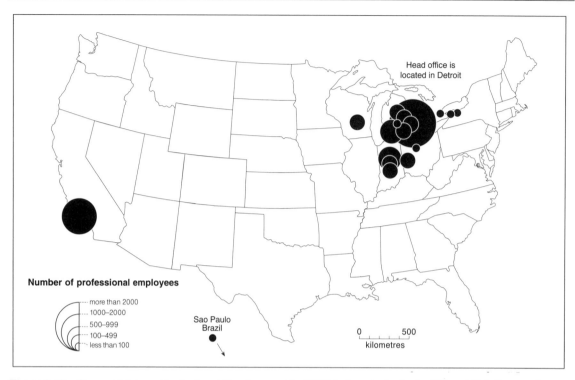

Figure 8.12   General Motors: Professional Employment in R&D Laboratories in the US, *c*. 1986. Source: Jacques Cattell Press

(CBDs) of inner cities. It has long been recognized that such behaviour facilitates person-to-person contact among strategic decision-makers and the related services they require, that is, the close proximity of head-office and business services facilitates the operation of dense networks of information which are characterized by uncertainty, secrecy and trust (see Chapter 4).

Yet it is also empirically transparent that in the US the head-offices of large industrial companies can viably exist in 'peripheral' regions, such as Washington State, and medium-size metropolitan areas, such as Seattle. Moreover, perhaps most notably in the case of New York, the importance of suburban areas and surrounding cities as head-office locales for large industrial companies, has increased. To some extent, such a development reflects inner city problems, and also suggests that industrial companies are perhaps gradually losing their power to bid for

the most expensive and accessible sites, in New York's case from the organizations which are the heart of the the continuing growth of the city as a global financial and communication centre (Mitchelson and Wheeler 1994; see also Ley 1985).

There is no doubt that the form and nature of information networks are significant constraints on head-office locations that need to locate to facilitate face-to-face communication and that recent rapid developments in communication technology are so far not undermining established concentrations of control functions. Whether or not the distribution of control, as reflected by the head-office locations of the largest 100 companies in the US (Figure 8.11), is optimal from either private or social viewpoints, is another matter. The 'efficiency' of such distributions is difficult to assess and it may be that a more decentralized pattern is equally or even more viable. If so, in countries such as the UK and Japan, where decision-making concentration is even more formidable, this point may have more relevance.

Similarly, the R&D establishments of major corporations favour relatively few centres. In the case of GM, as of the mid-1980s, its R&D centres and workforce in the US are strongly concentrated in cities of the Manufacturing Belt, generally within Michigan and Ohio and not far from its head-office in Detroit (Figure 8.12). GM does have an important R&D group in California but no other region stands out. It might also be noted that GM does little R&D in Canada or Latin America, although it does have an important European R&D centre. GM's R&D efforts are unusual in terms of number of centres and overall size: in the mid-1980s GM employed over 7000 professionals, a total bigger than most university faculties and another indication of the power of this giant (even if the power is misdirected from time to time and not to be equated with efficiency).

The R&D strategies of large industrial corporations do vary and inherently reflect, and reinforce, firm-specific circumstances. Even so, increasingly during the 1950s and 1960s, R&D centres were established by many corporations, including GM, in separate locations from manufacturing facilities and, to a lesser extent, from corporate head-offices. With respect to the latter trend, in the US at least, many R&D centres, but not all, remained in the same metropolitan area as corporate head-offices but were more likely to be located in the suburbs (Malecki 1979, 1980). Three related groups of reasons have been put forward to explain this behaviour. First, the location of R&D activities reflects a search for particular combinations of location factors, notably access to scientists, engineers and skilled labour, and, in turn, access to a variety of cultural, educational and environmental amenities thought conducive to attracting such professionals (see Chapter 4). A second argument emphasizes the advantages arising from the physical separation of R&D from manufacturing operations, notably with respect to allowing

R&D groups to focus on long-range problems without the interference of requests to become embroiled in short-term troubleshooting activities at particular manufacturing plants. A related point is that some distance between head-office and R&D centres reduces disruptions caused by too frequent contacts and 'interference' by head-office executives. Third, as Clark (1981) notes, the locational separation of R&D professionals from manufacturing workers facilitates managerial preferences to 'segment' each group according to different employment policies (see Chapter 7). While these rationales are important, the growing recognition of the loopy nature of R&D is encouraging a re-integration of R&D within manufacturing and marketing processes, and this re-integration has potential implications for the location of R&D (Figure 2.8; Hayter 1996).

Core–periphery effects According to the SDL thesis, the tendency of international firms to locate head-offices, R&D and related control functions in selected centres and regions simultaneously defines these places as 'cores' or 'metropoles' in two main related ways. First, within the context of the operations of international firms, it is in the cores where decisions are made about the allocation of investment, jobs and technology in the periphery. Second, it is in the cores that the highest paid and most stable jobs are provided. In addition, core functions generate a variety of spin-off jobs, many of which also involve high-income occupations in a variety of business services such as legal, advertising, computing, engineering and financial services. Moreover, the local wealth generated by core functions can in turn help sustain various 'high order' levels of cultural, entertainment and sporting amenities, which in turn help to define the core status of regions. Any assessment of corporate control functions should also appreciate their size. Individual head-offices of the largest corporations can employ thousands of people in buildings which occupy the most expensive real estate in the world. R&D programmes can also be huge, and comparable to large-scale factories in size. In 1993, for example, General Motors (GM) and IBM invested US$7.4 billion and US$5.6 billion respectively in in-house R&D and employed more scientists and engineers than most universities. Collectively, head-offices and R&D centres provide enormous concentrations of wealth and power.

On the other hand, according to the SDL and NIDL theses, the tendency of MNCs to disperse branch plants producing mature products and commodities among small, isolated rural places simultaneously defines these places as peripheral. Such branch plant operations emphasize manufacturing jobs that are less well paid and less stable than head-office jobs (or more innovative manufacturing operations). In addition, any local spin-offs generated by branch plants are determined by overall corporate policies and will typically involve similar functions. At the same time, within the context of peripheral areas, branch plants may offer relatively high-paying jobs.

The tendencies of international firms to help create core–periphery contrasts through the *internal* distribution of corporate functions and the *internal* division of labour is constrained in practice by a variety of considerations. Thus, the impacts of the traditional sectoral division of labour are still evident as some very large corporations retain parent head-offices and control functions in the regions where they were first established. The locational strategies of international firms vary; for example, occasionally head-offices are located in suburbs rather than downtown; R&D laboratories may be established in isolated places; and not all low-skilled work is inevitably decentralized to peripheral regions. In addition, it does need to be emphasized that these arguments refer to the internal division of labour and immediate spin-offs. There are many firms and jobs that are not controlled, directly or indirectly, by large

corporations, while the public sector, including the functions of governments themselves, is also an important source of high-income jobs and decision-making power which need not necessarily respond to the same principles of locational hierarchy as the large corporate sector.

Conclusion

The geography of firms and their local development implications is enormously varied. The Hakanson (1979) model is one attempt to summarize this geography and the SDL/NIDL theses are attempts to summarize the local development implications. The SDL/NIDL theses are particularly important themes within the literature based on 'matching' the locational hierarchies of large firms with core–periphery contrasts. In practice, it is reasonable to argue that the SDL and the NIDL have both captured important tendencies and grossly over-simplified reality. The locational hierarchies of the largest firms and core–periphery contrasts have proven more dynamic than anticipated in the original formulations of the SDL and NIDL. For example, some cores are now peripheries and some peripheries are now cores, or at least 'semi-peripheral'. As Gereffi (1989) has noted, the location of the realization of profit, and distribution of value added, varies from production chain to production chain in a manner which is not easy to geographically categorize among the pre-established core, periphery and semi-periphery.

Indeed, the difficulty in summarily generalizing the geography and local impacts of firms, reinforces the need to investigate the structure and strategies of firms in different segments of firms, while recognizing the variability that exists within segments. The next chapter continues this investigation by focusing on new and small firms.

9

FORMATION AND FUNCTION OF NEW (AND SMALL) FIRMS

Some time ago, Marshall (1922: 315) likened business enterprises to "the young trees of the forest" – they start as seedlings, grow slowly, then decay and die. Some firms, to be sure, have demonstrated remarkable resilience over decades, even centuries and appear more as long-established members of an old growth forest. Stora Kopparberg, for example, has been in continuous existence as a manufacturing company for over 400 years. Large corporations in any event can reproduce themselves in various ways, expand, rationalize and mutate without apparent reference to age. Indeed, the very invention of the modern corporation was designed to remove powerful limitations to growth, size and vulnerability associated with the individual entrepreneur. The vast majority of manufacturing firms, however, are small and will remain so. For these firms, Marshall's life cycle analogy has more immediacy, even if particular paths of evolution are highly variable.

This chapter examines various aspects of the formation and function of new and small firms, particularly with respect to their locational characteristics and their relationship to local development. For a considerable period of time, and with a few exceptions, small firms (SFs) were peripheral to the concerns of industrial geography. Typically, SFs were perceived, implicitly at least, as marginal units of operation surviving only because of limited market possibilities or technical limitations on the realization of economies of scale. In the conventional view, the 'ideal' unit is the large firm and factory exploiting economies of scale. Over the past 15 years or so, however, there has been a surge of interest in new firms, SFs and the associated topic of entrepreneurship (Cross 1981; Storey 1982, 1994; Curran and Blackburn 1991, 1994; Karlsson et al 1993). The reasons for this interest relate to the extent of the downsizing among large firms, the rapid growth of new business formation and an ideological shift celebrating the role of initiative and entrepreneurialism in international competition (Stevenson and Sahlman 1989: 96). Within geography, there has been a growing focus on the contributions of new firms and SFs in local development, itself stimulated by the demise of the traditional form of industrial location policy with its emphasis on large branch plants and its replacement, or at least modification, by policies which have tried to stimulate local entrepreneurship.

This chapter focuses particularly on the characteristics of new firms, the formation of new firms, especially why new firms locate where they do, regional variations in new firm formation rates, the challenge of survival of new firms, the characteristics of different types of new firms (and SFs), and some of the associated policy implications. These themes are introduced by a discussion of new (and small) firms as entrepreneurs.

New (and small) firms as entrepreneurs

The idea and definition of new and small firms are inextricably bound up with one another since new firms are almost inevitably small and the essence of both new and small firms is their 'entrepreneurial' nature. That is, new and small firms differ from multinational corporations not simply in terms of size. Rather, new and small firms offer a distinct 'entrepreneurial' form of business organization which stands in sharp contrast to the bureaucracies or technostructures that administer large corporations. Moreover, this distinction has important implications for industrial location and local development.

Given the rationale for paying special attention to the new firm as representing increments to society's entrepreneurial pool, in many ways the notion of entrepreneurship remains an elusive, ambiguous concept (Stevenson and Sahlman 1989). Kilby (1971), for example, likens attempts to define entrepreneurship with "hunting the hefflelump" which although (supposedly) a large, important animal, hunters are never sure they have caught it! Yet the hunting goes on and there have been numerous attempts to define the personality characteristics of entrepreneurs and the behavioural features of entrepreneurial firms (McLelland 1961; Birley 1989; Dewhurst 1989; Stevenson and Sahlman 1989). It is also often said that the societal 'hero' of neoclassical theory is the individual entrepreneur and that the entrepreneurial firm is the practical expression of firms as black boxes (Table II.1). This model provides an appropriate point of departure for thinking about entrepreneurship.

In a neoclassical landscape of perfect competition, firms have some clear characteristics. In particular, neoclassical black boxes *qua* entrepreneurs have no market or social power; they are

independent, decision-making and risk-taking are conflated and personalized, and they pursue economic self-interest on the basis of perfect information and perfect rationality. Even if in the real world entrepreneurs do not have perfect information or rationality, a neoclassical landscape is a highly competitive one in which market forces eliminate the misinformed and irrational. Indeed, models of the dual economy such as that outlined by Taylor and Thrift (1983; Figure 8.3) also emphasize that small firms, and by implication entrepreneurial firms, must necessarily be highly responsive to market forces, so that the 'power' of the small firm is limited to the effectiveness of these responses.

There is therefore agreement in the neoclassical and dual economy models that entrepreneurial firms operate in markets that are strongly regulated by competition. In addition, the neoclassical model indicates that entrepreneurial firms have several other features. First, such firms have little or no power to change market forces; that is, they are 'small' in two senses: they are 'relatively' small in terms of market share so that by themselves each firm cannot determine prices or output levels; and they are 'absolutely' small in an absolute sense of 'few' employees or sales, so that the opening and closure of each firm has virtually no measurable social impacts. Second, entrepreneurial firms are independent in the sense that input and output transactions are at 'arms length' and closely regulated by market (external) forces. Third, in entrepreneurial firms ownership and management is combined and personalized: decision-making is of the direct, 'hands-on' or personal contact variety. In this view, decision-makers are also risk-takers who have strong personal incentives to improve efficiency or market performance, i.e. to be innovative in some sense. Indeed, some observers place particular stress on the criterion of innovativeness as the heart of 'entrepreneurialism'.

Broadly speaking, small (absolute and market) size, independence, owner-management and risk-taking provide the defining characteristics of the entrepreneurial firm. Simultaneously, these defining characteristics of the entrepreneurial firm reflect its ambiguity. Thus, the meanings of small size (*qua* lack of power), independence and owner-management are themselves problematical and inevitably involve some kind of judgement. In terms of relative size, for example, what constitutes smallness in terms of market share? One firm may have a tiny share of a regional market but a dominating position in a community within a region. Another firm may have a significant share of a regional market but face the constant threat of competition from imports and from the formation of new firms because 'barriers to entry' into the business are so small. That is, defining smallness in a relative sense requires judgements concerning shares of the market in relation to judgements about the geographic size of markets and barriers to entry in the sense of the various difficulties facing the formation of a new firm. Moreover, if small size in absolute terms implies that an individual firm's opening or closure has negligible community impacts, judgements have to be made regarding what is meant by 'negligible' and 'community'.

There are also difficulties in defining the characteristics of independence and owner management. In the case of independence, for example, one legally incorporated small firm may serve thousands of customers at 'arms length' while another may sell its entire output to a giant corporation and operationally is not independent of its market. Between these two cases, which represent realistic polar situations, when does independence become dependence? With respect to the criterion of owner-management, even among new and small firms owners have the option of appointing professional managers and can make choices as to how close and regularly the managers are managed. These options in turn raise questions as to how owner-management is to be defined. Finally, while there is inevitably some risk in creating and operating a small firm, it cannot be assumed that all small firms are innovative.

Given the judgements that have to made regarding the interpretation of size, independence and owner-management (or innovativeness for that matter), it should be noted that some new (small) firms are strongly entrepreneurial in some but not necessarily all of these characteristics. Indeed, this observation underlies Taylor and Thrift's (1983) disaggregation of the small firm segment (Figure 8.3). Thus, the (market) independence of 'loyal opposition firms' which have established their own customer base in competition with large corporations is substantially different from 'satellite subcontractors' which service large firms. In turn, the independence and autonomy of the 'satellite franchise' is likely to be less in many respects than that of the satellite subcontractor. Similarly, small firm 'laggards' differ from small firm 'leaders' in their pursuit of efficiency and product improvements, or innovativeness.

In practice, the absolute size of firms, particularly with respect to employment, is used to define the SF sector and countries make different judgements as to the upper limits of what constitutes SFs (or small and medium-sized firms) in this context (Table 8.1). Whatever limits are chosen, such a definition uses absolute size as a surrogate for a range of entrepreneurial characteristics. Such an assumption is practical since information on employee size is often available while information on other entrepreneurial features is rarely systematically collected. Moreover, measures of absolute size probably correlate to some extent with other characteristics associated with entrepreneurial firms. At the same time, this correlation is by no means perfect as some larger firms are owner-managed and behave in an entrepreneurial manner, while some SFs have bureaucratic structures and characteristics such as risk-taking and innovativeness seem scarcely present.

Over time, the members and perhaps even the overall size of the entrepreneurial pool of firms changes. On the one hand, some firms are lost to this pool as a result of failure, internal growth and change, or acquisition by large corporations. On the other hand, new firms are constantly added to the entrepreneurial pool. Indeed, for some observers, the rate of formation of new firms, considered as a supply of entrepreneurship, provides a vital measure of the health of a local economy (George 1970; Firn and Swales 1978).

The formation and characteristics of new firms

Stories of new firm formation and entrepreneurship are incredibly varied and individual case studies can never hope to fully represent the range of behaviours and processes. With this important caveat in mind, case studies provide concrete illustrations of new firm formation which help to understand and link the previous relatively abstract discussion of new firms as entrepreneurs and subsequent summaries of empirical surveys of new firm populations (Exhibits 9.1 and 9.2).

Exhibit 9.1

Pacific Emergency Inc. (and Pacific Body Armour)

The founder of the firm, Bradley Field, was on the ski patrol at Crystal Mountain in BC's interior and he was waiting for an opening so that he could become an ambulance attendant. However, as a ski patrol member he had difficulty carrying emergency medical supplies in his conventional backpack. The problem with the old tackle box type of carrier was that it was awkward to handle and breakage was a problem. Field wanted to develop a more convenient system that could meet individual needs. Subsequently, "He developed an adjustable elastic system that can fit around bandages, bottles, needles and other equipment. It can be velcroed on anything from an ambulance wall to a paramedic's leg to hold items within easy reach" (Shaw 1993: D4).

So, in 1982, with his wife, Lori, "Field, who didn't know a backstitch from a buttonhole, set up a rented sewing machine in a spare bedroom and created his own custom made backpack to carry emergency medical equipment" (Shaw 1993: D1). When they started, Bradley and Lori Field had to cut the patterns and do the sewing themselves and they "learned everything the hard way" (Shaw 1993: D4). Initially, they made the backpack for themselves but people on the ski hill kept asking to buy one. They had no start-up capital except for an uncle who gave them $1000 and a Mastercard. Both kept their existing jobs for a

while, Lori at an insurance company and Bradley as ski manager on Crystal Mountain. After all, sales in the first year were just $20 000. But sales increased and both quit their old jobs, and they moved production from the spare room to the garage, then to rented premises behind someone else's factory, and finally to their own factory.

The Fields eventually established two companies. Pacific Emergency Inc. makes 100 different emergency medical products from under $10 first aid kits to $400 + sophisticated equipment for paramedics and ambulances. Most ambulances in BC now carry some of their products. The second company, Pacific Body Armour, makes bulletproof vests "a sideline that started when US customers said their paramedics needed protection as well as medical products" (Shaw 1993: D1). They were already making vests for paramedics but not bulletproof ones, until requested by US customers. Although they knew nothing about body armour, they developed such vests which combined comfort with protection. They vests costs $600–1400 and Canadian police departments are also buying them.

By 1993 the two companies employed 30 people, sales were projected at $2.5 million, of which about 30–40% comprises exports, principally manufactured by Pacific Emergency for the US, especially the East coast, with additional sales to Europe and around the Pacific Rim. The Fields no longer do any sewing.

Source: Shaw (1993: D1 and D4).

Exhibit 9.2

Madge Networks

Richard Madge founded Madge Networks in 1986 after a brief career as an architectural journalist and then as a partner in a computer game company which started and failed in the early 1980s. He then created Madge Networks by mortgaging his farm in Buckinghamshire, England, hiring four people and targeting networking technology. Madge Networks focused specifically on token rings which allow computers across networks to communicate with one another, a high-tech product, 70% of the global market for which was supplied by IBM in the mid-1980s. Notwithstanding IBM's dominance, Madge Networks, from its Buckinghamshire base, sold its first products in 1988, and in, 1990 achieved noteworthy sales of $20 million and profits of $1.1 million. Then the firm 'took off'. In 1992, sales amounted to $78 million and 1994 sales reached $213 million. In 1995 Madge Networks employed 200 computer scientists in a newly built research centre in Buckinghamshire and is currently investing about $7 million in R&D. The first has successfully developed a technological lead in token rings and it is estimated to have about 15% of the world market (while that of IBM has dropped to 55%). Madge Networks is currently researching products that will connect video and telecom facilities into computer networks.

To obtain crucial venture capital, and after failing to get sufficient offers from British sources, Robert Madge got $1 million from 3Com, a Silicon Valley computer company, in return for a 10% share in his company. In 1993, Robert Madge listed his firm on Nasdaq, the American stock market, because other specialist computer networking firms are in the US, while at the same time registering the company's head-office in the Netherlands. Then early in 1995, Madge Networks formed a strategic alliance with Cisco Systems, a leading computer networking company in the US. Each will market the other's products and Cisco, previously a partner with IBM, will also license some of Madge Network's products. Sales for 1996 are forecast at $377 million.

Source: Lynn (1995: 2.4).

New firm case studies

Pacific Emergency Pacific Emergency Inc. (and a subsequent company, Pacific Body Armour Inc.) was started by Bradley and Lori Field in Kelowna, in the Okanagan Valley region of British Columbia, Canada, a resource-rich peripheral region which is growing rapidly (Exhibit 9.1). This firm manufactures custom-made backpacks containing emergency medical equipment and bulletproof vests, and several summary points can be usefully drawn at the outset from this example of a 'successful' new firm.

In many ways, Pacific Emergency offers a good example of the principal features of an entrepreneurial firm. As a new firm in 1982, it was an extremely small ('micro') operation and, even after several years of rapid growth, by 1993 it employed just 30 people and had sales of Can$2.5 million (about US$2.0 million). While informa-

tion on market share is not available, Pacific Emergency (and Pacific Armour) serve a wide variety of consumers over whom it has no control and, if it does enjoy a significant market share for some products, the barriers to entry facing potential competitors are clearly not intimidating. Its small size also means that the firm's fortunes do not affect the prosperity of Kelowna and its population of 100 000. Moreover, the firm is owner-managed, the Fields pay close attention to all aspects of the firm's operations and they enjoy considerable independence and autonomy.

In addition, in this particular case, the competitive basis of Pacific Emergency (and Pacific Armour) rests on innovativeness – the creation of a new product to meet a market need – while additional new products since start-up have been designed to exploit related market niches. These innovations are not 'path breaking', 'radical' or 'major' and no new scientific, technological, engineering, marketing or organizational principles were involved. Rather, Bradley Field identified a very specific market opportunity and he and Lori Field were able to learn the sewing skills – established skills – necessary to make the opportunity a reality. As a new and small firm, Pacific Emergency provides one example of 'a leader' (small) firm (Figure 8.3).

There is great diversity in the origins and characteristics of new firms and Pacific Emergency (and Pacific Armour) is one particular example (Cross 1981). Indeed, as new manufacturing entrepreneurs, the Fields' background would appear to have some unusual characteristics and in some ways their venture occurred 'by chance': they identified a commercial opportunity virtually by accident; neither Bradley and Lori were apparently even thinking about creating a business; and neither were working for manufacturing firms or even employed by other SFs (he was in the tourism sector and she worked in the insurance industry). At the same time, there are features about the Fields' story which are more typical of new firm formation. Bradley Field, for example, was already a manager and his idea for a new firm stemmed directly from his work experience. Moreover, the process of creating and developing Pacific Emergency reveals some other themes that are consistently highlighted in the new firm literature. First, in this case as in many others, the owners, a husband–wife team, did virtually 'everything' in the early years, including manufacturing, designing, financing, purchasing and marketing. Second, the two owners had little or no prior knowledge of these activities and their learning of appropriate skills primarily occurred 'on the job' on a trial-and-error basis. Third, both owners kept their previous full-time jobs while developing their new firm and the associated skills. Fourth, they had no money of their own for investment. Basically, as do many new firms, they relied on personal credit and a relative, and on the income from their existing jobs (fortunately, they were able to generate cash flows to cover costs as they expanded). Fifth, 'location' is not an explicit part of the story. As is typical for many new firms, the Fields did not make a location decision in an explicit sense; they simply started working in their home. Subsequently, their relocation to rented premises and then to a factory occurred within Kelowna where they lived. Sixth, in developing Pacific Armour, the Fields obviously faced significant uncertainty, both in the sense of true uncertainty, and of knowledge gaps (see Chapter 5). Thus, they could not know the market potential of their products and they were unfamiliar with many of the key activities involved in manufacturing their specific products and in running a small business in general. They also had little money. Indeed, in response to such uncertainty, new entrepreneurs, like the Fields, often maintain existing jobs for a while and live in communities which they know well.

The Pacific Armour story in particular also reveals well-known difficulties in defining exactly when a new firm is formed (Mason 1983: 53). Thus, did the firm originate when the Fields started experimenting in a part-time way at home, when they legally incorporated their firm, when they established the business on a full-time basis and/or when they established production in

a separate location? The answers to these questions probably involve different dates, a not uncommon problem when identifying the origins of new firms. There is also the question of whether Pacific Body Armour should be regarded as a new firm. While it is small and owner-managed, it is also affiliated to an 'established' firm with which its operations are inextricably mixed and it does not represent an additional supply of entrepreneurship. Moreover, this kind of affiliated development frequently occurs even among relatively new firms. These problems may be simply definitional but they are worth noting because they have to be solved in a statistical analysis of samples or populations of new firms which, for example, compare new firm formation rates over time or across regions. Moreover, such comparisons are inevitably complicated if underlying definitions are different.

Madge Networks While Pacific Emergency has carved out a niche with a conventional product using long-established processes, Madge Networks is a high-tech manufacturer which has developed a sophisticated component for computers (Exhibit 9.2). In many respects, Madge Networks reflects conventional thinking about new firm formation, notably with respect to its location near the home of its founder, Richard Madge, the hands-on management provided by Madge, the financing problems he faced and the significant uncertainties associated with the firm's start-up and development. Madge Networks was not the first firm created by Madge but this is not unusual for new firm founders either. In other respects, Madge Networks is unusual, especially with respect to its growth and degree of internationalization. Thus, Madge Networks made its first sale in 1988 and by 1994 had sales of US$145 million. With respect to international connections, by 1994 Madge Networks had obtained venture capital from the US, was listed on Nasdaq, a US stock exchange, shifted its registered head-office to the Netherlands, and had achieved global sales in the face of competition from IBM. Then in 1995 it entered a strategic alliance with a leading US medium-sized firm (see Table 10.1). Madge Networks, like Pacific Emergency, is distinguished not simply by its growth and internationalization but by its survival. Many new firms fail within a year or two after formation and many that survive do not grow much and never participate in the international economy.

New firm characteristics: survey evidence from Japan

With a few notable exceptions, such as Japan, national data on the formation of new manufacturing firms, pertaining to date of start-up, location, employee size or product mix, are not systematically collected. Occasional data that are available are soon outdated and limited in content. Consequently, basic statistical information on new firms has depended on field-work and while this work has been extremely valuable, it is selective in focus and research designs vary considerably, for example, in terms of time period, spatial scale, size of sample, knowledge of underlying populations (of new firms) and definition of new firms. In Japan, however, organizations such as the Ministry of Trade and Investment (MITI) and the Kokumin Kinya Koko (People's Finance Corporation) regularly publish information on the characteristics of new and small firms. Several of these surveys are reported by Patchell (1991) and his summary provides the basis for the present discussion. While the distinctive nature of the Japanese economy has to be considered, including with respect to the unusually significant role played by small firms (Patchell 1992), these data provide a comprehensive framework for assessing trends and a frame of reference for other studies (e.g. Cross 1981).

Motivation to begin a new business While neoclassical theory emphasizes economic rationality and the importance of cost-minimizing or profit-maximizing behaviour, in practice, individuals begin their own firm for a wide variety of reasons. In this regard, the Kokumin Kinya

Koko publishes the results of a survey every five years of all new firms it has helped finance. In the case of its 1990 survey, 2891 firms responded, 12% being manufacturing firms, and each firm was able to make three choices (Figure 9.1). Essentially, the agency asked firms to indicate their motivations according to three main general categories, labelled income, work satisfaction and lifestyle gratification reasons. (An 'other' category is also identified.) Each of these categories is further subdivided. Income motivations, for example, are expressed as three related but separate categories.

Clearly, motivations are complex. Thus, income motivations, quality of life and work satisfaction reasons are all important, with the latter two categories being more important than the income category. Within these three categories, there are eight subcategories which received significant response. Further generalizations are difficult. Bearing in mind the categories are preordained, the responses reflect the individualism of new entrepreneurs and the importance of being able to have control over work, to develop ideas to their fullest potential, to have the ability to balance work and leisure as they see fit, and to be financially rewarded in a way that reflects their true worth. It might also be noted that western-based surveys of new firm founders consistently reveal complex motivations, which include non-economic considerations and the high value placed on independence and 'being one's own boss' (Cross 1981; Stevenson and Sahlman 1989: 100–101). Citing US-based data, for example, Stevenson and Sahlman (1989: 100) give particular emphasis on the degree to which self-employment satisfies personal values.

Entrepreneur's age at formation of business An interesting trend revealed by data collected by MITI on manufacturing firms established since the late 1940s, is that new firm founders in Japan are getting older (Table 9.1). Thus, until 1986, most Japanese new firm founders were in their 30s. Since 1986, over 46% of Japanese entrepreneurs have been in their 40s and since

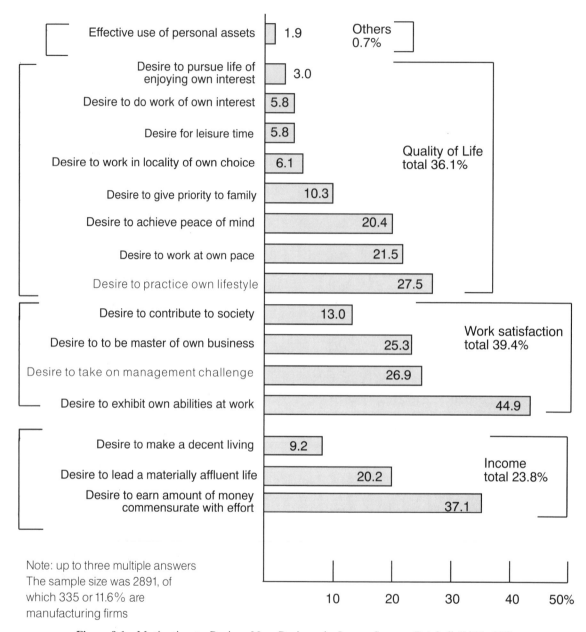

Figure 9.1   Motivation to Begin a New Business in Japan. Source: Patchell (1991: 107)

1986 this age group has been the single most important category. Previously most new firm entrepreneurs had been in their 30s. Moreover, entrepreneurs over 50 years old accounted for about 10% of new firm starts in the 1946–1955 period and over 25% since 1986. Some British

studies (Cross 1981: 209; Lloyd and Mason 1984: 215) also reported that the majority of new firm founders (in the 1970s) were in the 30–45 year age group – data that potentially corresponds with Japanese trends. One set of related ideas that have been put forward to

Table 9.1 Entrepreneur's Age in New Firms in Japan

Period	Age cohort (years)					Average age (years)
	< 30	30–39	40–49	50–59	60+	
1946–55	29.1%	41.6%	19.7%	9.4%	0.9%	35.0
1956–65	22.9%	32.8%	28.2%	12.2%	3.8%	38.0
1966–75	21.5%	33.3%	31.2%	11.3%	2.7%	37.9
1976–85	11.8%	34.1%	29.4%	16.5%	8.2%	41.3
1986–	2.7%	18.1%	46.4%	25.6%	7.2%	46.2

Source: Patchell (1991: 109).

Table 9.2 Pre-establishment Occupation of New Firm Founders in Japan

Period	Pre-establishment occupation					
	None or student	Employee of SME	Manager of SME	Employee of large firm	Manager of large firm	Other
1956–65	4.4	51.5	24.3	2.2	8.0	8.6
1966–75	2.1	45.1	18.8	3.7	5.2	18.8
1976–85	2.3	41.9	22.1	0.0	11.6	22.1
1986–	1.3	40.3	38.3	0.7	9.4	10.1

Source: Patchell (1991: 111).

explain such a trend is that more new firm founders in recent years have required more technological knowledge, greater financial capability and a greater need to have strong connections with the previous workplace before creating new ventures.

Founder's previous occupation In fact, the reasons suggested for the increasing age of new entrepreneurs are supported by changes that have occurred in the previous work histories of entrepreneurs (Table 9.2). In Japan, workers employed by other SFs have traditionally been the dominant source of entrepreneurs and since 1986 they have continued to be the dominant category. Yet, former employees of SFs are a relatively declining source of new firms. Indeed, according to these Japanese data, managers of SFs are now almost as large a source of new firm founders. MITI believes this trend reflects the increasing diversification of SFs and their advances into new manufacturing fields.

As a partner of a small (manufacturing) firm, Richard Madge's previous occupation fits this profile, and neither he nor the Fields were previously employees of small firms. In terms of comprehensive survey-based research there is strong evidence from British regions that by the 1970s the majority of new firm founders previously occupied managerial positions, particularly within the manufacturing sector (Cross 1981: 214; Fothergill and Gudgin 1982: 120; Gould and Keeble 1984: 197).

Relationship of new firms with founder's previous workplace MITI's information also indicates the various ways new firms are related to the founder's previous workplace and the fact that such relationships are becoming more important (Figure 9.2). Thus, since 1966 the percentage of new firms with no links to their founder's previous workplace declined from around 50% to about 28%. As more ex-managers are becoming new firm founders in Japan, perhaps such a trend is to be expected. These relationships are varied. In 1986, for example, the most important forms of 'co-operative' relationships were the receipt of subcontracted orders, loans of personnel, financial help and receipt of some preferred customers. Other types of relationship exist, however, including competitive relations. Bradley and Lori Field's venture is also a good illustration of the process whereby a previous employer becomes a customer, in this case an important one to the development of the firm. Such connections have been documented in other studies.

The seed-bed hypothesis

With respect to the location of new firms, MITI's data on motivations indicate that 'the desire to live in one's own locality' is not of particular significance. Yet the geographical literature has long assumed a strong geographic rationale underlying where new firms are created. This assumption has a strong basis. Various studies in North America and the UK report that between 68 and 90% of new firm founders located in their local area (Taylor 1970; Gudgin 1978: 109; Johnson and Cathcart 1979: 14; Cross 1981: 228; Khan and Hayter 1984: 5; Lloyd and Mason 1984: 214). Bradley and Lori Field's 'decision', apparently made without much thought, to locate in their home town is therefore quite typical. And, even if Robert Madge believes his firm to be a 'global business' (Lynn 1995: 2–4), he also 'automatically' located 'at home'.

The rationale for the tendency of new firm founders to locate in their local or home regions is expressed in terms of the seed-bed hypothesis. As Hill (1954: 185) noted some time ago, a new firm:

> "on the face of it . . . is highly mobile. The capital outlay is probably small, labour demands are not difficult to satisfy, raw materials may be generally available throughout the country at fixed prices . . . such a unit, in fact, more often than not turns out to be completely immobile; if

it is not established where the originator wants it, it will not be established anywhere."

The behavioural rationale for such decisions is the fact that new entrepreneurs are thoroughly familiar with their home locales, and within these locales they are likewise known. Thus, potential entrepreneurs are aware of possible premises in which to locate, know about possible workers or can assume labour will be provided by their families, and they will at least understand the characteristics of local labour. They may well have contacts with local financial institutions and knowledge of local markets; they may know of available equipment and suppliers. The entrepreneur's home provides a ready-made head-office and may even provide space for manufacturing for a while. In other words, local entrepreneurs 'inherit' considerable knowledge about their home environment as part of their birthright. To locate elsewhere would involve all the costs and uncertainties in collecting and understanding information on unfamiliar places. Indeed, for the most part, new founders do not even contemplate moving elsewhere. In this sense, location is not a decision for them. Moreover, new firms face considerable uncertainty in establishing themselves as a business, as was the case with Bradley and Lori Field and Robert Madge. If the Fields and Madge had located their manufacturing plants elsewhere they would have compounded all those uncertainties of starting a new business with the uncertainties of operating in unfamiliar locations.

In countries such as Israel, Canada and Australia, where there have been high levels of immigration, and immigrants are an important source of entrepreneurship, questions can be raised about how new firm founders compensate for the costs and uncertainties of not locating in their seed-bed. One strategy adopted by immigrant entrepreneurs is to live in their new home for a number of years in order to acquire the same level of local geographical know-how as native inhabitants before creating a new firm. In the case of Vancouver metro, a random sample of new firms found that new firm founders who were born outside of Canada lived at least 11

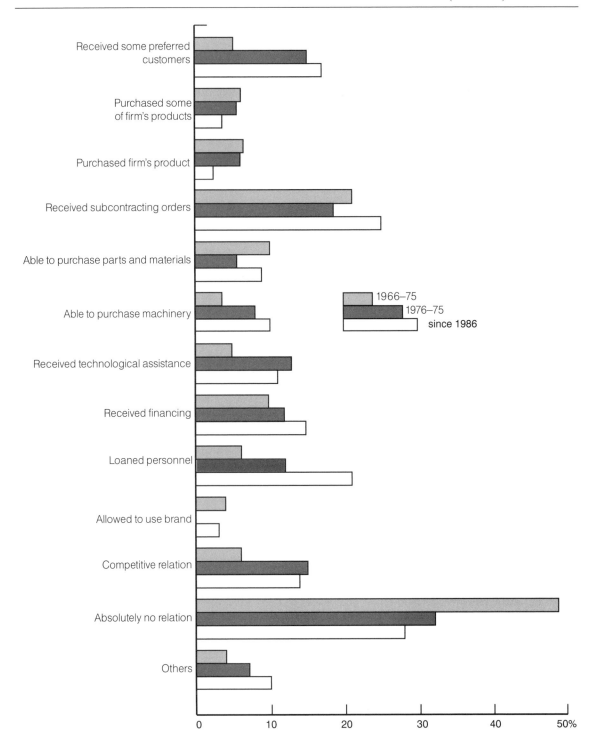

Figure 9.2 Relationship of Newly Established Firm with Founder's Previous Workplace. Source: Patchell (1991: 113)

years here before investing in a manufacturing enterprise (Khan and Hayter 1984). Clearly, such behaviour reinforces the rationale underlying the seed-bed hypothesis. In Canada, however, current immigration policy is favouring immigrants willing to invest on or soon after arrival, which inevitably increases the risks of such investments. In some situations whereby immigrant-entrepreneurs contemplate investment more or less immediately on arrival, some new firm founders can compensate for the strictures of the seed-bed hypothesis by joining established, closely linked cultural groups which serve to nurture their new firms as they try to come to grips with unfamiliar surroundings.

The incubator hypothesis

The idea of the incubator hypothesis has been developed in different contexts. Thus, local ethnic groups that help to nurture immigrant entrepreneurs are incubators; business organizations or educational institutions which spin-off entrepreneurial ventures through the initiatives of former employers are incubators; while the original rationale for the incubator hypothesis in the geographical and planning literature stressed the importance of long-established industrial cores within metropolitan areas as incubators. This last rationale is based on the argument that the rate of new firm formation is positively influenced by the supply of agglomeration economies and was developed by Vernon and Hoover (1959) to help explain the location of industrial activities within the New York metropolitan area. In this view, the incubator hypothesis states that new firms "will be attracted to areas offering services essential to their operation that they, because of their small size and limited resources, would be unable to provide internally" (Struyk and James 1975: 109).

In particular, the original geographical version of the incubator hypothesis argued that the old industrialized sections of metropolitan areas offered new firms with a supply of buildings of various sizes and costs as well as close access to cheap accommodation, suppliers, markets and a variety of ancillary business services such as transportation, legal, consulting, advertising and typing services. By concentrating together in downtown areas, new and small firms create external economies of scale by buying and selling among one another and sharing close access to transportation and storage facilities, as well as wholesalers, which facilitate exports and imports. Such downtown locations also provide access to labour pools, while providing various employee-related services such as public transportation and shops. Considerable support for this argument can be found in the literature. In particular, although there are exceptions (Gudgin 1978: 118), the markets and sources of supply of new and small firms in metropolitan areas do tend to be highly localized (Taylor 1970; Lever 1972; Taylor and Wood 1973; Gilmour 1974; Steed 1976a,b,c; Lloyd and Mason 1984; Peck 1985). In addition, the supply of old buildings in metropolitan areas has often been a great advantage to new firm formation, even if the buildings are dilapidated and breaking various planning regulations. In such cases, costs of acquisition or rental costs have often been low, or even zero whenever new firms have simply 'squatted' in them without the knowledge of owners.

Old industrial cores of metropolitan centres doubtless continue to constitute incubator functions to new firms by providing a supply of agglomeration economies. The extent to which metropolitan centres act as incubators, and whether or not they are superior incubators compared to other locales, is more debatable. One reason is that old industrial cores are disappearing. Over the past 30 years or so, many cities have undertaken extensive urban renewal schemes which have literally replaced existing industrial land uses with other forms of land use, including retail, transportation and residential land uses. In Vancouver, for example, since the late 1960s the False Creek area, once the city's dominant industrial district, has been 'gentrified' into a residential complex, with associated entertainment, retail and personal services, for professionals (Ley 1980, 1986; Ley and Mills 1990). False Creek's transformation was completed by World Expo 1988 which occurred on a nearby site. Admittedly, in the Vancouver metropolitan area, the growing importance of inner city areas as high-income residential zones housing professionals is distinctive in a North American context. In many US cities, it is not a lack of physical space *per se* that has created problems for new manufacturing firms. Rather, it is the growing external diseconomies of scale associated with problems of crime and obsolescent physical and social infrastructures, in association with long-standing trends of suburbanization and non-metropolitan industrialization, that have reduced the attraction of inner cities for new manufacturing firms.

In the UK, the 'constrained location hypothesis' has been put forward to suggest that the shift in manufacturing activity from urban to rural areas results from the lack of physical space available to manufacturing firms within cities (Lever 1972; Fothergill and Gudgin 1982; Keeble 1984). In some cases, as Rowley's (1985) documentation of the decline of cutlery manufacturing in parts of inner Sheffield suggests, industrial space has been reduced by urban renewal schemes that have led to the closure and replacement of factories. By the mid-1980s, however, the de-industrialization of industrial districts in the UK, including Sheffield's main industrial district, had *created* substantial spaces that have facilitated commercial and other revitalization efforts, including plans for re-industrialization, such as the provision of industrial parks which contain buildings to help incubate new firms. Such plans assume that the decline of inner cities as industrial incubators is at least not absolute.

The incubator functions of technology-oriented complexes The relative decline, and in some cases total eclipse, of the 'natural' incubator functions of the industrial districts of inner cities is paralleled by the emergence of incubator functions elsewhere. In general terms, as will be discussed shortly, in advanced countries new firm formation rates over the past several decades have been higher in areas outside of long-established industrial cities, particularly in rural areas (e.g. Davidsson et al 1994: 398). In addition, an important feature of Scott's (1988a) 'new industrial spaces' (e.g. 'third Italy') and high-tech centres or 'technology oriented complexes' (TOCs) such as those in Silicon Valley and Boston 128 in the US, is their supposed attraction to entrepreneurs.

In the case of TOCs, Silicon Valley (Saxenian 1983) and Boston 128 (Malecki 1986; Markusen, et al 1986) are widely regarded as the prototype models. Silicon Valley grew around a formally designated science park created in 1951, and Boston 128 along a highway constructed in the 1950s. Both may be regarded as the result of 'spontaneous' development arising from individual and corporate initiatives which featured the establishment of R&D facilities by major corporations as well as small, specialized operations run by individual entrepreneurs. With respect to the latter, the primary attraction for new firms is the shared access to the large-scale research activities of universities, research institutes and other firms, the large employment talent pool available in adjacent urban agglomerations, various support services and access to venture capital (Malecki 1991: 222; Saxenian 1994). Interestingly, the sources of new firms differ in the two TOCs. While faculty from MIT, and nearby universities, have been important creators of new firms for Boston 128, in Silicon Valley new firms have frequently been spun-off from existing firms, in many cases by former engineering students of Stanford University (Saxenian 1994: 3).

Moreover, in the setting up of new firms in Silicon Valley and Boston 128, venture capital, and local venture capitalists, which function as financial intermediaries providing capital to new firms from such sources as pension funds, corporations, foundations and even families (Florida and Kenney 1988b), have been important. As noted in Chapter 4, the San Francisco–San Jose and the Boston–Hartford regions are second only to New York as locations for venture capital offices (Figure 4.5(a)) and they dominate the distribution of venture capital investments in the US (Figure 4.5(b)). In both Silicon Valley and Boston 128, local venture capitalists invest most, but not all, of their capital locally, while also importing capital from other centres. In these TOCs, local venture capitalists developed gradually to serve the highly specialized and highly uncertain needs of high-tech firms. Typically, venture capital is provided on an equity basis and close proximity has allowed locally based venture capitalists to gain an intimate knowledge of industry in the TOCs and "to identify, monitor, supervise, and assist with [highly uncertain] investments" (Florida and Smith 1993: 441). In Silicon Valley, in particular, local venture capitalists have become 'embedded' in local technology networks and serve to reduce the uncertainties of investments and to compensate for ambiguous information (Florida and Smith 1993).

More generally, while there is evidence from the US and the UK that many small high-tech firms prefer to use profits as a source of investment capital (Oakey 1993: 229), venture capital markets have expanded considerably and more new firms in new and risky high-tech activities are relying on venture capital funds (Kenney 1986; Florida and Kenney 1988a; Malecki 1991: 334–342). In theory, venture capital might be expected to be highly mobile. In practice, the supply of venture capital is both "highly mobile and highly local" (Florida and Smith 1993: 449). On a global scale, the largest and most sophisticated venture capital markets are in the US. This capital is available to non-US firms, as the Richard Madge example reveals (Exhibit 9.2). At the same time, for Madge, the implication of his deal was a US-based partner and a US-based location. Within the US, there are substantial interregional flows of venture capital which compliment the highly localized flows (Florida and Kenney 1988a; Florida and Smith 1993). At the same time, these flows reveal a distinct geographical structure as the major destinations of venture capital are concentrated in relatively few places, including Silicon Valley and Boston 128 (Leinbach and Amhrein 1987; Florida and Smith 1993).

In contrast to the situation in Silicon Valley and Boston 128, peripheral regions face difficulties in accessing capital (Ewers and Wettmann 1980). Shaffer and Pulver (1985), for example, found that new firms in the more remote part of northern Wisconsin and in rural areas experienced 'capital stress', or insufficient capital funds, to a much greater degree than new firms in the more industrialized southern parts of Wisconsin. Indeed, some regions of the US are virtually without venture capital (Leinbach and Amhrein 1987), and in the UK, Thwaites (1982: 378) concluded that firms in peripheral regions experienced greater difficulty in accessing venture capital than firms in the core regions.

New high-tech firms have a distinct, highly uneven geography which in turn contributes to regional variations in the formation of new firms in manufacturing in general.

Regional variations in new firm formation rates

It has long been recognized that the supply of new firms varies among regions. Indeed, this variation is an implication of the traditional geographical incubator hypothesis. From a more explicit comparative perspective, Chinitz (1961) and George (1970), in US and Canadian contexts respectively, emphasized regional variations in new firm formation. Chinitz's classic comparison argued that new firms were far more likely to be created in New York than in Pittsburgh primarily because of variations in supply conditions, notably industrial structure. For Chinitz, Pittsburgh's economy, dominated by heavy industry and a few large corporations, is a structure that does not encourage the start-up of new entrepreneurial firms. New York's economy, on the other hand, enjoys a more diversified composition comprising a large population of small firms which Chinitz saw as providing more appropriate conditions to stimulate a supply of new firms. George was more concerned with the effect rather than the cause of regional variations in entrepreneurship; thus he contrasted the economic performance of Ontario

Table 9.3 Regional New Firm Formation Rates in the UK: Two Measures

Region	Employment in new manufacturing firms per 1000 in manufacturing employment, 1966–75	Net registrations per 1000 of working population, 1981
North	7–12	3.8
Yorkshire	7–12	4.5
East Midlands	14–19	4.6
East Anglia	> 25	5.0
South East	20–25	5.4
South West	7–12	6.2
West Midlands	< 7	4.7
North West	< 7	4.2
Wales	20–25	5.2
Scotland	14–19	3.4
Northern Ireland	20–25	3.9

Sources: Pounce (1981: 59) and Whittingham (1984).

(the 'leader') and Nova Scotia (the 'laggard') and explained the difference primarily in terms of the greater entrepreneurial vitality of the former.

Mapping new firm formation rates

In the UK, there is considerable statistical evidence of regional variations in new firm formation rates (Firn and Swales 1978; Cross 1981; Pounce 1981; Storey 1981, 1982; Gould and Keeble 1984; O'Farrell and Crouchley 1984). By way of illustration, the results from two studies reveal this variation among the UK's major regions for the mid-1960s to the mid-1970s and for 1981 (Table 9.3). In Pounce's (1981) analysis (column 2 in Table 9.3), the manufacturing employment generated in new manufacturing firms (per 1000 total manufacturing employees) in the 1966–1975 period is at least three times greater in the highest ranked region (East Anglia) compared to the lowest ranked regions (the North West and the West Midlands). While regional variations are less in Whittington's (1984) calculation of new firm registrations in *all* sectors in 1981 per 1000 total employees (column 3 in Table 9.3), the leading region (the South West) has almost double the rate of the lowest ranked region (Scotland).

Broadly speaking, the patterns revealed by the two surveys are similar in that new firm formation rates are higher in the growing regions of the South and East, and lower in the more peripheral regions. In Pounce's survey, Wales and Northern Ireland depart from this observation. Other studies at more localized geographical scales tend to reinforce the general pattern. For example, Lloyd and Mason's (1984) study reveals the higher birth rates in South Hampshire, part of the rich South, compared to northern metropolitan areas, thus confirming Storey's (1981) earlier results which noted the low rate of new firm formation in Cleveland, part of the poor North. It might be noted that subregional variations in new firm formation rates during the 1970s are greater than those revealed at the interregional scale (Gould and Keeble 1984).

A recent study has comparatively analysed new firm formation rates in the 1980s for all economic sectors, including manufacturing, for several advanced countries (Reynolds et al 1994). In several European countries and the US, the difference in the birth rates of manufacturing firms among regions is substantial (Table 9.4). Among the European countries, the regions with the highest birth rates generate from 2.7 (Germany) to 6.5 (Sweden) as many new firms

Table 9.4   Manufacturing Firms: Birth Rate for Selected Countries

| Country | Annual firm births per 10 000 population | | | |
	Regional average	Regional minimum	Regional maximum	Maximum/minimum
Germany (1986)	6.8	4.5	12.0	2.7
Ireland (1980–1990)[a]	22.3	10.7	42.7	4.0
Italy (1987–1991)	26.8	12.7	51.0	4.0
UK (1980–1990)[a]	27.5	10.0	59.5	6.0
Sweden (1985–1989)[b]	10.3	4.4	28.7	6.5
US (1986–1988)[a]	16.8	2.4	114.0	47.5

Source: Reynolds et al (1994: 449).
[a] Manufacturing workers used as denominator.
[b] Population 16–64 used as denominator.

(per 10 000 population or 10 000 employees) as the regions with the lowest rates. In the US, the variation is many times larger. While there are differences in time periods and methods of calculation, these data also suggest that there are national variations in birth rates. Although any comparisons have to be cautiously stated, the relatively high values for the UK are unexpected, given the relatively lesser role small firms have played in the UK in the past. Several observers do suggest that birth rates of firms in the UK are increasing although, as of the late 1980s, small firms were still relatively less important in the UK economy than in, for example, Germany (Table 8.1).

In the specific case of the UK, regional variations in new manufacturing ('production') firm formation rates between 1980 and 1990 generally confirm previous studies (e.g. Pounce 1981; Storey 1982) and support Martin's (1988) two-nation hypothesis (Figure 9.3). Thus, regions in south-east and south-west England and East Anglia are all in the highest three birth-rate categories, and most of the top two categories are located in these regions. On the other hand, the most peripheral regions of northern England, most of Scotland and Northern Ireland are in the lowest two categories. The main exceptions to this overall pattern are found in the rural areas of mid-Wales and northern Scotland. As Storey (1981) points out, the regions with the longest traditions of government assistance for industry

perform the worst in terms of new firm formation. In the case of Sweden, there are also variations in new manufacturing firm formation rates (Figure 9.4). The authors claim that the highest rates are located in regions that are "with few exceptions, small in terms of population", while the populous centres of administration and education tend to have lower rates of new firm formation in manufacturing (Davidsson et al 1994: 398). These observations, however, are *not* supported by their statistical analysis (Davidsson et al 1994: 401).

Statistical explanations   Various attempts have been made to statistically explain regional variations in new firm formation rates in terms of variables (or 'indicators') representing various regional characteristics, such as population, demand and employment conditions, and urban industrial structure. Storey (1982), for example, suggested that new firm formation rates correlated with six main indicators, namely the percentage of small firms, the percentage of the population that are managers, the percentage of the population with degrees, savings per head of population, the percentage of owner-occupied houses, the percentage of the workforce in low-entry barrier industries, and the regional income distribution. Recently, Reynolds et al (1994) have summarized the results of surveys in six countries which have sought to (statistically) associate

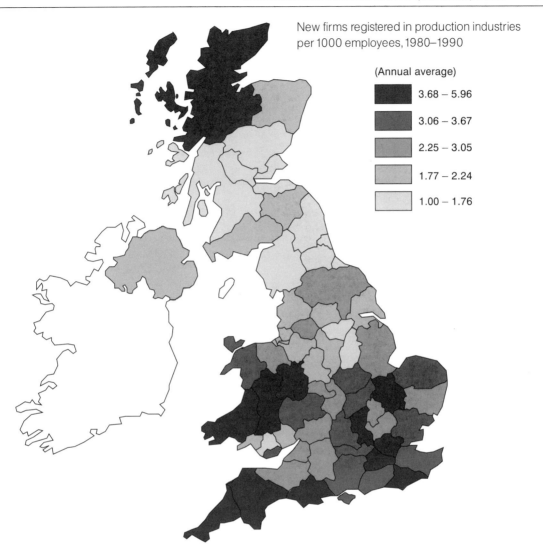

New firms registered in production industries
per 1000 employees, 1980–1990

(Annual average)

■	3.68 – 5.96
■	3.06 – 3.67
■	2.25 – 3.05
■	1.77 – 2.24
■	1.00 – 1.76

Figure 9.3 UK: Spatial Variations in New Production Firm Formation Rates, 1980–1990. Source: Keeble and Walker (1994: 417)

birth rates among manufacturing firms with indicators of regional characteristics (Table 9.5; see also Table 9.4). This study, it might be noted, also reported on non-manufacturing firms.

There are two immediate, general observations that can be made regarding these results. First, the variables employed in the six surveys account

for between 37% (Sweden) and 76% (Ireland) of the variance in new firm formation rates. In the 'best' Irish case, Hart and Gudgin (1994) found that the main determinants of new firm formation at the regional level are the proportion of small firms in the area, growth in local industrial demand, the proportion of professional and

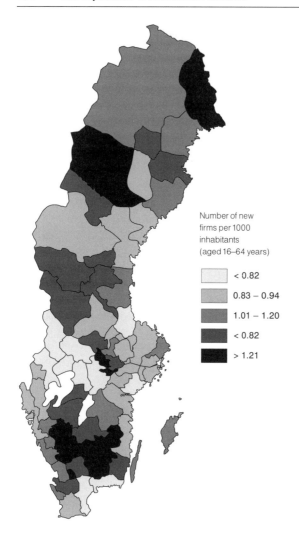

Number of new
firms per 1000
inhabitants
(aged 16–64 years)

▢	< 0.82
▨	0.83 – 0.94
▩	1.01 – 1.20
▤	< 0.82
■	> 1.21

Figure 9.4 Sweden: Spatial Variations in New Manufacturing Firm Formation Rates 1985–1989. Source: Davidsson et al (1994: 399)

managerial occupations in the population, and the impact of government policy. On the other hand, in four of the six national cases, at least 38% of the variance is statistically unexplained. Possibly, the variables used do not adequately capture the regional characteristics they repre-

sent; there may be other characteristics and factors that are important but have not been included; the regions chosen may be internally highly differentiated; and there may be some peculiarities in the data on new firm formation in the particular time periods chosen.

With respect to potentially important factors not included, the statistical models outlined in these surveys assume that regional variations in new firm formation rates can be explained solely by internal regional characteristics. Yet the surveys themselves note that new firm formation rates vary internationally so there is the possibility that factors operating at the international scale are ignored as countries are implicitly treated as closed systems. Also, not all relevant regional characteristics may have been included. Reynolds et al (1994), for example, found that regions in the US with more flexible labour relations (for all industries, not just manufacturing) were more likely to spawn new firms. It is also worth noting that the maps of new firm formation often reveal that isolated regions with small populations, such as northern Scotland (Figure 9.3) and northern Sweden (Figure 9.4), have high rates of new firm formation. In absolute terms, however, the number of new firms in such regions is likely to be quite small; the vast majority of new firms are elsewhere. In addition, there may be intangible considerations that are hard to quantify but which have some bearing on new firm formation rates. In this context, for example, Karlsson et al (1993: 8) raise the possibility that within European countries at least, there are variations in 'entrepreneurial spirit', which they in turn relate to the extent of networking by new and small firms with other actors in local economies (Johannisson 1986; Birley et al 1991; Curran and Blackburn 1994).

Second, among the variables that are important, one stands out above all the others; namely, the positive association between the existing proportion of small firms in a region and rates of new firm formation. Indeed, it is the only indicator that is important in all six countries. As such, these results confirm the insights of

Table 9.5 Regional Characteristics and Manufacturing Firm Birth Rates: Generalized Statistical Relationships

	Firm birth/year/10 000 population					
	Germany	Ireland	Italy	Sweden	UK	USA
Demand growth						
In-migration/population growth	0	(−)	(+)	+	+	+
Growth in gross domestic product	0	+	0	0	0	−
Urbanization/agglomeration						
Percentage 25–44 years old	0	NI	0	NI	NI	−
Population density	+	−	(+)	0	+	−
Percentage secondary housing	NI	NI	NI	NI	NI	NI
Percentage managers in workforce	0	+	0	0	+	+
Percentage with higher education	NI	−	0	0	0	0
Unemployment						
Unemployment level	−	0	(−)	0	0	0
Change in unemployment	0	0	0	0	0	−
Personal, household wealth						
Household income	NI	NI	0	NI	0	NI
Percentage owner-occupied dwellings	0	0	0	0	0	0
Dwelling prices	NI	0	0	NI	0	+
Land prices	0	NI	NI	NI	NI	NI
Small firms/economic specialization						
Proportion of autonomous workers	NI	NI	(+)	NI	NI	NI
Proportion of small firms	+	+	+	+	+	+
Industry specialization index	0	0	+	NI	NI	0
Political ethos						
Socialist voters	NI	NI	0	+	(−)	NI
Government spending/policies						
Local government expenditures	0	NI	NI	NI	0	0
Government assistance programmes	NI	+	0	0	0	NI
Number of regions	74	27	84	80	64	382
Variance explained – adjusted (%)	51	76	58	37	70	62

Source: Reynolds et al (1994: 451).
Notes: + = statistically positive influence, ( ) marginal significance.
– = statistically significant negative influence, ( ) marginal significance.
0 = measure included, not statistically significant.
NI = no measure included in the analysis.

Chinitz's (1961) study. Although of lesser importance, the demand growth characteristics, especially the in-migration/population growth variable, may be noted. In three countries, this variable is significant and in another it is marginally significant. That is, although a large number of characteristics are thought to be relevant to understanding spatial variations in new firm formation rates, the industrial structure and dynamism of regions are the most noteworthy (Gudgin and Fothergill 1984; Mason 1989).

The effects of industry structure

At a regional scale, since Chinitz (1961), a number of studies have argued that variations in industrial structure, especially as measured by plant size distributions, are associated with variations in new firm formation rates (Cross

1981: 221; Storey 1981; Fothergill and Gudgin 1982). In this context, the main argument is that regions dominated by large-scale plants are less conducive to new firm formation than those dominated by small firms. Simply put, this hypothesis says that small firms begat small firms, while large plants replace them. There is a related argument that regional industrial structures dominated by large foreign or externally controlled branch plants likewise dampens new firm formation (Pred 1974; Firn 1975; Britton and Gilmour 1978; Thwaites 1978; Britton 1981).

The basis for the view that large plants and firms are less likely to incubate new firms compared to small firms rests on differences in employee experience and labour relations. Thus, in a large plant it is argued that employee experience is more likely to be strictly specialized and even demarcated. Until recently, large plants have also offered greater employment security and much more attractive non-wage benefits in the form of pension schemes, vacation rights and pay, and medical insurance related benefits. Such characteristics are likely to be particularly strongly felt in union plants where there is additional 'peer' pressure on workers to conform to union standards. In contrast, a small firm potentially offers much more relevant experience to employees thinking about starting a business. Thus, employees in small firms are more likely to perform a wide variety of tasks; they will be familiar with the owner and manager and be in a position to watch, and learn from, their activities. In other words, there is greater opportunity for small firm employees to become relatively knowledgeable about a firm's total business. Moreover, job security in small firms is likely to be less than in large plants and employees are likely to be less inhibited about leaving because of accumulated benefits (e.g. pension benefits) and because of union seniority.

An additional argument, consistent with this view, is that large plants that are foreign or externally controlled are highly likely to rely on the head-office for various services and on affiliated plants for various parts and processes. This argument, has been expressed in British

(Firn 1975; Thwaites 1978), Canadian and US (Pred 1974) contexts. It needs to be emphasized that this thesis emphasizes that regions *dominated* by a population of plants, especially if externally controlled, will be less likely, *on the whole,* to generate the same per capita levels of new firms than a region *dominated* by a population of small firms. To the extent that branch plants automatically rely on their head-offices to provide them with services and affiliated factories for component parts, the potential for making demands for these same goods and services from locally based small firms is accordingly reduced. Moreover, since foreign-owned branch plants are, on average, typically larger than domestic plants, the effects of large plant size and ownership combine to dampen locally based entrepreneurial activity. Such demands are likely to be particularly dampened in the secondary manufacturing sector if branch plants are producing goods in the late stages of the product life cycle (Figure 4.6). In this stage, workers are relatively unskilled and management functions relatively limited to labour supervision and machine maintenance. In recent years, there has been particular concern that high levels of foreign ownership undermine the potential for high-tech new and small firms (Britton 1991; MacPherson 1994) and for business services, particularly high-order business services which are concentrated in major metropolitan 'control' centres.

It is important to bear in mind that the relationships between plant size, ownership and new firm formation need to be explored in particular contexts. Two broad points can be made to underline this statement. First, as noted in Table 8.1, the size distribution of firms varies internationally and, the stronger role played by small and medium-size firms in the Japanese and German economies than those in the American and British economies directly reflects the stronger tradition of large firms and plants in the former economies to spawn and co-operate with new and small firms. Second, changes occur over time. Indeed, in the contemporary period of restructuring, the creation of leaner and flatter organizations is a significant theme, perhaps

Table 9.6 Paulcan Enterprises, Chemainus: Source of Plant and Equipment, 1985–1994

Item	Source
Main planer	MacMillan Bloedel, Chemainus (sawmill rebuilt)
Small planer	Fletcher Challenge, Youbou (sawmill rationalized)
Lumber sorter	Northcoast Lumber, Coquitlam (sawmill closed)
Strapper	MacMillan Bloedel, Nanaimo (sawmill rebuilt)
Beams/sawmill	City of Vancouver, Lion's Gate Bridge (maintenance)
Trimmer set	Champion Lumber, Libby, Montana (sawmill closed)
Beams/trusses	Canadian Pacific, Gold River, Tahsis cedar mill
Roof/main mill	Mayo Lumber, Nanaimo (sawmill rebuilt)
Boilers for kilns	Alberta Liquor Control Board, Edmonton
Kilns	Welland, Marysville, California
Tilboy system	Vancouver

Source: Author's research files.

especially in North American and British contexts, and there has clearly been significant re-thinking about the relative roles of large and small firms in favour of the latter (Storey 1994). Contemporary corporate restructuring, for example, often involves decisions to contract out ('vertically disintegrate') work formerly done in-house. Lyons (1993), for example, reveals evidence that in the engineering sector of the UK there has been a net tendency for firms to increase subcontracting; while one-third decreased subcontracting, over half the surveyed firms increased subcontracting, during the late 1980s. In addition, contemporary corporate restructuring has featured the permanent lay-off of managers as well as workers, thus creating a potential pool of entrepreneurs. In a related development, management buy-outs of firms in the UK increased rapidly during the 1980s (Wright et al 1990). Both these processes can operate together.

An interesting case in point relates to the Chemainus sawmill in British Columbia (Barnes et al 1990; Barnes and Hayter 1992). MacMillan Bloedel, the parent company, closed its old sawmill at Chemainus and built a new one, reducing its payroll from over 700 to 125 managers and workers. However, the new mill does not plane or kiln dry the sawn wood. These activities are now contracted out, partly to a new firm, Paulcan (Table 9.6), which was set up by a

former manager of MacMillan Bloedel who also hired several former employees and purchased the planer from the old Chemainus mill (see Chapter 12). The extent to which vertical disintegration and lay-offs are contributing to new firm formation in an aggregate sense, however, is not clear, in British Columbia or elsewhere. There is also the question of the geographical implications of vertical disintegration. While in the case of wood processing in Chemainus, vertical disintegration has encouraged a localized concentration of activity, trends elsewhere are more varied (Storper and Christopherson 1987; Milne 1991). This issue, which is complicated by the fact that vertical disintegration as well as vertical integration are reversible processes, will be addressed in Part IV.

The effects of regional dynamism

In addition to the effects of industrial structure, there has long been the suggestion that regional dynamism underlies variations in new firm formation rates (George 1970; Firn and Swales 1978). In this context, regional dynamism refers to absolute increases in levels of economic activity and regional income and/or per capita increases in regional income. In this view, those regions which sustain growth over a long period of time are more likely to generate higher rates of new firm formation than slow growth or stagnating regions. Given that causes and effects between regional dynamism and new firm formation are likely to be interrelated, this argument suggests that sustained growth encourages entrepreneurship because it implies that threshold levels for an increasing number of economic activities are passed; economies are more diversified in terms of skills and occupations, which provides more opportunities for new firms; bigger populations in turn increase the chances for innovative behaviour; growth is likely to fuel its own spirit of speculation and optimism; and growth typically implies that net in-migration and interregional migration, at least within advanced countries, is a selective process emphasizing better-educated, higher-skilled and higher-income individuals. Conversely, slow growth or stagnating regions might be expected to reveal the reverse of these effects.

The influence of regional dynamism, which captures the effects of a number of factors on new firm formation rates, is strongly supported by a comprehensive statistical analysis of firm births across the industrial spectrum in 382 labour market regions in the US during the highly volatile early 1980s (Reynolds et al 1993). Thus, this study determined that the single most important ('major') influence on new firm formation (and for branch plant formation) in all (manufacturing and non-manufacturing) industries is local economic diversity, and that population growth and greater personal wealth revealed 'strong' positive associations. Interestingly, the previously noted surveys of new firm formation in manufacturing, and for different time periods, found the extent to which regions are specialized (and thereby diversified) to be unimportant (Tables 9.5 and 9.6). The surveys agree, however, on the influence of population growth and migration. In this regard, perhaps more attention needs to be paid to the effects of migration. Storey's (1982) index of entrepreneurship, for example, stresses that the employment of professionals is positively correlated with new firm formation and such individuals are often important components of economically motivated migration streams. More generally, in-migration potentially diversifies the range of skills, ideas, contacts and investment opportunities in a region. Out-migration, on the other hand, has a narrowing effect.

The 'obvious' relationship between regional dynamism and entrepreneurship becomes more complicated when consideration is given to business cycle effects. Thus, the business cycle may encourage 'perverse' effects. The 'desperation' (or 'recession push') hypothesis suggests, for example, that new firm formation rates increase during recessions as people who are laid-off are encouraged to invest in their own businesses (S Johnson 1991: 90–94). While some support for this hypothesis has been found in the UK (Harrison and Hart 1983; Hamilton 1989; S Johnson 1991), although not consistently (Table 9.5), in the US, unemployment during business downturns across a wide spectrum of industries is strongly negatively correlated with new firm formation (Reynolds et al 1994: 100). Rather, the Reynold's et al (1994) study found economic diversity, wealth and population growth, along with the nature of labour relations, to be the most important variables during recession and subsequent recovery. If lay-offs are temporary or perceived to be temporary, however, and a social assistance programme is in effect, then this relationship may not be important. It could also be supposed that during an unusually rapid growth cycle, the leading firms and sectors may be in a position to pay high wages, salaries and prices and thus 'crowd out' smaller firms and pre-empt new firm formation, at least for a time. Bourque (1971), noted that between 1965 and 1968, the Washington State economy experienced a boom, led by the growth of the Boeing Company, one result of which was that the economy became less rather than more diversified. Although Bourque did not assess new firm formation, the implication of his analysis is that during this particular period, entrepreneurial opportunities declined. Of course, in such a situation eventually in-migration might be expected to help off-set the effects of local bottlenecks. Another slant on this issue is provided by Vianen (1994) who found that during the 1980s, SMEs in the Netherlands were more stable than large firms during recessions as they were willing to accept lower labour productivity during periods of declining demand, i.e. they were willing to hoard labour.

New firms: the challenge of survival

It has been recognized for some time that new firms experience the highest death rates among firms of all ages, and that death rates are high (Steindl 1947). Recently, Bates (1995) has noted that among American firms formed between 1984 and 1987, 35% of franchises and 28% of non-franchises had failed by 1991. Across a variety of US sectors, including manufacturing, evidence from the 11 year period between 1976 and 1986, showed that over half of new firms failed to survive, a pattern similar to UK experience (Storey 1994: 95). From this perspective, Pacific Emergency (Exhibit 9.1), Madge Networks (Exhibit 9.2) and Paulcan (Table 9.6) are already exceptional in that they have all overcome the critical challenge to survival facing new firms in their first few years. In fact, the owner of Paulcan had to delay start-up until he had cleared bankruptcy problems from an earlier venture.

As Marshall (1922: 315) intimated, the problems facing new firms can be usefully reviewed within the context of a life cycle model of small firms (Figure 9.5). This model portrays a (successful) firm moving through five stages. In stage 1 (Introduction), new products are introduced, at which time the firm encounters consumer ignorance and possibly resistance. Sales are typically low, growth is slow for a while, and profitability is hard to achieve. In stage 2 (Take-off), the product becomes accepted, sales grow and profitability is achieved, although the higher the profits, the greater the possibility of competition. In stage 3 (Slow Down), growth typically slackens for a combination of reasons such as saturation of demand, competition and attainment of a size that the owner is 'comfortable' with – the so-called 'comfort level'. Profits might slip. In stage 4 (Maturity), sales are static and may simply be replacement sales. There is a

need for product differentiation. In stage 5 (Decline), in the absence of product innovation, decline may set in and the firm may die.

This model is clearly an 'idealization' and in practice the life-cycle characteristics of real firms vary considerably. As noted, many new firms never 'take-off'. Many new firms quickly reach a comfort level of say one employee and survive modestly for decades without noticeable further change. Others, such as Pacific Emergency, Madge Networks and Paulcan, grow rapidly after take-off and are able to innovate new products and processes. At this point, there is no indication that Pacific Emergency, Madge Networks or Paulcan have reached their 'comfort level'. We do know, however, that very few firms grow to a large size. Given the wide range of trajectories that new firms follow, an important lesson of life-cycle models in this context is the recognition that the nature of the business challenges facing new firms are difficult and they can evolve rapidly.

Organizational metamorphosis and new firms The thesis that organizational structures are adjusted to evolving organizational strategies was developed in the context of the long-term growth of firms, from the very small to the very large (Figure 7.1). But new firms often face significant, interrelated problems of strategy formulation and organizational adjustment in a short period of time. The first two stages of the life-cycle model, for example, raise a considerable number of problems facing new firms (Burns 1989: 43; Storey 1994: 121). According to this model, in the first phase, the Introduction Stage, new firms are preoccupied with survival: in terms of management, owners do everything including direct supervision of employees; the marketing problem is that of accessing a sufficient number of customers and developing a market niche; the accounting problem is one of establishing a cash flow; while the financial problem is typically overcome by use of personal funds and borrowed money from friends, relatives and perhaps banks. Once survival is ensured, the policy issues facing new firm founders change. In particular,

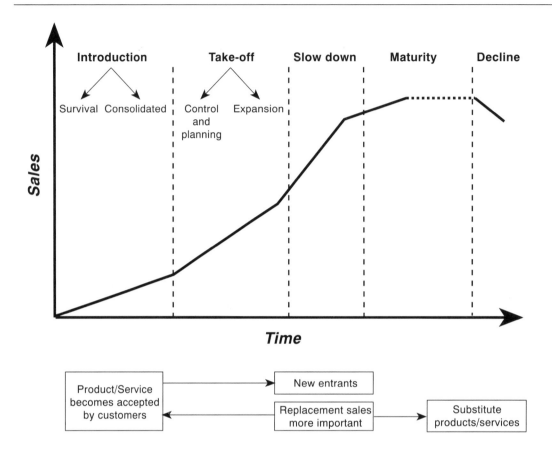

Figure 9.5 The Life Cycle Curve for Successful New Firms. Source: Burns (1989: 41)

emphasis shifts towards consolidation. Similarly, if the Take-off Stage is reached the nature of the management, marketing, accounting and financial problems continue to evolve. Thus, the firm has to develop proper procedures for control and planning, and solve the problems necessary for rapid expansion. In the case of marketing, the firm has to access more consumers on a more permanent basis. During take-off, a more systematic approach to marketing (e.g. administration, files, cost allocation, and hiring of sales people) is needed, while expansion requires a strategy to reach new customers (e.g. advertising, identification of customers, personal contact with new markets, circulars, and innovation of new products).

Employment practices and relations, which interestingly are not incorporated in Burns' (1989) model, can also change rapidly for a firm such as Pacific Emergency which within 10 years of start-up expanded their labour force from just themselves to 30 people. Such an expansion involves decisions to hire, train, supervise, establish pay scales, hours of work, and non-wage benefits, while the owners have no doubt had to handle a variety of personnel matters in the absence of a formal grievance procedure. Moreover, as employment expands, an increasing amount of paperwork and record-keeping is required in order to meet tax and business laws and to meet medical and insurance requirements. Continued growth also requires increasing delegation of supervisory responsibilities and there may come a time when the workforce wish to unionize, which requires an entirely different system of labour relations for the firm.

New firms therefore face a wide range of critical problems, each of which can change substantially as the firm expands. Typically, new firms have to face these problems with little prior experience. Regardless of market forces, new firms face considerable uncertainty, which helps explain the high death rates among such firms. The nature of the challenges facing new firms also provides insights into the concept of 'comfort level' which essentially defines the size of operation that is optimal from the perspective of the owners. For the majority of firms, this size is very small.

New (and small) firms in local development

New firms, considered as a supply of entrepreneurship, are an index of the health of regional economies. Indeed, there is a growing recognition of the distinctive contributions that new (and small) firms play in local development. Various contributions can be identified.

First, it is now widely argued that even in countries such as the US and UK, where large firms have dominated, new and small firms have increased in importance as job creators. Especially since Birch's (1979) US-based analysis, a number of studies have emphasized the growing role of new and small firms as a supplier of jobs in a variety of contexts in Ireland (O'Farrell 1986), the UK (Storey and Johnson 1987; Storey 1994), the US (Reynolds and Maki 1991) and Sweden (Davidsson et al 1994). In practice, it is difficult to sort out the contributions, net or gross, of new firms or of small firms to employment change in a region over a period of time. Storey's (1994: 168–173) recent review of evidence from the US and the UK suggests that from 1985 to 1990 firms employing between 5

and 20 workers created relatively more jobs than firms employing more than 500 workers. The contributions made by new firms (births), expansions and contractions within different size categories of firms varied according to time period and country. Thus the formation of new, small firms was far more important to (net and gross) job creation in the UK than in the US in the late 1980s. Even so, in the US small firms (5–19 employees) increased employment levels by 18.3% between 1988 and 1990, while large firms (> 500 employees) declined.

Second, according to Fredriksson and Lindmark (1979), an appreciation of the relative contributions of small versus large firms begins with the recognition of the differing abilities of the two types of firms, first, in utilizing the internal human and capital resources available to them and, second, in influencing external organizations. Large firms, in this view, are particularly effective in influencing rivals, consumers, suppliers and governments. Entrepreneurially run firms, on the other hand, have high potentials for fully utilizing their own resources in terms of their ability to effectively use available labour, space, management and equipment. The ability to make decisions quickly is another example of such potential.

Third, new and small firms are said to increase the capital:output ratio to the extent that new firms utilize the existing stock of 'old' machinery and buildings. In this regard, the tradition of new firms using available space in old industrialized areas or even in the home of founders, as was the case with Pacific Emergency (Exhibit 9.1), is well known. New firms are also very effective scavengers of old machinery. A good example is Paulcan Enterprises which, as noted above, began business as a subcontractor for a nearby sawmill at Chemainus owned by a forest product giant, MacMillan Bloedel (Table 9.6). Its principal equipment comprises several kiln dryers, a planer and a trimmer, and these pieces of equipment were all purchased second-hand, mostly at auctions. Indeed, most of the firm's equipment (including such relatively small items as welder's pipes), the firm's roof and some of its booms were purchased in this manner. In fact, the only new technology are the firm's computers. For virtually everything else the firm is constantly scanning for auctions and it is prepared to travel as far west as Manitoba and as far south as California to fetch equipment, even if this means hiring labour and trucks. The reason behind this is cost. In August of 1994, for example, Paulcan had just bought a trimmer at an auction in Libby, Montana, after a big lumber mill had closed. The cost of the second-hand trimmer, including delivery, was about Can$50 000, while a new one made a few kilometres away would have cost Can$450 000. Apparently, the firm usually buys second-hand items at around 10% of the purchase price.

Fourth, new and small firms are believed to play distinctive roles in the innovative process (Oakey 1984; Britton 1989a; MacPherson 1994). Indeed, some studies suggest that the innovativeness of small firms is superior to large firms. Pavitt and Townsend (1987), for example, claim that in the UK, firms with less than 1000 employees account for 3% of R&D expenditure but one-third of significant innovations. Admittedly, 1000 employees is an unusually high limit for defining small firms and there is evidence that small firms of less than 200 employees, in part because of the expense involved, have been poor adopters of new technology such as CAD-CAM systems (Kelly and Brooks 1988; Schoenberger 1989). Indeed, the scavenging activities of firms like Paulcan are consistent with this view. At the same time, Paulcan has an ability to integrate new computerized control systems with old equipment in a manner that would not be contemplated by most large firms. Although the majority of new and small firms do not engage in 'significant' innovations, Britton (1989a,b, 1991, 1994) has emphasized their role in incremental innovations which occur as more or less continuous, minor improvements to production by managers, engineers and workers (MacPherson 1987, 1988, 1994). Although not dramatic, cumulative impacts of incremental technological change can result in impressive productivity improvements over periods of time (Freeman

and Perez 1988). Indeed, by virtue of their specialization and narrow focus, small firms are a particularly important source of incremental innovations, the benefits of which can be passed on to large firms through subcontracting linkages (Patchell 1993a,b).

Fifth, it is widely argued that new and small firms enhance local multiplier processes by their proclivity to purchase local supplies of inputs of goods and services. Indeed, the direct implication of the seed-bed hypothesis is that new firms will rely on highly localized suppliers (Table 9.7). Thus, over half of a small sample of new manufacturing firms located in the Vancouver metropolitan area purchased at least 75% of their inputs from this area, and the main exceptions were provided by immigrant entrepreneurs.

Sixth, the marketing linkages of new and small firms are often local (Table 9.7) and as such these firms play an important role in improving levels of self-sufficiency in a local economy, reducing reliance on imports or at least acting as a relatively cheap form of learning as to whether or not the particular good they are manufacturing is actually viable within a local economy. In this regard, high death rates among new and small firms is not simply a waste of resources as the entrepreneur and locality can learn from such disappointing experiences.

Seventh, in a more general way, new and small firms contribute to local development because they imply local control which enhances a region's potential to determine its own future and realize its own development potential. Local control of an economy may also have implications for the cultural, social and educational vitality of a region. To the extent that the local control of an economy is important for political and social, as well as economic reasons, new and small firms are critically important. As noted, new and small firms are not inevitably 'entrepreneurial' in all characteristics and some may be classified as laggards (Figure 8.3). In addition, the opportunism of entrepreneurs may generate socially undesirable outcomes in the form of hyper-competition, labour exploitation and

Table 9.7   Material Linkages of New Firms at Start-up in the Vancouver Metropolitan Area in the Early 1970s

| Percentage distribution | Location of inputs and markets | | | | | | | |
| | Vancouver Metro | | Rest of BC | | Canada | | Imports | |
	Inputs	Markets	Inputs	Markets	Inputs	Markets	Inputs	Markets
0%	12	0	33	25	25	25	27	32
1–25%	3	3	3	11	6	11	6	6
26–75%	3	14	1	4	6	4	2	2
76–100%	22	23	3	0	3	0	5	0

Source: Khan and Hayter (1984).
Note: Markets for rest of BC and Canada are the same. Total size of sample = 40.

environmental degradation. Any implications of new firms for local development does of course depend on their establishment and survival.

Small firms and regional development policy

With increasing momentum since the early 1970s, new firm formation, and the small firm sector in general, has acquired a higher profile in regional and local development polices in many countries. In the US, UK and Canada, as well as elsewhere, stimulation programmes in support of small firms have generally emerged out of the ashes of more traditional forms of industrial stimulation programmes that favoured the attraction of large-scale branch plant operations. In several cases, long-established designated regions such as northern England, central Scotland and Atlantic Canada that since the 1950s had attracted branch plant investments, were found to be deficient in the supply of new firms (George 1970; Firn 1975; Storey 1981). Bearing in mind that, traditionally, industrial incentives were designed to change the locational preferences of branch plant investors, these policies encouraged industrial structures, particularly with respect to large size and external ownership, now considered inimical to the creation of new firms (Storey 1981; Hayter 1982).

In contrast to traditional forms of industrial incentives which seek to change location preferences, policies to stimulate small firms generally take location as given, and generally assume the actual or potential presence of locally based entrepreneurship. Broadly speaking, the goals of small firm stimulation programmes have been to encourage the creation of new firms and the consolidation and expansion of small firms, and often applied to non-manufacturing as well as manufacturing industries. The policies themselves are highly varied and in different forms and mixes have provided financing, information and advice, services, access to foreign markets, and infrastructure, while in recent years there has been growing concern for promoting networks of small firms, especially high-tech firms (Britton 1991; Oakey 1991; Johannisson 1993; Storey 1994: 253–303). One spatial strategy that has been widely attempted is the idea of science parks which seek to help incubate high-tech firms (Oakey 1981, 1985; Amirahmadi and Saff 1993).

In practice, many early policies in support of new firms addressed financial problems and were justified in terms of a so-called 'financial gap' theory. According to this theory, new firms have distinctive financial problems, notably a lack of capital at start-up, as was the case with Fields' and Madge's firms; and then a lack of capital to finance growth, e.g. in moving from a spare room into rental premises with equipment and then into a wholly owned factory. Subsequently, the firm may more easily participate in the capital market or sell itself to a bigger firm. From this theory's perspective, mainstream banks are highly conservative in giving loans to new firms, perhaps rightly so given the track record of such firms and their high death rates. Banks consider a new / small firm a potential customer if its accounts reveal high profitability, high liquidity and a low gearing ratio. Such characteristics may be particularly hard to demonstrate for a new firm which does not have a track record, while banks also have difficulty with applications that reveal high profitability but low liquidity.

Thus, financial support for new firms is justifiable to overcome a financial gap that is only temporary and because the banks are too conservative to help. Such policies, whether in the form of grants, interest-free loans or low-interest loans, are controversial. The counter argument, in part, is that financial markets are there and any financial gap reflects a lack of knowledge and / or financial competence. Another concern is that financial help is limited to new firm owners and does not extend to employees who are also tax-payers. In addition, there is the question of the efficacy of large public sector bureaucracies trying to arrange very small loans to very small private sector firms. The actual administration of financial packages often faces difficulties simply because the emphasis of bureaucracies on procedure, qualifications and forms is often inconsistent with the *modus operandi* of individual entrepreneurs. In any case, many new firms die.

In addition to financing, government policies to help new firms have become more varied and geared to providing a range of advisory and support services, in addition to financial help, on how to run a small business. In many countries, an important theme of these policies is to provide assistance to small firms in penetrating export markets (Bilkley 1978; LeHeron 1980; Hayter 1986a,b; Erickson 1989). More recently, especially in the context of high-tech firms, there have been pleas for policies encouraging richer and more varied small firm information networks. As Britton (1991: 201) notes, information access can be a problem to small firms located in information-rich metropolitan areas, such as Toronto, even if the problems are greater for firms located in peripheral areas. Thus many small firms in

Toronto lacked the information necessary to effectively innovate and "given the social importance of the innovation process some form of policy is required to assist firms overcome their search and linkage problems" (Britton 1991: 201). Britton's suggestion is for a government-sponsored agency which would act as an intermediary linking firms to markets and appropriate sources of knowledge. A related idea that has already been widely implemented is the establishment of centres specifically aimed at incubating new, preferably high-tech firms.

Planning on incubators

Over the past decade or so, regional and community policy-makers have actively sought to facilitate the incubation of new firms. Industrial parks and estates have, of course, been a long-established feature of regional development policy (Bale 1977; Barr 1983). From the 1950s to the 1970s, however, the main purpose of industrial parks / estates was to provide the physical infrastructure to attract large-scale branch plants of already established firms. While planners often hoped that branch plant investments would encourage linked developments, such linkages were not an explicit thrust of policy. With the shift in regional and local development policy towards small firms, however, efforts have been made to design industrial parks which hopefully will act as incubators to new high-tech firms (Oakey 1985). Thus, in addition to factory space, contemporary industrial parks frequently contain incubator buildings which offer various services such as typing, faxing, computing, seminar rooms, video equipment, business advice centres and such social services as restaurants and day care, which can be shared among new firms. These services may be provided free or at low cost, at least for a particular time period to help firms overcome the difficult period of start-up. In some cases, new firms can pay for such services by providing landlords with equity shares (Malecki 1991: 348).

Incubator policies and parks have been mandated to attract any kind of firm. Since the

Table 9.8 Number of Science Parks: Selected Countries

US	285
UK	65
France	43
Canada	38
Japan	29
Australia	22

Source: Fusi (1991: 610).

early 1980s, however, there has been an extremely rapid growth in TOCs (also called innovation centres and science parks), an important rationale for which is the impetus they provide for new firm formation in high-tech activities (Amirahmadi and Saff 1993: 108). By 1990, for example, the US alone had 285 such parks, including 31 in California, 22 in Florida and 17 in Michigan (Table 9.8). Many of these parks offer some form of incubator help, in some cases including venture capital help. Admittedly, science parks, and their incubator functions, have often experienced great difficulty in meeting expectations. In the UK, for example, Massey et al (1992) note that approximately 27–29% of science park tenants are new firms and a further 8–9% are new branches of existing firms, while the majority of facilities are the result of

relocations by small and large firms (Table 9.9). They also note that the importance of new firm formation varies among science parks. The Aston Science Park, near Birmingham, for example, has been successful in attracting new firms and in developing local sources of venture capital, while in the more famous Cambridge Science Park, new firms are exceptional (Massey et al 1992: 32).

In addition to the debate over whether or not science parks incubate new facilities which might not otherwise have occurred, questions have been raised regarding the effectiveness of science parks in promoting university–entrepreneurial links (Monck et al 1988: 70–97) and linkages among park tenants (Howells 1986; Gillespie et al 1987; Malecki 1987; Keeble 1989). Oakey (1991, 1993) has also consistently raised doubts about the job potentials of new high-tech firms and has been particularly critical of the growth performance of new British companies in relation to US companies.

In practice, available evidence suggests that planning for incubation, whether with respect to manufacturing firms as a whole or high-tech firms in particular, is problematical. At the local level, there are successes and failures, and perhaps in a high-tech context, attempts to clone Silicon Valley have been predominantly frustrating exercises.

Table 9.9 Science Parks in the UK: New Facilities and Relocations

Status	UKSPA-OU-CURDS 1986 Survey Plus Number (%)	UKSPA Tenants Directory 1987 Number (%)
New facilities		
independent	49 (27)	58 (29)
non-independent	17 (9)	16 (8)
Relocation		
independent	86 (47)	74 (37)
non-independent	33 (18)	52 (26)
Total	185	200

Source: Massey et al (1992: 31).

Conclusion

Since 1980, a large body of literature on the subject has led to a much better appreciation of new and small firms. Debate continues to occur regarding the importance of new and small firms to local economies, e.g. with respect to job generation. The basic point, however, is that the small firm sector remains a significant segment within capitalist economies, and in some countries it is probably becoming somewhat more important. Fears for the death of small firms have proven unfounded. At the same time, the new-found vitality of small firms has brought mixed reactions. In general terms, while some authors celebrate the renaissance of small firms as the most appropriate way of promoting local development (Piore and Sable 1984), others suggest that the growth of small firms is part of a wider trend toward an 'enterprise culture' which gives too much priority to individualism and not enough to community values (Heelas and Morris 1992; Marquand 1992).

10

MEDIUM-SIZED FIRMS, BIG FIRMS LOCALLY

Within geography, it is common to interpret the size distribution of firms (Figure 8.1) as a duality between the two 'polar cases' of very large and very small firms, or between planning system and market system firms (Figure 8.3). Thus, the geography of the firm is based around well-established theories and concepts of the formation of new firms and the behaviour of small ('micro') firms (see Chapter 9) and the growth, structure and behaviour of multinational firms (see Chapters 11 and 12). To some extent, medium-sized firms (MSFs) have been incorporated within the new and small firm segment under the rubric of 'SMEs', i.e. small and medium-sized enterprises (Rothwell and Zegweld 1982; OECD 1993; Bagnasco and Sabel 1995; Pfirrmann 1995; Vaessen and Keeble 1995). In practice, this literature implicitly assumes that the behaviour, structure and impacts of MSFs are more or less the same as small firms – MSFs are simply on the 'big' side of the range of small firm sizes.

The presence of a size continuum of firms, however, is a reminder that the reality of enterprise behaviour and structure is highly varied. As Oinas (1995a: 178) states, in an observation specifically directed at the variations that exist *within*, as well as between, the small and large firm segments, "Enterprises differ from each other along innumerable dimensions". This chapter seeks to further underline this variability by an explicit focus on MSFs. In particular, the chapter draws on a few studies that interpret MSFs as an important segment of firms which exhibit distinctive characteristics. Most notable in this context is Nakamura's (1990) pioneering analysis which argues that MSFs play a vital role in the Japanese economy, an argument underlined by the literal interpretation of the Japanese word for MSFs, *chucken,* as 'backbone'.[1] In an American context, Kuhn (1982, 1985) further highlights the particular contributions of MSFs (*Business Week* 1993). Similarly, in Germany, MSFs are referred to as 'hidden champions' (Simon 1992), while Steed's (1982) Canadian study summarized the potential of MSFs as 'threshold firms'.

Geographically, the locational choices and spatial organization of MSFs have important implications for local development. Indeed, from this perspective, the potential significance of MSFs is that they are 'big firms locally.' Like small firms, MSFs typically have strong attachment to 'home'

[1]In writing this chapter, I am particularly grateful for the help of Jerry Patchell, especially with regard to the translation of Nakamura's (1990) work and for the suggestion of interpreting MSFs in Japan as backbone firms. Kevin Rees has also provided valuable help in compiling material for this chapter.

localities where ownership, control and investment are largely, if by no means completely, concentrated. If MSFs are 'hidden' outside of their market niches in terms of national or global awareness, locally they are well known. Moreover, in a local context, the employment and output levels are sufficiently big that the strategies of individual MSFs have noticeable local impacts. At the same time, the growth of MSFs requires a geographical extension of activities, most obviously with respect to markets, and MSFs are typically aggressive exporters and may even engage in direct foreign investment. As such, MSFs have a special role in linking the fortunes of localities with the forces of globalization.

This chapter is in two main parts. The first part discusses the distinctive nature of MSFs as organizations which combine selected characteristics of entrepreneurship and corporate power. A model is outlined which explores these characteristics in terms of corporate strategy, especially in relation to technology, marketing and employment. Particular emphasis is given to the role of MSFs in developing market niches, which are frequently international in scope, through a strong commitment to innovation and entrepreneurial initiatives. Different types of MSFs are then identified, based on the categories of small firms employed in Taylor and Thrift's (1983) dual model of business segmentation (Figure 8.3). In particular, key types of MSFs are identified as (market and technological) leaders, subcontractors and loyal opposition firms, and case studies are outlined to illustrate each type. The second part of the chapter examines geographical dimensions of MSFs, especially with respect to locational behaviour and as big firms locally in the mediation of local/global forces.

Towards a theory of medium-sized firms

Any definition (and explanation) of medium-sized firms must be made in relation to small and large firms. Although precise definitions vary, the small firm segment is often summarily defined as comprising firms which employ less than 100 people and have sales of less than US $25 million. In many cases, such a definition is stretched slightly to incorporate (small) medium-sized firms, meaning firms employing between 100 and 200 or 300 people, and with sales in the US $25–50 million range. At the other end of the size continuum, large multinational corporations (MNCs) are typically seen as firms employing in excess of 10 000 and with sales measured in billions of dollars.

But what about the firms between these size ranges, i.e. firms that employ between 300 and 10 000 people and have sales in the US $50 million to US $2 billion range? Even given some arbitrariness in statistically defining the boundaries of business segments in terms of firm size, there are many firms that occupy this (large) size range. These firms may be regarded as incorporating the 'true' MSFs. *Business Week*'s (1993) recent list (selectively) provides 50 US-based examples of (relatively large) MSFs which range in size from $236 million to almost $1.4 billion in sales (Table 10.1). It might be noted that *Business Week*'s interest in these 'little giants' stemmed from their impressive growth and profitability performance, especially in relation to the real giants. These firms also specialize in specific product-market niches, all have major exports and most have international operations. Indeed, in the context of the Japanese economy, Nakamura (1990: 2) argues that since at least the 1960s such MSFs have constituted an important 'third' group or segment of firms. Initially cast to critically assess the once prevailing view of the optimality of large firms and the marginality of small firms, Nakamura's argument stresses the qualitative differences of MSFs, in relation to small and large firms. Support from this perspective is also evident in the US (Kuhn 1982, 1985) and Germany (Simon 1992).

Medium-sized firms as backbone firms: a model

For Nakamura, MSFs are vitally important to the economy as a whole: they are literally 'backbone' firms (Figure 10.1). In backbone

Table 10.1 50 'Little Giants' Based in the US

	Firm/headquarter location	1992 Sales ($mil)	% International	Product niche	Foreign manufacturing
1	Advanced Technology Labs / Bothell, WA	323.7	40.0	Medical ultrasound systems	
2	Ametek / Pooli, Po.	769.6	30.0	Small motors, precision instruments	yes
3	AMSCO International / Pittsburgh, PA	498.2	20.0	Sterilization equipment, surgical tables	yes
4	Applied Materials / Santa Clara, CA	751.4	61.0	Semiconductor production	yes
5	Beckman Instruments / Fullerton, CA	908.8	55.0	Laboratory equipment for biological research	yes
6	W.H. Brady / Milwaukee, OR	236.0	30.0	A small version of 3M	yes
7	Cherry / Waukegan, IL	266.2	50.0	Auto parts, computer keyboards	yes
8	Cirrus Logic / Fremont, CA	354.8	60.0	Controller chips	yes
9	Cisco systems / Menlo Park, CA	339.6	36.0	'Internetworking' gear	
10	Continental Can / Syasset, NY	511.2	58.0	Packaging	yes
11	Dexter / Windsor Locks, CT	951.4	38.0	Speciality materials and coatings	yes
12	Dynatech / Burlington, MA	528.0	33.0	Intercommunications systems	yes
13	EMC / Hopkinton, MA	349.1	35.0	Data storage technology	yes
14	Ferro / Cleveland, OH	1097.8	57.3	Plastics, chemicals	yes
15	John Fluke Mfg. / Everett, WA	271.8	36.0	Testing and measuring equipment	yes
16	General Datacomm Ind. / Middlebury, CT	197.9	38.0	Multimedia and telecommunications gear	
17	Guardian Industries / Northville, MI	1200.0	50.0	Automotive and architectural flat glass	yes
18	Harnischfeger Industries / Brookfield, WI	1390.8	50.0	Heavy manufacturing equipment	yes
19	Haworth / Holland, MI	650.0	20.0	High-tech office furniture	yes
20	IDEX / Northbrook, IL	708.2	48.6	Material handling, conveyors	yes
21	Interlake Corp. / Lisle, IL	277.1	28.5	Fluid-handling products	yes
22	Intl. Flavors & Fragrances, New York	1126.4	70.0	Flavours and fragrances	yes
23	Invacare / Elyria, OH	305.2	35.0	Home medical equipment	yes
24	Keystone Intl. / Houston, TX	528.4	57.7	Valves and actuators	yes
25	Loctite / Hartford	608.0	60.0	Adhesives and sealants	yes
26	Lord / Erie, PA	242.2	32.0	Industrial adhesives, fasteners	yes
27	Medtronic / Minneapolis	1328.2	40.6	Heart valves and medical implants	yes
28	Millipore / Bedford, MA	777.0	63.0	Scientific instruments, filters, separators	yes
29	Mine Safety Appliances, Pittsburgh, PA	502.4	42.0	Gear used in mining	
30	Molex / Lisle, IL	776.2	72.0	Electronic components	
31	Nordson / Westlake, OH	425.6	67.0	Adhesive application equipment	yes
32	Pall / East Hills, NY	685.1	62.0	Maker of filters	yes
33	Perkin-Elmer / Norwalk, CT	911.1	53.0	Analytical instruments	yes
34	Pioneer Hi-Bred Intl. / Des Moines	1261.8	29.0	Sophisticated seed hybrids, affiliated products	
35	Read-Rite / Milpitos, CA	389.4	85.0	Recording heads for disk drives	yes
36	RPM / Medina, OH	625.7	22.0	Speciality coatings	yes
37	Sealed Air / Saddle Brook, NJ	446.1	31.0	Protective packaging	yes

Continued

Table 10.1 – *continued*

	Firm/headquarter location	1992 Sales ($mil)	% International	Product niche	Foreign manufacturing
38	Sensormatic Electronics / Dearfield Beach, FL	309.9	55.0	Electronic security devices	
39	Silicon Graphics / Mountain View, CA	866.6	47.0	Workstations	
40	Snydergeneral / Dallas, TX	750.0	45.0	Air filtration	yes
41	Standard Microsystems / Happauge, NY	250.5	44.0	Local area network hardware	yes
42	Stryker / Kalamazoo, MI	477.1	31.0	Special surgical products	yes
43	Sybron / Milwaukee, OR	382.6	30.0	Laboratory equipment	yes
44	Symbol Technologies / Bohemia, NY	344.9	35.0	Hand-held laser scanners	yes
45	Synoptics / Santa Clara, CA	388.8	31.0	Computer networking products	
46	3COM / Santa Clara, CA	617.2	50.0	Personal computer networking gear	yes
47	Thermo Electron / Waltham, MA	947.0	23.3	Energy equipment	yes
48	TJ International / Boise, ID	400.5	20.0	Manufactured wood	yes
49	Vishay Intertechnology / Malvern, PA	664.2	50.0	Electronic resistors	yes
50	Western Digital / Irvine, CA	938.3	43.0	Computer disk drives	yes

Source: *Business Week* (1993: 44–45).
Note: These companies have sales of at least $200 million and at least 20% of their sales off-shore.

firms, the corporate priorities and culture of MSFs are comprehensively shaped by the values of dominating entrepreneurs. In turn, these entrepreneurs closely co-operate with progressively skilled and participatory workforces as backbone firms seek to achieve superiority in the recruitment and effective utilization of personnel. Within backbone firms, R&D and marketing are key, interrelated functions which sharply focus this co-operation on innovation, with respect to technology, work organization, marketing and, ultimately, product markets. Indeed, backbone firms are highly innovative and continually seek to enhance the penetration of highly specialized market niches, often on an international scale. For backbone firms, the refinement and extension of particular product market opportunities frequently, if not invariably, implies a strong commitment to exporting and to foreign investment. Typically, a corollary of product specialism is geographic diversification of markets and production. In turn, market size and growth helps backbone firms realize economies of scale at the level of the product, factory and firm.

MSFs typically evolve from small firms but they can originate in other ways, including as spin-offs from large firms. MSFs may remain medium sized, grow to become large firms themselves, be acquired by existing large firms or decline back into the small firm sector or even fail (Figure 10.1). As Nakamura (1990) and Kuhn (1985) emphasize, MSFs differ from small firms in terms of organizational structure and technology as well as with respect to size. Thus, as is the case in small firms, MSFs are strongly influenced by the abilities, values and even personalities of individual entrepreneurs. On the other hand, the scale and often speed of growth means that MSFs deal with a range of new managerial challenges as they learn to functionally decentralize to an important degree; co-ordinate multiple plants, often in different countries; manage large labour forces which may be unionized; and secure substantial financing, often from equity markets. In terms of

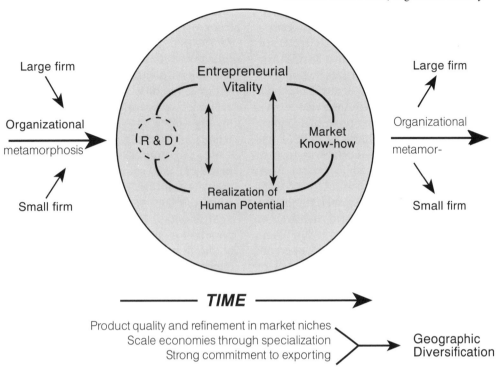

Figure 10.1   Medium-sized Enterprises as Backbone Firms

labour relations, for example, a continuous innovation strategy based on the realization of human potential throughout the firm necessarily implies the empowerment of employees and managerial responsiveness to employee wishes in order to gain the maximum amount of creativity from the workforce. Moreover, in MSFs, but not in small firms, the realization of firm-level economies of scale, notably with respect to R&D and marketing functions, is important.

In contrast to corporate giants, MSFs are more entrepreneurial (less bureaucratic, fewer managerial layers) so that decision-making is a faster process and key decision-makers are in closer contact with the workforce. Indeed, just as labour relations in MSFs differ from small firms, a key difference between MSFs and the giants is the greater capability of the former to fully engage and develop human resources in terms of

manual and intellectual contributions and in generally promoting interdependencies between employer and employee (Simon 1992: 122). The efforts of MSFs, in contrast to large firms, are also typically more focused in highly specialized product markets, even though, like large firms, they may be global in scope. For both Nakamura and Kuhn, MSFs are better able than large firms to escape the restraints of already saturated markets by more effectively capturing the latent value of employees to develop market niches.

The highly specialized, global market focus of MSFs is well illustrated by the 50 little giants listed by *Business Week* (Table 10.1). Kuhn (1982, 1985) describes scores of other US-based examples pursuing narrowly defined product markets, often on a global scale, as does Nakamura (1990) for Japan. Indeed, as Nakamura (1990: 3) notes, MSFs with focused innovation efforts can achieve dominant national and even global positions in highly differentiated markets, a characteristic which distinguishes them from most small firms. Simon (1992) similarly emphasizes the market strength of Germany's MSFs. He cites 25 German-based MSFs, out of a group of 39 interviewed, that were either the world's leading or second leading producer of particular products, c. 1990 (Table 10.2). One firm (Gerriets) was the only producer in the world of its specialty, while several firms enjoyed global market shares several times larger than that of its nearest rival. Korber/Hauni, for example, had 90% of the world market for cigarette machines throughout the 1980s. Some summary statistics of the 39 firms interviewed are also worth noting: on average, they had sales of US $303 million, 2904 employees, a world market

share of 22.6% and a European market share of 31.7% (Simon 1992: 116). Simon further underlines five 'common practices' of Germany's hidden champions, notably product market specialization with geographic diversity; an emphasis on customer value; an emphasis on blending technology with customer needs; considerable self-reliance on technical competence; and a strong sense of mutual interdependence between managers and employees. These characteristics are consistent with the model of backbone firms (Figure 10.1).

MSFs may well become attractive acquisition targets for large firms (or other MSFs). In addition, while some 'dominant' entrepreneurs may wish offspring or other individuals to inherit 'their' firm, others may wish to sell. On

Table 10.2   The Market Position of 25 Hidden Champions in Germany

Company	Main product	World market rank[a]
Baader	Fish-processing machines	1
Bamberger Kaliko	Bookbinding textiles	1
ECH Will	Cut-size sheeters for paper products	1 (4.0)
Erhardt & Leimer	Web-handling technology	1 (8.0)
Gehring	Honing machines	1–2
Gerriets	Theatrical stage horizons; stage decorations	1
GKD	Metal filters and fabrics	1
Heidelburger Druckmaschinen	Offset printing machines	1
Heidenhain	Measurement and control instruments for lengths and angles	1 (3.0)
Hymmen	Continuous pressing and heating machines	1
J M Voith	Machines for paper and cardboard industry	1
Korber/Hauni	Cigarette machines	1 (10.0)
Krones	Labelling machines for beverage industry	1 (4.0)
KWS	Seeds for moderate climates	1 (1.3)
Leybold	Vacuum technology	1 (1.7)
Markin & Cie	Model railways	1
Piffler	Front-operated lathes	2–3
Prominent Dosiertechnik	Pumps; gas and liquid measurement apparatus	1
Starck	Special plastic products	1
Stihl	Chain saws	1 (1.3)
TetraWerke	Food for tropical fish	1 (5.0)
Trumpf	Bending, sawing and punching machines	1
Webasto	Sunroofs for cars	1 (2.5)
Weinig	Automatic moulding machines	1 (4.0)
Zinser Textilmaschinen	Textile machines for spinning and twisting	2 (0.9)

Source: Simon (1992: 119).
[a]Refers to market strength of based on ratio of market share of company in relation to market share of strongest rival.

the other hand, in dynamic MSFs continued internal growth, possibly from medium to giant size, raises questions about organizational metamorphosis for the MSF. In this regard, it might be noted that firms can achieve substantial size while combining entrepreneurialism with more decentralized organizational structures. Anderson and Holmes (1995), for example, provide a fascinating case study of Magna International, a Canadian auto parts supplier which has experienced extremely fast growth as its sales grew from $302 million in 1983 to $2.6 billion in 1993. According to the statistical definitions adopted in this chapter, Magna has shifted from a MSF to a giant, and in response to its growth strategy it has developed a highly decentralized organizational structure. At the same time, its decentralization has been innovative, notably by retaining an entrepreneurial perspective.

Thus, the founding entrepreneur still exercises a dominant influence on overall corporate priorities and culture. More specifically, Magna has created a set of operating groups, co-ordinated by an executive management headed by the founding entrepreneur, and each group comprises a geographically concentrated network of small, specialized plants (Anderson and Holmes 1995: 664). Each group has its own R&D and engineering staff, marketing responsibilities and human resources management; each plant is a profit centre, has a profit-sharing scheme and is responsible for training, tooling and the identification of markets and product development. Most plants are highly automated, flexible and have less than 100 employees, and the plants within each group constitute what Magna calls an 'industrial concept' in which 10–20 small plants provide the basis for shared social and recreational facilities, in some cases in Magna-owned industrial parks. As Anderson and Holmes (1995: 657) point out, Magna's organization illustrates a hybrid version of Piore and Sabel's (1984: 265) concepts of solar firm and workshop factory. It also reveals that the organizational metamorphosis from MSF to giant is open to innovation and one which does

not simply involve MSFs copying the giants. Rather, MSFs such as Magna may be providing prototypes for existing giants.

The social rationale of medium-sized firms
Within the overall size distribution of firms, MSFs contribute to national economic development by promoting competition and efficiency, by responding to consumer demands and by enhancing innovativeness. For Kuhn (1982: 2–3), MSFs are simultaneously essential to national economic efficiency "by thwarting monopolies" and to political pluralism "by disrupting the political hegemony between 'big business' and 'big government'". Indeed, Kuhn (1985: 70) argues that "true" MSFs directly "compete against the giants" and that the public benefits of this competition largely result from the greater efficiency (or less inefficiency) of MSFs. Thus, Kuhn (1985: 114) suggests that "X-inefficiency", defined as the excess of unnecessary cost as a proportion of total cost, increases with firm size as large size (and power) in the market-place protects bureaucracies, sluggish behaviour and managerial perks while maintaining high prices. A related argument notes that labour relations become more difficult with large size, as a result of both supervisory difficulties and worker satisfaction. Although measurement difficulties are considerable, there is statistical support for the view that MSFs define the most efficient levels of operation and that MSFs are big enough to fully exploit economies of scale and as a group are often more efficient than giant firms (e.g. Blair 1972; Shepherd 1979; Scherer 1980).

A significant thrust of Nakamura's (1990) argument is the dynamic role of MSFs in the Japanese economy since the 1960s. In particular, Nakamura's analysis identifies three stages in the evolution of MSFs from the 1960s to the 1980s. First, in the 1950s and 1960s, MSFs provided a key role in introducing new organizations and labour relations strategies to facilitate the introduction of new technologies. In a variety of industries, a few pioneering MSFs, including Honda and Sony, revolutionized existing SF management thinking by focusing on products with growth potential; specific market niches achieving the benefits of mass production and quality improvements (gaining benefits over both SFs and giants); and rapidly introducing new process technologies (often reverse engineered from models elsewhere) and introducing meritocracy and egalitarianism in the workplace. According to Nakamura (1990: 8), the leading MSFs of the 1960s distinguished themselves from SFs by achieving scale economies and high wages, while providing critical support for the advances made by large firms in the assembly and new materials industries. These new entrepreneurial-driven MSFs explored different market strategies: machine-tool MSFs focused on developing internationally competitive products; MSFs in many regionally diffused industries (e.g. laundry, raincoat, miso, ham and sausage, furniture, etc.) developed national identities; and MSF subcontractors in auto and other assembler industries emerged by improving the precision and standardization of their products as well as improving delivery dependability.

Second, in the 1970s, Nakamura (1990: 12–17) argues that MSFs led diversification and economic restructuring processes in the economy as a whole. In this period, leading Japanese MSFs developed new applications and products in related or entirely new directions, either through original R&D or by introducing technology developed elsewhere. In many instances, a key strategy involved developing high-variety, small-lot production as existing mass production/ distribution systems no longer sufficed in saturated markets. In this context, Nakamura (1990: 17) suggests that the ability to deepen R&D in order to transfer technology to and from other fields and to develop high-variety, small-lot production constitutes a type of R&D-based scope economy. The experience in developing high-variety, small-lot production also helped prepare the way for the factory automation systems of the 1980s.

Third, in the 1980s and 1990s, Nakamura (1990: 19–29) argues that MSFs have led the shift in the Japanese economy from its heavy industry and mass production bias towards integration in a more international division of labour in which Japanese firms elaborate niche markets, increasingly bridge manufacturing with primary and service activities, and create entirely new businesses. Within the general context of the information and communication techno-economic paradigm, there has been tremendous growth of highly differentiated markets as a result of the breakdown of the mainframe computer which created demands for a wide variety of peripheral devices and software, and a shift towards flexible mass production which created demands for a wide variety of automated products and machines. Information technology also opened up new markets in such fields as records, ticketing and music/video rental. In addition, Nakamura argues that MSFs in Japan are leading the way in terms of offering Japanese consumers new lifestyle choices in clothing, record rentals, computer communications and corporate vision.

Nakamura (1990: 27–29) further suggests that the leading edge MSF in the 1980s differs from that in the 1960s in several ways. While MSFs throughout the period depend on entrepreneurial ability to create businesses, the entrepreneurs of the 1960s were more likely to be 'lone wolves', whereas the entrepreneurs of the 1980s are more willing to collaborate, merge and network with other firms of similar size and power. Japanese MSFs have typically been closely involved in subcontracting relationships (and in a social division of labour) – far more so than their counterparts elsewhere. In the 1980s, however, MSFs are engaging in alliances with other firms that are proactively defining strategic directions. In this regard, Japanese MSFs have become more interested in diversifying their activities and markets by bringing together related technologies within strategic alliances (see Chapter 11) or by internally using existing technological expertise to promote new products. In addition, the leading-edge MSFs in the 1980s allow employees more input in the direction and organization of the firm, so that the firm is becoming more than the entrepreneur's vision. Finally, for Nakamura, the MSFs that become the giants of the 1990s will draw their competitive advantage less from economies of scale than from economies of scope and networking with other firms to enhance their innovativeness.

While there are distinctive dimensions to the MSFs in Japan, Nakamura's analysis of the dynamic role of MSFs has considerable implications elsewhere, including the increasing tendencies of MSFs to participate in strategic collaborations and to engage employees in decision-making processes more effectively (Kuhn 1985; Simon 1992). Similarly, the argument that MSFs are not only a source of the giant firms of the future but that they offer models of organization to which existing giants must adjust, has been widely noted and is, for example, at the heart of the idea of 'intrapreneurship.'

Case studies of medium-sized firms: alternative strategies

In practice, MSFs exist in all sectors of modern economies and exhibit a variety of market, technological and location strategies. As noted in the introduction, the terms used by Taylor and Thrift (1983) to classify different types of small firms, may be applied to MSFs (Figure 8.3). In particular, dynamic MSFs may be thought of as *leaders*, *subcontractors* and as *loyal opposition* firms (and the possibility of MSFs becoming *laggards* should be recognized).

Leaders Leader MSFs are innovative, growth-oriented firms which enjoy significant, possibly dominant, market shares. Giant companies are typically not present in the particular market niches being served, or are minor players. Three examples of leader MSFs are offered. Invacare, a US-based firm (Table 10.3), and Murata, a Japanese-based firm (Table 10.4), are large MSFs and MacDonald Dettwiler is a much smaller Canadian-based MSF, which (until 1995) fits Steed's category of threshold firms (Table 10.5). Invacare manufactures health care products, especially wheelchairs, Murata is a manufacturer of machinery including textile machinery and robots and MacDonald Dettwiler is primarily a

designer, developing software for geo-information systems.

The differences among the three case studies reflect variations in the characteristics of backbone firms (Figure 10.1). Thus, the oldest of the firms, Murata, began as a small family enterprise, Invacare began as a spin-off division of a giant MNC, and MacDonald Dettwiler was created by two 'boffin-businessmen' as a small firm. All three companies have been controlled by dominant entrepreneurs who created distinctive corporate cultures. In the case of Invacare, its founding entrepreneur, Mal Mixon, has emphasized aggressive marketing and the importance of beating rivals, a philosophy frequently communicated in militaristic rhetoric (Butler 1995). The founding entrepreneurs of Murata and MacDonald Dettwiler have given particular priority in developing skill and creativity among employees: 'freeing people for creative work' in the case of Murata (Patchell 1991: 756) and 'creating wealth from knowledge' in the case of MacDonald Dettwiler (*BC Business* 1995).

All three firms have been strongly innovative in research and developing new products. Invacare relies on a separate R&D department for product development while Murata more closely integrates R&D and manufacturing by diffusing engineering know-how throughout its operations. MacDonald Dettwiler basically employs highly qualified engineers and its emphasis has been on developing software, with most manufacturing processes contracted out. The three companies tightly relate R&D efforts to specific market opportunities. All three companies originally focused on specific market niches and have been able to use established technological and marketing expertise to diversify into new but related products. Invacare, for example, has focused on products, beginning with a standard wheelchair, that serves 'home-based' (rather than hospital-based) patient needs, while MacDonald Dettwiler has progressively developed its expertise in communications systems to diversify beyond (custom-designed) land-based satellite systems to include defence, space and commercial aviation projects. In Murata's case, the firm has

Table 10.3   Selected Characteristics of Invacare Corporation

Characteristic	Comment
Origins	Founding company opened in 1885. Various ownership changes and in 1979 Invacare spun-off as part of a small division of Johnson & Johnson created by a leveraged buy-out led by Mal Mixon, a sales executive in a hospital products company
Headquarters	Elyria, a suburb of Cleveland, Ohio
Size	1994 (1992) sales of $411 million ($305 million); 1994 employment: 3200
Manufacturing	Elyria, Ohio (three plants); Bridgend, Wales (1984); Bad Oyenhausen, Germany (1984); Reynosa, Mexico (1986); Montreal and Toronto, Canada (1991); Tours, France (1992); Basel, Switzerland (1995); Auckland, New Zealand; Birmingham, UK (1995)
Product-mix	Standard and power wheel chairs, patient aids (e.g. crutches); respiratory core products (e.g. oxygen concentrators)
Entrepreneurship	Close personal direction by Mal Mixon and a small group of 'functional' VPs. Particularly strong emphasis on aggressive marketing; corporate culture heavy with military symbolism (Mixon is a Vietnam veteran)
R&D	1994 R&D budget: $7 million
Innovativeness	In 1982, introduced motorized wheelchair with industry's first computerized controls; 50 new products in 1994. Also pioneered extensive distribution system and close contact with independent home health care dealers.
Work organization	Strong job demarcation prior to 1990
Market position	Since 1979, replaced Everest and Jennings as global market leader in home health care equipment. Dominant producer in N America; second in Europe
Exports	By 1994, sales in 80 countries including Russia; 26% of sales outside US; power wheelchairs most important product line
Organizational metamorphosis	In 1978, Invacare was a tiny division of J&J with sales of $19 million. After buy-out by Mixon, Invacare has grown rapidly and stated ambition is to become a Fortune 500 company by 2000

Source: compiled from various sources including Standard and Poor's Registrar of Corporations, Butler (1995) and Palmeri (1993).

translated existing technical strengths to pursue a more substantial horizontal diversification strategy (Patchell 1991: 321–322).

Murata originally produced Jacquard machines to mechanize the weaving of elaborate textile patterns in Kyoto, the centre of Japan's kimono industry. Murata broadened its base in textile machinery and became a pioneer in its automation. This knowledge of automation provided the basis for the firm to re-enter the production of machine tools in 1962 and to develop a strategy for the manufacture of fully automated manufacturing systems (including robots). Its subsequent growth in the 1970s in communications technology, which stemmed from the initiative of the founder's son when he

took over the business, is focused on facsimile machines and was designed to reduce Murata's reliance on machinery industry. While this development was a substantial diversification, which required learning new technology and developing new marketing systems, Murata's did draw on its expertise in automated systems to internally create a facsimile production system which eventually became profitable in the face of fierce competition from giant firms (Patchell 1991: 779).

Murata provides a good example of Naka-mura's backbone firms and how they have evolved from the 1960s to the 1990s. As such, it offers some interesting differences with Invacare. First, as noted, the extent of its horizontal

Table 10.4 Selected Characteristics of Murata Machinery Company

Characteristic	Comment
Origins	1935 as the Nishijin Mfg Co. partnership (textile machinery). In 1945, renamed as Murata Textile Machinery Co.
Headquarters	Kyoto
Size	1993 (1987) sales approx. $1 billion ($900) million; 2300 employees
Manufacturing	Kyoto (two plants, 1951 and 1960); Inuyama (1962); Kanto (1981); joint venture, Charlotte, NC (1970); São Paulo, Brazil (1973); Düsseldorf, Germany (1984). (Korean factory opened on 1940 but since closed)
Product-mix	Textile machinery (1935); machine tools (1940s and 1961); automated systems (1963); communication equipment (1972)
Entrepreneurship	First Murata President replaced by his son, J. Murata. Personalized corporate philosophy represented as 'freeing people for creative work by delegating to machines the work that machines do best' (Patchell 1991: 756–7)
R&D	Located at Kyoto
Innovation	Original emphasis on reverse engineering of foreign technology has shifted to development of world-leading unique products. In the automated systems division, innovation has focused on fully computer integrated manufacturing systems. Since 1950, when employment was 100, firm has recruited engineers on annual basis
Work organization	Functional flexibility. Strong links between design and production. Strong reliance on suppliers organized by JIT
Market position	In 1987, textile machines still over half of sales; growing share of facsimile market (7% in 1986); leading Japanese supplier of physical distribution systems and automated warehouses
Exports	Exports important since at least the 1950s (over half sales in 1956)
Organizational metamorphosis	Divisional re-organization in 1986; each of four divisions became directly responsible for its own growth and development. Foreign sales also re-organized as reliance on trading firms replaced by independent manufacturing and sales firms in Europe, North America, Asia and Africa

Source: drawn from Patchell (1991, especially pp. 321–43, 752–82); see also Patchell (1993b)

diversification is greater. Second, Murata's growth has been based largely on internal growth, notably by building, rationalization and expanding its own factories and by learning new technologies and markets. While Invacare has also developed new products, its growth has been substantially realized by acquisition, especially of rival firms in foreign countries. Third, especially with respect to manufacturing automated systems for a variety of industries, Murata constantly demonstrates economies of scope in R&D. Fourth, although Murata's sales are twice those of Invacare, employee totals are similar. This difference reflects Murata's reliance on subcontractors, or on a social division of labour, for a wide variety of components and processes

(Patchell 1993b). Invacare's production is more vertically integrated, i.e. it relies more on an internal division of labour (Figure 10.1; see Chapter 13 for a full discussion of the internal and social divisions of labour). Fifth, as Nakamura (1990) anticipates, Murata has begun to engage in joint ventures with foreign firms, its most important co-operation being with the US-based Warner and Swassey Company, a well-known machine tool company, which was established in the early 1970s to manufacture automatic lathes. It has also signed an agreement with a US firm to manufacture facsimile receivers.

Both Murata and Invacare have become market leaders and manufacture large outputs of specialized products which generate important

Table 10.5   Selected Characteristics of MacDonald Dettwiler and Associates

Characteristic	Comment
Origins	Founded 1969 by two colleagues (John MacDonald and Vern Dettwiler) in the electrical engineering department of the University of British Columbia in order to bid on a contract with Landsat 1
Headquarters	Richmond, a suburb in the Vancouver metropolitan area, British Columbia
Size	1994 (1992) sales: Can $110 million ($72 million); 1994 employment: 1000
Manufacturing	Richmond, BC. Company also wholly owns subsidiary 'consulting' and 'service' companies in the US, UK, Australia and Canada (Ottawa) and has majority and minority control of two other UK firms
Product-mix	Large-scale systems engineering and software development for geo-information systems (satellite stations), aviation systems (e.g. air traffic control systems, pilot briefing systems), and defence and space programme projects (e.g. radar jamming equipment), communication equipment for combat zones
Entrepreneurship	Both founders actively involved in running company (MDA), especially John MacDonald who has consciously sought to create wealth from knowledge and to link creative employers with challenging tasks
R&D	Company conducts significant R&D funded through commercial contracts
Innovation	Since building a receiving box for radio signals for first civilian satellite, MDA has innovated three generations of earth-imaging technology and developed market niches in aviation systems (since 1987) and in space / defence areas (since 1988)
Work organization	Highly qualified workforce engaged in sophisticated design and engineering work. Teams and participation required. Employees own 25% of MDA
Market position	Leading global supplier of land-based systems for retrieving, analysing and mapping satellite data
Exports	Global sales with a particular thrust towards Asian markets, including China and Japan, and North America. In 1994, half of MDA's sales comprised geo-information systems
Organizational metamorphosis	The founders' backgrounds were 'technical'. By 1980, MDA faced managerial problems and a CEO with considerable business experience was hired (while John MacDonald remained president). In July of 1995, MDA was acquired by Orbital Sciences of Virginia which will now shape its future strategy

Source: compiled from various sources including MacDonald Dettwiler.

economies of scale from their R&D and marketing networks, as well as in the production process itself. Invacare, for example, has specialized its models and component production in different factories and has begun to take advantage of an international division of labour. Its factories on the Mexican–Texas border and in Spain, for example, focus on low end products, while its Elyria (Ohio) and Bridgend (Wales) plants manufacture the newer, higher-value products. In the UK, its newly acquired plant in Birmingham supplies aluminium tubing to Bridgend.

MacDonald Dettwiler is much smaller than either Invacare or Murata, in part reflecting the fact that MacDonald Dettwiler is primarily a service company whose growth potential is less than the other two. Nevertheless, MacDonald Dettwiler's growth is impressive and its competitive advantages are increasingly based on the economies of scale and scope of its highly qualified workforce as they rely on established expertise, creative potential and tremendous flexibility in forming teams to develop leading-edge communication technology in response to a variety of customer needs. Until 1995, this firm illustrated a threshold firm, defined as "Canadian owned enterprises with 100–2499 employees in Canada . . . belonging to one of the more technology intensive sectors" (Steed 1982: 47).

However, in July, it was acquired by Orbital of Richmond, Virginia, a giant space-based firm. Indeed, the acquisition of Canadian firms by foreign giants is a major reason why there are so few threshold firms in Canada (Britton and Gilmour 1978; Steed 1982). Admittedly, it is an operation that would be very difficult to relocate since its expertise is not embodied in equipment but is defined by its employees who are likely to want to stay in British Columbia. At the same time, the loss of its entrepreneurial characteristic may at least limit its future growth. From the perspective of high-tech development in BC, MacDonald Dettwiler is likely to remain important, even if its future is less certain.

Subcontractors The vast majority of subcontractors are relatively small and it is possible in some industries for such firms to transform themselves and become MSF leaders. It is also possible for highly innovative and specialized MSFs to develop as subcontractors. Indeed, in Japan, numerous such firms exist and two

examples are outlined: F-tech (Table 10.6), a supplier to one core firm in the auto industry; and THK, a supplier of linear ball bearings to several core firms in several industries, but particularly robotics and machine tools (Table 10.7). F-tech is a highly specialized supplier of suspension systems to Honda. Its principal plants in Japan and in Ontario are located within 90 minutes drive of Honda assembly operations and its R&D operations are adjacent to Honda operations. Depending on the component, F-tech's delivery schedule to Honda varies from one hour to one week. In turn, F-tech has also developed its own supplier system and each of its two major factories in Japan rely on about 20 subcontractors. Core auto assemblers in Japan, such as Honda, typically rely on a large base of subcontractors (see Chapter 14). Honda's reliance on F-tech is unusual, however, in that F-tech is its sole supplier of suspension systems. On the other hand, F-tech is a high-quality, highly innovative manufacturer which keeps a close watch on suspension system developments

Table 10.6 Selected Characteristics of F-Tech Corporation

Characteristic	Comment
Origins	Began as toy manufacturer at Soka city and changed to a component supplier for motor cycles before becoming an auto supplier for Honda
Headquarters	Saitama, northern Kanto (Tokyo metro area)
Size	1995: 1616 employees
Manufacturing	Plants in Saitama and Tochiga prefectures; Ontario, Canada; Philippines; US plant planned for 1997. All plants, except Philippine facility, close to Honda assembly
Product-mix	Suspension systems for Honda (since 1985)
Entrepreneurship	Hands-on running of company by President Fukuda; 20% owned by Honda
R&D	83 R&D professionals in two laboratories by Honda factories
Innovation	Highly focused and developmental. F-tech is Honda's sole supplier of suspension systems – an unusual situation. On 6-year R&D product cycles
Work organization	Functionally flexible, 1992: each of its Japanese plants uses about 20 subcontractors and the Canadian plant uses 5. JIT
Market position	1993: 93% of sales are to Honda. Some sales to Mitsubishi. Trying to diversify away from Honda; the aim is to have 66% of sales to Honda
Exports	Indirectly through Honda. Canadian sales from Canadian plant
Organizational metamorphosis	They are anticipating more threats to *keiretsu* system and its apparent "closed" relationship with Honda

Source: interview 1992; translations kindly provided by A. Takeuchi.

Table 10.7   Selected Characteristics of THK

Characteristic	Comment
Origins	1971; Toho Seiko KK became THK Co. in 1973 (THK: toughness, high quality and know-how)
Headquarters	Tokyo
Size	1987 sales: approx. $370 million; 1780 employees
Manufacturing	Tokyo, Kofu, Yamaguchi, Nagoya, Osaka, Mie (Osaka region)
Product-mix	Custom-made bearings, machine tools and parts, electronic and electric parts, hydraulic and pneumatic devices
Entrepreneurship	Haku Teramachi, founding entrepreneur and owner, exercises strong 'hands-on' leadership. Emphasizes both high-quality production and price leadership
R&D	'Techno-plaza' built in Kofu in 1987. Strong reliance on sales workforce who receive engineering training to link market needs with innovation
Innovation	THK created (and patented) the linear motion bearing that greatly improved the speed, precision and energy-saving performance of robots, machine tools and other types of equipment. Constant innovation. By 1988, 69 patents in Japan, 170 in the US and 300 in Europe
Work organization	The precision of the bearing has required heavy investment in specialized and integrated equipment. Even so, THK uses about 100 suppliers (for drilling, plating, finishing, etc.) and has a co-operative association
Market position	A subcontractor with a highly diversified customer base. Three or four rivals but THK has dominating market position in Japan. Supplies all robot-makers and has 90% of machine tool market for bearings. Some exports to US and EC
Sales	Late 1980s: linear motion bearing is 70% of sales – almost half to robot-makers, almost half to machine tool makers and rest to semiconductor makers. 23 sales offices in Japan and offices in US, and Germany. Strong engineering sales support services
Organizational metamorphosis	Effective adjustment to rapid growth, including creation of a supplier system

Source: Patchell (1991: 473–474, 501–504, 865–869); see also Patchell (1993b).

around the globe (it has examples of virtually every major suspension system developed by other companies in its Saitama factory) and constantly seeks to improve the quality and cost of its product for Honda. While Honda has 20% equity, F-tech is run in a 'hands on' way by its President who is in constant interaction with his employees. Although F-tech's relationship with Honda is extremely close, both firms talk about diversifying and in recent years, for example, F-tech has established a small relationship with Mitsubishi. Its plan is to reduce its dependence on Honda to 66% of sales, a plan which reflects a broader gradual erosion of *keiretsu* relation-ships, which essentially comprise stable, mono-poly/monopsony business relations between firms, across the Japanese economy (Yaginuma 1993).

THK was founded by Haku Teramachi, who had previously been working in another firm in the same industry, to manufacture and sell a significant innovation, the linear ball bearing system. This new system, in comparison to conventional ball bearings, substantially reduces friction and resistance and permits robots and other NC (numerically controlled) and CNC (computer numerically controlled) machines, such as lathes and machining centres, to perform more accurately, with greater productivity (speed), with less wear and tear, and with power cost savings of up to 90% (Patchell 1991: 869–870). In this last regard, the growing energy crisis in Japan in the early 1970s helped stimulate demand for Teramachi's innovation. The linear ball bearing system is also simple to install and allows the construction of more flexible robotic machines by increasing movement and carrying-capacity possibilities. In contrast to F-tech, THK is completely independent and supplies several core firms; for example, in the robot industry it sells to all the important manufacturers. For individual robot manufacturers this can be a source of tension and, in Hirata's case, for example, technological co-operation occurs on a specific product basis. However, THK has control of the patents and its willingness to work closely with Hirata in Kyushu has ensured that the relationship between the two is stable and growing. Because of the very high quality and precise nature of its product, THK has invested considerably in specialized and integrated factories. Even so, it has cultivated a stable base of over 100 suppliers, many of which it has included in a co-operative association.

Teramachi provides THK with strong 'hands on' leadership. He has been particularly emphatic on installing engineering know-how in the work-force and among sales staff to facilitate communication between consumers and the firm. THK pays constant attention to innovation and is an acknowledged world leader in ball bearing systems. The firm sells its technology to the West and has taken out many patents in the US and Europe as well as Japan (see Table 10.7). While promoting innovation and high-quality products, Teramachi has emphasized that THK should be a price leader. Thus, one of his directives issued in the 1980s is that if the products of rivals are only 10–20% more expensive, then THK must find ways to cut costs. THK has been a growth-oriented firm, and within 16 years its sales had reached over $370 million. While most of these sales are within Japan, THK has established permanent sales offices in Chicago, New Jersey and Los Angeles in the US, and Düsseldorf in Germany. Export markets, however, remain to be developed.

Loyal opposition MSFs Kuhn's interpretation of MSFs gives particular emphasis to the role of MSFs as competitors to giants. In a general sense, MSFs inevitably operate "under the shadow of giants" (Edwards 1979: 73), at least because of their potential role as competitors, and threats as acquisitors. In practice, many leader SMEs have pioneered market niches which are not exposed to the direct competition of the giants, while subcontractors deliberately seek to be complementary. At the same time, in some markets, MSFs clearly directly compete with the giants, and in others MSFs have sought to meet this competition more obliquely by developing competitive strengths not evident among existing giants operating within the same industry.

In the case of the US soft-drink industry, Dr Pepper for a long time offered loyal opposition to Coca Cola and Pepsi, two giant MNCs (Table 10.8). Beginning as a family enterprise in Waco, Texas, in the 1930s, Dr Pepper grew particularly rapidly in the 1960s and 1970s by emphasizing market distinctiveness in terms of taste and geography, enjoying a particularly strong market position in Texas (Kuhn 1986: 291–309). Indeed, Dr Pepper represented a conventional loyal opposition firm whose rivalry with the giants is strongly regionalized and rooted in particular consumer loyalties. Its above-average rate of growth, however, encouraged Dr Pepper to develop the national market and to confront the giants more directly. Although this strategy experienced considerable difficulties, by the early 1990s, Dr Pepper had merged with 7Up and market share expanded to 11.6% of soft-drink sales by 1994 (Table 10.8). In 1995, however, Dr Pepper / 7Up was acquired by Cadbury Schweppes to create the third largest giant in the industry. By 1995, the three largest giants accounted for 88% of the US soft-drink market.

In the case of the Japanese robot industry, the world's largest, rapid growth since the late 1970s has been organized by a large number of firms (Patchell 1991). A few of these firms are true giants and of these, Matsushita, a highly diversified electrical product MNC, has emerged as the largest (Table 10.9). In most cases, robot manufacturers in Japan, including Matsushita and MSFs such as Murata, Hirata and Yaskawa, were well established in other industries and entered robot manufacture, typically in innovative ways, to further develop their existing specialisms or to help diversify into a related

Table 10.8 The Size and Role of Dr Pepper in the American Soft Drink Market

Company	Market share (%)				1995 Sales ($ million)	No. of employees 1995
	1989	1991	1993	1994		
Dr Pepper	9.9	10.6	11.4	11.6	769	952
Coca Cola	40.1	40.7	40.4	40.7	16 200	31 312
Pepsi Cola	31.8	31.5	30.9	30.9	28 500	423 000
Schweppes	5.0	5.0	4.9	4.8		
Royal Crown	2.7	2.4	2.2	2.0		

Source: Standard and Poor Industrial Survey.
Notes: In February 1995, Dr Pepper/7Up was acquired by Cadbury–Schweppes for $1.7 billion. Pepsi's 1995 sales and employment totals include Kentucky Fried Chicken.

Table 10.9 Giants and Loyal Opposition: The Japanese Robot Industry

Company	1988 Sales ¥ 100 million	Robot specialization
Giant Firm		
Matsushita	816	Material handling, especially Cartesian robots, 50% for external markets
Loyal opposition		
Yaskawa	180	Multi-articulated playback machines for cutting, grinding, welding, etc. (especially for auto industry); robots incorporate its leading-edge motion control technology
Hirata	32	Most robots (90%) sold as part of manufacturing systems; arm base and machine base, especially Cartesian robots designed for assembly purposes
Murata	90	automated guided vehicles which incorporate robots and sold to many industries

Source: based on Patchell (1991). This source lists all important robot manufacturers in Japan

area (Patchell 1993b). Thus, Matsushita primarily manufactures Cartesian robots, which are robots used in material handling. The robots are either incorporated in its own products or sold externally. Its rivals are basically the electronic product giants of the world.

With respect to MSFs, Murata produces robots that are included as part of its automated systems, including automated vertical warehouses, dangerous-substance handling systems and a variety of automated guided vehicles with various loading and movement characteristics. Hirata manufactures manufacturing systems and has specialized in robots which perform assembly work, most of which are also sold as part of manufacturing systems. Yaskawa, on the other hand, is the fifth largest of Japan's electric machinery manufacturers and it has sought to produce a distinctive product by emphasizing robot production, especially for the auto industry (Patchell 1992: 576). Yaskawa sought to develop market niches by innovating highly sophisticated robots used in welding, cutting, grinding operations and which incorporate its established expertise in motor control technologies. Yaskawa's arc-welding robots soon dominated the world market, and the firm has developed expertise in the manufacture of peripheral devices used in robots and in related software packages. Robot equipment is now almost as important as electrical machinery to the sales of the firm.

Each of the loyal opposition firms, as well as Matsushita, has developed its own network of suppliers (Patchell 1993b). Indeed, to a significant degree, and underlining an important trend in the Japanese economy, the suppliers in the robot industry have played vital roles in innovating products which enhance performance and reduce costs. Among some of the suppliers, the level of expertise of the suppliers equals or surpasses that of the core firms, to create products not conceived by the core firms. Thus, loyal opposition firms compete with the leading firm by incorporating the expertise of others. These rival networks are not entirely independent or entirely localized, as the case of THK, which supplies components to all robot manufacturers, demonstrates (Table 10.7). Yet the rival networks are strongly localized and core firms that use non-exclusive suppliers are careful in the organization of co-operation.

Geographic perspectives on medium-sized firms

While there are few systematic analyses, there are reasons to believe that the geographical behaviour, structure and impacts of MSFs are distinctive, in relation to large and small firms. In contrast to giant firms which have increasingly concentrated their control functions in the central cities of major metropolitan areas, there is evidence that the origins and headquarters of MSFs reveal more diverse location patterns and that peripherality appears to pose no significant disadvantage to their evolution. The location preferences of the branch plants established by MSFs are also diverse, if anything favouring non-core locations. With respect to mutual dependence between firms and localities, a framework recently outlined by Oinas (1995a) strongly implies a distinctive and important role for MSFs.

The location preferences of MSFs

Support for the view that the headquarters of MSFs are relatively dispersed is provided by the 50 MSFs listed by *Business Week* (Table 10.1).

While these firms are not randomly chosen, they were not selected with any geographic purpose in mind either. In any event, the location pattern of head-offices revealed by these firms is interesting. Just three of the firms had their head-office within a city of more than a million people, and just three more were located in cities of between one-half and one million people. The remaining 44 were located in smaller places. On the one hand, the regional distribution of the head-offices of these firms emphasizes traditionally important industrial regions. On the other hand, the intraregional pattern clearly reflects distinct preferences, ranging from the suburbs of metropolitan areas to small, even rural places. Invacare, for example, is headquartered in Elyria, a suburb of Cleveland, Ohio, while Ametek is based in Pooli, Pennsylvania, a town with a population of 5600, and Vishay Intertechnology is based in Malvern, Pennsylvania, a town of 2900 inhabitants.

Invacare has also chosen branch locations in relatively small places (Table 10.3). Its major overseas facility, for example, is Bridgend, Wales, a town of about 50 000 people, while its recent acquisition of France (and Europe's) largest wheelchair manufacturer is based in Tours. Its choice of location for its maquilladora is Reynosa, a remote community just inside Mexico and located near Mcallen, Texas, where Invacare has an office. Its new Spanish plant is in a similarly remote location. Indeed, the only 'large' city in which it operates is Birmingham, England, where it acquired a small firm which supplies it with tubing supplies.

The location choices made by the other 'little giants' have not been identified. However, Simon's (1992) analysis of the German situation reinforces the thesis that MSFs pursue distinctive location strategies. According to Simon (1992: 122):

"The typical hidden champion is located in a small town or village rather than a big city. Few can be found in urban centres like Hamburg, Munich, or Cologne, but many are scattered in places like Neutraubling, Harsewinkel, Tauberbischofscheim, Melle, and Stockdorf – towns most Germans have never heard of".

In the German case, Simon notes that owner-managers of MSFs were often born and raised in the small towns where they established operations and that such locations facilitate a close interdependence between employers and employees. Indeed, in contrast to the view that remote locations offer opportunities to firms to tap low-waged, passive labour supplies (see Chapter 6), he argues that in many of Germany's rural-based MSFs, labour relations are characterized by forms of 'enlightened patriarchy' (Simon 1992: 122). In support of this view, he notes the very low rates of turnover experienced, and the limited number of strikes, arguing that "internal problems absorb much less managerial energy than in large companies" (Simon 1992: 122). On this last matter, Simon's rule of thumb is that while management in large companies typically spend about 65% of their time "overcoming internal resistance", in MSFs in Germany the figure is closer to 25%.

Even in Japan, where head-office patterns and industrial activity has traditionally been highly concentrated in the Tokyo and Osaka metropolitan regions, it is nevertheless clear that MSFs, including highly innovative 'leader' firms, do originate and thrive in peripheral locations. Patchell's (1993b) analysis of the Japanese robot production system, for example, revealed the significance of MSFs in the development of geographically diverse patterns of control and innovation, as well as in production. In particular, these firms play vital roles in fully incorporating peripheral areas within a national system of robot production. Thus, the locational origins and structures of the giant firm, Matsushita, and of the MSFs, Yaskawa, Murata and Hirata, are all different (Table 10.9). On the one hand, Matsushita has anchored its robot production system in the Osaka metropolitan area. Some branch plants are now in peripheral locations, but control, innovation, production and supplier firms remain concentrated in the Osaka region. In contrast, Murata retains its R&D and control functions in Kyoto and has shifted its production functions towards industrial core regions, albeit in small towns.

Both Yaskawa and Hirata have developed extremely sophisticated production systems in remote locations, from where they manufacture products exported throughout the world. In Hirata's case, for example, its home town is Kumamota in central Kyushu, a small rural town in a remote agricultural region, where it continues to build some of the most complex manufacturing systems in the world, based on a local network of subcontractors it has built up and nurtured over recent decades. Hirata has established a titular head-office in Tokyo, as well as sales offices and many connections. Even so, its base, the home of its controlling family, its principal manufacturing facilities and its subcontracting system remain in Kumamoto. Similarly, Yaskawa has concentrated its robot production system in the remote region of northern Kyushu, especially around its home town of Kurosaki (Patchell 1991: 602). Its new state-of-the-art robot plant where robots make robots, built in 1991, is located at Yahata, adjacent to its old factory in northern Kyoshu and drawing on the older plant's established network of subcontractors (Patchell 1991: 573).

MSFs as big firms locally

Among industrial countries in the world, including the three most important (the US, Japan and Germany), the control functions of innovative MSFs can be found dispersed throughout the country in remote regions and in small places. From a political economy perspective, Kuhn's (1982) point that MSFs pre-empt a hegemonic coalition between big business and big government, has geographical expression by spatially diffusing important private sector decision-making functions, which in turn enhances the resources and influence of peripheral regions and communities. Local control adds to local self-determination.

Within the localities in which they are principally based, MSFs are big economically as well as politically. This observation is self-evident when MSFs locate in small rural localities. Even within metropolitan areas, however, MSFs typically are not 'hidden' but an important part of the local economy. Indeed, it may be argued that MSFs offer distinctive contributions to local economic development. This distinctiveness is revealed by reference to a framework recently developed by Oinas (1995b) to help explain the relationships between the strategy and structure of firms and locality. As Oinas argues, decision-makers vary in their willingness to take into account the long-term interests of local development and to enter into local political affairs. Some firms are interested in long-term local consequences, some are not. The unintended consequences of enterprise strategy and structure for local development also vary.

To help understand the implications of enterprise strategy and structures for local development, Oinas constructs a simple typology linking enterprise dependence and non-dependence on each other (Figure 10.2). Thus, firms are either dependent or non-dependent on localities. While some firms depend on specific localities for specialized, skilled and loyal supplies of labour, a variety of

business contacts which have been moulded by close personal relationships and supportive community attitudes other firms are not tied to any particular place in terms of specialized labour, localized inputs and outputs and personal preference. Similarly, localities are either dependent or non-dependent on firms. In particular, locality dependence increases with the size of operations and the extent to which firms are locally linked.

In Oinas's terms, 'committed agents' occur in situations where there is a high level of inter-dependence between firm and locality. Committed agents employ large numbers of people locally, have developed strong local integrative functions (in terms of personal contacts, business transactions, etc.), participate extensively in local economic functions (e.g. industry associations, chambers of commerce, economic development associations) and social and cultural activities (e.g. sponsorships and endowments of various kinds), and are locally controlled, entrepreneurial and have a close identification with the community. On the other

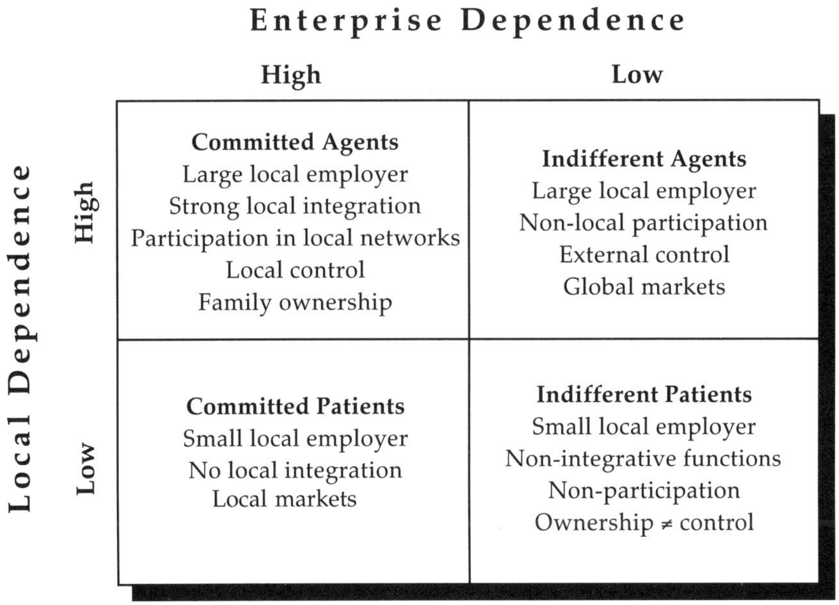

Enterprise Dependence

	High	**Low**
High (Local Dependence)	**Committed Agents** Large local employer Strong local integration Participation in local networks Local control Family ownership	**Indifferent Agents** Large local employer Non-local participation External control Global markets
Low (Local Dependence)	**Committed Patients** Small local employer No local integration Local markets	**Indifferent Patients** Small local employer Non-integrative functions Non-participation Ownership ≠ control

Figure 10.2   Relations of Dependence Between Enterprise and Localities. Source: Oinas (1995a: 191)

hand, 'indifferent agents' are typically large branch plants which are externally controlled and are better connected globally than locally. Enterprises for which localities have little dependence are typically small and are labelled by Oinas as either 'committed' or 'indifferent patients' depending on the extent to which they participate in local networks.

From the perspective of this framework, MSFs offer examples of 'committed agents,' especially in 'home' locations where control and major manufacturing facilities are concentrated and where owners have strong local attachments. The contributions of MSFs to local development are particularly evident in peripheral locations, including agricultural regions where innovative manufacturing has not been traditionally expected. In these regions, MSFs have had the capabilities to develop globally competitive industrialization. Thus, it may be argued that to the extent they act as committed agents, MSFs are important for defining global roles for localities in a way that conflates enterprise and local interests. At the same time, internationalization means that communities are more vulnerable to outside influences, to which they must adapt. The growth of MSFs can bring mixed blessings for local development. Thus, the growth of MSFs in one area may threaten smaller rivals based in other areas. A classic example in Japan is soya sauce, which for centuries was locally produced. When Kikkoman developed a national brand, production in many areas was threatened. In addition, the foreign investment activities of MSFs can incorporate low-wage countries, as Invacare's maquilladora investment reveals (Table 10.3). While such a strategy may complement high-wage locations, over time relationships may change and firms may become 'less committed' to high-wage home regions.

In addition, if MSFs should be acquired by another firm based elsewhere, the shift to subsidiary status may sooner or later re-define local impacts in a way that reduces the firm's local commitments. Such a shift is not inevitable, of course, as one of the advantages of acquiring a MSF may be the extent of its local networks. Some MSFs also become giants themselves, and with such growth, changes in established organizational structures and location preferences can arise, particularly with respect to the movement of control functions to big metropolitan centres. Again, such a shift is not inevitable. It may well be, for example, that in the 1990s it is the MSF that is the organizational model for the giants, rather than the other way around.

Conclusion

Because MSFs are important locally their behaviour in relation to the pressures of globalization is particularly fascinating. Thus, the evolution of MSFs plays a crucial role in mediating global forces in a manner that has direct implications for local development. On the one hand, localities can provide MSFs with access to location advantages which help the internationalization process, and in turn MSFs can help localities, even remote ones, participate in highly competitive global environments. On the other hand, relationships between MSFs and localities are subject to changes which may be less mutually reinforcing. In any event, how MSFs respond to contemporary global competitive pressures is likely to be a sensitive indicator of the relationships between capital and community.

Traditionally, it has been assumed that foreign investment has been dominated by giant firms. In present times, however, MSFs (and even small firms) clearly have internationalization options beyond that of exporting, and they are clearly capable of directly investing in foreign countries. The next chapter addresses the international growth of firms.

11

The Growth of Multinational Firms

This chapter focuses on the growth of multinational (or international) manufacturing firms, defined as firms which control manufacturing establishments in at least two national jurisdictions. In particular, the chapter offers an explanation of why a firm that is based in one nation – its home or donor economy – can establish factories in foreign, more distant countries, i.e. in host economies. The same question can be raised with respect to how firms based in one region can establish factories in more distant regions. Indeed, the multinational firm can be seen as a special case of the interregional firm.

The chapter is in three main parts. The first part offers a brief, broad perspective on the internationalization process by outlining general trends in direct foreign investment (DFI), i.e. the establishment of branch plants in foreign nations, and by recognizing alternative ways in which firms can internationalize their operations. The second part of the chapter reviews what is termed the theory of locational entry of multinational corporations (MNCs) which directly draws on the pioneering work of Hymer (1960, 1972) and other economists, notably Caves (1971, 1974) and Dunning (1973, 1977, 1980). At the heart of this theory is the idea that firms wishing to locate a factory in a foreign country must have some competitive or entry advantage in order to overcome the various problems of

doing business in unfamiliar environments. The third part of the chapter offers two rival hypotheses about how foreign firms behave in host countries following entry. It might be noted that the chapter does not distinguish 'international' from 'multinational', but 'transnational corporations' are seen as representing a new type of development. In addition, the chapter focuses on DFI, or equity investment, while portfolio investments, involving loans from institutional investors to (other) institutional borrowers, are ignored.

Perspectives on international firms

International companies which control facilities in more than one country have a long history and can be traced at least as far back as the 17th century and the development of the big, British-based trading houses such as the Hudson Bay and the East India Companies. Even in the manufacturing sector, international companies have a long history, particularly in the textile and iron industries which were key industries in the first Kondratieff. By 1770, for example, John Wilkinson and his brother, leading iron masters of the Industrial Revolution in England:

"were in possession of three important ironworks – at Broseley, Bersham and Bradley.

He gradually extended the Broseley works and connected them with the Birmingham canal. There he built one after another five or six blast furnaces, obtained coal from deposits which he owned and worked himself. He had interests in foundries in South Wales and was a shareholder in Cornish tin mines. He owned a big warehouse in London with five or six landing stages on the Thames. His activities were extended to France, where in 1977 he set up ironworks at Indret, near Nantes, and where in 1778 he built furnaces for the Creusot foundry. The whole made up a *kind of kingdom, an industrial state*, which Wilkinson governed with a strong and autocratic hand. This State, more important and much richer than many Italian or German principalities, enjoyed a credit which they might well envy and, like them, coined its own money." (Mantoux 1966: 308) (Italics added)

This large, integrated empire or 'kingdom' provides an interesting historical precedent to the corporate giants, such as GM, Ford and Exxon, which comprise Galbraith's (1967) *New Industrial State*. However, in Wilkinson's time, and for decades to come, ownership, profit-receiving, risk-taking, decision-making and often innovation were concentrated among a few individuals. The shift from the owner-managed kingdom of Wilkinson to the giant corporations led by 'captains of industry' (Veblen 1932) and then to the even bigger publicly owned corporations controlled by 'technostructures' (Galbraith 1967) was still to come. Yet the Wilkinsons' firm does raise the central theme of the theory of location advantage: they were able to contemplate establishing an ironworks far from their home base in a country with a very different culture because they had competitive advantages in the form of expertise in iron-making and considerable power and resources.

At the beginning of the Industrial Revolution, the possibilities for international manufacturing firms were severely limited by the still small-scale nature of production, the customized or local nature of technology, financing problems, owner-management, and the difficulties of transportation and communication. With gathering momentum in the 19th century, and since, a variety of technological and institutional innovations facilitated the growth of international organizations (Dicken 1992). In an increasing number of industries the scale of production expanded significantly, while improvements in transportation networks and communication systems allowed access to larger market areas and information between distant places to be more readily exchanged. Of equal, if not greater significance to the emergence of giant firms and MNCs, was the innovation of limited liability and the public corporation. Indeed, according to Marris (1968), it was to overcome the disadvantages imposed by managerial diseconomies of scale and the restrictions of internal financing to the growth of firms that:

"the social architects of the nineteenth century invented the public, joint stock, limited liability company, and thus invented modern capitalism: the managerial restraint on scale was overcome by resort to collective ownership and delegated control, while the financial restraint was handled by the issue of marketable shares carrying limited liability."

Above all else, these social inventions permitted the separation of ownership from control, the hallmark of the modern corporation (Berle and Means 1932). Organization and the managerial division of labour, in effect, became a factor of production in which firms could invest to facilitate growth – and if General Motors is used as a guide, as of 1995, firm sizes of US$168 billion were possible and there is no reason to believe that this size is some defining limit. Indeed, it is important to recognize that the MNC is an evolving form of organization: the MNCs of the present time are different from those of 100 years ago, not simply in terms of size and scope but also with respect to organization. Several Sheffield-based speciality steel manufacturers, for example, established branch plants in the US in the late 19th century and the first two decades of the 20th century, following the imposition of a tariff by the US government to protect domestic competition from exports from Sheffield (Table 11.1). As Tweedale (1986; see also 1987a and b) notes, these branch plants were created by the relocation of skilled workers and managers, and operated in a largely autonomous manner with only limited forms of communication and integration with their parents back in Sheffield. Tweedale concludes that the Sheffield parents received few benefits from these ventures, several of which were closed or sold within a decade or so of opening. Primarily since 1950, however, branch plants have typically been closely integrated within the overall parent company strategy. Thus, while forms of organization vary, including with respect to the degree of decision-making decentralization, the contemporary MNC is distinguished from its predecessors by the centralization of policy-making and closely knit integration of facilities across national boundaries (Behrman 1969; Barnet and Muller 1975).

DFI as a form of internationalization

If internationalization is broadly defined as "the process of increasing involvement in international operations" (Welch and Luostarinen 1988: 36), individual firms internationalize their activities by exporting to foreign countries, selling licenses of a product they have developed to foreign firms and participating in strategic alliances with foreign firms, as well as through DFI. In addition, firms can internationalize by procuring foreign sources of raw materials and other inputs which may or may not be transformed into products for export. To some extent, the export, licensing, strategic alliance and DFI alternatives substitute for one another. These alternative forms of internationalization, however, also complement each other and these complementarities are significant, especially those between trade and DFI.

Exports From a product market perspective, firms typically first internationalize by exporting. Many manufacturing firms do not export and among those that do, the relative importance of exports as a percentage of sales (the export : sales ratio) and the commitment to exporting varies substantially (LeHeron 1980; Hayter 1986a,b). For some firms, exporting can be an occasional, 'one-off' type of sale which is arranged through an export wholesaler, export association or foreign-based sales agent in which the exporting firm itself has little or no direct contact with the foreign market. Traditionally, in fact, many small and medium-size firms (SMEs) have developed relatively significant export sales without making direct contact with foreign consumers. In other cases, foreign consumers visit potential exporters to establish contacts. However, exporting in this fashion, necessarily limits exporting firms' understandings of foreign market dynamics such as those which arise from changes in taste and new competition. Indeed, an important feature of

Table 11.1 Sheffield High-Grade Steel Firms in America 1860–1940

Sheffield firm	Product	American location	Start-up	Product	Comment
1. W&S Butcher	Crucible steel; tools	Philadelphia	1867	Steel castings	Reorgan. 1870
2. Sanderson Bros	Crucible steel	Syracuse	1876	Crucible steel	Sold 1900
3. Thos Firth & Sons	Crucible steel; arms	Mckeesport, PA	1896	Crucible steel; arms	Sold 1949
4. Wm Jessop	Crucible steel; castings	Washington, PA	1901	Saw steel	Sold 1920
5. Edgar Allen	Manganese steel castings	Chicago Heights	1910	Manganese steel castings	Sold 1920
6. Hadfields	Steel castings, arms	Bucyrus, IL	1917	Projectiles castings	Failed 1927

Source: Tweedale (1986: 79).

greater export commitment is the establishment of direct contacts in foreign markets and more control of sales channels.

The initiation of exports is often a difficult process as firms must make contact with distant and unknown customers, arrange some form of financing and insurance, complete the necessary protocols and documentation demanded by the importing nation, arrange for transportation and distribution, and absorb the risks of costs incurred prior to payment (Bilkley 1978). Nevertheless, as firms grow in size, markets must be enlarged and in most countries such growth quickly implies exports. In the Sheffield region of the UK, for example, medium-size and large cutlery, tool and steel firms virtually all developed substantial exports, as measured by the export : sales ratio (Figure 11.1). On the other hand, the export behaviour of the smallest firms in these industries is far more variable: some firms have high export : sales ratios and others do not export at all. The narrowing of the 'envelope' of export : sale ratios as firms increase in size, however, indicates that the development of an export trade is an important challenge typically facing the growth of manufacturing firms.

To some extent exporting and DFI are alternatives available to firms in accessing foreign markets. They are often complementary processes. Thus, exports can prepare the way for DFI by establishing business contacts and sources of information; once established, DFI allows firms to gain a much better understanding of foreign markets, thus facilitating exports. It can also be argued that as firms grow they will need to extend their purchasing patterns. Moreover, as with sales, purchasing links can prepare the way for DFI. However, DFI to secure inputs is almost always designed to institutionalize established trade links or create new ones.

Direct foreign investment

Frequently, if not inevitably, DFI occurs in places where the firm already has export or sourcing connections (Welch and Luostarinen 1988: 44–46; Yoshihara 1978). The establishment

of foreign operations, by acquisition of existing facilities or investment in new ones, typically involves firms in complex learning and bargaining processes comprising long-term strategic decisions. In terms of control, international firms may choose to establish wholly owned and controlled subsidiaries and branch plants, or share ownership and control with local firms and governments or other international firms. Joint ventures typically involve agreements between two or more firms or government organizations to build a new factory in which ownership and control is shared, often according to the share of financial commitment. Joint ventures may also share marketing responsibilities.

For the individual firm, the most costly, uncertain and sophisticated form of internationalization is DFI. Even so, international firms have grown massively throughout the 20th century (Dunning 1958, 1993; Wilkins 1994). In fact, although estimates are crude, by 1914 accumulated DFI was already substantial (Table 11.2). Comparing the major sources or host economies of DFI in 1914 with those in 1988, two major points may be made. First, the origins of DFI in 1988 remained highly concentrated among leading industrialized countries, if slightly less so than in 1914. Second, among leading industrialized countries, the origins of DFI remained concentrated in 1988 but notably less so than in 1914. In 1914, four countries (the US, the UK, France and Germany) accounted for almost 87% of accumulated DFI, and MNCs in the UK (Jones 1986) and the US (Wilkins 1970) were especially important. By 1988, the leading four countries (the US, the UK, Germany and Japan) accounted for 65.6% of accumulated DFI, and between 1914 and 1988 the US replaced the UK as the world's leading supplier of DFI. Indeed, in the 1950s and 1960s the US accounted for an astonishing 40–50% of DFI. In 1988, the US was still easily the largest donor economy and US-based MNCs continue to invest aggressively in different parts of the world (Schoenberger 1990). Since the 1960s, Germany and Japan have emerged as major donors. The growth of Japanese MNCs has been particularly noteworthy

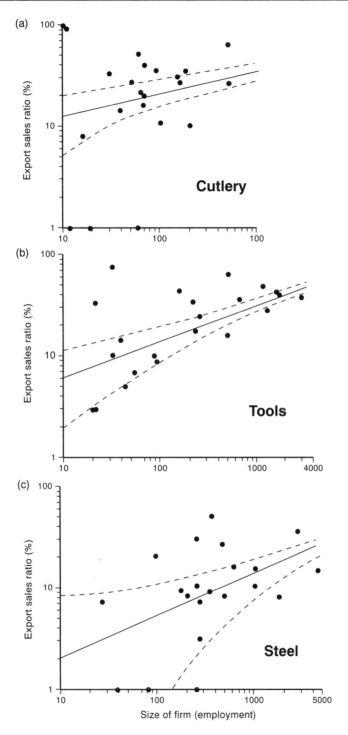

Figure 11.1 Export Performance and Size of Firm in Sheffield's Manufacturing Industries, 1979, Source: Hayter (1986a: 790)

Table 11.2 Estimated Stock of Accumulated Direct Foreign Investment by Country of Origin

Country	Percentage distribution					
	1914	1938	1960	1973	1983	1988
USA	18.5	27.7	48.3	48.1	39.6	30.5
Canada	1.0	2.7	3.8	3.7	5.1	4.4
UK	45.5	39.8	16.3	12.8	16.7	16.2
Germany	10.5	1.3	1.2	5.7	7.0	9.1
France	12.2	9.5	6.2	4.2	5.2	5.0
Netherlands			10.6	7.5	6.4	6.8
Switzerland			3.5	3.4	3.5	3.9
Japan	0.1	2.8	0.8	4.9	5.6	9.8
Developing countries			1.0	2.9	2.6	2.8
Total ($B)	14.5	26.4	66.1	210.5	572.8	1140.5

Source: Dunning (1988: 74), except for 1988 data which are extracted from Dunning (1993: 17).

Table 11.3 Estimated Stock of Accumulated Direct Foreign Investment by Recipient Country or Area

Country	Percentage distribution					
	1914	1938	1960	1973	1983	1988
USA	10.3	7.4	13.9	10.4	23.2	27.0
Canada	5.7	9.4	23.7	16.8	10.2	
Western Europe	7.8	7.4	22.9	36.5	35.8	36.5
(UK)	(1.4)	(2.9)	(9.2)	(8.9)	(9.1)	(9.8)
(Germany)						(6.8)
Japan	0.2	0.4	0.2	(0.8)	(0.7)	0.9
Latin America	32.7	30.8	15.6	(12.5)	(12.6)	9.4
Africa	6.4	7.4	5.5	(2.9)	(2.2)	2.5
Asia	20.9	25.0	7.5	(4.8)	(7.2)	9.3
Total ($B)	14.1	24.3	54.5	166.7	552.6	1219.3

Source: Dunning (1988: 75), except for 1988 which are extracted from Dunning (1993: 20).

in recent years and has been particularly focused on Australasia, North America and Europe (Yoshihara 1978; Edgington 1987, 1990; Dicken 1988).

With respect to the destination of DFI, patterns are more diverse and developing countries are important, especially prior to 1939 (Table 11.3). In 1988, developing countries accounted for about 20% of accumulated DFI. While the amount of DFI destined for developing countries is significant, the majority of DFI occurs among already industrialized countries. Since before the turn of the century, Canada has been a significant host country for

its size (Ray 1965; Britton 1980; Hayter 1981; McNaughton 1992a,b; see Chapter 15). Western Europe, especially the UK, has a long tradition of receiving DFI, and after 1950, stimulated by the Marshall Plan, western Europe received a considerable amount of DFI by US-based MNCs (Dunning 1958; Servan-Schreiber 1968; Blackbourn 1974). DFI in the EC continues to be important and involves much intra-European flows, such as from European countries to the UK (Watts 1979, 1980b, 1982). Since 1970, DFI in Asia, including China and the ASEAN countries, has grown very rapidly in relative as well as absolute terms (Leung 1993; Eng and Lin 1996). Latin America's relative share of DFI continues to decline but in some countries, notably Mexico, there has been massive growth in recent years (South 1990; Kenney and Florida 1994). In addition, in the 1990s, since the break-up of Soviet-controlled eastern Europe, there has been a flood of investment there in a variety of industries (Murphy 1992; Michalak 1993). Finally, if in the 1980s Europe was the most important destination region for DFI, the US has emerged as the single most important national destination (McConnell 1980, 1983; Ó'hUalla-cháin 1985, 1986; MacPherson and McConnell 1992; Ó'hUallacháin and Reid 1992).

The location conditions affecting DFI on an international scale are varied. In a highly generalized way, however, at this scale access to markets has arguably been the most important concern. In primary and primary manufacturing sectors, access to resources is a more important location condition. In addition, DFI has occurred in low-wage countries to access cheap, pliable labour. Even so, DFI in poor countries is relatively small compared to flows occurring in the developed countries; and among developing countries, DFI tends to be highly concentrated among relatively few countries. Moreover, some DFI in poor countries, including China, is market-driven as MNCs have established branch plants with an eye on the country's vast market potential. It might also be noted that historically firms based in OECD countries have been the most important foreign investors, although in recent decades firms from other parts of the world have established foreign operations. While in many cases these operations are small, there are examples of newly emerging corporate giants, such as Daewoo of South Korea which has invested extensively in foreign countries, most recently in the automobile industry of eastern Europe (Kraar 1996b).

International firms: a theory of locational entry

There are many explanations for the international firm (Dicken 1992: 120–145). There are theories of

internationalization which are abstract, focusing on aggregate capital shifts rather than the strate-gies of individual firms from conventional and radical perspectives (Palloix 1977). There are theories which focus on how international firms function, rather than on how they grow, and these in turn comprise conventional approaches (e.g. McManus 1972), which stress the efficiency of MNCs in allocating resources, and radical approaches, which stress how MNCs underpin spatially uneven development (e.g. Frank 1967). Another set of theories explain how firms grow internationally. Some of these theories are 'partial' (by highlighting one particular dimension of the internationalization process), such as the Scandi-navian model which interprets internationalization in terms of market learning processes (Johanson and Vahlne 1977; Hakanson 1979; Figure 8.4). Another partial approach interprets internationa-lization as a bargaining process (Gregerson and Contreras 1975; Krumme 1981; Soyez 1988b; Figure 7.6). From the perspective of how indivi-dual firms grow to become MNCs, however, the most influential tradition is provided by the so-called theories of oligopolistic advantage pioneered by Hymer (1960), Caves (1971), Dunning (1973) and others (Vernon 1966) primarily in the context of the massive expansion of US-based MNCs in the 1950s and 1960s. In this chapter, this approach is summarized as a model of locational entry (Figure 11.2).

Basically, the locational entry model argues that firms internationalize their operations to exploit some internally generated entry or competitive advantages, for example, in the form of marketing, production or technological know-how. These entry advantages are of sufficient strength to overcome various spatial 'barriers to entry,' defined ultimately by the problems of competing with local firms in foreign markets (Hayter 1981). The oligopolistic advantage / loca-tional entry model has several advantages for geographical enquiry. First, if this approach is not a macro theory of international capital movement, it does provide a relatively general framework within which to understand the internationaliza-tion of individual firms. Thus, the idea that

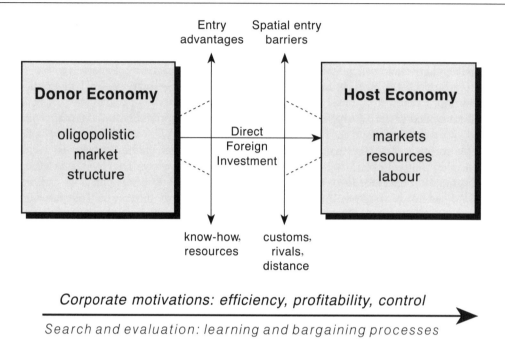

Figure 11.2   Entry Advantages and Spatial Entry Barriers

internationalization is a learning or bargaining process is readily incorporated within its terms of reference. Second, it makes no *a priori* assumptions about the contributions of the international firm to local development. Alternatively, this approach allows MNCs to be recognized as efficient resource allocators or as exploiters reinforcing local dependency. Third, the locational entry model imparts to studies of industrial location studies, especially those from a behavioural and institutional perspective (see Chapters 6 and 7), greater significance than is commonly supposed in the geographical literature (Hayter 1981). Fourth, the model explicitly incorporates alternative forms of internationalization, notably the export behaviour of firms (Hirsch 1976). The locational entry model is briefly and simply summarized in this chapter in terms of two basic dimensions: entry advantages and entry barriers. It might be noted that a much more complicated variant of this approach is provided by Dunning's (1973, 1977) eclectic theory of international production (see Edgington 1987).

Spatial entry barriers

The essence of the location entry model is that firms contemplating investments in foreign environments must have some entry or competitive advantage *vis-à-vis* potential local competitors in order to compensate for various spatial 'barriers to entry' (Caves 1971; Hayter 1981). Local entrepreneurs, for example, do not face communication problems across national boundaries and they typically 'inherit' without cost much information pertaining to local legal, cultural, political, economic and physical conditions which in one way or another impinge upon the viability of plants. In addition, they have a better idea of where relevant information can be found. In other words, spatial entry barriers pertain to the additional managerial costs and uncertainties incurred by foreign investors over and above those facing (potential) local entrepreneurs during the selection of regions, communities and sites.

To use the terminology introduced in Part II, spatial entry barriers are defined for behaviouralists by the costs, time and uncertainties involved in attempts to reduce the 'knowledge gap' in making locational choices (see Chapter 6). Alternatively, for the institutionalists, spatial entry barriers are defined by the costs, time and uncertainties in negotiating locational choices (see Chapter 7). In practice, both information search and bargaining processes are closely intertwined and to some extent they can be measured, for example, in terms of managerial costs and the price of consultant studies. However, given that location conditions comprise intangible features, that knowledge gaps cannot be defined precisely and because the behaviour of others cannot be completely controlled, locational search and bargaining processes are inherently uncertain. Consequently, any *a priori* assessment of spatial entry barriers is necessarily judgemental. Certainly, these barriers should not be under-

estimated. Studies in the geographical (Stafford 1972; Townroe 1971; Hayter 1978) and non-geographical (Ricks et al 1974; Tweedale 1986) literature provide numerous examples of industrial problems that have occurred when firms unfamiliar with local conditions have made decisions based on inadequate and inappropriate information (see Chapter 6).

A case study by Soyez (1988a) systematically analyses the concept of spatial entry barriers that faced the Swedish firm, Stora Kopparberg, when it established a pulp mill in Nova Scotia, Canada (Table 11.4). Nova Scotia offered a fundamentally different host environment, in terms of its physical environment, as well as its economy and culture, compared to Stora Kopperberg's home or donor environment. These differences implicated the firm in years of collecting information and negotiating with local interests. Stora found that it could not simply transfer its

Table 11.4 Important Differences in Locational Environments for the Company in Sweden and Nova Scotia, 1960

Sweden	Nova Scotia
Extensive company-owned forests	Leased crown land
All forest operations within the company's own responsibility	All forest operations to be approved and controlled by provincial authorities
All forest operations planned and executed by the company's own personnel and machinery	Logging and delivery made by contractors and their equipment
Good knowledge of existing species and of many operational effects on the forest, based on long experience with sustained-yield forestry	Lack of familiarity with local species and with effects of any operation
Woodland consisting of managed forests	Woodland consisting of high-graded secondary or untouched natural forests
Detailed forest inventories	Only reconnaissance inventories
Integrated supply linkages to private woodlots	Supply linkages to small woodlots must be developed
Sustained-yield forestry even in privately owned woodland	Use of privately owned woodland mostly as exploitation, with almost no silvicultural input
Clear legal framework provided by appropriate forest legislation	Rudimentary forest legislation
Strong and competent forest authority, with many local branches controlling appropriate implementation of existing regulations	Insufficiently manned forest authority, with almost no tradition regarding intensive silviculture and only rudimentary regional and local structures controlling implementation of existing regulations
Close working relations with a single labour union	Numerous sectoral trade unions
Limited number of merchantable tree species, forest management being focused on pine and spruce	Great diversity of tree species, with a significant proportion of hardwood

Source: Soyez (1988b: 136).

'state-of-the-art technology', for example, concerning silviculture, from Sweden in order to 'improve' local practices. The problem for Stora was that Swedish practices were not entirely appropriate for local conditions, while local interests did not regard Swedish practices as necessarily constituting improvements.

Psychological distance Countries constitute different kinds of cultural, institutional and economic environments in which to conduct business and these differences have physical, social, political and economic foundations which are reinforced by varying degrees of geographic separation. These differences (or distances) form the basis of Johanson and Vahlne's (1977: 24) concept of 'psychological distance' which they define (in the context of exporting) as "the sum of factors preventing the flow of information from and to the market. Examples are differences in language, education, business practices, culture and industrial development." In the 19th century, geographic distance alone constituted major problems for intra-corporate communication and control. In recent decades, as these problems have been reduced if not eliminated, other forms of 'distance' have become more important, e.g. the distances created by different languages and business culture. On the whole, the greater the psychological distance between countries, the greater the size of spatial entry barriers. From this point of view, corporate preferences for investment in adjacent, similar environments are not only 'conservative' but economically rational. This helps explain, for example, the timing, extent and location of American corporate penetration in the Canadian economy (Marshall et al 1936; Aitken 1961; Ray 1965; Wilkins 1970) and the preference of Canadian firms to locate in the US (Litvak and Maule 1981; Niosi 1985). Similarly, the often-observed corporate support for such policies as free trade, common market agreements, metrication and related measures, which have the effect of facilitating movement and reducing regional differences, are explicable as attempts at reducing the costs of adjusting business practices to

particular local circumstances (Levitt 1970; Christopherson 1989; Schoenberger 1994).

Even between countries considered close in terms of psychological distance, such as Canada and the US, spatial entry barriers exist and can be significant. Gates (1992: U2), for example, warned Canadian firms contemplating investment in the US that, even following the Free Trade Agreement of 1989, "rules can be different and harder". This report acknowledged the growth of non-tariff barriers since the Canada–US Free Trade Agreement of 1989, while also paying particular attention to differences in tax laws. The general, and correct, view is that income tax in the US is lower than in Canada and so in this respect the US is more attractive to firms. Yet, in the US, there is a "minefield of other taxes" (Bianchi 1992: U3) which are different to those in Canada (Exhibit 11.1). For two countries typically considered as relatively close in terms of 'psychological distance', the differences in the US and Canadian tax systems alone are impressive.

Exhibit 11.1
Tax Differences in the US from a Canadian Perspective

1. A payroll tax paid by both employee and employer and an additional levy on higher paid workers
2. A federal unemployment tax which may be supplemented by (varying) state taxes which may or may not be credited against the federal unemployment tax
3. A federal corporate income tax which is supplemented by (varying) corporate income taxes which are imposed by 43 states (Connecticut has the highest) while some cities, including New York and Detroit, also charge corporate income taxes
4. Taxes imposed by some states on new equipment, especially states that do not have a sales tax
5. Taxes imposed by some states and counties on personal property such as cars and refrigerators
6. Luxury taxes recently imposed by some states and counties on cars, planes, furs and jewellery
7. A federal environmental tax introduced in 1987 on all businesses with sales in excess of US$2 million

adjustable taxable income. There is also a petroleum oil spill tax and a petroleum super fund tax
8. A battery of taxes specifically targeting foreign investment. As of late 1992, these taxes included a 15% US non-resident withholding tax on dividend payments to foreign shareholders; a 10% withholding tax on royalty payments from a US company to a foreign company (plus an additional 10% on profits in excess of C$500 000 returned to foreign countries); and a 15% US non-resident withholding tax on interest payments from the US to foreign countries. In California the unitary tax allows for a subsidiary's tax to be computed from the foreign company's tax base, and in 1992 the Internal Revenue Service planned to clamp down on subsidiaries who file late. Bianchi (1992: U3) further notes that "In a further attempt to crack down on foreign investors, two influential congressmen have introduced a proposal that, if passed, would mean foreign-owned subsidiaries would have to pay US taxes – even those reporting net losses"

Source: Bianchi (1992: U3).

More significant differences in language, culture and business philosophy that exist between nations (and regions) widen the psychological distance and therefore the size of spatial entry barriers. Many western corporations, for example, wish to export and invest in the Chinese and Japanese economies. However, both countries pose formidable entry barriers. In the case of Japan, its chosen path of industrialization, which has given priority to reverse engineering and the development of indigenous enterprise, has traditionally restricted opportunities for foreign firms, confirmed by policy restrictions on DFI (Freeman 1988; Kudo 1994). While these policy restrictions have softened in the 1980s and 1990s, Japan's distinctive culture, language and market preferences mean that most foreign firms face significant learning and bargaining costs as a prerequisite to selling or investing there (Encarnation 1992; Safarian and Dobson 1995). Moreover, western companies complain about high tariff and non-tariff barriers, including deeply embedded inter-firm relations and complex distribution systems controlled by *keiretsu*

(powerful enterprise groups) and affiliated *sogo shosha* (giant trading companies). For most western firms, substantial participation in the Japanese economy requires considerable patience and effort (Hayter and Edgington in press). In the case of China, while it has welcomed DFI since the economic reforms of the late 1970s, most western firms find the political, social and economic system, and the ways business transactions are conducted, to be extremely difficult (Ho and Heunemann 1984). Indeed, these problems have encouraged the Singapore government to establish an agency to help western firms and Chinese authorities negotiate with each other (Kraar 1996a). For the western firms the Singaporeans provide an understanding of Asian business culture, and for the Chinese they provide considerable experience in developing industrial parks and attracting DFI. For Singapore, such services are a way of further exploiting their know-how of the industrialization process in an Asian context which is consistent with a high-wage economy that is less able to attract low-wage industry itself.

Method of entry Even within a given industry and given host and donor environments, the spatial entry barriers facing internationally expanding firms may differ. Firms vary in competence and willingness to substitute a higher (lower) degree of uncertainty for lower (higher) costs of collecting information on new environments. Some firms may accept the higher risk associated with not spending time and money on locational choice by mistake or to avoid 'procrastinating' and possibly losing business opportunities. Spatial entry barriers facing firms also vary depending on whether entry is realized by investments in wholly owned new site branch plants or as a joint venture with local firms, or by acquisitions of existing firms. Acquisition as a method of entry is particularly attractive since it involves fewer uncertainties than building new plant as foreign firms inherit both existing locations (accumulated capital resources) and existing management and workers (accumulated human resources).

In other words, acquisition offers an 'instant' and possibly cheap way of understanding local conditions, especially if local owners of capital underestimate the importance of such geographical knowledge. Given its inherited and implicit nature, local know-how may well be underestimated – a tendency that may be reinforced by fears among management and workers over job security following the take-over. Acquisition also facilitates relatively rapid expansion and from this point of view local firms may constitute not so much a rival to the foreign firm as an opportunity to gain advantage over, or react to, its 'real' international competitors. Acquisition is not always an option, however, and foreign firms may have to build new plant. In this case, foreign firms can sometimes arrange joint ventures with local firms and in so doing obtain the necessary geographic know-how and circumvent foreign ownership criticisms. Otherwise, the foreign firm must accept the costs and uncertainties of establishing a manufacturing plant in a new environment. From the point of view of local development, the benefits to be derived from acquisition, if they exist, are typically much less obvious than entry by internal growth.

The importance of acquisitions and joint ventures in the internationalizing process is well demonstrated by the interregional and international expansion of MacMillan Bloedel (MB), a forest product giant based in British Columbia (Table 8.3). These expansions have been dominated by acquisition as a method of entry while new mills were only built as part of joint ventures. Subsequently, MB did acquire full control of some of the joint ventures once they had proved successful. Similarly, the numerous foreign firms entering the forest sector of British Columbia prior to 1980 did so either by acquiring an existing operation or by building a new plant as part of a joint venture with firms already operating in the province (Hayter 1981).

Host government policy Governments, even among OECD countries, vary in their openness to DFI, especially as regards acquisition. Thus,

Canada, Australia, the UK and the US have favoured relatively liberal 'wide open door' policies to DFI. In contrast, countries such as Sweden, South Korea and especially Japan have limited the possibilities for DFI, especially in the past. Government policies towards DFI can change and in recent times there have been several dramatic turn arounds in this regard. The most significant changes have occurred in still Communist China and in former USSR and its satellite east European countries. Thus, China in 1979 completely reversed its existing attitudes and began to welcome foreign firms, in selected economic zones at first and in more and more regions of the country since then (Sit 1986). The majority of the firms that have invested in China are in fact Hong Kong based and are of Chinese origin, while firms from elsewhere have been slower to participate (Leung 1993; Eng and Lin 1996). Clearly, the entry barriers for the Hong Kong based firms are less than for western firms. Similarly, in eastern Europe, the leading foreign firms participating in the newly liberalized economies are from other European countries and, to further compensate for their lack of familiarity with local situations, these firms have typically formed joint ventures with local partners (Michalak 1993). Other examples of developing countries radically changing their policies in favour of more liberal attitudes to DFI include Brazil in the 1960s and Chile following the overthrow of the Allende regime in the 1970s.

Differences in host-country policies towards DFI constitute formal, institutional variations in entry barriers facing internationally expanding firms (and simultaneously reveal the controversial nature of DFI in the local development process). While some countries have deliberately restricted DFI, others have been far more welcoming, not only by allowing DFI but by offering monetary incentives and other forms of inducements. Similarly, many regional and local governments actively seek to welcome and induce DFI to their jurisdictions. From this perspective, the incentives made available to entice foreign firms, such as tax breaks, cash grants, low resource royalties, favourable profit repatriation schemes and attractive investment depreciation allowances, and the range of services offered to investing firms, including the provision of low-cost buildings and more commonly much free information on the local economy, are specifically designed to reduce entry barriers. At the same time, since incentive schemes reduce the entry barriers of foreign firms it may be argued that they simultaneously discriminate against local entrepreneurs by reducing their natural advantage *vis-à-vis* international competitors (Hayter 1982).

Apart from specific host-country policies affecting DFI, firms contemplating international expansion are widely reported to be influenced by questions of 'political stability'. In general, it is argued that firms seek out areas which are politically stable. As a location condition, 'political stability' is not without ambiguity since the principle of 'free enterprise' is closely associated with the principle of democracy and yet the very nature of democracy invokes the right to change government. Yet, what appears to be crucial for international business in its interpretation of political stability is that rules and regulations affecting business will not change in unanticipated ways. Issues of individual freedom and human rights, on the other hand, are typically seen as beyond the legitimate concerns of international business. Thus, *ceteris paribus,* countries with military dictatorships can be attractive to firms because they guarantee 'stability'. On the other hand, countries which democratically elect a government that introduces new ('anti-business') laws may well be regarded as unstable and not a suitable place to invest.

Size of firm According to Caves (1971: 13), spatial entry barriers are greater for small compared to large firms. His argument is that international investment involves relatively high and to a large extent fixed planning costs and that larger firms are better able to bear this burden. In addition, it may be argued that the high level of uncertainty associated with international investment in new plant can be more effectively borne by the already large firm. Moreover, large firms are powerful simply because they are big. Thus, they enjoy 'economies of size', i.e. the bargaining advantages associated with size and which include the ability to locate somewhere else. They also enjoy what Penrose (1959) calls the 'economies of expansion', which are resources the firm has available for planning purposes, notably under-utilized managerial resources. Consequently, large firms are in a better position to negotiate the conditions of entry compared to small firms (Krumme 1981; Soyez 1991; see Chapter 7). In general terms, in contrast to small firms, large corporations already dominant in domestic markets, typically enjoy broader spatial planning horizons, are able to draw on past experience in adapting organizational structures to growth, may perceive geographical concentration itself to be a source of uncertainty, or at least a constraint on expansion plans, and, by no means least, are more likely to have acquisition (and joint venture) possibilities (McNee 1974; Taylor 1975a; Hayter 1976).

Historically, the evidence indicates that aggregate patterns of DFI have been dominated by large firms while the size of branch plants also tends to be much larger than industry averages (Britton and Gilmour 1978). As Caves (1971) notes, small firms able and willing to internationalize their operations may find the planning costs associated with establishing licensing arrangements more compatible with their resources and abilities to withstand uncertainty. There are, however, many small firms that have invested in foreign countries. Large numbers of Japanese SMEs in the textile industry, for example, established operations in various Asian countries in the 1970s. Even so, in this case, the small firms were aided in financial, marketing and planning terms by the huge Japanese trading companies or *soga shoshas* (Yoshihara 1976). Another example is provided by the numerous Canadian SMEs who have invested in the US. In this case, it seems that Canadian entrepreneurs have frequently found the spatial entry barriers of establishing a branch plant at nearby locations in the US less than establishing a more distant branch plant in another Canadian province. In addition, as the previous chapter noted, medium-sized firms within the contemporary global economy frequently internationalize operations by exporting and by DFI. Their behaviour may well denote a new trend of the 1980s and 1990s in the organizational structure of DFI.

Spatial entry advantages

In a general sense, firms expand internationally to meet strategic motivations which relate, on the one hand, to profitability and efficiency considerations, and on the other hand, to control of markets and resources and related security reasons (see Chapter 7). More specifically, given the constraints of important, even formidable entry barriers, firms pursue international expansion strategies which further extend established entry (competitive) advantages. In this regard, Caves (1971) distinguishes between horizontally and vertically integrating firms (Figure 7.1).

Horizontal and vertical integration   Horizontal integration occurs when firms expand in their existing or closely related line of business. In this case, entry advantages comprise some internally developed asset or expertise related to technique, product organization, marketing, financing and/or human skill which can be invested in a new host economy without the need to incur much or any of the fixed costs associated with its original development. In other words, the size of the horizontally expanding firm's entry advantage is fundamentally the fixed costs and uncertainties of acquiring or imitating its distinctive asset by local entrepreneurs (Vernon 1970).

Vertical integration occurs when firms grow by investing in facilities which provide a market for existing products (forwards vertical integration) or in facilities which supply inputs to existing activities (backwards vertical integration). The entry advantages associated with vertical integration, whether forwards into markets or backwards into raw materials, emphasize the (cost) advantages of supplanting the market mechanism. By definition, vertical integration

means that corporations gain control over technically linked stages in the production process in terms of decisions about the timing, quantity and quality of flows of goods and services. In part, the advantages of vertical integration have a technical base where there are clear economies of continuous flow operations such as exists between pulp and paper or iron and steel. But corporations often vertically integrate physically separate production stages. One important theory argues that firms vertically integrate for efficiency reasons, specifically to reduce the 'transactions costs' of utilizing markets, i.e. independent suppliers and consumers (Coase 1937; Malmgren 1961; McManus 1972). These transaction costs relate to the costs of searching for information about markets and supplies, the costs and uncertainties of negotiating contracts and the vulnerability that potentially arises from failures by independent firms in meeting the terms of contracts for a variety of reasons. Another theory argues that firms vertically integrate to reduce uncertainty in supply and marketing chains and threats from rivals, i.e. to realize greater stability and security of operations (Galbraith 1967). MB's first international venture to acquire paper-box plants in the UK which were markets for its pulp and paperboard, for example, was stimulated when a rival threatened to acquire these same plants (Table 8.3; Hayter 1976). In like manner, vertical integration may provide firms with bargaining power in order to ensure that prices of inputs and outputs are 'fair' and to compete with equally large vertically integrated rivals. In principle, the size of the vertically integrating firm's entry advantage is 'measured' by either the savings in 'transaction costs' or by the advantages of greater security in lines of supply and markets (see Chapter 13).

In practice, FDI typically incorporates elements of both horizontal and vertical integration. In summary, branch plants can be interpreted as a bundle of integrated production, marketing, financial, planning and technological assets and expertise which collectively can comprise extremely powerful entry advantages in a foreign country. In addition, corporations planning on international expansion may be able to draw upon Penrose's (1959) 'economies of expansion', and economies of size. The former refer to under-utilized resources within the firm (especially managerial resources) which may disappear once the firm has grown, and the latter to the power of already large firms.

However precisely defined, entry advantages or, in the vernacular, 'head starts', that firms accumulate over time both permit, and are reinforced by, the building of new branch plants or the acquisition of new subsidiaries. Thus, branch plants add to the rate of return on established entry advantages. In addition, branch plants ensure that these entry advantages remain under the control of the firm while at the same time adding to them by acquiring knowledge of the host economy. Moreover, over the past four decades the literature suggests a change in the defining characteristics of leading corporate exemplars of entry advantages. This shift may be summarized in terms of the (fordist) 'American Challenge' and the (flexible) German/Japanese Challenge.

The (fordist) American Challenge: technology and marketing From a European perspective, Servan-Schreiber (1968) interpreted 'the American Challenge', in a book with that title, as a result of the power, leadership and innovativeness shown by US-based MNCs during the 1950s and 1960s when US-based MNCs dominated aggregate patterns of DFI (Table 11.2). He saw the US-based MNC as the model form of corporation and for Servan-Schreiber the American challenge was how to stimulate similar types of European companies and similar types of business practices throughout European industry. For Servan-Schreiber, the defining core assets of US corporations lay in their technology and marketing. Indeed, two well-known models, the product cycle and market linkage models, both emphasize technological and marketing expertise as the guiding forces underlying international investment by US-based MNCs. The product cycle model gives particular priority to technological advantages (Vernon 1966; Hirsch 1967) and the market linkage model to marketing advantages (Kolde 1972).

The product cycle model interprets the entry advantage of horizontally expanding firms primarily in terms of technological expertise or 'head starts' (see Chapter 4). Thus, Vernon's (1966) original formulation of the product cycle model emphasizes that US corporations, specifically manufacturers of consumer durables, were able to pioneer new product developments in part because of access to large pools of scientists, engineers, skilled labour and related external economies, and in part because of access to a huge domestic market comprising wealthy consumers willing to try new products. Thus US corporations were often the first to innovate new products and the profits gained by such 'head starts' further reinforced investments in corporate R&D to provide *firm-specific* technological advantages and to ensure a continuing stream of products. As individual products mature, however, input conditions change as firms increasingly rely on mass production technology and relatively unskilled labour. According to this model, US MNCs internationalize in search of unskilled labour with relatively low labour costs.

Kolde (1972) recognizes the technological strengths of the US corporation while stressing marketing know-how, networks and power. Thus, for Kolde, in the first instance, market knowledge directs the R&D process with critical information about consumer needs. Second, investment in marketing channels, distribution channels, advertising, product differentiation and brand names seeks to influence and control consumer demands to ensure that the firm's technological advantages are extended as far as possible and investments in R&D (and marketing efforts) are profitable. Initially, these investments concentrate in the US. Subsequently, US firms expand internationally as an extension of their marketing tentacles. Over time, firms develop exports and related marketing infrastructure (from agents and occasional personal contact to sales offices, warehousing and distribution systems) and as exports increase, the associated development of market power and market information prepares the way for international investments.

In their basic forms, as just reviewed, the product cycle and market linkage models suggest alternative paths of internationalization (Figure 11.3). In the product cycle model, for example, exports from the donor economy (the US) primarily to other rich countries in stage 1 are replaced by DFI in poor (low-wage) countries in stage 2, and exports from these 'platforms' supply the donor (the US) and rich host markets (Figure 11.3(a)). In the market linkage model, on the other hand, DFI follows (stage 2) and replaces previously established exports (stage 1) and both exports and investments predominantly favour rich countries (Figure 11.3(b)). The existence of tariff barriers protecting host countries from exports further reinforces this trend.

However, several 'hybrid' cases can be readily identified in which product cycle and trade linkage mechanisms complement each other. Thus, in one hybrid version, firms select host countries on the basis of market linkage, but within the host countries firms choose regions and communities that offer low labour costs (and government inducements) for a product which is mature (Figure 11.3(b)). Foreign firms in the US auto industry have recently pursued this kind of strategy (Krumme 1981). Another hybrid is that MNCs choose low labour cost 'platforms' which are nevertheless closer to the markets they wish to serve (Figure 11.3(a)). European and Japanese MNCs that locate in the Mexican maquiladora zone to supply US markets (Kenney and Florida 1994) or US and Japanese MNCs locating in Spain (and increasingly the UK) to serve European markets provide examples of this hybrid strategy. Indeed, as trade barriers decline, particularly within continental trading blocs, such strategies have become increasingly important.

In general terms, both the search for low labour costs and market access remain important to the pattern of MNC evolution. If the spectacular growth of DFI in low-wage economic zones in Mexico, China and other countries gives weight to product cycle mechanisms, market access also provides an enormous pull on DFI. Schoenberger (1990), for example, has documented that at least until the 1980s, the predominant national pattern of DFIs by American multinationals has reflected a market orientation and has favoured Europe as a host region. At the same time, DFI is no longer as dominated by US-based MNCs as it once was. Indeed, for many observers, in the 1980s and 1990s, German and Japanese MNCs have taken on the mantle as 'model' corporations.

The (flexible) German/Japanese challenge: firms organized as learning systems   As the hegemony of the US has experienced relative decline, the primacy of the US-based MNCs has also been challenged, particularly by German and Japanese MNCs. Technology and marketing expertise remain vital parts of the entry advantages of MNCs, wherever they are based. However, the defining characteristics of this challenge relate to work organization and labour relations. In particular, German and Japanese corporations, in admittedly different ways, have given priority to the continuing and comprehensive development of worker skills and to the more effective integration of workers in the operation of factories and, particularly in Germany, in corporate level decision-making.

The basis of the German and Japanese challenge is that firms are explicitly and comprehensively organized as learning systems. Streeck (1989, 1992) and Kioke (1988) model German and Japanese corporations respectively as learning systems in which skill formation among the workforce is a deliberative strategy in order to enhance productivity, willingness to think innovatively and a commitment to high-quality work. Moreover, in Germany and especially Japan, this same strategy is extended by large corporations to incorporate the activities of subcontractors (see Chapter 14). There are differences between German and Japanese corporations, *qua* learning organizations. In Japan,   worker–management   participation

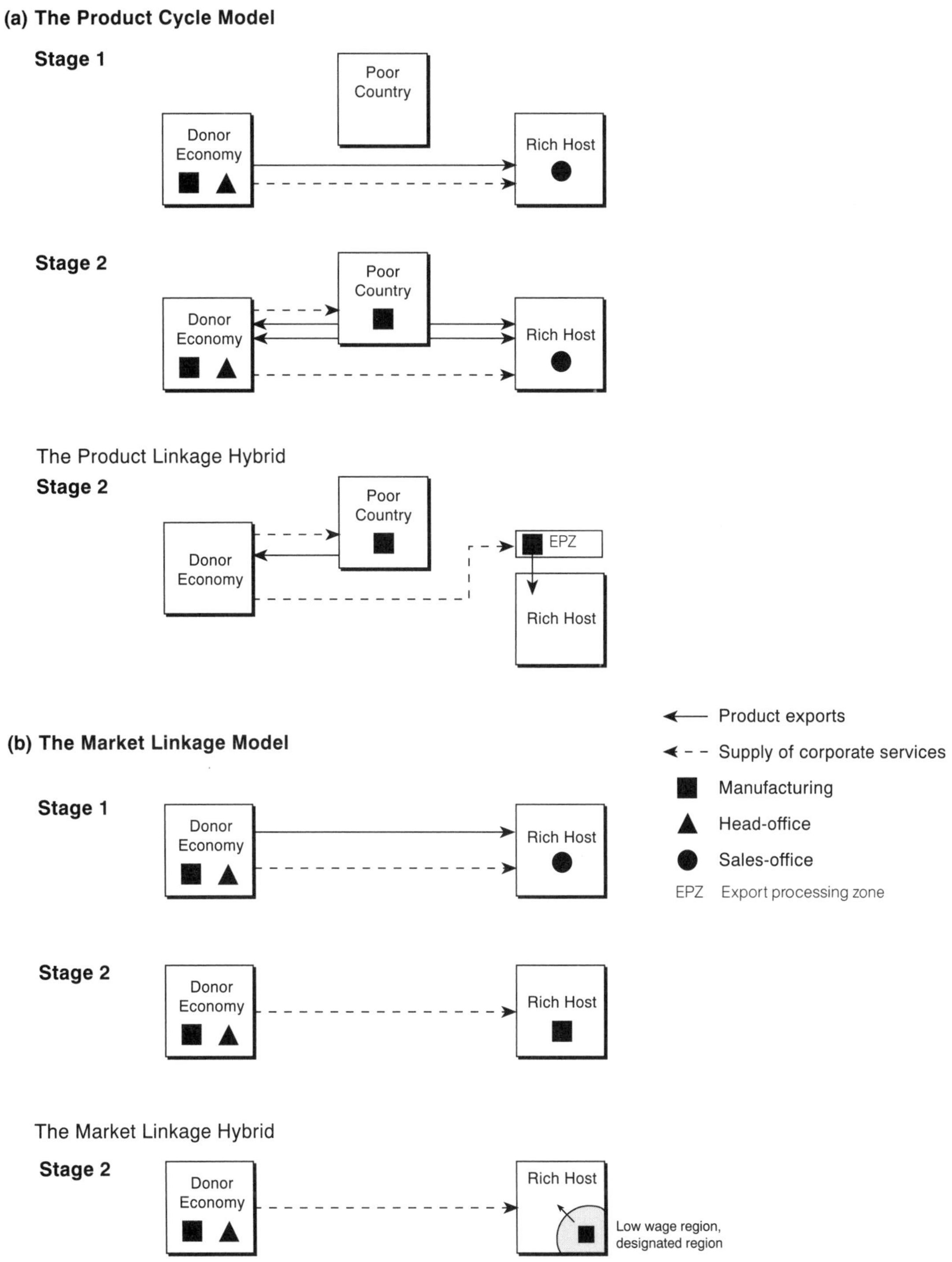

(a) The Product Cycle Model

Stage 1

Stage 2

The Product Linkage Hybrid
Stage 2

(b) The Market Linkage Model

Stage 1

Stage 2

The Market Linkage Hybrid

Stage 2

— Product exports

◄ - - Supply of corporate services

■ Manufacturing

▲ Head-office

● Sales-office

EPZ Export processing zone

Figure 11.3   Two Paths to Internationalization: The Product Cycle and Market Linkage Models

primarily occurs within factories on the 'shop floor', while in Germany unions have representatives on boards of directors (Krumme 1981). With respect to skill formation, Germany has developed an extensive apprenticeship system while in Japan on-the-job training is more critical (Kioke and Inoke 1990; Marshall 1994; Patchell and Hayter 1995).

Nevertheless, the development of sophisticated forms of work organization, along with technological and marketing strengths, has provided Japanese and German MNCs with formidable entry advantages which other corporations and regions are seeking to emulate. Thus, an important reason stimulating the big US auto manufacturers to enter into joint ventures with Japanese corporations is to learn more about the Japanese approach to work organization and labour relations (and related subcontracting practices). Similarly, the arrival of a Nissan branch plant in Sunderland, northern England, has been welcomed for its new approach to management and labour relations in a region characterized by 'entrenched union attitudes'. At the same time, the success of this plant has given more credence to the growing view in the UK that British de-industrialization has more to do with managerial failure than unions (Williams et al 1989).

In the US, Marshall and Tucker (1992) argue that corporations organized along Taylorist lines are not as competitive as firms organized as learning organizations. Marshall and Tucker (1992: 37) argue that US manufacturers have to reject Taylorism, which seeks to restrict thinking and learning to management, in favour of the German model in which work is organized 'around highly skilled, well paid workers, using high performance work organizations'. In this view, the key to corporate competitiveness with high wages (and living standards) is technological change in new products and processes and an associated commitment to product quality (Figure 11.4). In firms organized for learning, emphasis is placed on labour–management interaction and a high level of worker involvement in supervision, monitoring and design; the organi-

zation of workers in teams involving the development of polyvalent skills and continuous training; and managerial attitudes which promote participatory styles, quality, the continual up-grading of their own and worker skills and 'bench marking' which requires constant learning about markets and rivals. The expected advantages of such interactiveness are thin (lower cost) management ranks; improved quality because of better co-ordination of myriad functions and because of fewer mistakes; improved design; enhanced worker motivation and morale; and improved ability to serve more quality-conscious and differentiated markets.

It should be emphasized that there exists a considerable variation around national corporate 'models' within and among countries and corporations also adapt to local conditions. Moreover, in the fifth Kondratieff, competitive conditions are intense and sources of DFI increasingly wide. Even Japanese and German leading-edge MNCs are having to adjust to new challenges. For example, for the first time since 1945, a major car assembly factory was closed in 1995, by Nissan in Tokyo. Indeed, both Japanese and German MNCs are also establishing branch plants in low-wage platforms around the world as they relocate factories to places where labour costs are consistent with skill levels. Even for 'nationally' oriented and loyal German and Japanese MNCs, the pressure to innovate in order to maintain a high-skill/high-wage base in their home or donor economies is becoming more intense.

Exporting: a comment

The locational advantage model of the international firm, as articulated by Caves (1971), can be readily accommodated to explain export behaviour (Hirsch 1976; see also Hayter 1986a,b). Thus, as in the case of DFI, the essence of the theory is that firms contemplating exporting must have some entry advantage over local competitors in order to compensate for various spatial barriers to entry. That is, the advantages enjoyed by domestic firms, in the form of knowledge about local conditions and informa-

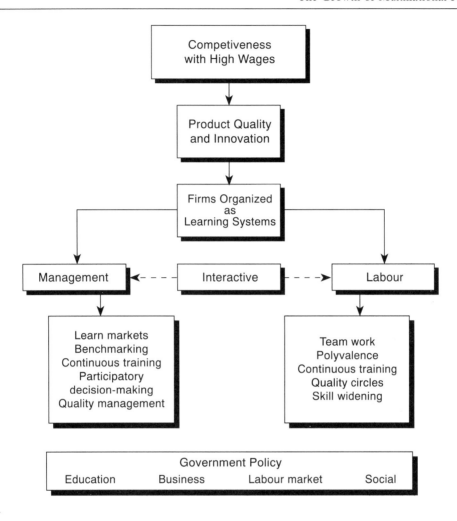

Figure 11.4 Strategies for High Performance: The De-Taylorization of Business. Source: based on Marshall and Tucker (1992)

tion sources and proximity to customers, may be viewed as barriers to potential exporters. Thus, for exporters, the spatial entry barriers are the additional transportation, information, communication and transaction costs and uncertainties associated with exporting compared to domestic sales.

'National entry barriers' that arise from the (potential) existence of locally based entrepreneurs are often supplemented by problems created by competitors based in other countries.

These 'international' barriers to exporting are naturally greater whenever foreign rivals enjoy advantages as a result of specific institutional arrangements. For example, in the case of western Canadian secondary manufacturers wishing to export around the Pacific Rim, strong barriers exist as a result of bilateral and even multilateral trade agreements that do not include Canada and, most important, the control of trade by multinationals, primarily of American and Japanese origin (Hayter 1986b: 27–28).

Thus, a Japanese multinational that establishes a manufacturing plant in the Pacific Rim country will eliminate opportunities for the small Canadian exporter. If the purchase of capital equipment from Japan is part of the investment package, then exports from Canadian equipment manufacturers are also blocked. In addition, the sourcing of inputs by these subsidiaries to affiliated plants further reduces Canadian export potentials. Indeed, given the significance of affiliated trade within international firms, these observations have widespread implications.

Given the existence of various kinds of spatial entry barriers, companies wishing to export must possess some entry advantages. These entry advantages are similar to those already discussed in relation to direct investment and refer to firm-specific revenue-producing factors, such as access to, or control over, some scarce natural resource or product, or to some technological and marketing expertise the firm has developed over time.

For most individual manufacturing firms, the difficulty of exporting varies considerably by national market. Foreign markets are different to domestic markets and the most vital of these differences relate to language and culture, including differences in business culture. During the 1970s and 1980s, for example, many Sheffield-based cutlery, tool and steel firms, which in some cases had been exporting to Commonwealth countries and the United States for over 100 years, found the penetration of EC markets, characterized by different business practices, consumer tastes and language, to be extremely difficult and often impossible (Hayter 1986b). Indeed, these difficulties are consistent with Johanson and Vahlne's hypothesis that firms export, *ceteris paribus,* first to the psychologically closest country, and subsequently to countries that are psychologically more distant. This hypothesis is systematically confirmed by Luostarinen's (1980) analysis of Finnish firms and by Johanson and Wiederscheim-Paul's (1975) analysis of Swedish firms.

Licensing As a substitute for, or complement of, exports (and DFI), firms can internationalize by selling (and procuring licenses) to foreign-based firms and by entering into cross-licensing arrangements with foreign-based firms. For innovative firms, including small firms, licensing of production to other companies (for payment) is an attractive option whenever the difficulties facing exports or DFI are substantial (Malecki 1991: 197–200). Such situations arise, for example, when small firms innovate a relatively bulky product which is expensive to transport and export, and when the firms have neither the ability nor the desire to establish branch plants. Small firms, even if they have patented their product, may also be concerned about copying by foreign competitors, which in turn raises the spectre of costly and uncertain court cases in foreign countries. In the early 1980s, a small firm based in Edmonton, Alberta, designed and patented a new loading platform and while local sales grew rapidly the product is costly to distribute and could have been 'reverse engineered'. Consequently, as an alternative to exporting the product, the firm chose to sell licenses to firms in the United States and contemplated the possibility of selling licenses elsewhere (Hayter 1986a). In addition, it might be surmised that the reason why Madge Networks is considering licensing some of its products to foreign firms is a pre-emptive move to ensure some return on its R&D before competitors copy or develop a similar product (see Chapter 9).

For similar reasons, SMEs may engage in cross-licensing arrangements, especially in situations when firms in the same industry innovate products which serve distinct but related market niches. SMEs that manufacture machinery, for example, may wish to concentrate design and production resources on a limited product range that primarily serves a regional market but has potential for sales in export markets. The cross-licensing of products in this context creates a return on technological expertise and widens each partner's product range in domestic markets without incurring additional R&D and patent costs. Many small Canadian firms in various industries have engaged in cross-licensing with foreign partners (Hayter 1988: 73; Ahern 1993b).

The post-entry behaviour of foreign-controlled activities

Foreign-controlled manufacturing operations have several characteristics which distinguish them, as a group, from the domestic population of firms (Caves 1971; Dicken 1976; Britton and Gilmour 1978). On the whole, foreign-controlled activities tend to be larger and are more likely to be found in either capital-intensive industries or technologically sophisticated industries than is true for the domestic population of firms. Industries that often feature DFI occur where entry advantages can be best developed and include large-scale resource processing, such as petroleum, aluminium, pulp and paper, and copper, and in more research-intensive industries such as autos, pharmaceuticals, machinery, electronics and computers. In comparison to domestic firms, foreign-controlled activities, by definition, are ultimately controlled by a parent company whose head-office and technostructure is located in a foreign country. This control has significant implications for the actual and potential structures and strategies of foreign-controlled firms in comparison to domestic firms. Within the population of foreign-owned activities, structures and strategies also vary considerably.

Foreign branch plants: does strategy follow structure?

For parent or independent companies, as Chandler (1962) argues, organizational structure typically follows strategy (Figure 7.2). For foreign-controlled subsidiary companies and branch plants, on the other hand, strategies frequently depend upon organizational structures. Thus the structure of a branch plant or subsidiary is typically defined by parent companies, including with respect to autonomy and mandate which respectively define limits on decision-making discretion (autonomy) and

function (mandate). The scope of decision-making within the subsidiaries then follow from these limits. Autonomy and mandate are closely related in that a branch plant that has an extremely limited mandate or purpose is likely to have limited autonomy with respect to such functions as marketing, procurement, labour relations, planning, investment decision-making, and research and development. Moreover, from the perspective of post-entry behaviour, the nature of local autonomy and mandates is crucially important. In general terms, branch plants and subsidiaries that have greater degrees of autonomy and a wider range of functions are more likely to plan strategies than subsidiaries with limited autonomy and functions. The structure of local subsidiaries can of course be changed if parent companies so wish.

Clearly, the parent company's motivations in implementing investments play a decisive role in shaping the behaviour of foreign-owned companies. Thus, highly specialized, export-oriented operations set up by resource-based companies seeking additional supplies of raw material differ from highly specialized, domestically oriented operations set up by a secondary manufacturing company seeking additional national markets for an established product line. In this instance, different mandates imply different sales patterns (and potentially different implications for the balance of payments of the host economy). However, in both cases, the market mandate is closely prescribed by parent company policy: the former does not have the mandate, and therefore the capability, to 'add value' prior to exporting, and the latter does not have the mandate, and therefore the capability, to export (or at least to export to the particular market region designated by the parent company). Within the given mandates, the two subsidiaries may have varying degrees of autonomy for the actual market process. Thus, the parent company has the option of performing all market functions itself from the head-office or delegating them to the subsidiary. Similar observations can be made regarding other corporate functions.

Decision-making autonomy of branch plants
Several factors affect the degree of autonomy of a foreign-owned operation (Watts 1981). The first factor is the method of entry and the nature of equity ownership. In this regard, an important distinction is between wholly owned subsidiaries and joint venture companies. Joint venture companies, particularly those established on a 50 : 50 basis between two partners, at least one of which is foreign-owned, are typically established to serve a specific purpose and are managed by representatives from both parent companies. Such companies usually exist simply to realize the initial motivation for the investment and they have no particular 'growth dynamic' of their own; further growth of the joint venture is likely to occur sporadically and according to the needs and agreement of two parent companies (Hayter 1981). Given the importance attached to joint ventures by host governments, as a way of alleviating foreign ownership criticisms, this limitation is potentially important. In general, wholly owned subsidiaries have more potential to grow, so long as such growth is consistent with parent company plans.

The second, possibly most important factor affecting local autonomy is the size of the foreign-owned subsidiary. Generally speaking, bigger companies have more autonomy; indeed, for big, diverse operations, local decision-making capability is likely to be critical for effective performance. In this regard, international firms frequently establish a large and diverse presence on entry into a foreign country through the acquisition of existing operations. In addition, some foreign-owned subsidiaries may have the mandate to pursue active policies of growth within their host economy and thereby grow and become increasingly diverse. For these 'active' subsidiaries, a cause and effect relationship between growth and decision-making autonomy is to be expected.

A third factor thought to affect the degree of autonomy is the nationality of the parent company. Several observers suggest that US-based multinationals control their subsidiaries differently from British or Japanese MNCs. It is argued that US firms are more likely to prefer to establish wholly owned subsidiaries and closer levels of integration between parents and subsidiaries than Japanese or British firms. Whatever the validity of such 'national' differences, it might be expected that the 'style' of individual corporations over matters such as local autonomy for subsidiaries will vary considerably. Michelon, for example, has a reputation for extremely tight, centralized control of operations combined with high levels of secrecy to the point where even the names of its senior executives are difficult to publicly access.

Several other factors are thought relevant to understanding the degree of a local subsidiary's autonomy, including the distance between the parent company head-office and the subsidiary location, the degree of product specialization of the subsidiary within the parent company, and the stability of product markets (Watts 1981). In general, it is argued that, *ceteris paribus,* the local autonomy of subsidiaries and branch plants increases with distance from parent company head-offices, with the degree to which their roles are specialized, and with increasing instability of product markets. However, it is difficult to assess the importance of these effects with precision.

For parent firms, too much centralization may lead to the dampening of initiative, the discarding of local opportunities and a failure to recognize key trends, while excessive delegation may lead to duplication, contradictory decisions and even threaten the integrity of the corporation as whole. Consequently, no matter how large and geographically diverse they are, subsidiary companies will be under some form of control and integration by parent companies. At one level, this integration occurs as regular flows of information in monthly, quarterly and annual reports, in person to person meetings, phone calls, faxes and e-mail. At another level, senior personnel may occasionally be moved among various subsidiary companies in several countries to help establish a corporate culture. At yet another level, capital budgets, investment and R&D programmes of foreign subsidiaries are typically closely scrutinized and ultimately subject to parent company control and priorities. That is, the strategies of subsidiaries need to be placed within a parent company context to make sense and these strategies inevitably have limited mandates in one way or another. Subsidiaries, even those with considerable autonomy, may not have the mandate to export, and/or to add value, and/or to engage in R&D, and/or to diversify their product range, while in most cases subsidiaries do not have a mandate to invest in foreign countries themselves.

Over time, changes can occur in the nature of subsidiaries and their relation with parent companies: they can grow and decline, they can be bought and sold, and the process of foreign ownership even reversed as well as increased in any specific host economy. They also may be subject to the so-called 'obsolescing bargain'.

The obsolescing bargain

The obsolescing bargain hypothesis was developed in the context of the bargaining processes that take place between multinational corporations and host-country governments (HCGs) during and following entry (Vernon 1971). In fact, this idea was principally developed in relation to resource-based multinationals and developing countries (Pinelo 1973; Moran 1975; Sklar 1975; Auty 1985; Shafer 1983) although Grieco's (1982) study of Indian experience dealt with the international computer industry. The essential assertion of the obsolescing bargain hypothesis is that bargaining power between multinationals and host countries initially favours the multinational during the negotiations about entry but gradually shifts in favour of the host country following entry.

According to this hypothesis, the HCG and the MNC each strive to maximize benefits from investment (Figure 7.6). To varying degrees, their goals are congruent and in conflict. Moreover, the bargaining process involves risks and uncertainties and these risks and uncertainties change over time. Prior to the investment, negotiation typically favours the MNC since the MNC has more information about the relative profitability

of the investment and alternative locations. The MNC can potentially use this advantage to bargain for concessions from the HCG, especially if the host country is capital-scarce and information-poor and there is limited competition among MNCs. Subsequently, once the investment has been established, the bargaining position of HCGs becomes stronger to the extent that individuals and organizations within the host country gain information and experience as regards the operation of the investment and to the extent that the operations are capital-intensive and difficult to relocate, which is typically the case in resource-based manufacturing or primary activities. In addition, as the operation becomes profitable, uncertainties over markets and technology are dissipated, not only for the MNC but also for the HCG. Consequently, for a variety of reasons, HCGs may be in a position to demand concessions from the MNC, rendering the initial bargain obsolescent, at least in part. Indeed, the HCG may threaten and implement a nationalization strategy.

Even if nationalization is not invoked, the obsolescing bargain hypothesis predicts that the initial bargain favouring the MNC will eventually be renegotiated over time, implying a decline in the MNC's entry advantage. Several studies have supported this prediction and have noted trends towards tightening the early beneficial terms granted by HCGs after new operations proved successful. These renegotiations have included demands for higher taxes, greater local processing, joint marketing, a greater commitment to hiring locals in managerial positions, and a higher share of domestic ownership by host country nationals or the HCG itself (Pinelo 1973; Moran 1975; Sklar 1975).

At the same time, even in resource-based manufacturing, the obsolescing bargain should not be assumed or the power of MNCs underestimated. The entry advantages of MNCs are embodied both within plant and machinery and in the world-wide facilities, networks, expertise and connections of the parent company as a whole. The former are often difficult to disaggregate and the latter inaccessible to HCGs. A plant or mine may be nationalized but it is far more difficult to acquire the built-in engineering know-how and to duplicate the MNC's world-wide marketing connections. Any one subsidiary serves a particular role within the parent company and these connections provide the MNC with sustained bargaining power. More-over, MNCs, particularly those based in the US, can lobby for help by donor governments which may be in a position to sanction 'unfair' HCGs. It might also be noted that initial bargains may have been excessively favourable and the MNC may even anticipate a degree of renegotiation after start-up.

Furthermore, it needs to be stressed that if a host country *may* move up a learning curve following DFI, in that it learns something about the nature of the MNC's operation, MNCs inevitably acquire knowledge and connections within the host country. That is, over time, the MNC learns about a host country and acquires the capabilities and characteristics of local residents. Moreover, MNCs can still threaten to withdraw investments at existing operations and allow the possibility of technological obsoles-cence. There are always investment alternatives for MNCs in a general sense. Indeed, it is perhaps surprising that so much attention should have been given to the obsolescing bargain hypothesis. There are strong arguments that DFI are good deals for MNCs – deals that imply accumulating bargains.

The accumulating bargain

While there may be debates as to whether in some cases the obsolescing bargain occurs, in which the MNC's entry advantage dissipates, there is little doubt that for many, perhaps most, established foreign subsidiaries and branch plants, the concept of spatial entry barriers becomes obsolescent. Thus, over time the MNC becomes familiar with local conditions and sooner or later acquires the local know-how and connections inherited by domestic firms. Once established, a subsidiary can expand within a host economy and while it still has access to

parent-company expertise and resources (i.e. entry advantages), post-entry expansions do not have to overcome the problem of spatial entry barriers.

Moreover, MNCs frequently negotiate finan-cial support as a condition of entry. Such financial help may still be available for subse-quent growth. Even if subsidies are not available, subsidiaries typically have access to domestic sources of funds through the normal channels, including banks and investments by local share-holders of an equity and non-equity kind. Indeed, the post-entry growth of subsidiaries and branch plants may be entirely financed by local earnings and domestic funds (Hayter 1981). In this way, it may be argued that domestic sources of financing fund increases in the size of foreign-owned companies. Yet parent companies still retain control of assets which can be sold for their market value. In fact, subsidiaries are a source of funds to parent companies in the form of parent-company charges for head-office services and as interest and dividend payments. In less tangible ways, the control of foreign operations also significantly increases the parent companies' geographic scope for adjusting to change.

For a variety of reasons, therefore, it may be argued that DFI is a cumulating bargain for MNCs. Available evidence strongly suggests that DFI has been extremely profitable to parent companies. Indeed, subsidiaries are potentially the ultimate cash cow for parent companies looking for large injections of financing for modernization in facilities elsewhere, including the donor economy. Given the profitability of DFI, divestment may not be considered a preferred strategy. However, in times of severe corporate crisis, the sale of subsidiaries can readily be converted into cash. In the severe recession of the early 1980s, for example, several forest product corporations sold foreign subsidi-aries for substantial sums (Hayter 1985). Inter-national Paper of New York, for example, sold its Canadian subsidiary, which it had acquired in the 1920s and which had been self-financing since then, for over \$1 billion which it used for modernizing its US-based facilities. Similarly, MB sold several of its foreign subsidiaries and a large mill in Atlantic Canada in the 1980s to help pay for large losses in the early 1980s (Barnes et al 1990).

As noted above, many US states charge a foreign corporation tax (Exhibit 11.1) and, given the nature of the accumulating bargain, other jurisdictions may give some consideration to this idea.

Conclusion

Within the radical literature on the international firm, a central theme is the enhanced mobility of capital and a more implied argument that this mobility has rendered the theory of locational entry redundant (Fröbel et al 1980). From this perspective, entry has already been accomplished around the world and MNCs can simply move from one known location to another; there is no unfamiliar territory. Yet, for many large MNCs, entry into new host economies continues to be an important phenomenon; for example, European and North American companies are trying to crack Japan, and MNCs from all over the world are seeking DFI in Russia, east Europe and China. In addition, there is an emerging stream of medium-sized firms that are internationalizing for the first time and there are a growing number of MNCs coming out of the developing world. DFI, in other words, remains a problematical process and the theory of locational entry remains a useful starting point from which to investigate the international evolution of firms.

At the same time, the tentacles of MNCs are continually expanding to remote parts of the globe, sometimes directly and sometimes indirectly through subcontracting patterns which literally reach right into the living rooms of people. This global reach (Barnet and Muller 1975) raises important implications for labour, some of which are discussed in the next chapter.

12

CORPORATE RESTRUCTURING AND EMPLOYMENT FLEXIBILITY

The long boom of fordism during the 1950s and 1960s was a period of sustained growth during which time multinational corporations (MNCs) emerged as a dominating controlling influence of the global economy. In contrast, since the early 1970s, the global economy has been volatile and many corporations have experienced profound restructuring of technology, production, organization, markets, location and employment (Taylor and Thrift 1982; Townsend 1983; de Smidt and Wever 1990). While these changes are complex, there is a broad consensus that a basic theme of corporate restructuring is a search for 'flexibilities' of one kind or another, as the global economy shifts, in Freeman and Perez's (1988) terms, from the fordist mass production techno-economic paradigm to the information technology (ICT) techno-economic paradigm (see Chapter 2). Yet the meaning and implications of flexibility are controversial. Flexibility is expressed in terms of the characteristics of machines, factories, firms and of society as a whole (Gertler 1988, 1992); it can refer to a variety of organizational, market, technological and employment characteristics (Schoenberger 1987; Sayer 1989), and societal shifts towards flexibility can be seen as a benign (Piore and Sabel 1984) or an exploitative process (Harvey 1989). In practice, corporations have consider-

able discretion in integrating the various facets of their operations that are in some ways 'flexible'.

In this chapter, corporate restructuring is examined from the perspective of the search for employment flexibility. The chapter is in three main parts. The first section notes various dimensions of corporate restructuring and then gives particular consideration to employment by comparatively reviewing fordist with more flexible labour markets. The second and third parts of the chapter relate to different types of geographical strategy, specifically new location strategies and *in situ* change strategies, with different types of employment flexibility, specifically between core and peripheral workforces.

Corporate dimensions of restructuring

Globally, the winds of economic change gathered momentum during the 1970s. Energy crises, inflation and high interest rates increased costs significantly. New technologies, particularly those developed in micro-electronics, transformed production structures across the industrial spectrum, as computer-assisted design (CAD), computer-assisted manufacturing (CAM) and numerically controlled (NC) machines became critical measures to assess the innovativeness of factories, firms and regions (Ewers and Wettman 1980; Oakey et al 1982; Thwaites 1982; Rees et al 1984). Enabled by these technologies, firms sought to create new products for markets which were becoming more volatile, differentiated and competitive as American MNCs were challenged by European firms and even more so by Japanese firms. Gradually, it also became apparent that the strength of the Japanese challenge was not based on temporary advantages of low labour costs or unusually hard-working and loyal labour but was more deeply rooted in the nature of production organization (Fruin 1992).

For western economies and firms, declining competitiveness was publicly expressed in the form of large-scale lay-offs. Moreover, these lay-offs were massive and permanent, frequently occurring within the context of factory closure, rationalization and modernization (Massey and Meegan 1982). Permanent lay-offs affected core as well as peripheral regions and were especially pronounced after the mid-1970s in the US and the UK (Bluestone and Harrison 1982; Townsend 1983), although they occurred elsewhere as well (de Smidt 1990; Fuchs and Schamp 1990). In the UK, for example, between 1976 and 1981, British Steel reduced jobs by 67 000, British Leyland (cars) by 38 500, and Courtaulds (textiles) by 24 700 (Townsend 1983: 74). Between 1974 and 1980, ICI, the giant chemicals firm, reduced its world-wide workforce by 57 800, including 44 700 in the UK (Clarke 1982: 99). Moreover, the recession in the early 1990s signalled another round of job losses. By the mid-1990s, downsizing had bitten deeply in North America and Europe (and had touched Japan) and had been dominated by planning system firms. No corporation in North America and Europe seemed immune. General Motors (GM), for example, experienced a considerable buffeting from the winds of change, which took a substantial toll on its employment. Thus, between 1980 and 1995, GM reduced its net global workforce by over 200 000 people and its actual lay-offs, bearing in mind it built several new factories, were greater still. Even so, GM still lost over US$16 billion in 1991 and 1992,

indicating that the problem of competitiveness remained.

In recessionary situations of declining market performance and competitiveness, firms have traditionally sought to reduce costs, especially labour costs (Fredriksson and Lindmark 1979). The scale, extent and permanence of recent job losses implemented by so many major corporations, however, have typically been part of wider restructuring processes. The nature of these restructuring processes can be summarized by reference to the fordist firm, as the 'ideal' type in the fordist techno-economic paradigm. In the fordist firm, specialized managers do the 'thinking' and a unionized workforce does the 'operating' while production utilizes dedicated machines to manufacture standardized items in order to exploit economies of scale (Marshall and Tucker 1992; Storper and Scott 1992). In the more competitive economy of the ICT techno-economic paradigm, however, the fordist firm is considered uncompetitive for several reasons. In particular, fordist management is considered top heavy, bureaucratic and slow to respond to change; fordist workers are judged to be insufficiently versatile and self-reliant to cope with a variety of new tasks, resulting in problems of labour productivity; and fordist technology is too specialized, sacrificing quality control for volume at a time when markets are demanding product innovation, differentiation and reliability.

The interrelated faces of restructuring

In recent decades, corporate restructuring designed to lower costs, enhance productivity and improve market positions has taken on multiple dimensions. While restructuring plans inevitably differ among corporations, several common themes and possibilities are emphasized in the industrial geography and related literature. These themes may be briefly identified in relation to labour, production and technology, organization and product-markets. First, with respect to labour, firms can immediately reduce costs by lay-offs and can increase productivity by intensifying work practices, i.e. by maintaining existing functions and tasks with a smaller workforce (Massey and Meegan 1982). Over the longer term, firms may prefer to hire workforces who can be more easily hired and fired. Alternatively, or in a complementary way, firms can increase labour productivity by automation, by providing workers with more sophisticated capital equipment, and/or by increasing the skills of the workforce (Patchell and Hayter 1995).

Second, in the context of production and technology, firms can close down facilities completely or rationalize operations in which either capacity or functions or both are reduced. Alternatively, or in conjunction with selective closure and rationalization, existing facilities may be modernized, in whole or in part, and/or firms may choose to introduce new technologies, including CAD–CAM, at new locations. Third, with respect to organization, large integrated firms can restructure operations by (selective) vertical disintegration, i.e. by contracting out, or outsourcing, functions previously conducted internally (Scott 1986). Vertical disintegration may be motivated by desires to gain (indirect) access to cheaper supplies of labour. Alternatively, vertical disintegration may be an attempt by firms to specialize on core activities to which it can give more focused attention and to gain access to equally specialized and focused suppliers. At the same time, vertical disintegration is by no means an inevitable consequence of restructuring and it should be recognized that large corporations can combine elements of both vertical integration and disintegration. In addition, firms can buy and sell parts of their operations and engage in joint ventures and strategic alliances. Another organizational implication of restructuring is the creation of flatter decision-making structures to the extent that more flexible labour needs fewer managers, and greater specialization allows thinner managerial layers.

Fourth, the ultimate purpose of corporate restructuring is the definition of profitable market roles (Peck and Townsend 1984; Hayter 1986a). From this perspective, a general theme of corporate restructuring is to shift towards more valuable, design-intensive, higher quality products and towards higher income market areas. Such a shift typically involves a commitment to R&D, quality control and servicing so that competitiveness is not simply based on cost and price. For many UK manufacturing corporations in the 1970s and 1980s, for example, restructuring emphasized both adding value and shifting from former Commonwealth markets to EC markets (Hayter 1986a). Similarly, for many lumber firms in coastal British Columbia, restructuring in the 1980s and 1990s has involved a shift from the mass production of a limited range of low-value commodities for the US housing market to a wider range of products for the Japanese market (Grass and Hayter 1989; Edgington and Hayter in press). If the value-added shift is a general theme of corporate restructuring, it is not a universal response, however. Some high-quality producers, have sought to establish the production of medium-quality products to their marketing mix in order to realize economies of scale. Moreover, even in high-income locations, opportunities for low-cost operations exist.

The relative importance of the various dimensions of restructuring, how they are related to one another and how they evolve over time vary considerably from corporation to corporation, even within the same region and industry. Corporations develop different restructuring plans and some firms inevitably make mistakes (Schoenberger 1994). Moreover, restructuring plans are often implemented under the exigency of financial crisis, even for large corporations, and the immediate solutions undertaken to resolve financial threats to survival in turn shape subsequent restructuring plans.

Whether successful or not, corporate restructuring typically has many faces. Case studies that have detailed the comprehensive nature of the restructuring process by large MNCs include IBM in Europe (Kelly and Keeble 1990), ICI (Clarke 1982), Fletcher Challenge (LeHeron 1990), Philips (de Smidt 1990), three UK firms (Peck and Townsend 1984) and an unnamed Asian-based firm (Clark 1993). GM, Ford and

IBM, each of which reduced their global work-forces by at least 200 000 are probably the largest-scale examples of corporate restructuring of the 1980s and 1990s. Indeed, the context for the search for employment flexibility is provided by (a) complex, multifaceted restructuring processes and (b) massive employment down-sizing by major corporations.

Core firms and segmented labour

Ideas about labour market segmentation or dualism were developed in the 1950s and 1960s as a critique of conventional or neoclassical economic thinking (Morris 1988; Hayter and Barnes 1992). While neoclassical theory inter-prets labour – including the hiring, mobility and firing of labour in terms of laws of demand and supply – dual labour or segmentation market theory argues that the rules and conditions governing groups of workers vary according to the segment they occupy. The principles under-lying forms of segmentation are dynamic (Clarke 1982; Gordon et al 1982), while at any particular time there are considerable variations from place to place (Peck 1992, 1993). Taylorism or scientific management, which became incorporated within fordist firms, began to emerge around the beginning of the 20th century, replacing more idiosyncratic forms of work organization based on the power of managers and foremen. In the 1980s, at different speeds and along different lines, the fordist–Taylorist principles of labour market segmentation became threatened by more flexible ones.

Labour markets in the fordist firm Fordist labour markets are an expression of the idea of a dual economy, particularly as developed by Averitt (1968) and Galbraith (1962, 1967) in the context of the US economy in the 1950s and 1960s (see Chapter 2). In particular, the dual economy is organized by a dominant fordist sector comprising a few large capital-intensive, oligopolistic firms that engage in mass produc-tion and serve relatively stable markets, and a competitive sector made up of many small firms primarily supplying small, fluctuating markets. Kerr (1954) and Doeringer and Piore (1971) elaborated the idea of economic dualism by suggesting the complementary thesis of the dual labour market. Kerr (1954) spoke of 'balkanized' labour markets and Doeringer and Piore of 'segmented' labour markets.

According to Doeringer and Piore, there are two main labour market segments, primary and secondary, which in turn correspond to the fordist and competitive sectors (Figure 12.1). In broad terms, the primary segment is character-ized by "high wages, good working conditions, employment stability, chances of advancement, equity, and due process in the administration of work rules". Doeringer and Piore (1971) further subdivide the primary, or fordist, segment into a primary independent segment (management, research and development workers) and a primary subordinate or dependent segment (production workers, tradespeople, office workers). The former generally enjoy higher levels of remuneration, greater employment stability and better non-wage benefits than the latter. In contrast, jobs in the secondary segment tend to have "low wages and fringe benefits, poor conditions, high labour turnover. Little chance of advancement, and often arbitrary and capricious supervision" (Doeringer and Piore 1971: 165). In comparison to the primary sector, workers in the secondary sector are more likely to be non-union, female and to belong to a visible minority. Moreover, workers in the secondary sector are hired and fired according to competitive condi-tions, i.e. along the lines prescribed by neoclas-sical theory.

The primary segment, comprising the internal labour markets of large firms, are typically highly structured, often by formal agreements between management and unions governing working conditions and levels of remuneration of the blue-collar workforce. Under fordism, labour markets are structured by two dominant princi-ples: seniority and job demarcation (Lester 1967). Seniority imposes structure by limiting entry into the firm to the lowest grades and positions so

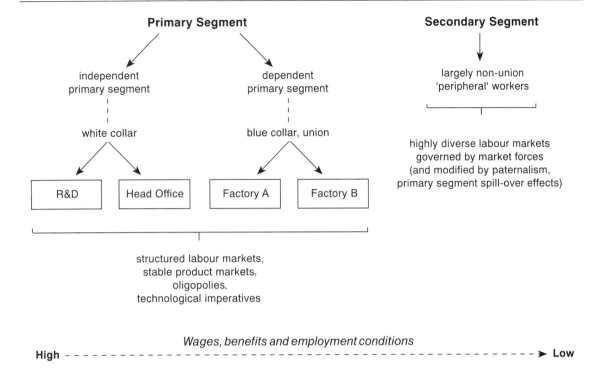

Figure 12.1   Labour Segmentation under Fordism. Source: based on Hayter and Barnes (1992)

that higher level vacancies have to be filled by internal (vertical) mobility according to the principle of seniority which begins to accumulate once workers have fulfilled any probationary period and are considered 'full time' (Lester 1967). Wage rates, promotion and lay-offs are then regulated by seniority. Job demarcation provides structure by precisely defining the tasks of workers.

Under fordism, the principles of seniority and job demarcation offer significant advantages to both (fordist) management and unions. For unions, these principles address a central philosophical concern, namely the need to reduce competition among workers and the implications of such competition for degraded working conditions and declining levels of remuneration. Thus, for unions, seniority and job demarcation create order, stability and dignity in the workplace as each worker has clearly defined job roles and expectations. For management, seniority and

demarcation have two main benefits. First, they encourage a stable workforce as seniority, and the associated accumulating wages and non-wage benefits, progressively 'lock-in' workers to the firm while job roles are precisely demarcated. Second, job demarcation is the mirror image of Taylorism, which emphasized narrowing jobs into the simplest possible tasks that could be repeated rapidly (and therefore efficiently). Moreover, under Taylorism the promotion of workers on the basis of seniority rather than merit is not a major problem as obedience and discipline are more important than intellect and initiative.

Given a mutual commitment to seniority and job demarcation, fordist labour relations were further structured by regular contract bargaining, which in many industries occurred every two or three years (Holmes 1992). As part of job definitions, grievance procedures were negotiated to allow formal challenges to management decisions. Typically, collective bargains became more complex with respect to wages, non-wage benefits and working conditions to create what Clark (1986) calls 'cascading benefits' which, rooted in law, have proven resilient to changing managerial demands.

The fordist collective bargains negotiated in the primary dependent segment typically occurred in industries, (e.g. the auto industry) characterized by oligopolistic market structures in which a few, large firms invest in mass production technology and enjoy market power and product market stability (Hayter and Barnes 1992). In the case of the North American auto industry, the size and technological sophistication of their factories encouraged GM, Ford and Chrysler to establish increasingly well paid, secure workforces who in return provided stability and productivity. During the 1950s and 1960s, lay-offs of primary dependent workers were mainly temporary and organized according to seniority. Generally, the primary independent segment were untouched by recessionary conditions.

In contrast, Doeringer and Piore argued that labour markets in the secondary segment were largely non-unionized and that wage levels and employment stability were governed closely by market forces (Figure 12.1). Particularly in those segments populated by small firms operating in highly competitive markets, jobs were relatively low paying and unstable, and principles of seniority and job demarcation were dependent on local circumstances, i.e. specific arrangements between management and labour. Between the 'polar cases' of the unstable, low-waged industries of the secondary segment and the highly structured internal labour markets of, for example, the auto firms, the nature of labour markets varied. Even in the secondary segment, forms of managerial paternalism were possible, while part of the 'giant shadow' cast by the large fordist firms of the primary segment over the small firms in the secondary segment involved labour relations. Thus, in some cases, for example where small firms desired skilled, stable labour, the standards set by collective bargains in

the primary dependent segment established goals for the secondary segment.

Labour segmentation in the flexible firm   The kind of labour segmentation process that developed under fordism, especially in North America, depended on a set of conditions peculiar to the 1950s and 1960s. Since the beginning of the 1970s, however, these conditions have been threatened, even in North America. Thus, product markets in the core industries have become less stable, oligopolistic power has been diffused by international competition, new technologies have required different employment levels and skills, and labour's bargaining power has lessened. In this more competitive economy, the bases for labour market segmentation began to change towards more 'flexible operating cultures' of the kind pioneered by successful Japanese and German firms (see Chapter 11).

In response to recession, technical change and the market challenge posed by foreign competition, Atkinson (1985, 1987) argued, in a UK context, that an increasing number of firms were pursuing a policy of employment flexibility (Figure 12.2). His basic argument is that firms are seeking to develop flexible 'core' workforces and flexible peripheral workforces in which the characteristics of flexibility differ in the two segments. Thus, core workers are 'functionally flexible', i.e. highly skilled and able to perform different functions as part of team work, responsible for decision-making, committed to developing further skills as needs arise, and willing to accept benchmarking (i.e. to study and if necessary incorporate the best practices developed by competitors). In contrast to fordism, core workforces in the flexible firm are not narrowly structured by job demarcation or by seniority. If job demarcation is not eradicated, in the flexible firm job categories are far fewer and more broadly described. The advancement of workers depends upon performance and the ability to perform tasks, rather than solely by seniority. In addition, in some types of flexible firms at least, in contrast to fordist firms, workers are recognized as human resources capable of

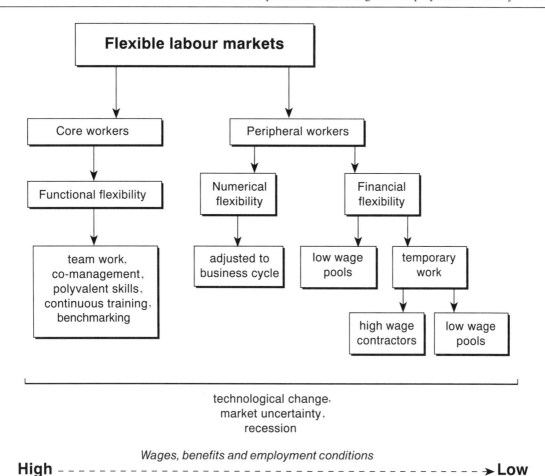

Figure 12.2   Labour Segmentation based on Flexibility. Source: based on Atkinson (1987)

initiating productivity improvements and self-supervision. These workers are rewarded with high wages, non-wage benefits and employment stability.

In contrast, peripheral workers can be either numerically and/or financially flexible. Numerically flexible workers are hired as needed and they include part-time and temporary workers and 'permanent' workers whose hours can easily be varied by adjustments in shifts, additions, contractions to overtime and by lay-off if necessary. In addition, numerically flexible workers may be employed directly by firms or indirectly through subcontracting. Financially,

flexibility is achieved by hiring workers traditionally associated with lower wages, such as females; they are used on a part-time basis as needs arise within the firm or via subcontracting. Firms can also achieve financial flexibility by subcontracting high value work to suppliers comprising highly skilled workers and professionals who have expertise which firms want occasionally (Christopherson 1989).

In summary, Atkinson's (1985: 16) argument is that because firms "have put a premium on achieving a workforce which can respond quickly, easily, and cheaply to unforeseen changes" a new form of labour market segmentation has arisen between a functionally flexible core labour force, and a numerically/financially flexible peripheral one. Clearly, the characteristics of core workers in the flexible firm are fundamentally different from the characteristics of core workers in the fordist firm. Similarly, in the flexible firm, management is articulated with workers in a fundamentally different way from the fordist firm as supervision and worker discipline are replaced by co-management and initiative. In addition, flexible employment has significant implications for unions by its direct confrontation with the principles of seniority and job demarcation. Indeed, the general shift towards flexibility is associated with a general decline in the power of unions, most notably in the US and the UK. On the other hand, the characteristics of flexible workers in the periphery are similar to the characteristics of workers in the secondary segment of fordism. Moreover, in the flexible firm, established polyvalent skills and accumulating experience, perhaps reinforced by structured, unionized labour relations, provide barriers to ready movement of workers from the periphery to the core.

Flexible labour and geographical strategies In order to create a flexible labour force, firms have two basic geographical strategies. The first strategy is the new location strategy in which a new, flexible labour bargain is created at a new location. The second *in situ* strategy occurs at existing locations where firms seek to replace fordist with flexible labour relations. Simultaneously, firms, perhaps in conjunction with labour, must choose a particular flexibility option. Again, two polar strategies can be identified. The first strategy aims to emphasize the creation of core workforces based on principles of functional flexibility, and the second emphasizes the creation of a peripheral workforce based on principles of numerical and financial flexibility. Each of these flexibility strategies may be associated with either a new location or an *in situ* geographical strategy.

Thus, the geographical implications of the transition from fordist to flexible employment may be explored in terms of four basic strategies: new locations with functional flexibility; new locations with numerical/financial flexibility; *in situ* adjustments incorporating functional flexibility; or *in situ* adjustments incorporating numerical/financial flexibility. It does need to be reiterated that reality is more complex, and behaviour in practice is not restricted to these theoretical categories. Reality is more complicated as firms can and do select different bundles of flexibility characteristics in their workforces at new and existing locations, often in the same region (Morris 1988; Pinch et al 1989; Hayter and Barnes 1992). Moreover, in practice, it is not easy to distinguish between different types of flexibility. Flexibility of any kind is designed to ensure that workers are always working, so flexibility can imply the simultaneous intensification of work practices and the development of multiple skills as workers perform various tasks. At the same time, the performance of multiple tasks does not automatically equate with multiple skills if the tasks involved are simple. The notion of skill is itself problematical.

New locations and flexible workforces

New locations provide firms with significant bargaining advantages (see Chapter 7). For firms, it is easier to develop new labour relations in new locations. Such a strategy allows the firm to literally 'start again' and to avoid the difficulties of trying to change established attitudes and contractual relations. This advantage applies whether the firm is seeking functionally flexible core workforces or numerically flexible peripheral workforces.

Perhaps the most significant experiments in labour relations in recent years have occurred in the auto industry at new locations. The most radical of these experiments are GM's Saturn plant in Tennessee and Volvo's short-lived Uddevalla plant in south-west Sweden (Figure 7.7). Toyota's recently built plant on Kyuyshu in southern Japan, its first plant in Japan built outside of its core region around Toyota Town, is another attempt to redefine labour relations. In each case, the firms emphasized the creation of new functionally flexible core workforces in the new plants in the hope that the new practices could then be transferred to existing operations.

Creating core workforces in new locations: GM's Saturn plant

GM's Saturn project at Spring Hill, Tennessee (Figure 6.5), originated in the recessionary crisis of the early 1980s as part of GM's corporate-wide restructuring plans. In 1982, an internal study concluded that GM was unable to compe-titively manufacture a small car with existing work practices (Rubinstein et al 1993: 341). In December 1983, GM's senior management and the UAW created a study centre to initiate the Saturn project and subsequently, the 'group of 99' was formed, comprising managers and union members representing wide-ranging interests (Berggren and Rehder 1993: 194). The 'group of 99' divided itself into work teams which travelled over two million miles and visited over 100 ('benchmark') companies to determine appropriate state-of-the-art practices. Saturn's mandate was to manufacture a small car in the US that was a world leader in cost and quality, and to become a source of 'know-how' for GM as a whole (Berggren and Rehder 1993: 195).

In practice, Saturn has become a significant innovation in the history of industrial relations in the US (Rubinstein et al 1993). Saturn was designed explicitly to depart from fordist work principles and to embrace the full implications of functional flexibility, with a particular emphasis on the integration of traditionally distinct management–worker roles. At Saturn, union and management agreed to virtually eliminate job demarcation and supervisors in the tradi-tional sense and to organize work in the form of teams, modules and decision rings (Table 12.1). In particular, self-directed teams of 6–15 members, which enjoy considerable authority and responsibility over day-to-day operations, are organized into modules which are in turn integrated into three business units. Management

and labour are further integrated by the weekly meetings of several decision rings related to, for example, corporate long-range planning, manu-facturing at Spring Hill and labour relations issues. In addition, co-management features a 'partnering programme' in which union members are partnered with managers in a growing number of functional areas including sales, service and marketing; finance; industrial engi-neering; quality assurance; health and safety; training; maintenance; product and process development; and corporate communications. According to Rubinstein et al (1993: 346), this partnering initiative "represents the most far-reaching innovation in the Saturn governance system", which in their view has taken labour's role in management further than any other organization in the US. They also point out that since the start-up of the Saturn plant in 1990, further developments in co-management have occurred, notably 'problem solving circles' which are teams created to solve off-line problems.

The Saturn project is a major experiment in industrial relations developed in a new location. Saturn employs over 7 000 workers, including about 5 300 UAW members. It is a 100% union operation in a non-union state. As of 1993, investment in Saturn amounted to about US$5 billion, including product development costs. Whether Saturn will succeed remains to be seen. From a sales and marketing point of view, Saturn's new automobile has clearly been a success. However, Saturn does have cost and productivity problems. Berggren and Rehder (1993: 195), for example, report that Saturn lost $700 million in 1992. Apparently, the operation is simply not as lean as rival Japanese factories and there are quality problems with suppliers, all of which are UAW operations and were chosen jointly by union and management. In the mean-time, Saturn offers to GM a highly valuable lesson in marketing and in non-adversarial co-management.

Saturn and Uddevalla compared   In the mid-1980s, as GM was planning its Saturn project, Volvo decided that within Sweden it needed both

Table 12.1 GM Saturn's Partnership Structure: Key Initiatives

Initiative	Comment
Work teams	Self-directed teams of 6–15 members, elected leaders. Broadly shared authority over daily operations regarding matters related to producing to budget, quality housekeeping, safety and health, maintenance, material and inventory control, training, job assignments, repairs, scrap control, vacation approvals, absenteeism, supplies, record-keeping, personnel selection and hiring, work planning, and work scheduling.
Modules	Comprise teams interrelated by geography, product or technology. Modules have common 'partnered' advisors integrated into three Business Units (Body Systems, Powertrain, and Vehicle Systems).
Decision rings	Joint union–management groups which meet weekly. Important decision rings include the Strategic Action Council (long-range planning and issues related to dealers, suppliers and community, etc.) and the Manufacturing Action Council (manufacturing issues at Spring Hill). Each Business Unit has a decision ring, also organized at the module level.
Partnering	The one-to-one partnering between union members and 'middle' management providing for on-line co-management by the union (and replacing foremen, superintendents and their assistants).
Problem resolution circles	Labour–management committees to address 'off-line' problems.

Source: based on Rubinstein et al (1993, especially pp. 342–5).

new, integrated auto assembly capacity and to restructure its labour relations (Alvstam and Ellegard 1990; Berrgren and Rehder 1993). This decision necessarily took into account Volvo's existing operations which in Sweden principally comprise Torslanda, west of Gothenburg, where an assembly plant is integrated with a body shop and paint shop (Figure 7.7). This plant was opened in 1964 and another assembly plant was opened in Kalmar in south-eastern Sweden in the early 1970s. Both these locations are supplied from other Volvo plants including Skövde (engines) and Köping (transmissions), as well as from a large number of subcontractors (Fredriksson and Lindmark 1979). As Alvstam and Ellegard (1990: 188) note, a new body shop, paint shop and assembly operation could have logically been added to the Kalmar site, at that time only an assembly operation, or at Olofström which had a press shop. In either case, new integrated facilities would have increased the efficiency of existing flows of components and processes within the Volvo system.

In practice, Volvo eschewed its Kalmar and Olofström options in favour of a new location, specifically Uddevalla, to the north of Gothenburg (Figure 7.7). While there were several factors underlying this choice, including local development issues, as was the case with GM's choice of a Tennessee location, a critical factor underlying the choice of a new site reflected a desire 'to develop new ideas of work organization' (Alvstam and Ellegard 1990: 189). Similar to GM's strategy at Saturn, Volvo management involved the union from the beginning of the planning of the new plant. At its Uddevalla plant, Volvo's motivation was to replace the traditional, fordist system of assembly in which 'core' workers perform jobs narrowly demarcated according to short-cycle, highly repetitive tasks, a system it employed at its Torslanda plant, with more flexible 'core' workers performing longer-cycle and more varied tasks. Alvstam and Ellegard (1990: 193–194) offer several factors that encouraged Volvo to create more interesting, varied and responsible work experiences. In particular, Volvo was keen to offset the problems of increasing rates of turnover and accident rates associated with short-cycle repetitive work, and also to offer jobs to

Swedish youth who constitute a declining cohort in the Swedish population. Moreover, Volvo recognized that better educated youth would demand more challenging and varied work experiences and would not be attracted by narrowly defined 'blue collar' jobs.

The Uddevalla plant that opened in 1988 represented a truly radical approach to work organization in the auto industry. While auto assemblers traditionally performed repetitive, specialized tasks lasting one to two minutes, at Uddevalla work cycles lasted one to four hours. The factory itself comprised six, separate product shops for assembly plus a materials store, several areas for final test and completion, and an administrative building (Alvstam and Ellegard 1990: 197). Each of the product shops comprised several small teams of (10) workers and each team assembled a complete car. Prior to the start-up of the plant, in 1986 Volvo completed a training centre where workers were extensively trained in the building of actual cars. The general goal was to provide each worker with the knowledge to build one-quarter of the car. The training centre also played an important role in testing new techniques and new forms of organization. In theory, just four workers would be needed to build a car at Uddevalla, although when the plant first began production, workers were grouped into teams of ten. In addition to the traditional type of assembly work, each team was responsible for structuring and checking materials and components before assembly, quality control, and making adjustments where necessary. Each team was also responsible for tasks previously performed by foremen and managers, which not only included controlling their own work day, but also involvement in longer-term planning, including with respect to such matters as recruitment, budgeting and production technique (Alvstam and Ellegard 1990: 198). Almost half of the original workforce comprised women.

In comparison with GM's Saturn venture, Volvo's Uddevalla plant represented a more radical 'humanistic' way of building cars at the factory floor level. The transition from very short to long work cycles, the novel approach to training, the breadth of the training of assembly workers emphasizing functional understanding and a holistic perspective, and the plant's innovative ergonomic design and lay-out are the particularly noteworthy features of the Uddevalla experiment (Berggren and Rehder 1993: 195). If Volvo's redesign of the assembly process and of worker skills at Uddevalla was more profound than at Saturn, from a broader perspective GM's Saturn venture is more radical than Uddevalla. In particular, the integration of manufacturing processes with innovative product development, supplier relations, industrial relations and marketing has been more explicit and comprehensive at Saturn than at Uddevalla. GM is clearly committed to the Saturn project; despite large corporate losses in the early 1990s, which included losses at Saturn, the plant has continued to manufacture cars. In contrast, when Volvo suffered corporate losses, in 1993, the Uddevalla plant was closed

According to Berggren and Rehder (1993), the Uddevalla plant did not fail because of its pioneering forms of work organization. Indeed, they suggest that in terms of productivity, quality, cost-effective technical solutions, cost savings related to health, lower tool and training costs, and in reducing delivery times to customers, the Uddevalla plant recorded rapid improvement after 1990, which either met or surpassed the performance in the Torslanda plant. Rather, they relate the closure of the Uddevalla plant to a substantial downturn in Volvo's markets, sluggishness in Volvo's pace of innovation, the failure of Volvo's senior management to fully embrace the Uddevalla experiment and, ultimately, the power of vested management and union interests in the Torslanda plant. Thus, when Volvo felt it necessary to rationalize capacity, it was easier to close the newer (least senior) and small Uddevalla plant than the bigger and older Torlsanda plant. At least for Volvo, any immediate reform of work organization will have to be *in situ*.

Peripheral workforces in new locations: the importance of women

It has been intimated since the 1960s that firms choose new locations, in part, to access a peripheral workforce, i.e. workers who are passive, low-waged, hard working and stable. To varying degrees, trends towards suburbanization and non-metropolitan industrialization as well as offshore leaps to poor countries have sought labour with these characteristics. To an important degree, hiring preferences in these contexts have implied a search for female workers and in some cases for child labour. In the very heyday of fordism, many firms gave emphasis to the hiring of what is now referred to as flexible, often female labour. Peripheral and exploited workforces have never been confined to females, and core and privileged labour markets are not confined to males. Even so, the process of industrialization marginalized women in relation to paid employment.

Traditionally, female labour has received lower wages than male labour, even for similar work, and firms widely believe that female labour is less likely to be unionized and easier to control than male labour. Indeed, the original dual market based labour segmentation theory suggested that women are over-represented in the peripheral secondary sector (Figure 12.1), while the recent flexibility based labour market segments similarly argue that women play a proportionately bigger role in jobs that are considered numerically and financially flexible (Figure 12.2). Across most sectors of advanced countries, for some time women have provided important shares of peripheral workforces (Christopherson 1989; Saso 1990). More recently, women have greatly expanded participation in the core workforces, especially within the service sector but also in manufacturing. Within British Columbia's private sector, for example, in manufacturing, wholesaling and services, women professionals were the most rapidly growing segment between 1981 and 1986, even in industries experiencing employment decline, although their share of these jobs remains small (Hayter and Barnes 1992).

Married women and the suburban option Historically, World War II reminded many industrialized countries that women could perform the same tasks as men in a variety of industries with at least equal efficiency under extremely difficult circumstances (and for low wages). The war did not seem to change existing family values, particularly with respect to gender roles and relations, and women were comprehensively replaced by men in the manufacturing sector. Within 10 years or so, however, an increasing number of firms in 'light' industries, such as various branches of metal-working and tool manufacture, began to seek out married as well as single women workers at a time when married women in growing numbers were looking to return to the job market. The 1950s and 1960s was a period of rapid growth, with, by contemporary standards, extremely low unemployment rates in most advanced countries, and firms often faced labour supply shortages. At the same time, factory-skilled and low-cost women were available, including married women who already had children and who were looking to supplement family incomes. Moreover, society's values still supported the idea of a male 'breadwinner' earning a 'family wage', i.e. a wage sufficient to maintain a family, so that women's wages were widely viewed as secondary, perhaps even temporary. Such values provided tacit support for firms to continue to pay low wages to married women as well as to unmarried women whose traditional roles in the workforce had been temporary. At that time, within families, arguments centred as much on whether women should work as on the question of pay.

Geographically, in many countries, including the US, the UK, Canada and Australia, one aspect of the suburbanization of manufacturing in the 1950s and 1960s included the matching of female labour demand with female labour supply, married and single. Such matching may or may not have been important location factors: for many firms the move to the suburbs was motivated by land costs, land availability, building design considerations, taxes, congestion in inner cities, and labour-related reasons. Nevertheless, to some extent it was at least assumed that by locating near residences, married women could find jobs within walking distance or within a short commute. Such assumptions were not necessarily validated in practice. One anecdotal story in this regard is provided by a Sheffield-based cutlery firm (Case B in Hayter and Patchel 1993) which relocated in the 1950s to a site near to the Sheffield Wednesday football ground, adjacent to new housing estates which the firm assumed would provide female labour. Instead, most of its existing female labour force chose to commute across town – a pattern which has remained in evidence.

In recent years, more rigorous attempts to investigate female labour markets in the suburbs have been conducted within the context of the 'spatial entrapment hypothesis' (Hanson and Pratt 1988, 1991, 1995; England 1993). Hanson and Pratt (1995), for example, in the context of suburban areas in Massachusetts, suggest that women are typically 'entrapped' within peripheral labour markets and 'spatially entrapped' within distinctly female labour markets, which in turn exist within a patriarchal capitalist society. This hypothesis focuses on suburban-based, white, married women with children, who wish to work close to home in order to maintain domestic obligations. Such women, the hypothesis argues, are willing to forego high wages in favour of local, but low-income jobs in order to give more priority to domestic roles (England 1993). This hypothesis predicts, for example, shorter journey-to-work distances for women, especially married women, compared to men. Hanson and Pratt's (1991) study supports this thesis and they further note that since most women (93%) found jobs after choosing a residence in comparison to two-thirds of men, residential location decisions are primarily affected by male choice of jobs. Hanson and Pratt also emphasize that variations in female labour markets in terms of income, race and age attract different kinds of activities. Manufacturing firms, for example, favoured the lower-income suburb containing Puerto Rican workers. In general, employers deliberately sought female labour markets by locating in particular areas, by using particular recruitment methods and by arranging job schedules more attractive to

women than men, e.g. by emphasizing the part-time and temporary nature of employment.

Other studies suggest a more complex view of suburban female labour markets than implied by the spatial entrapment hypothesis. Indeed, England (1993: 239) cautions strongly against a too literal interpretation of the hypothesis, which she sees as "an overgeneralization and over-simplification" (see also McLafferty and Preston 1991). England's investigation of journey-to-work distances in a suburb of Columbus, Ohio, reveals only minor differences between males and females, while she finds bigger differences between married and single women. Moreover, contrary to expectations (of the spatial entrapment hypothesis), married women with children have longer commutes than single women (England 1993: 232). Rather, she notes that the relationships between choice of residence, choice of work, marital status, number of children, age of children and family situations are extremely varied and that women (and men) have various ways of 'coping', to use Dyck's (1989) term, with their various obligations, including journey to work. For firms, a suburban location, even if targeted at women workers, may or may not mean that labour supply is restricted to that suburb (Hanson and Johnston 1985; Villeneuve and Rose 1988; McLafferty and Preston 1991). Similarly, if suburban women can work downtown or elsewhere in the metropolitan area, the large supplies of low-wage immigrant (female and male) workers that exists in many inner city areas is to varying degrees available to new plants in suburban locations (Scott 1988b).

Single women and the export processing zone option   The dispersal of manufacturing activities to peripheral regions and non-metropolitan areas within advanced countries and to low-wage countries has frequently targeted female labour markets. Within advanced countries, job opportunities were often limited for women in rural areas and small communities, where availability, low wages and low levels of militancy among women have provided advantages to firms. Data on the gender structure of workforces were not

Table 12.2   Female Workers in Export-Processing Zones

Country	Women in EPZ industries (%)	Women in non-EPZ industries (%)	Percentage of women EPZ workers by age group [a]
Hong Kong	60	49.3	85 (20–30)
India	80	9.5	83 (<26)
Indonesia	90	47.9	83 (<26)
Korea	75	37.5	85 (20–30)
Malaysia	85	32.9	(average: 21.7)
Philippines	74	48.1	88 (<29)
Singapore	60	44.3	78 (<27)
Sri Lanka	88	17.1	83 (<26)
Mexico	77	24.5	78 (<27)
Jamaica	95	19.0	(average: 20s)
Tunisia	90	48.1	70 (<25)

Source: Dicken (1992: 186)
[a]Age group (in parentheses) is in years.

often collected in early surveys, although Townroe (1975: 54) notes that in one UK region females were more important in externally owned branch plants than in locally owned plants (see also Thwaites 1978), and Morgan and Sayer (1988) observe the tendency of electronics manufacturers to hire females in assembly-line operations. In the US, rural locations have provided manufacturing firms in various industries with large contingents of female workers (Schmenner 1982).

In developing countries the availability of female labour is even greater and the wages and non-wage benefits (if in existence) even lower. Indeed, in such countries, most obviously in export processing zones (EPZs), which provide low-cost labour and tariff-free imports for export activities, the seeking out of female labour is particularly pronounced (Elson and Pearson 1981; Fernandez-Kelley 1989; Tiano 1994). For some observers, the shift towards more flexible forms of production inevitably implies continued growth in the female labour force (Standing 1989). A survey by the International Labour Office (1988), in export processing zones in 11 different countries, found that the percentage of women employed never dropped below 60% and only two cases were below 74% (Table 12.2). This level of female employment is significantly

higher than national averages found outside the export processing zones, and in general the women workers are young and single. In some instances, young single women have been specifically targeted; Nelson (1989: 14) says that some employers have fired women who married and others "have actually awarded prizes to women who undergo sterilization, which is considered the sign of a dependable company employee". Largely unprotected by unions, the rights of workers in export processing zones (and elsewhere in poor countries) are limited. Management's concern can be highly focused on immediate productivity. As Schoenberger (1988: 116) notes, "Management often provides pep pills and amphetamine injections to keep the women awake and working; some of the women become addicts" (see also Fuentes and Ehrenreich 1987).

Moreover, with low opportunity costs and no union representation, the wages paid to workers in EPZs are extremely low. In the early 1990s, for example, a cutlery firm in the Shenzhen economic zone in China paid its workers, mostly women, 17 cents an hour (Hayter and Patchell 1993). A survey of Japanese-owned maquilas in Mexico reported that in late 1991 average wages amounted to US$1.10–1.25 per hour, i.e. about four times less than the US minimum wage, and the range was between 50 cents and $2.00 (Kenney and Florida 1994: 31). Kenney and Florida also note that the non-wage benefits were relatively small, and varied from free bus services and cafeteria food to company-sponsored parties. While this study did not provide details on the extent of female employment, Florida and Kenney (1994: 34) observed that gender ratios varied considerably by plant and that distinct gender divisions of labour did exist; for example, while television cabinet manufacture was done almost entirely by men, circuit-board stuffing was done almost entirely by women (see Sklair 1989).

The workers hired by the Japanese maquilas represent classic types of numerical flexibility. Most production jobs, regardless of gender, are low skilled and routine, and according to Kenney and Florida, there are few signs of training. They report that absenteeism and turnover rates are extremely high, but since hiring and training costs are negligible, workers are readily replaced. They conclude that in their maquilas "Japanese firms are managing their operations in ways that resemble their usage of temporary and part-time workers in Japan" (Kenney and Florida 1994: 35).

Tapping child labour   Although illegal, in several countries there are large numbers of children working for extremely low wages. India is estimated to have the largest amount of child labour, the estimates varying from the low official count of 15.5 million to between 50 and 100 million (Bramham 1996: D3). China has an estimated 5 million child workers, and there are about 20 million children in Pakistan who work under a bonded system in which the children work to pay off loans made to their parents. Many of these children stitch carpets. Wages can be as low as US$8 per month. Children who stitch soccer balls in the Sialkot region of north Pakistan, the world's principal source of hand-stitched soccer balls, do better, receiving about US$1.00–1.50 per hour (Lees and Hinde 1995: 6). Children begin work at the age of six and are paid on a piece-work basis so that wages primarily depend on the number of balls stitched per day.

Children, for the most part, are probably directly employed by local organizations, including government agencies. Established MNCs, even if they have policies which ostensibly prevent the use of child labour, nevertheless do tap into child labour through subcontracting linkages. The soccer balls stitched in the Sialkot region, for example, are marketed by such well-known firms as Mitre, Adidas and Reebok. Similarly, garment manufacturers in the US tap into cheap labour, including child labour, in Asia and several Pacific Islands. Although illegal and widely condemned as immoral, child labour is likely to remain for some time. The children and their families need the money, no matter how paltry, and their nimble fingers literally offer physical flexibility which is prized and which can perform in ways older fingers cannot. At the same time, in southern Asia at least, child labour is a brake on adult wages and helps contribute towards already high levels of unemployment.

Nike's overseas leaps   Nike, the sports footwear company, provides a good illustration of a company that has pursued a geographical strategy in pursuit of numerical and financially flexible labour primarily through subcontracting production to low-wage regions (Donaghu and Barff 1990; see also Barff and Austen 1993). The company originated as a US-owned distributor of athletic shoes produced by a Japanese manufacturer, Japan then being a low-wage region. Faced with a threat of takeover by its Japanese supplier, in 1971 Nike began a partnership with a Japanese *sogo shosha* or trading company which arranged alternative lines of supply. In the meantime, Nike concentrated its research and development in the US, at Beaverton, Oregon. However, further battles over control, and rising costs in Japan, encouraged Nike to establish its own partnerships in a variety of countries (Figure 12.3). Most of the partnerships Nike has created have been on a contract basis although it did acquire its own factories in the US, the UK and Eire. During the

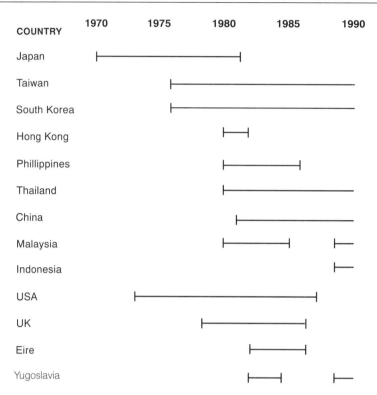

Figure 12.3   The Duration of Nike's Athletic Footwear Production and Plant Ownerships, by Country. Source: Donaghu and Barff (1990: 542)

1980s, however, Nike divested these connections and in recent years virtually all of Nike's production has been subcontracted.

Most of Nike's subcontracting relationships are in south-east Asia and take on one of three forms (Figure 12.4). *Developed partners* are the most sophisticated producers within the system, and are primarily located in Taiwan and South Korea. They manufacture the most recent, design-intensive shoes while subcontracting the price-sensitive components to other suppliers, including the developing sources, in even lower-wage countries. Developed partners are exclusive to Nike, which tries to ensure as even a flow of orders as possible. The *volume producers*, mainly based in South Korea, are non-exclusive to Nike and manufacture large quantities of standardized models. These suppliers experience considerable fluctuations in output. Finally, there are the *developing sources* in low-wage countries (parti-

cularly Thailand, China and Indonesia), which Nike is trying to develop on an exclusive basis, presumably initially by contracting out relatively small job lots of relatively low-value components. The developed partners and developing sources are closely integrated through subcontracting arrangements and by Nike's 'expatriate programme' in which engineers from Beaverton spend time with suppliers to transfer skills and quality-control concerns.

As a result, Nike has developed a highly flexible system which seeks to promote stability among its most valued suppliers and actively promotes skill development while also minimizing labour costs and maintaining quality (Donaghu and Barff 1990; Barff and Austen 1993). Even the favoured suppliers are in relatively low-wage locations, however, and they, in turn, contract out the more standardized components to even more peripheral locations.

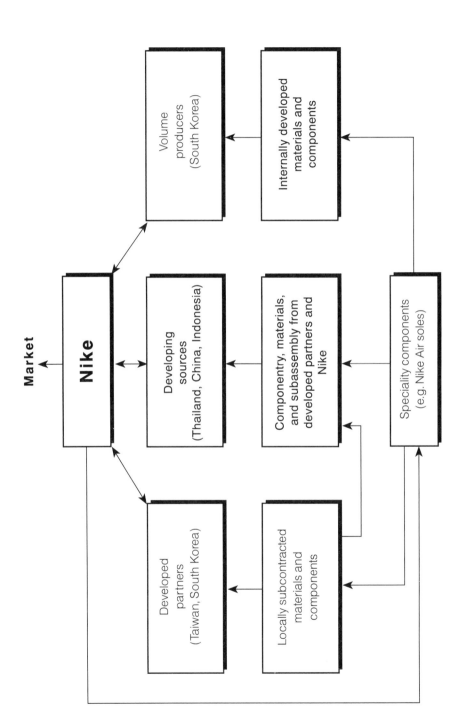

Figure 12.4 Nike's Search for Employment Flexibility. Source: Donaghu and Barff (1990: 545)

The developing sources provide a stream of alternative as well as complementary suppliers, while the volume producers are turned on and off with demand. Herbert (1996) argues that Nike has organized an exploitative sweat-shop system in which Nike constantly switches its production in search of ever lower labour costs. He notes that sweat shops with Nike contracts in Indonesia pay $2.20 a day, and the firm is currently contemplating Vietnam, where labour costs are $30 a month. On the other hand, Herbert (1996) reports that Nike has paid Michael Jordan $20 million for advertising.

In situ change and flexible workforces

The development of various kinds of flexibilities at new locations has placed increasing pressure on established factories to adapt to similar labour bargains to the extent that the former are more productive, innovative and value-oriented than the latter. The *in situ* search for employment flexibility is an inherently more difficult process than at new locations, especially in unionized factories where fordist labour relations are entrenched in tradition and law. Nevertheless, geographical options are not always readily available and in all industries existing factories comprise immovable capital and inherited human know-how and may still offer appropriate location factors which encourage firms to renegotiate labour bargains in place.

Not surprisingly, within established fordist contexts, such as the North American auto industry, the most radical changes towards functional flexibility have occurred at new locations, such as GM's Saturn plant and the new Japanese transplant locations. Changes towards more flexible work practices have been made at existing plants, and at NUMMI, the joint venture between GM and Toyota in Freemont, California, these changes have been far reaching in incorporating Japanese practice (Womack et al 1990: 82–84). NUMMI occupies a factory that had been closed down by GM in 1982; however, although located on an existing site, NUMMI's

management were able to insist on a new contract with the union which, for example, fully accepted team work and only two job categories: assemblers and technicians. In other existing factories the shift towards flexibility has been more problematical (Clarke 1989).

Renegotiating core workforces in situ: MacMillan Bloedel in British Columbia

The British Columbia (BC) forest industry is a good example of a strongly fordist industry that has been struggling to shift towards more flexible operating cultures (Grass and Hayter 1989; Barnes et al 1990; Barnes and Hayter 1992; Hayter and Barnes 1992). These struggles have largely occurred at existing sites. Since the early 1980s, MacMillan Bloedel (MB) has been the leader of the corporate unionized sector in pioneering new forms of work organization. The best-known example is its Chemainus sawmill (Barnes and Hayter 1992), where (like NUMMI) changes in employment practice at this mill were facilitated by the permanent closure of an existing mill and its replacement by an entirely new operation.

The Chemainus sawmill In 1982, MB closed down its sawmill at Chemainus, which employed 600–700 workers in 1980 but was obsolescent (Barnes et al 1990; Barnes and Hayter 1990, 1992, 1994). The old sawmill at Chemainus was entirely torn down and replaced on the same site, two years later, with a new mill, which was quickly fully computerized and able to cut lumber precisely to a wide variety of dimensions, particularly for the Japanese market. From its start-up in 1985, the new mill has been fully committed to the principles of functional flexibility with its emphasis on team work, the development of multiple skills, pay for knowledge scheme, ongoing training and close management–worker interaction. In addition, the apprenticeship programme has been reintroduced for trades occupations, and profit-sharing has recently been experimented. Bearing in mind its favoured tidewater location and access to high-quality logs, the Chemainus sawmill has been consistently profitable, and it is has operated without lay-off since its re-opening, including through the recession of 1991.

In the terms of Marshall and Tucker (1992), the Chemainus sawmill has successfully shifted from Taylorized mass production principles, to flexibility principles based on the high-performance strategy model. The International Wood-workers of America (IWA), which represents most sawmill workers throughout British Columbia, accepted the comprehensive introduction of flexibility principles at this mill and the local union of the IWA at Chemainus operates the sawmill according to a sub-agreement within the master contract negotiated between the IWA and most sawmill employers. (For a general discussion of union organization, with specific reference to North America, see Clark 1989.) Admittedly, in prior negotiations, management had a powerful negotiating advantage from closing the old mill and laying off all the workers. A clause in the master contract tied the firm to established seniority rights only until two years after closure, at which time the firm had the discretion to hire whom it wanted when opening the new mill. Nevertheless, the Chemainus example does demonstrate that functional flexibility can be readily incorporated within a union environment. Difficulties in developing flexibilities in existing mills arise, however, because customs and contractual relations are deeply embedded.

The Powell River papermill In practice, among operational union mills, attempted shifts from fordism to flexibility principles have occurred within the context of employment downsizing and increasingly tough negotiations at the factory level between management and the union locals which represent workers in specific mills. These negotiations, which can be lengthy, typically feature specific deals or quid pro quos. A summary of the experience of MB's Powell River paper mill, which has been detailed elsewhere (Hayter and Holmes 1994; Hayter forthcoming), provides an illustration (Table 12.3). Thus, in

December 1973, the mill employed 2600 people, including 233 relief workers, all of the hourly staff belonging to one of two union locals. In December 1981, employment still amounted to 2335, but by 1985, following recession, over 500 jobs had been lost and a fundamental turning point in employment conditions had been reached at the mill. For the first time in the mill's history, salaried personnel were affected as badly as the hourly workers, and jobs had been permanently lost. While a number of jobs came back in the market boom of the late 1980s, another 700 jobs had been lost by 1994, most occurring following the recession of 1991. By January 1994, the mill's employment was down to 1275 jobs (including 235 relief workers) and, according to management estimates, which are believed by the union, employment will continue to decline in the near future until it stabilizes at a level of no more than 945 jobs, probably lower.

By 1994, employment levels were less than half of peak levels (and declining), but production capacity was still 80% of former levels (and may increase without any employment increment) and compromises a wider range of higher-value papers. The job loss, and the associated increase in labour productivity, clearly has been essential in order to maintain the competitiveness of the mill (Holmes and Hayter 1994). To a significant degree, the job loss can largely be explained by the effects of technological change and rationalization. To a much lesser, if not precisely known degree, the introduction of more flexible work arrangements has also become a factor since 1990, when the first set of flexibility concessions were agreed upon. Indeed, employment flexibility is considered vital to improving productivity, increasing skill and lowering costs, and to ensuring that Powell River remains competitive with greenfield paper mills, which are typically organized around highly flexible working cultures. At Powell River, flexibility has principally concerned modifying job demarcation lines among the trades, and between trades and production line workers, to speed up maintenance work completion times; flattening out the organizational structure of the mill to increase

Table 12.3 Powell River Paper Mill: Employment Levels, 1965–1994

Year	Salaried	Hourly	Relief	Total
1965	216	1730	130	2076
1966	242	1724	187	2153
1970	308	1915	246	2475
1973	325	2032	233	2596
1974	339	1948	240	2527
1980	330	1665	315	2310
1981	332	1720	283	2335
1982	271	1435	238	1944
1983	286	1393	286	1943
1984	265	1325	273	1863
1985	265	1290	272	1827
1986	265	1297	271	1833
1987	303	1328	268	1899
1988	306	1338	346	1990
1989	314	1349	303	1966
1990	303	1318	285	1906
1991	270	1309	249	1828
1992	230	1065	205	1497
1993	204	860	233	1297
1994	182	857	235	1275

Source: Powell River Mill records.
Note: The data pertain to the situation on 31st December except in 1994 when the data are for 31st January.

managerial efficiency; facilitating labour–management interaction, notably by assigning more responsibility to labour; making entry requirements more rigorous to ensure that any new recruits are (mentally and physically) able and willing to be functionally flexible; and through contracting out. Job rotation and comprehensive reductions in job demarcation are also presently matters under discussion and work flexibility is very much an ongoing issue.

At an existing mill such as Powell River, this search for flexibility is inevitably a difficult and at times emotionally harrowing experience. In this regard, four summary points may be noted. First, in an existing union workplace, flexibility, regardless of the underlying model driving it, is bound to be contentious, not only because it demands previously 'hard fought' concessions from workers, but because flexibility strikes at the central principles of modern unions, namely job demarcation and seniority. For unions, these principles serve to eliminate wage competition among workers, constrain managerial autonomy, provide security and discipline among workers, and prevent arbitrary job intensification. Contracting out is also a potential threat to these values as well as to the lowering of employment standards. Thus, even forms of functional flexibility that stress enskilling and job satisfaction, along with productivity gains, raise legitimate concerns, and any shift towards flexibility demands difficult trade-offs for unions to make.

Second, flexibility concessions in unionized workplaces require formal negotiations between management and unions, whether the negotiations concern individual jobs and workers or are across the board. At Powell River, for example, workplace flexibility has so far been negotiated in two agreements, in 1991 and 1992, involving specific quid pro quos. Thus, following a wild cat strike in 1988 over a contracting-out issue, a subsequent court action required the union to pay MB over $4 million. Instead, in 1991, the union agreed to flexibility concessions in lieu of the fine. In 1992, the union agreed to a further set of flexibility concessions in return for an early retirement package. Since then, although they have tried, management has been unable to offer workers another acceptable quid pro quo in return for more flexibility.

Third, the negative effects of relentless downsizing on worker morale and trust (and potentially on productivity) are likely to be reinforced and complicated by flexibility discussions, especially if protracted (which they are virtually bound to be). In the case of Powell River, the fact that agreement to be more flexible has not provided job security for the surviving workforce has become a problem. Indeed, the mill's workers became embittered when shortly after the 1992 flexibility concessions another round of lay-offs occurred, and the jobs the unions thought saved, disappeared. For the unions, agreement to an early retirement package and increased job flexibility was traded for job security. For management, job flexibility was traded for early retirement with no implications for job security. It might also be noted that even among the surviving workers at Powell River, flexibility is seen as personally distressing because it creates job loss among friends.

Fourth, in an established Taylorized mass production workplace, many managers and workers will not have the appropriate attitudes and skills (let alone formal qualifications) for a more flexible operating culture. Managers can be replaced but new staff require time to learn their new positions. At Powell River, for example, there is some evidence that new, more flexible managers are not highly regarded in terms of their knowledge about the mill, and there are problems in implementing a flatter organizational structure.

A critical issue in the transition to flexibility is the question of training. In high-performance strategies, emphasis is on the development of the 'core', stable and well-paid worker with polyvalent skills. Education and training are essential in helping to qualify for 'core' jobs; in turn, 'core' jobs require an ongoing commitment to education and training. In the particular case of Powell River, labour flexibility – in particular, the elimination of traditional job demarcation lines,

including with respect to management functions – is associated with this idea of the polyvalent worker. The nature of the production changes from being a commodity producer to being a higher value producer of a range of higher quality papers further reinforces this association.

Yet, in an old mill like Powell River which is downsizing, the question of 'skill formation' is a problematical one. The apprenticeship programme, has been a victim of downsizing. The team concept in the wood room experienced problems because of difficulties in training everybody to the level necessary to practise job rotation. Extensive job bumping has disrupted traditional on-the-job training (OJT). The lack of articulation between the new managers and the workforce poses problems for effective, interactive learning. For the workforce, commitment to ongoing training and education raises the spectre of 'testing'. For senior management, training and education is costly and it takes people away from their jobs so that inevitably training is selective and a potential source of dismay for people not chosen. In the context of rapid downsizing, who is to be trained? And what about training for workers (and managers) who are laid-off, particularly those people who lack the skills required by society outside of the mill? Clearly, 'training' is not some 'packaged' option or magic wand that can be purchased or waved to move the mill from mass production to high-performance strategies. Rather, training itself involves significant costs, uncertainties and negotiation.

Maintaining peripheral workforces in situ

Even in high-wage countries, it should not be automatically assumed that firms are generally pursuing value-added strategies based around functionally flexible workforces. Indeed, the spirit of the rise of neoconservatism in the US and the UK, respectively associated with Reaganism and Thatcherism, as well as privatization and deregulation, is to provide individual firms with greater latitude in decision-making. Many firms have translated this spirit to impose greater control over the workforce, cost-cutting and a greater reliance on a peripherally flexible workforce. General economy-wide increases in part-time and temporary workers in the US and the UK, for example, illustrate such a shift (Christopherson 1989). Certainly, in unionized environments, attempts to significantly degrade working conditions are likely to be strongly resisted. Yet union power in manufacturing industries is in decline, notably in the US and the UK (Clark 1989; Martin et al 1992). Indeed, the shift towards flexibility is both a cause and an effect of the decline in union power (Hayter and Barnes 1992). Flexibility is a cause of declining union power because through the threat of unemployment, firms are able to introduce new flexible work practices that potentially undermine the ability of unions to control job task; and an effect because once the power of the unions is weakened, other firms begin to introduce flexible practices.

In communities with weak or non-existent union traditions, and a history of poor working conditions, the possibilities for entrenching numerical and financial flexibility increase. Such a trend has characterized the Sheffield cutlery industry, which over the past 25 years has declined drastically in the face of foreign competition from advanced and developing countries (Hayter 1985, 1986b; Hayter and Patchell 1993). The Sheffield cutlery industry has always been characterized by low wages and poor working conditions, and during the present restructuring, firms are emphasizing the pursuit of low labour costs, and numerical and financial flexibility. Thus, the most successful firm in the industry is a kitchen-knife manufacturer which expanded its employment base from 160 to 500 between 1980 and 1989, was consistently profitable and won export awards (Case A in Hayter and Patchell 1993: 1434–1436). Apart from a small core group of 30 male engineers and technicians, however, the firm pays relatively low wages and its production workforce entirely comprises young, non-union, female workers who operate as many as six machines. Training is minimal and seasonal upsurges in demand are

met by hiring casual female labour. This firm combines the strictures of Taylorism – notably the separation of managers and workers, strong job demarcation of the latter and deskilling through automation – with non-union, numerical flexibility.

Skill formation of doubly peripheral workforces

In developing (peripheral and semi-peripheral) countries, skill formation can occur and the conditions of a peripheral workforce are enhanced. Kenney and Florida (1994) stress that Japanese (maquila) branch plants in Mexico have hired workforces that are numerically flexible. Yet even these plants are planning for wage increases. For example, they cite a company that was paying $1.00 an hour in October 1991, but which was planning to increase wages to $1.25 an hour in December 1991 and anticipated wages of $2.50 an hour by 1996. All the Japanese managers felt that wages were being driven up in Tijuana (Kenney and Florida 1994: 31). To some extent, such wage increases reflect increasing demand for labour. The wages are still low and the Japanese maquilas are not planning on formal off-the-job training, job rotation or working circles, and unusually high levels of worker turnover obviously reduce on-the-job training (OJT). However, Kioke and Inoke (1990) suggest one possible way in which peripheral workforces in peripheral countries can improve their employment conditions.

Enterprise-specific skills   In the context of Japanese branch plants in Asia, Kioke and Inoke (1990) emphasize the skill formation of workers over time through OJT to create what they term 'enterprise-specific skills'. Thus enterprise-specific skills are acquired by workers over time, and bind firms and workers together according to the unique characteristics of these skills (Patchell and Hayter 1995: 344). From this perspective, increases in worker skill are defined in terms of increases in worker productivity using given machinery. According to Kioke and Inoke (1990), workers increase their efficiency by

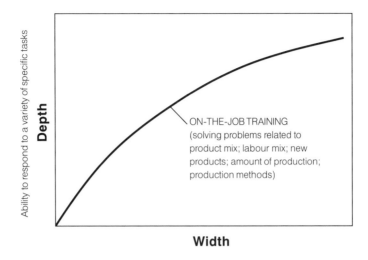

Figure 12.5 Enterprise Specific Skill Formation. Source: Patchell and Hayter (1995: 345)

improving their intellectual capability to deal with both routine problems and more significant changes that occur on the shop floor (within the factory). In turn, the intellectual ability of workers is gained by the width (horizontal) and depth (vertical) of factory experience and OJT (Figure 12.5). As workers deal with increasing complexity through OJT, they create enterprise-specific skills.

Both of these abilities can be displayed separately and concurrently as workers switch amongst their range of usual or routine tasks (width or horizontal experience) and as they deal with unusual or difficult situations within each of those tasks (depth or vertical experience). For example, when the *product mix* changes, the firm benefits if its employees have the horizontal skills necessary to adapt to the changes required on a production line. Workers display vertical or depth of experience by choosing and calibrating equipment that will allow those changes to be made faster and allow quality manufacture to proceed unhindered. Similarly, width and depth of experience enables workers to be rotated to ensure the unimpeded operation of production when a preferred *labour mix* breaks down due to absenteeism or fluctuations in experience at some positions. In addition, the introduction of *new*

products is enhanced if workers understand the logic of the production process and can assist in the changeover and calibration of machinery. Koike and Inoki claim that worker experience also proves valuable when the *amount of production* varies and when *production methods* are changed.

The ability to deal with problems is another major advantage that the skilled worker brings to their firm, and perhaps best exemplifies the intellectual depth that any worker can bring to the tasks they perform. Dealing with the inevitable problems that arise in production requires that a worker is able to detect the problem, diagnosis it and rectify the problem. Fixing major problems may be out of a worker's depth of abilities, but "knowledge of the structures, functions, and mechanisms of equipment, products, and the production process itself" (Koike and Inoki 1990: 9) will enable the worker to perform some maintenance and repair. Koike and Inoki also found workers' intellectual contributions to increase with automation.

On-the-job training (OJT) is the key to the improvement of workers' width and depth of job performance and their intellectual ability to deal with change. The character of most of the tasks that a worker performs are indefinable. The worker learns the intricacies of the machines used, the characters of fellow workers, the interrelationships between people and machines, and the process of production in a hands-on manner (Koike and Inoki 1990: 10). Workers learn different tasks in a step-wise manner by following an instructor, performing the task under supervision, working independently and then showing results to superiors, and seeking assistance only when necessary; when primary training is completed, workers continue to improve their skills incrementally by moving on to slightly more difficult tasks. The increase of the horizontal and vertical span of a worker's intellectual skills takes time, and OJT is primarily an enterprise-specific learning process. Off-the-job training can supplement OJT in an effective manner but theoretical instruction cannot match the direct transmission of skill offered by the combination of theory and application in OJT. OJT requires an investment by the firm in sustained salary increases and promotion, and although cost is incurred by the slowdown in production as the worker develops skill, the development of intellectual enterprise-specific skills (ESS) using OJT costs less and is more effective than taking workers away from production entirely for off-the-job-training.

In the context of a stable workforce, even once-unskilled workers can become more functionally flexible if provided with width and depth of experience. As Atkinson (1987) stresses, adaptability to change and the sustained reproducibility of that capability within the firm are the essential characteristics in creating core workforces. The co-operation necessary for learning skills that are often indefinable, however, needs to be nurtured by appropriate institutional arrangements (Peck 1993). Koike and Inoki emphasize the transmission of skills within firms. Skill formation is also possible through co-operative learning between firms, including between core firms and subcontractors (Patchell and Hayter 1995). Nike, for example, does not simply rely on its Asian network to reduce costs, but has taken the time to ensure quality improvements in its first-tier suppliers,

especially its developing partners (Figure 12.4). In this way, core workforces can occur in the periphery.

Conclusion: The illusion of the spatial division of labour

Since Hymer (1960), there has been increasing recognition that large corporations spatially sort out their internal labour markets to create distinct corporate locational hierarchies. However, the internal employment geographies of individual corporations can take on a variety of forms and are constantly subject to change within the context of a broader division of labour. Corporate restructuring can involve a search for numerical flexibility and this search can focus on inner cities, the suburbs, non-metropolitan areas and developing countries. Corporate restructuring can involve a search for functional flexibility, which can also occur in geographically diverse circumstances. Japanese branch plants in Mexico seem to be reinforcing numerical flexibility; in parts of Asia, functional flexibility. Within Asia, Nike relies on different workforces with different types of flexibility characteristics.

If labour markets are examined from a local rather than a corporate perspective, the impacts of the shift towards flexibility on employment differentiation and variability is confirmed. Within the UK, for example, the nature and extent of shifts towards functional flexibility varies within and among regions (Morris 1988; Pinch et al 1989). Within Canada, studies of provinces, resource communities in general and individual resource communities reveal that employment flexibility is developing along different lines in the same place and between places (Hayter and Barnes 1992; Norcliffe 1994; Randal and Ironside 1996). In the US, in metropoles such as New York, labour markets are remarkably diverse; downtown sweat shops contrast with the suburban world-scale R&D centres of firms such as IBM and International Paper. The precise forms of flexibility are finely variegated, varying geographically across region and nation, sectorally across industry type, and by market type and segment. The local economic consequences of flexibility are equally diverse. In some areas there is a burgeoning of high-paid, high-skilled functionally flexible jobs; in other areas, well-paid semi-skilled occupations are replaced by low-paid, numerically flexible ones; in yet other areas, employment of all types is lost as firms move to practise flexibility elsewhere; and in still other areas, there is a mix of these trends.

The spatial division of labour is therefore a problematical concept. Similarly, Arrighi and Drangel (1986) suggest that traditional cate-gories of core and periphery (and semi-periphery) countries are now less valid as production is organized globally in increasingly complex ways. Using evidence from the garment and shoe industries, Gereffi (1989, 1995) emphasizes that the growth in interna-tional supply networks and production chains is leading to changing distributions for value-added activity and the realization of profits that do not conform with these established cate-gories. He argues that core–periphery distinc-tions are better applied to different nodes within specific geographically complex produc-tion systems and chains.

PART IV

PRODUCTION SYSTEMS AND LOCAL DEVELOPMENT

Within the past 15 years, industrial geographers have become increasingly fascinated with the interrelationships between forms of industrial organization among populations of firms and the dynamics of industrial location and local development (Hayter and Watts 1983: 174; Scott 1986; Dicken and Thrift 1992). This research has focused on how firms, large and small, relate to one another, particularly in the context of industrial agglomerations or industrial districts but also with respect to more geographically dispersed inter-firm (and intra-firm) relationships. Part IV examines the geography of inter- and intra-firm relations within the context of particular 'production systems', and the implications of these relations for local development. Chapter 13 defines and discusses the basic concepts of production systems, industrial district, geographic multipliers and flexible specialization. Chapters 14–16 focus on particular types of production system in particular

places. Thus, Chapter 14 analyses Japanese-style flexible mass production systems, and in particular Toyota Town, arguably the most efficient auto industrial district in the world by the 1980s. Chapter 15 examines the geographical implications of direct foreign investment for home or donor economies, using the Tokyo metropolitan area as its principal example. The chapter then examines host or branch plant economies, focusing on the Canadian industrial system (which in national terms may be regarded as the biggest branch plant economy in the world), and on the special economic zones in China. Chapter 16 focuses on the once great and dominating industrial districts of the world (in the UK, the US and Germany) which have been de-industrialized and are now seeking rejuvenation. Chapter 17 concludes the book by briefly commenting on contemporary job dilemmas and industrial geography's contribution to understanding them.

13

PRODUCTION SYSTEMS AND INDUSTRIAL DISTRICTS

In a pioneering paper, Fredriksson and Lindmark (1979) define production systems as a set of linkages of material goods, services and information occurring within and among firms who are directly and indirectly integrated to produce a particular final demand such as a car, a TV set, an airplane, a robot, a paper machine, a spoon or a door. If, in practice, production systems have been primarily discussed in terms of linkages between manufacturing firms, conceptually links with primary sector firms, retailers, wholesalers, trading companies, business services and financial organizations can be readily incorporated (Gereffi 1989). Manufacturing firms themselves typically engage in 'non-manufacturing' activities.

Production systems comprise populations of interacting firms producing and supplying a related set of goods. In the case of a sophisticated product, such as a motor car, there are literally thousands of separate operations or functions involved in its manufacture and considerable choices as to how these tasks are allocated among a set of firms (and among a set of locations). Even in the case of the manufacture of a simple product, such as a spoon, there may be 20–30 different tasks and potentially as many firms. In practice, whether in the case of a 'complex' car or a 'simple' spoon, there are tremendous organiza-

tional variations in the nature of production systems over time and across space. Thus the same industry can be organized by relatively few (horizontally and vertically) integrated firms or by a plethora of small specialized firms (and by any number of combinations in between). These variations reflect different choices about the internal and social divisions of labour, which in turn have important implications for industrial location and for the character and potential of local development.

Production systems exhibit widely varying geographic structures. On the one hand, there are production systems that are highly dispersed and globally integrated within the framework of multinational corporations (Donaghu and Barff 1990). On the other hand, there are production systems, or substantial parts of production systems, that are geographically concentrated in dense agglomerations to create 'industrial districts' (Scott and Storper 1990; Scott 1996). Between these two polar cases, a variety of possibilities may be conceived (e.g. Leung 1993). There is also considerable controversy regarding the relative importance of industrial districts, in relation to more dispersed forms of production systems, in contemporary industrial change. Historically, industrialization has been a profoundly regional phenomenon (see Chapter

3), and industrial districts – formed by interacting populations of firms – have been a well-established, important feature. However, particularly within the context of fordism, increasing levels of corporate concentration and the growth of multinational corporations, encouraged a dispersal of production away from established concentrations. During the present ICT techno-economic paradigm, some observers have related increasing globalization with increasing dispersal of production and the fragmentation of regional economies (Amin and Robins 1990; Lovering 1990). Other observers, including Piore and Sabel (1984), argue that a defining feature of contemporary trends towards flexibility is the re-agglomeration of activity within industrial districts, including in new industrial spaces (Scott 1986, 1992; Storper and Scott 1989; Sabel 1989; Storper 1989).

This chapter explores the concepts of production systems and industrial districts in three main parts. First, production systems are explained in terms of the social and internal divisions of labour. Second, the implications of the linkages between firms for local development are illustrated by reference to the concept of the geographic multiplier which is explained and distinguished according to 'quantity multiplier' and 'quality multiplier' effects. Finally, the concept of the industrial district is examined and, particularly with respect to the related notion of 'flexible specialization', contemporary forms are briefly elaborated (Storper and Harrison 1991; Harrison 1992).

The internal and social divisions of labour

An internal division of labour occurs when related activities required to produce a final demand (or indeed some intermediate product) are entirely controlled *within* the firm. A social division of labour occurs when related activities required to produce a final demand occur *among* a population of small firms performing highly specialized functions. Theoretically, the various possibilities for organizing production may be

conceived, in the first instance, as existing between two polar types of 'governance'; namely, the extremes of all production being organized and controlled by one huge firm, and all production being organized and controlled by an infinite number of tiny firms.

The choice between the internal and social divisions of labour is often summarized as the 'make or buy' decision; an internal division of labour occurs when a few firms decide to make themselves the parts and perform the functions they need, while a social division of labour occurs when firms decide to buy most of their goods and services from other firms. In this latter case, subcontracting occurs when firms buy, outsource, externalize or contract-out part of their production needs to other firms (Taylor and Thrift 1983; Holmes 1986; Patchell 1993a). The two polar cases underlying the make and buy decisions reflect production systems respectively organized according to (integrated corporate) *hierarchies* or (small-firm-dominated) *markets*. Subcontracting, however, permits a range of possibilities between these two cases. In particular, subcontracting by large core firms links the internal and social divisions of labour.

Two approaches to explaining production systems are outlined. The first approach is the transaction cost model which, as originally articulated by Coase (1937) and recently elaborated by Williamson (1975, 1985), explains choices of markets or hierarchies as a (neoclassical) exercise in maximizing efficiency (usually minimizing costs). The second approach recognizes greater flexibility in these choices, focuses more on the varying rationales for subcontracting, and adopts the dual model of business segmentation as its point of departure (Figure 8.3).

The transaction cost model

The transaction cost model is a neoclassical extension of the theory of the firm. Simply stated, whether markets or hierarchies dominate depends on their relative efficiency in reducing transaction costs. Coase's (1937) fundamental observation, and the starting point for this analysis, is that the use of the 'market mechanism' for the buying and selling of goods and services is not free. Rather, markets involve transaction costs. In particular, the search for information regarding the possibilities for exchange and the negotiation and arrangement of the details of exchange (for example, with respect to price, quantity, quality, insurance, transportation, timing and the preparation of documents, including legal contracts) involves expenditures of time and money. In addition, there are uncertainties involved in buying and selling goods and services in the market. Judgements have to be made regarding the extent of the search for information; negotiation may be difficult and details hard to confirm; buyers and suppliers may be unreliable and fail to deliver on time and/or in the right quantity and/or in the right quality, or they may even go bankrupt or be acquired by a competitor. Legal contracts can have loopholes and in any event there are costs and uncertainties in enforcement. Moreover, market transactions may give rise to opportunistic behaviour.

The transaction cost model therefore recognizes the costs and uncertainties of using the market mechanism. In the terms of this model, integrated firms replace market-based transactions with hierarchically based or internally administered transactions. In the neoclassical view, hierarchical markets exist whenever they provide a more efficient alternative to market-based transaction costs and uncertainties. Potentially, administered trade between affiliated (i.e. controlled) plants can reduce the costs and uncertainties of information search and negotiation; indeed, information flows and many types of decision-making processes can be facilitated by routines and informal access. Such gains may be realized in situations in which buying or selling involves large volumes of raw materials, commodities or components, especially if firms are fearful of disruptions to such flows if left to the market. Firms may also prefer hierarchies in which to administer those goods and services which are highly specific to their operations and/

or which represent an asset that is critical to their specific competitive advantage. For research and development (R&D) services related to the firm's particular operations and market roles, for example, the market is likely to be a high-cost, uncertain supplier and to generate problems of proprietary rights and secrecy. Indeed, the firm may not be able to find exactly the kind of R&D it desires, at the time it desires it, in the market. According to the transaction cost model, in such a situation, firms invest in internal supplies of R&D because this significantly reduces the costs and uncertainties of relying on markets. Moreover, internal R&D allows the accumulation of expertise which can be used again in branch plants or sold in the form of licenses, franchises or exports.

In the transaction cost model, limits to hierarchies occur when markets provide goods and services more cheaply. There are costs in establishing and operating hierarchies, and with increasing size hierarchies may become excessively bureaucratic. In situations characterized by small-scale, highly differentiated and rapidly changing demands, for example, firms may prefer to participate in markets rather than form hierarchies. That is, the central issue is the relative efficiency of markets versus hierarchies in producing and delivering goods and services. The basic principle of this model is that firms produce goods and services themselves if the additional costs and uncertainties of administration (governance) are more than offset by reductions in the costs and uncertainties of using the market mechanism, and vice versa. In essence, in this view, corporate control is incidental to achieving efficiency.

Given that transaction costs are important, some criticisms of the transaction cost model might be noted. In general, the relationship between control and efficiency is more complicated than anticipated by the model (Ahern 1993a; Patchell 1993a). Thus, the model suggests that firms minimize the costs and uncertainties of using the market. Yet a significant implication of uncertainty is that firms cannot know optimal solutions *a priori,* so that among a population of firms facing similar situations, judgements are likely to vary. While one company may believe that transaction costs can be minimized by control of branch plants, distribution systems and transportation carriers, another company may prefer market solutions – and both may turn out to be viable. Similarly, the organization of subcontracting linkages by core firms has varying rationales with differing implications for control and efficiency (Patchell 1993a,b). That is, there is discretionary scope as to whether markets or hierarchies exist. In reality, production systems, even within the same industry, vary in their organization, including by various subcontracting configurations.

Subcontracting: linking the internal and social divisions of labour

In practice, 'subcontracting' (or contracting out) is frequently used to classify decisions by firms to buy rather than to make goods and services. However, many of these purchases are not actually underpinned by 'contracts' in a formal sense (Holmes 1986). Indeed, formal contracts are noticeable only by their absence in some 'classic' industrial districts organized around

SMEs, such as the cutlery district of Sheffield (Townsend 1954). While formal contracts are more likely when large firms are involved within the production system, subcontracting relationships among firms are widely underpinned by tradition, trust, custom and power, as well as by law. Asanuma (1986), prefers to use 'inter-firm relations' rather than subcontracting, and 'suppliers' rather than subcontractors. However, since the term subcontracting is so widely used, conventional practice is followed here.

Subcontracting is organized in various ways and with varying geographical characteristics. Two basic forms of organization, for example, are hierarchical and lateral forms of subcontracting (Figure 13.1). In the hierarchical system, contractors maintain very few direct contacts with suppliers but allow their first-tier suppliers to organize second-tier suppliers. In a lateral system, to gain access to the same number of subcontractors the contracting firm would require far more direct contacts. Subcontracting relationships may or may not be organized on an exclusive basis and they may or may not be stable over the long term. Geographically, subcontracting may form dense highly localized interactions or involve extremely long distance

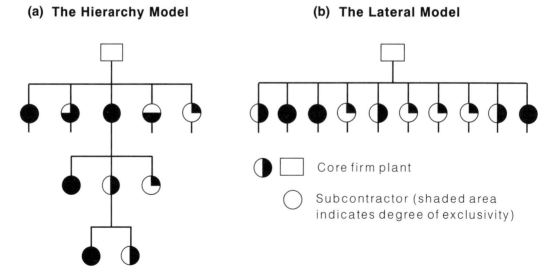

(a) The Hierarchy Model   **(b) The Lateral Model**

Core firm plant

Subcontractor (shaded area indicates degree of exclusivity)

Figure 13.1   Two Models of Subcontracting Organization Around a Core Firm

movements to remote places around the globe. Fredriksson and Lindmark (1979), however, note with respect to Volvo's production system, that more sophisticated, higher value suppliers tended to be closer to Volvo than suppliers of standardized goods which require little information exchange.

Subcontracting: types and rationale Firms use subcontractors for various reasons and two main types are recognized: the capacity subcontractor and the speciality subcontractor. The former duplicates an activity also provided by the contracting firm while the contracting firm, relies on the speciality subcontractor to provide a specialized product or service it does not provide itself (Holmes 1986). Firms may use capacity subcontractors to meet demand above some stable, minimum level of production of a good (or service) which is performed internally and can be sustained even during recessions. That is, capacity subcontracting allows 'production smoothing' by the contractor which subcontracts 'surplus' business during boom times and reduces subcontracting during recessions (while maintaining its own production). In this way, contracting firms pass on market risks to subcontractors. For large unionized firms, capacity subcontracting may be a way of accessing the lower-wage, non-unionized labour of SMEs. Speciality subcontractors, who may or may not employ lower-wage non-union labour, can provide considerable production, technological and marketing expertise and exploit economies of scale unavailable to firms relying solely on internal markets.

In the context of core-firm-dominated production systems, different views exist on the role of subcontractors. On the one hand, subcontracting relationships are interpreted as attempts by core firms to externalize costs and risks to SMEs and to indirectly exploit a low-wage labour force. Taylor and Thrift's (1983) study of business segmentation, for example, which is (implicitly) drawn from North American and British experience, gives particular emphasis to this view (also Holmes 1986). Such a view emphasizes the role of SMEs as buffers against business cycles and as low-wage reservoirs. An alternative view interprets subcontracting networks as attempts to gain access to the expertise and innovation potentials, as well as the efficiencies, of other firms (Patchell 1993a). In the case of core-firm-dominated networks, production systems may be seen as attempts to gain the benefits of both the internal and social divisions of labour while minimizing the disadvantages. Fredriksson and Lindmark's (1979) study of Volvo's production system in Sweden recognizes this interpretation of subcontracting, which is also at the heart of Sheard's (1983) and Asanuma's (1989) models of production systems which are derived from Japanese experience (see also Patchell 1993a,b). In this view, SMEs play a more dynamic role in wealth creation.

Fredriksson and Lindmark (1979) note that the internal and social divisions of labour offer different advantages and disadvantages as a form

Table 13.1 Internal and Social Divisions of Labour: Costs and Benefits

	Benefits	Costs
Internal division of labour	Co-ordination and control (reduce transaction costs)	Diseconomies of scale
	Strong external effectiveness (economies of scale and size)	Bureaucratic (low internal effectiveness)
Social division of labour	Competition (to reduce costs and to innovate)	Costs and uncertainties of market mechanism
	Strong internal effectiveness (flexibility)	Low external effectiveness (limited market power)

of governance (Table 13.1). Thus, the primary advantage of the internal division of labour is that it allows for control and co-ordination of production, while big firms also can exploit economies of scale, afford large-scale R&D and exercise power in the market-place. In Fredriksson and Lindmark's (1979) terms, large firms enjoy high levels of external efficiency. On the other hand, big firms are bureaucratic, which slows decision-making processes, weakens links between initiative and reward, and can potentially create different interest groups, and costly in-fighting among members. The primary strength of the social division of labour is its entrepreneurialism – the focused search for competitive advantage, the close relationship between initiative and reward, and the short-term flexibility of the small firm. In Fredriksson and Lindmark's (1979) terms, the benefits of the social division of labour rest on the high levels of internal effectiveness of SMEs in using available resources. On the other hand, SMEs do not exploit significant economies of scale, their opportunistic behaviour may threaten the stability of transactions, and their power in the market-place is limited.

Subcontracting potentially allows the advantages of the internal and social division of labour to be integrated while neutralizing their respective disadvantages. In this view, an 'ideal' type of production system is both stable (i.e. co-ordinated and secure for all participants) and competitive (i.e. committed to efficiency and quality improvement). Thus, stability, or co-operation, is important because it encourages long-term thinking and investments in risky projects including R&D and innovation, while competition is important because it ensures dynamism and helps regulate behaviour. At the same time, it is important to recognize that the process of defining and achieving an appropriate balance between the forces of competition and co-operation is problematical (Patchell 1996). However, Sheard (1983) believes that a key dimension of the astounding success of Japanese industrialization relates to the development of production systems which

have achieved, more effectively than elsewhere, this balance.

International variations in subcontracting The international variation in the size distribution of firms (Table 8.1) is paralleled by, and indeed related to, international variation in subcontracting. In particular, the greater role of small firms in Japanese manufacturing is reflected in the greater importance of subcontracting in Japanese production systems. Such variation again reinforces the notion that to some extent production systems reflect choices, and not simply mechanically defined technological imperatives.

In Japan, the percentage of small and medium-sized firms that are subcontractors varies by industry (Table 13.2). Between 1976 and 1986 this percentage tended to decline in more industries than not. Even so, in 1986, in 11 of 22 industries, at least 60% of SMEs were subcontractors. In 1986, the average for manufacturing as a whole was 56%. The only industry where subcontractors were a small share of SMEs was food. There is also a general tendency in Japan for exclusive subcontracting relations to decline as in most industries there has been an increase in the number of core firms supplied by individual subcontractors (Yaginuma 1993: 11). Yaginuma (1993: 12) reports that in 1990, just 16% of SME subcontractors still had at least 90% of sales to one core company, and a further 37% had at least 90% of their sales to between two and five core companies. Since at least 1976 the general trend across most industries has been for subcontracting firms to supply more core companies. For the manufacturing sector as a whole, and for individual manufacturing industries such as autos and electronics, Japanese core firms rely on subcontractors to a greater extent than US firms (Asanuma 1989; Fruin 1992; Shimokawa 1993; Yaginuma 1993). In the early 1980s, for example, a Japanese auto assembler in Japan relied on subcontractors for 75% of its production, in comparison to 52% for an American auto manufacturer (Yaginuma 1993: 24). As

Table 13.2   Japan: The Proportion of SMEs that are Subcontractors by Industry

	1966	1976	1986	1966–1986
		%		
Food	17	15	8	−9
Textiles	80	85	80	0
Clothing	74	84	79	5
Wood and wood products	35	43	22	−13
Furniture and furnishings	46	41	39	−7
Pulp and paper	51	45	41	−10
Publishing and printing	46	51	42	−4
Chemical manufacturing	40	37	22	−18
Petroleum and coal	30	27	18	−12
Plastics			69	
Rubber products	62	61	65	3
Leather	60	63	65	5
Ceramic, stone and clay	34	29	35	1
Steel	66	70	52	−14
Nonferrous metals	67	69	62	−5
Metal products	66	75	71	5
General machinery	71	83	75	4
Electric machinery	81	82	79	−2
Transport machinery	67	86	80	13
Precision machinery	72	72	70	−2
Other machinery		56	43	
Total manufacturing	53	61	56	3

Source: Yaginuma (1993: 9).

will be explored in the next chapter, the nature of these subcontracting relationships also differs between Japan and the US.

Geographic multipliers

From the perspective of local development, a vital characteristic of business transactions, whether organized by market or hierarchy principles, is their geography. The relationship between business transactions, particularly as expressed by *linkages* of goods (and services), and local development is captured by the concept of the 'geographic multiplier' (Hewings 1977). Simply stated, a geographic or local multiplier defines how growth in an activity (e.g. a particular factory or an entire industry) within a region stimulates growth in other activities within the region. Conventionally, the local multiplier is calculated quantitatively, usually on the basis of sales linkages (linkage multiplier), employment (employment multiplier) or income (income multiplier), although other variables are possible such as some measure of pollution. An important distinction in multiplier analysis is between the 'short term' and the 'long term'; a distinction that is based on whether or not structural change in the local economy is possible. In the short term, it is assumed that investment does not occur and a region's industrial structure, defined as its mix of industries and the linkages between industries, remains the same. In the long term, investment (and divestment) in public and private sector facilities, and thereby structural change, is possible. Long-term structural change is inherently complex and expresses how regions and societies 'solve' their problem of industrial transformation (see Chapter 2). The multiplier process in the short term nevertheless reveals the fundamental importance of business transactions and local linkage development.

Quantity (employment and income) multipliers

Local or regional multipliers essentially measure the impacts of some initial growth or change in an activity in a region on total or aggregate level of economic activity within a region. Simply put, a (short-term) multiplier (k) measures the ratio of initial growth to total impacts, i.e.:

$$k = \frac{\text{total economic impacts within a region}}{\text{initial economic impact}}$$

Because there are different ways to measure impacts (e.g. sales linkages, employment and

income) and because there are different definitions of 'total impacts', there are in fact a whole family of multipliers (Hewings 1977). The basic idea of the regional multiplier *process,* however, is that growth in one activity generates growth in other activities via industrial linkages. This idea can be simply illustrated (Figure 13.2). In this particular case, Factory EF has won an export order for $10 million. This export sale means that the region's economic activity has been increased by this amount and as such it constitutes the 'initial economic impact' or multiplicand. The multiplier process concerns how Factory EF's new exports 'spillover' and generate additional growth in other industries within the region. In this case, Factory EF, to produce additional exports, needs to purchase inputs (component parts, raw materials, services) from other firms in the region, including firms in industries A, B and C. These purchases are the equivalent of additional sales for firms in industries A, B and C and

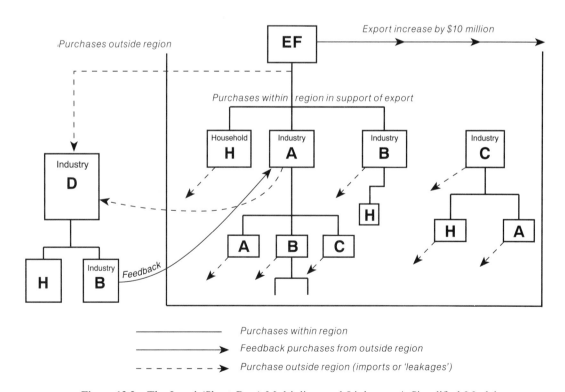

Figure 13.2   The Local (Short Run) Multipliers and Linkages: A Simplified Model

constitute 'direct impacts' that arise from the initial growth in Factory EF's exports. In addition, in order to increase their sales, these firms need to buy inputs from other firms so that another (second) round of sales increases ('indirect impacts') are generated. Indeed, additional rounds of sales increases (more indirect impacts) may be expected so that the initial increase in sales is now 'multiplying' as a result of these indirect impacts. It should also be noted that firms pay wages and salaries to employees (i.e. purchases from households in present terminology), and households may purchase from firms in the region, generating further multiplier effects. These household-generated effects are normally labelled 'induced impacts'.

Leakages and the size of multipliers The regional multiplier process is inevitably limited by 'leakages' of one kind or another. In this case (Figure 13.2), an example of a leakage is Factory EF's purchases from firms in industry D which is located in another region, so that these purchases represent imported goods and services. From the point of view of a local economy, imports are leakages which generate sales (and employment and income) elsewhere but not in the case study region. Given that firms in industries A, B and C also import with each successive round in the multiplier process, additions to sales within the region become less and less. Ultimately, if the initial increase in export sales of $10 million generates total additional sales of $20 million, then the sales multiplier, $k = \$20/\$10 = 2.0$. That is, the multiplier is two which means that, in the short term, for every $1 increase in exports (or 'final demands') for goods provided by Factory EF, $2 worth of sales are generated within the region.

This brief, simplified description of the multiplier process nevertheless reveals a fundamental characteristic of geographic multipliers: the size of the multiplier depends upon the extent of purchases of local goods and services. The greater the linkages among local industries, the bigger the geographic multiplier. Alternatively stated, the size of the multiplier is determined by

the extent of the leakages to local inter-industry linkages, and the more an industry imports, the smaller its multiplier effects. Admittedly, industries in other regions, such as industry D (Figure 13.2), may themselves require purchases, known as 'inter-regional feedback effects', from the case study region which will also generate multiplier effects. Nevertheless, the fundamental point is that leakages (i.e. removals of money from local circulation) dampen local multiplier effects. In this context, while imports are a significant form of leakage in practice, other types of leakages may be noted. Thus, savings, profits and dividends which are sent to other regions, and taxes paid to non-local governments, are potentially other important forms of leakage which, in the long term, may or may not return to the case study region in the form of investment or transfer payments.

Employment and income multipliers are conceived in the same way. In the case of employment, for example, if to meet additional exports of $10 million, Factory EF employs 100 people, then this employment number represents the initial economic impact. As the multiplier process works its way through the system (by direct, indirect and induced impacts), other firms also increase their employment as they expand their sales. If the total number of jobs generated is 250 then the employment multiplier is $250/100$ or 2.5. Similarly, an income multiplier can be calculated if information on wages and salaries, i.e. purchases of labour from households (Figure 13.2), is available. Thus, if the 100 people employed by Factory EF were paid $4 million and total wages and salaries in the region expanded by $8.8 million, the income multiplier would be $8.8 million/4.0 million or 2.2.

Input–output multipliers The most sophisticated method of calculating income, employment and sales multipliers is provided by input–output analysis (Miernyk 1965; Hewings 1977). A brief, non-mathematical introduction to how input–output multipliers are derived is provided in Appendix 2. In addition, the 'gross flows' input–output table, the basis for input–output analysis,

Table 13.3   Input–Output or Interindustry Gross Flows Table for Washington State, 1967 ($m)

Purchases From ↓	Sales to → Agriculture	Food products	Forest products	Metals machinery	Aerospace	Services	ALLWA industries	Final demands	Total sales
Agriculture	87	303	151	6	0	20	(585)	470	1055
Food products	34	72	0	0	0	28	(135)	1308	1442
Forest products	2	26	343	3	2	159	(555)	1276	1836
Metals, machinery	2	65	17	55	42	200	(414)	854	1268
Aerospace	0	0	0	0	59	5	(63)	2473	2536
Services	77	85	164	87	62	1074	(1671)	5833	8895
ALLWA industries	(238)	(577)	(708)	(161)	(173)	(1878)	(3952)		
Imports	130	400	298	578	1274	1503	(4787)	1699	8080
Value added	687	465	830	559	1089	5514	(9955)	2870	12733

Source: Morrill (1974: 97); see also Beyers et al (1970).
Notes: This table is a highly aggregated version of the original. The dotted line encloses the so-called intermediate sector although not all industries are identified. In this table, households are included in the value added and final demands sectors. In Appendix 2 imports and value added are included in the payment sectors.

for Washington State in 1967 is partially presented in Table 13.3. This table simply shows the 1967 dollar value of all business transactions for all activities in Washington State as a set of inputs (purchases) and outputs (sales), i.e. it is a form of double-entry bookkeeping which reveals the 'actual' nature of inter- and intra-industry linkages for 1967. For example, firms in the aerospace industry, principally the Boeing Company, had total sales in 1967 of US$2536 million, including US$59 million to other firms in the aerospace industry of Washington State, US$495 million to governments, and US$1979 million worth of exports to private sector firms. All of these outputs must be balanced with inputs and Table 13.3 shows that for aerospace firms these inputs included purchases of US$59 million worth of inputs from other firms in the aerospace industry of Washington State, and US$1274 million worth of imports and value-added payments, notably wages and salaries to households.

These gross flows data provide the basis for the calculation of different types of income multipliers (Table 13.4). An important observation is that different industries generate different multiplier effects within the same region. The aerospace industry in Washington State, for example, has a smaller income multiplier than the forest

product sector, so the same amount of export sales growth in the latter will create a bigger income impact than the former. Similarly, during recession, the downward or negative multiplier effects of the aerospace industry are less than that for the forest sector (Bourque 1969, 1971; Beyers 1972; Erickson 1974). The reason is that the forest sector has strong linkages with local suppliers, while in the aerospace industry the Boeing Company subcontracts many of its requirements out of the State.

An input–output table, such as the Washington State 1967 table, defines regional industry structure at a point in time in the form of inter-industry linkages. These linkages ultimately depend on the decisions of individual organizations (in the public as well as the private sector). Thus, at a point in time, *ceteris paribus*, business decisions which favour local purchases of goods and services increase local multiplier effects. Over time, investments in new activities facilitate the process of regional industrial diversification. Indeed, in regions that diversify over time, it is to be expected that income multipliers will increase and leakages will tend to decrease. In contrast, a feature of declining industrial regions is the fragmentation of industrial structures and declining multiplier effects.

Table 13.4 Income Multipliers in the Washington Economy: Selected Industries, 1963

	Income created per $1 of export sales		
	Direct income ($)	Direct and indirect income ($)	Direct, indirect and induced income ($)
Field crops	0.54	0.67	1.19
Vegetables	0.78	0.89	1.59
Livestock	0.34	0.52	0.92
Beverages	0.49	0.76	1.35
Logging	0.67	0.91	1.61
Sawmills	0.46	0.93	1.64
Pulp mills	0.38	0.82	1.46
Petroleum refining	0.24	0.42	0.54
Aluminium	0.38	0.46	0.86
Electric machinery	0.62	0.69	1.22
Aerospace	0.52	0.56	0.99
Shipbuilding	0.67	0.76	1.35
Construction	0.41	0.65	1.18
Insurance	0.85	0.97	1.72
Business services	0.76	0.89	1.58
Personal services	0.66	0.85	1.50

Source: derived from Bourque (1969: 6–8).
Note: In the terms of Appendix 2, the direct income effect is the technical coefficient for households; the direct and indirect income effect is the Type I income multiplier; and the direct, indirect and induced income effect is the Type II income multiplier.

The quality multiplier

Relationships among firms are, or at least potentially are, more than simply the purchase and sale of goods and services. Firms also develop and exchange know-how in the form of information, counselling, joint research and engineering initiatives, loans of employees including engineers, and co-operative training ventures, while firms can also observe the 'best practice' behaviour of other firms (LeHeron 1976). This exchange of know-how among firms within a region defines the quality multiplier (Thomas 1969). In this regard, the quality multiplier rests on an interpretation of production systems as learning systems and the development and transformation of skills (Patchell 1993a). An important way in which industrial skills are created and developed is through personal relationships in which learning takes place by the exchange of ideas, direct observation and by practice ('doing') where, for example, skilled workers teach unskilled workers and apprentices.

Such personal interactions and skill-formation processes are facilitated by the localization of related activities. Indeed, the most powerful geographical expressions of quantity and quality multiplier processes are found in industrial districts.

The nature of industrial districts

Scott (1992: 266) defines an industrial district as "a localized network of producers bound together in a social division of labour, in necessary association with a local labour market". That is, industrial districts feature (a) geographic concentrations of activities; (b) populations of small and medium-sized firms which are linked together in various ways; and (c) appropriately skilled and accessible labour pools. The economic rationale for industrial districts is provided by localized external economies of scale (Storper 1989; Scott 1992; Sunley 1992; see also Phelps 1992). Thus, firms in an industrial district gain benefits which are internal to the district but external to the firms themselves. In addition, economies of scope (i.e. the ability to perform different tasks with the same resources) are also an inherent feature of the 'flexibly specialized' nature of industrial districts.

According to Piore and Sabel (1984), industrial districts are 'flexibly specialized'. Within industrial districts, SMEs are individually highly specialized on a limited range of products and processes and collectively flexible because each SME has ready access to the (product, process and employment) specialisms of many other SMEs, some of whom compete and some of whom complement one other. In turn, this collective flexibility implies rapid response to highly differentiated consumer demands, highly differentiated input supplies, rapid absorption and diffusion of new technologies and market information, as well as effective use, training and redeployment of labour. Indeed, Morgan (1994) thinks of flexible specialization in terms of 'collective entrepreneurship' (see Cooke and Morgan forthcoming). In this view, high levels of market and supply uncertainty encourage a disintegration of production among SMEs, and a reliance on external economies of scale and scope as a means of coping with constantly changing demands. In fragmented and changing markets, for example, SMEs can "shift promptly from one process and/or product to another, and to adjust quantities of output up or down over the short run without any strong deleterious effects on levels of efficiency" (Storper and Scott 1989: 24).

Industrial districts as localized networks

A wide variety of products are manufactured in industrial districts that have existed in various forms since the Industrial Revolution (Scott and Storper 1990, 1992; Scott 1992: 267). Industrial districts also vary geographically in terms of institutional structure (Figure 13.3). A basic distinction is that between industrial districts solely comprising SMEs, and industrial districts in which SMEs are organized around core firms.

There are other possibilities related to these basic types. Inevitably, a size distribution of firms exists even in districts dominated by SMEs, while in core-firm-dominated industrial districts there may be several core firms of varying size and importance plus medium-sized firms which provide examples of secondary core firms. Piore and Sabel (1984) note further types of localized organization, including the 'solar firm' which comprises multiple branches of one firm or branches of affiliated firms. It should also be recognized that national and local government policy contexts vary in the degree of explicit support for industrial districts (Park and Markusen 1995).

It has long been recognized that core-firm and SME-dominated industrial districts, and related alternatives, create localized external economies of scale and economies of scope (Figure 13.3). As Sunley (1992) notes, Marshall's (1992) characterization of industrial districts of 100 years ago, emphasized three main forms of localized external economy, namely pools of skilled labour, the growth of subsidiary industries and the exchange of information. By agglomerating in particular places, firms in the same and related industries allow the development of pools of appropriately skilled and interested workers in which traditions and skills are transferred within families as well as through on-the-job training and specialized educational and training programmes, themselves made economic by the concentration of local demand. Second, a concentration of activity among a population of SMEs creates a range of economically viable demands for common, specialized suppliers of components, processes, machinery, wholesaling, business, marketing, research and development, financing and transportation services. Third, a geographic concentration of activity facilitates person-to-person contact through informal as well as formal channels and thereby the flow of information on all kinds of matters related to the industry. Marshall emphasized the importance of such information exchanges to the diffusion of ideas and innovations (Sunley 1992: 307; Glasmeier 1988).

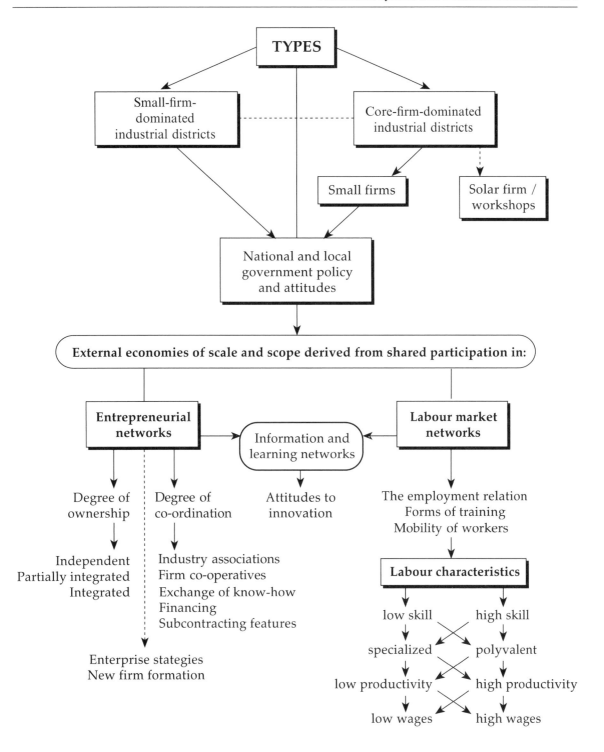

Figure 13.3 Industrial Districts as Localized Networks

The particular ways in which external economies of scale and scope are realized depends on the specific nature of the interrelated entrepreneurial networks and labour market networks that exist within industrial districts and how these networks are organized for learning and innovation (Figure 13.3). With respect to entrepreneurial networks, an important characteristic is the degree of ownership, i.e. the extent to which production is distributed among an independent social division of labour or, conversely, the extent to which production is concentrated within an internal division of labour. In a general sense, competition increases with the number of firms and independent decision-makers. Indeed, the flexibility of industrial districts as agglomerations of specialized activity rests to a substantial degree on the internal competition that exists among subcontractors for business with contracting firms and among firms competing for related final product markets inside and outside the region.

But the flexibility of industrial districts also rests on co-ordination among firms, i.e. the complementary behaviour required between contractors and subcontractors to manufacture products of the right quality, in the right quantity and on time, and among rivals who wish to develop external economies of scale (Storper and Scott 1989). Indeed, an important institutional question facing firms (and labour) in industrial districts is how to balance the forces of competition and the forces of co-operation. If competition drives efficiency, quality improvements and innovation, hyper- or cut-throat competition can lead to degraded labour conditions and reduce incentives to learn and innovate. Some degree of stability and co-operation in business transactions is important for firms and employees to invest in innovative and learning processes. In general terms, the degree of co-ordination within an industrial district is defined by industry associations, firm-level co-operatives, bilateral exchanges of know-how and financing arrangements, and subcontracting arrangements (Figure 13.3).

Co-operation and competition In practice, social divisions of labour are bound together by custom and tradition as well as by formal economic ties and contracts. Thus trust and co-operation are potentially important features which help bind industrial districts together, including among firms who are rivals for particular product markets. Indeed, according to Lorenz (1992: 195–196), the most prominent characteristics of industrial districts are:

> "a particular balance between cooperation and competition among the firms within them . . . Cooperation has two principal aspects. It takes the form of the provision of collective goods, notably training or education and research and development, but also medical care and unemployment insurance . . . through the auspices of some local institution. Cooperation also takes the form of adherence by producers to a set of norms of competition . . . their being embedded within 'communities', and the high level of trust among producers within them."

While the potential range of co-operative behaviour is considerable, some general points can be made. Thus, industry-wide co-operation is represented by the creation of associations by all or many of the companies in an industrial district to undertake responsibilities on behalf of all the members who provide funding. Industry associations provide goods and services, such as R&D, marketing, lobbying and representation, labour bargaining, forums for discussion, and information on industry matters, which are either beyond the abilities of individual firms or, if paid for by an individual firm, would be readily available to others. Typically, membership in industry associations is voluntary and an important task of associations is to ensure each member feels that participation is worthwhile. Co-operative consultation and co-ordination also occur in a wide variety of other ways involving two or a few firms in more limited ways; for example, involving two firms in the exchange of know-how and opinions, or in the provision of a variety of services by contractors to subcontractors, such as help with financing, space and planning (Leung 1993).

Moreover, as Lorenz (1992) indicates, firms within an industrial district typically share values as to the 'norms of competition'. While these norms are shaped by local culture, there are always possibilities for opportunistic behaviour in which firms seek to gain economic advantage over others by violating traditional, if not the legal bounds of behaviour, or the obligations of unwritten understandings. In all cultures, opportunism is at least informally constrained by the threats of loss of future business, loss of reputation and exclusion from exchanges of information and offers of help. That is, trust is motivated by economic incentives and beliefs that "anticipated benefits of future mutual co-operation are valued higher than the one-time rewards of defecting while others cooperate" (Lorenz 1992: 197). Opportunism, it might be noted, is not limited to western societies which emphasize individualism; it exists in Chinese and Japanese contexts where Confucianism is present (Asanuma 1989; Leung 1993). Indeed, as will be discussed in the next chapter, industrial districts in Japan are strengthened by powerful incentives rewarding reliability and trustworthy behaviour.

According to Takeuchi (1992), the resilience of industrial districts in Japan, including those organized by SMEs in the dynamic, high-cost environment of Tokyo, is greatly aided by high levels of inter-enterprise co-operation and trust, particularly with respect to the exchange of technological know-how and other forms of information exchange. For example, in Tokyo, knitting, tool-making and machinery manufacturers formed associations in the 1980s to collectively promote ways to improve the quality of their products, given that low-value items were relocating elsewhere in Japan or to foreign countries. Takeuchi (1992: 287–288) argues that the evident success of these associations lay in frequent meetings, in some cases on a nightly basis, a high level of trust among members, and "a strict adherence to the main goal", while also noting that "Absentees failing to provide notice to meetings are ousted. [and] Companies which are incapable of making suggestions or who fail to contribute to other companies must also

withdraw." In this way, trust relations among rivals in the same industry are sanctioned in the form of 'norms of co-operation'.

Co-operation involving joint learning, technology transfer and the development of new business requires face-to-face contacts and levels of trust that are facilitated by geographic proximity. Takeuchi (1992: 291) notes, that in Tokyo, "it is said that the necessary conditions for co-operation is that the parties involved should not be more than 15 minutes from each other for effective mutual exchange." In an interesting experiment, MITI and local governments in Japan began to offer financial support for co-operation regarding new business ventures and technology transfer among SMEs in different industries throughout Japan. Thus, between 1987 and 1989, the number of such groups increased from 700 to 1527, and the respective number of enrolled companies from 20 000 to 52 149 (Takeuchi 1992: 285). Typically, such groups meet monthly, much less than permitted by firms in close proximity to one another.

Employment perspectives   Principles of competition and co-operation similarly shape employment networks within industrial districts (Angel 1991). In general terms, the nature of labour market networks are based on the employment relations (including with respect to forms of training) negotiated between management and labour which over time reflect and develop particular cultures, traditions and 'norms of bargaining'. In this regard, unions provide a formal institution binding the interests of employees together across industrial districts, by bargaining and in other ways. Whether or not union representation exists, labour characteristics can evolve in many different ways, according to skill, worker mobility, job demarcation, productivity, wages and non-wage benefits, and there may or may not be industry-wide bargaining, including with respect to funding labour-training associations, pensions and apprenticeship schemes. On the one hand, there may be a commitment to linking productivity with improving wages and working conditions.

On the other hand, bargaining may be more inherently antagonistic and directed towards keeping wages and training costs low, and workers in competition with one another. Alternatively put, employment networks may emphasize principles of functional flexibility and the development of core, stable, well-paid and multi-skill workers or they may emphasize numerical flexibility and the entrenchment of a peripheral, low-paid, unskilled and easily replaced workforce.

The reality of local employment networks within industrial districts, however, is often complicated, combining elements of both functional and numerical flexibility. On the basis of detailed case studies, Rees (1993) found that in the newly emerging wood remanufacturing industrial district of the Vancouver metropolitan area of British Columbia, employment bargains are highly varied (Table 13.5). Thus, there are examples of firms employing highly paid, stable and multi-skilled workforces producing high-value goods, examples of poorly paid, high-turnover workforces producing low-value goods, and examples of firms employing workforces with mixed characteristics. The range in employment characteristics also seems to be a feature of the film industry in California (Storper and Christopherson 1987).

There is an ongoing debate about the implications of flexibly specialized industrial districts for employment. Based largely on the experience of Little Italy (Brusco 1986; Goodman et al 1989; Sabel 1989; Pyke et al 1990), Piore and Sabel (1984) stress the potential of industrial districts for mutually reinforcing and progressive relationships between entrepreneurial and employment networks. Others have raised issues of exploited and even degraded labour conditions, and stress that flexible specialization simply provides a new regime for capital accumulation (Harvey 1990). In this view, women are seen as especially vulnerable. Thus, Jensen (1989) argues that the conception of skill emphasized in flexible specialization is biased towards men in part because machines are designed by men for men, while Lever-Tracy (1988) and Walby (1990) emphasize the role played by women in part-time work.

Fujita's (1991) study of women workers in flexibly specialized industries in Tokyo since 1985 modifies these views. She suggests that it is inappropriate to conflate the general shift towards more female part-time work that has occurred since 1950 in the US and UK with flexible specialization, a trend only just beginning in these countries: Japan, where flexible specialization is most advanced, has the lowest rate of women part-time workers (Fujita 1991: 263). In addition, Fujita recognizes that the concern for ergonomics in Germany, Sweden and Italy, as well as Japan, has led to close design attention to the social aspects of machinery ('humanware relations'), i.e. to meet human needs of all workers, including female workers. Empirically, Fujita's (1991: 275) analysis of Tokyo's information industrial district reveals a rapid growth of female professionals, many of whom prefer high-wage part-time work over full-time work in order to maintain a wider range of family and community roles. Women have also been increasingly active in developing the skills and networks necessary to create their own businesses. Given that in Tokyo, as elsewhere, "child bearing and child care persistently hamper and interrupt women's continuous employment", Fujita (1991: 275–276) nevertheless argues that an increasing number of professional women working in a high-tech environment have achieved some autonomy over career paths while resisting patriarchal and traditional work roles. Women factory workers in Japan, on the other hand, remain in low-wage, low-skill jobs (Sosa 1990).

Workers, male and female, can also achieve more autonomy by forming workers' co-operatives. Workers' co-operatives have developed especially rapidly in Italy and Spain, where by the early 1990s they numbered around 10 000 and 20 000 respectively (Bartlett 1993: 57). Workers' co-operatives are typically small, specialized organizations, and in some cases they comprise important elements of local industrial districts. In the Basque region of Spain, for example, the Mondragon system of industrial co-operatives began in 1956, and by 1990 this group comprised

Table 13.5 Employment Characteristics of Remaining Firms in the Vancouver Metropolitan Area

Firm[a]	Full-time (temporary) employment			Union	Entry wages ($ hr)	Non-wage benefits	Flexibility characteristics
	81	86	91				
A	13 (3)	28 (10)	40 (10)	No	12.75	Full medical, dental & pension	Job rotation among skilled tasks; training; preference for qualified graders
B	6	10	15	No	8.00	Full medical, 80% dental, no pension	Job rotation but work unskilled; no qualifications; unstable workforce
C	16 (20)	10 (25)	11 (22)	No	14.50	Full medical, 50% dental, no pension	Strong job demarcation; temporary workers, relatively old; limited training
D	74	72	54	No	17.81	Full medical, 80% dental, 80% pension	Job rotation among skilled tasks; extensive training and qualifications
E	–	–	24	Yes	17.00	Full union package	Union sub-agreement permits shift scheduling and job rotation, extensive training
F	–	–	–	–	–	–	All work is subcontracted

Sources: Rees (1993) and Rees and Hayter (forthcoming).

Notes: Virtually all employees, apart from clerical, are male and white. Firm C's full-timers (but not temps) are mainly Indo-Canadians.

[a]A: medium-value independent producer; B: low-value contractor; C: low-value capacity subcontractor; D: high-value speciality contractor; E: high-value branch plant; F: low-value branch 'middleman'.

97 co-operatives employing 22 000 workers (Bartlett 1993: 63). For workers' co-operatives, it is efficiency, rather than worker exploitation, that is debated. A related question might well be how effective co-operatives are with respect to innovation.

The importance of innovation In Piore and Sabel's (1984) terms, the defining characteristics of industrial districts are the flexibilities provided by specialist firms in the form of speed, cost and quality of responses to highly variegated consumer demands and uncertain supply situations. In turn, these flexibilities are rooted in external economies of scale and economies of scope. Yet, flexible specialization is by itself no guarantee of long-term survival (Hayter and Patchell 1993: 1430). While Piore and Sabel (1984: 17) associate flexible specialization with a commitment towards innovation, they hint that innovation is not an inevitable result of specialization. Yet the social division of labour underlying industrial districts is dynamic and, in most cases, firms in industrial districts need to innovate to survive (Ide and Takeuchi 1980). In general terms, the supply of innovations, which may be organizational or technological and co-operatively introduced or originated by individual firms, is shaped by the skill of workers, entrepreneurial ingenuity and by more formal research, development and design activities (Figure 13.3).

Industrial districts vary in terms of the organization of worker training and innovation. Several studies have reported on the varying roles of local political parties, churches, local governments, trade associations, trade unions and chambers of commerce in supporting innovation in northern Italy (Brusco and Righi 1989; Zeitlin 1989), Baden-Württemberg in Germany (Herrigel 1989; Sabel et al 1989) and various small-scale industries in Japan (Ide and Takeuchi 1980; Friedman 1988). In some instances, markedly different attitudes towards worker training and innovation are evident. A case in point is provided by the cutlery industry districts of Solingen, Germany, and Sheffield, UK (Lloyd

1913; Hayter and Patchell 1993: 1442). Both districts developed as an export-oriented social division of labour; by 1900, the 12 000 workers in Solingen were almost three-quarters of the number in Sheffield, the leading centre of the 19th century. Yet Lloyd (1913: 374) reported that, by 1900, cutlers in Solingen were technologically more advanced than their deeply conservative Sheffield counterparts and that the workforce was better organized, had more influence and was more supportive of technical change than in Sheffield where cutlers were even slow to use stainless steel which was invented in Sheffield in 1913. In addition, the apprenticeship system in Sheffield relied entirely on on-the-job training by existing craftsmen who were not paid for this responsibility (and apprentices were paid very little), in contrast to the much greater level of support given to training in Solingen. Moreover, these differences have continued to the present day and a much fragmented Sheffield industry survives by emphasizing low costs (especially wages) and virtually no investment in new equipment, R&D or training. While Solingen's cutlery firms have also declined in number, the survivors are highly automated, high-quality manufacturers who pay high wages and maintain a strong apprenticeship and training system. A similar point can be made regarding the Tsubame cutlery district in Japan: its vitality relates directly to its strong commitment to innovation (Hayter and Patchell 1993).

Types of industrial districts

There is a large array of products and services that are produced in industrial districts. To select a few well-known examples, there are industrial districts that produce autos, such as Toyota Town in Japan; numerically controlled machine tools in Nagano, Japan; machine tools in Baden-Württemberg, Germany; shoes and knitwear in northern Italy; cutlery in Solingen, Sheffield and Tsubame; semiconductors in Silicon Valley; and aerospace manufacture in Los Angeles. In metropolitan areas such as London, New York and Tokyo, multiple industrial districts exist. A recent paper by Scott (1996: 312), for example, discusses five different types of production and service industrial districts (advertising, clothing, entertainment, furniture and jewellery) in the craft, fashion and cultural products industries of Los Angeles. In Japan, industrial districts are widely prevalent and have consistently provided the dominant framework for industrialization in the 20th century to a greater degree than in the West, including with respect to SME-dominated industrial districts.

Within the western literature, small-firm-dominated industrial districts have long been regarded as relic forms of industrialization. Over 80 years ago, for example, Lloyd (1913: ix) studied the Sheffield cutlery industry as a SME-dominated industrial district because it comprised "rapidly disappearing features which appear to be survivals from an earlier industrialization". In practice, even in the West, such districts proved more resilient than anticipated, and in support of Piore and Sabel's (1984) advocacy of flexible specialization there has been much celebration of the Emilia Romagne region of northern Italy, with its populations of SMEs manufacturing a wide range of products in the ceramics, clothing and footwear industries within flexibly specialized industrial districts (Brusco 1986, 1990; Goodman et al 1989). It is in Japan, however, where SME-dominated industrial districts are particularly prevalent (Ide and Takeuchi 1980; Yamazaki 1980).

In Japan, the term *jiba sangyo* or community-based industries (cbis) was introduced in the 1960s to refer to localized concentrations of small, owner-managed firms which employ a refined division of labour to manufacture specific types of products for sale, in many cases for export (Ide and Takeuchi 1980: 299). Ide and Takeuchi counted over 330 cbis located in both major metropolitan areas and in small, isolated rural communities which collectively accounted for about 10% of all small businesses in Japan (Yamazaki 1980). In Tokyo, in the early 1970s, cbis accounted for about 60% of its factories and

50% of manufacturing value added, the most important of which were in printing, metal pressing, book-binding, metal moulding, household furniture, paper goods, apparel for women and children, handbags, flat knitting, accessories, leather footwear, men's clothing, toys, circular knitting, precious metals, briefcases, cigarette lighters, rubber products and medical equipment. Other types of cbi (e.g. lacquer ware and metalworking) are found throughout Japan (Figure 13.4).

Traditional and modern industrial districts

Among OECD countries, there is an incredible variety of products and processes that characterize industrial districts. The extent of this variety is illustrated by reference to the Wajima *Shikki* industry in Japan (Figure 13.5) and the Los Angeles aerospace complex (Figure 13.6). Wajima *Shikki* is a handicraft industry producing wood products which are varnished and decorated, especially by lacquer and inlaid gold. Wajima *Shikki* involves about 60 processes which are dependent on artisan skill and experience, and performed within 11 categories of firms, including organizers (*nushiya*), subcontractors, decorators, joiners, hollowing-makers, turners, board-rounders, chopstick-makers and box-makers (Suyama 1996: 66–67). By 1989, there were 869 firms employing 2 869 workers in the Wajima *Shikki* industry; almost 90% of the firms employed fewer than five workers, and only three of the *nuishiya* had more than 51 employees (Figure 13.5(a)). These firms are involved in complex inter-firm relations. *Nushiya* B, for example, a small organizer employing 15 people which concentrates on varnishing and sales, relies on 26 outside firms within the region for a range of highly specialized processes, including differing types of surface decorating (polishers, lacquerers and gold in-fillers), wood-body-making (joiners, board-rounders, hollowware, specialist turners) and for other types of varnishing, especially priming and middle varnishing (Figure 13.5(b)). The firms of Wajima *Shikki* have also formed associations to purchase

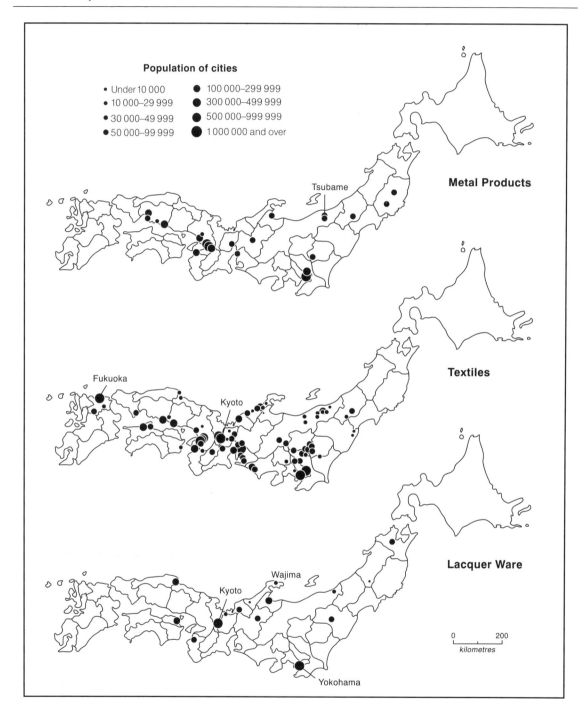

Figure 13.4   The Distribution of Selected Localized Industries in Japan. Source: Ide and Takeuchi (1980: 331)

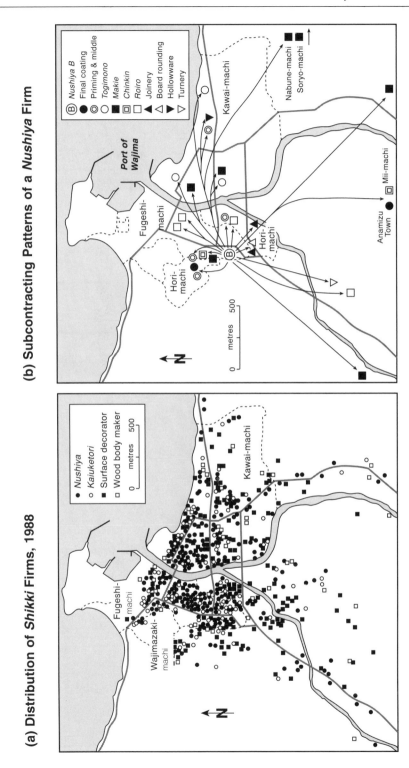

Figure 13.5 The Wajima *Shikki* Industry. (a) Distribution of *Shikki* Firms, 1988; (b) Subcontracting Patterns of a *Nushiya* Firm. Source: Suyama (1996: 69, 78)

specialized inputs not available locally, such as brushes and gold foil and powder, and to mine *jinoka*, a material vital to the varnishing processes. Other materials, such as wood, are bought independently or by groups of firms, and sales are primarily handled by the *nushiya* who sell to wholesalers and retailers.

As Suyama (1996) emphasizes, the technology and methods underlying the production of Wajima *Shikki* are essentially those established in the 18th century when it appears that the industry began in the region. The size and organization of the industry has changed considerably over the last 30 years, however. Thus between 1967 and 1989 the number of firms and

employees doubled, as did production. As the *shikki* industry in many areas of Japan collapsed, Wajima concentrated on high-value goods for which demand remained strong. In turn, the industry's expansion required organizational restructuring, notably by an increase in the number of independent subcontractors, the diversification of the sales links organized by the *nushiya*, the development of direct contacts between dealers and department stores, and by the organization of co-operatives to purchase materials. As a result, as Suyama (1996) documents, the Wajima *Shikki* industry has expanded and complicated its social division of labour while maintaining traditional methods of production.

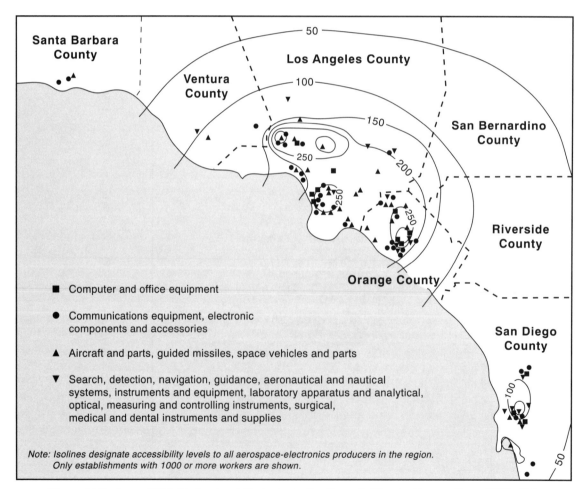

Figure 13.6   Large 'High-Technology' Factories in Southern California. Source: Scott (1992: 336)

An example of a very different, technologically sophisticated and dynamic industrial district is provided by the aerospace complex in Los Angeles (Figure 13.6). Scott (1992) suggests that in several high-tech industries, including aerospace, industrial districts are organized by large factories which he labels 'systems houses'. According to Scott, systems houses are complex organizations which operate R&D-intensive production processes which exploit various internal economies of scope in manufacturing a range of high-value products in small batches, often over a lengthy production period. While systems houses are internally extremely versatile (in terms of technology and worker skills, for example), they create highly articulated and geographically concentrated networks with other specialized R&D-intensive producers and suppliers. In particular, the systems houses of various high-tech industries generate huge, localized demands for highly specialized, customized inputs (Scott 1990).

While Scott recognizes that many of these inputs are supplied from distant sources, he offers evidence, specifically with respect to the subcontracting patterns of NASA prime contractors, of a "remarkable intra-regional network of economic transactions focused on large producers" (Scott 1992: 273). Thus, in 1989, five major contractors (California Institute of Technology, General Dynamics, McDonnell Douglas, Rockwell International and TRW) held 20 NASA prime contracts at eight factories (systems houses) in southern California. They awarded over half of the 4787 first-tier subcontracts of at least US$10 000 to other firms also located in southern California, while those first-tier subcontractors that engaged second-tier subcontractors found almost one-third of their suppliers in southern California (see also Scott and Mattingly 1989). With respect to subcontracting structure, while at least two tiers of subcontractors are recognized, the first tier has more members and therefore conforms more to the lateral model of subcontracting, rather than the hierarchical model (Figure 13.1).

Towards a classification of industrial districts

It is not easy to classify industrial districts by products. If cutlery evolved largely in the industrial districts of Sheffield, Solingen and Tsubame, in the US and Korea cutlery is manufactured within internal divisions of labour. Similarly, autos, electronics and shoes are produced in industrial districts and in geographically dispersed production systems around the globe. In addition, industrial districts themselves not only vary by product but in terms of how entrepreneurial and labour market networks are organized and integrated. Potentially, industrial districts can be classified on the basis of the manifold characteristics of these networks and how these characteristics evolve over time. A more partial and simpler alternative, which nevertheless illustrates the point that industrial districts vary in terms of their organization, is to classify industrial districts, and other forms of production system, according to a two-dimensional framework comprising the degree of ownership and degree of co-ordination (Figure 13.7).

Thus, on the vertical axis of Figure 13.7, increasing degrees of ownership imply increasingly stronger roles for core firms (and for the internal division of labour) as fewer firms are in control of total production. Along the horizontal axis, increasing degrees of co-ordination imply stronger forms of co-operation among firms. While it may be possible to measure these dimensions systematically, no such attempt is made here. Rather, the framework is presented as a heuristic device designed to illustrate the institutional basis of industrial districts (Langlois and Robertson 1995). Needless to say, the allocation of industrial districts and other forms of production system within the two-dimensional space of ownership and co-ordination is judgemental, but not arbitrary.

Thus, SME-dominated industrial districts share low degrees of ownership integration but vary according to co-ordination. Sheffield cutlery, at least until the 1960s, for example, represents a classic 'Marshallian' industry district which comprises large numbers of SMEs engaged

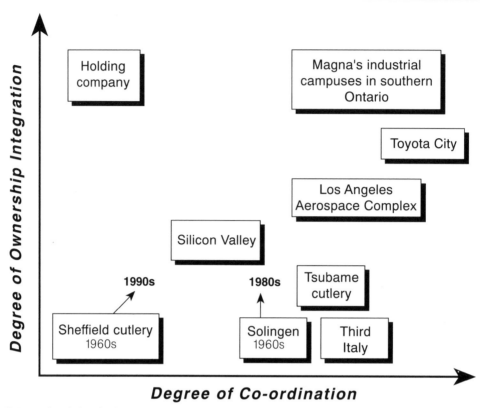

Figure 13.7 Industrial Districts Classified According to Degrees of Ownership and Co-ordination in Entrepreneurial Networks. Source: adapted from Langlois and Robertson (1995: 124)

in highly competitive relations and weak forms of co-ordination. Indeed, in Sheffield, even rival industry associations were established to lobby governments to support completely different trade policies (Hayter and Patchell 1993). Collective support for R&D and labour training is weak to non-existent and innovation is *ad hoc*, dependent on the initiatives of individual firms. In contrast, the various consumer goods' industrial districts found in Third Italy, and the Tsubame and Solingen cutlery districts comprise populations of SMEs which both compete and engage in coherent forms of co-ordination. Relationships between contractors and subcontractors are relatively stable, bilateral co-operation is common, firm-level co-operatives exist and effective industry-wide associations promote export marketing, R&D and information diffu-

sion. Firms in these districts typically strongly support skill enhancement and innovation.

With respect to industrial districts, such as Silicon Valley, the Los Angeles aerospace complex and Toyota City, which contain large core firms as well as populations of SMEs, the degree of ownership integration is higher than in SME-dominated industrial districts. However, degrees of co-ordination in these districts can be as high and, arguably in the case of districts like Toyota City, even higher than that achieved in Third Italy. Thus, in Toyota City, there is one dominant core firm which literally orchestrates the operation of the industrial district by closely co-ordinating and controlling its suppliers within its own co-operatives, kanban arrangements and long-term, stable relationships. This industrial district is analysed in the next chapter. Silicon

Valley and the Los Angeles aerospace complex define less well-controlled and stable districts but co-ordination and co-operation is evident, even if Silicon Valley co-operation is a matter of controversy. Another type of industrial organization which combines a high level of co-ordination with total control is illustrated by Magna, the Canadian-based auto parts giant, which has created several localized 'campuses', each of which comprises several, relatively small and specialized branch plants. Since a social division of labour does not exist, however, these campuses do not have all the attributes conventionally associated with industrial districts. Similarly, the companies that exist within holding companies may be localized but there is a common parent so that such groupings do not reflect industrial districts, while the geographically dispersed branch plant enclave is the antithesis of an industrial district.

The social division of labour as a source of dynamism

The social division of labour, including in SME-dominated industrial districts, constantly evolves, and it is the flexibility of specialized firms that provides the potential to adapt to changing circumstances. In so doing the nature of the social division of labour itself can change. In this regard, the literature has rightly emphasized the ability of firms within a social division of labour to cope with rapidly changing market and supply conditions. The social division of labour can also be a source of long-term structural adjustment. A 'small-scale' example is provided by Tsubame and its creation and development of a cutlery industrial district.

Cutlery manufacture in Tsubame   Traditionally, the Japanese people have not used knives and forks, and the chances are that in remote Tsubame, situated in Niigata Prefecture on the Japan Sea, local farmers and metalworkers engaged in (Japanese) pen, pipe and nail manufacture at the beginning of the 20th century had not seen western cutlery (Figure 13.4). Existing (hand-made) metalworking activities, however, were undermined by imported machine-made goods and the search for diversification focused on western cutlery, probably after a Tokyo-based wholesaler had provided samples (Patchell and Hayter 1992). In any event, the experiments in Tsubame with cutlery manufacture began in the first decade of the 20th century, and by the 1920s the region was an important cutlery exporter, albeit of low-value products, primarily to Asian markets. Following World War II, cutlery in Tsubame grew rapidly and the district became the leading world cutlery exporter by the 1960s. Even after the rapid growth of lower-cost cutlery manufacture in South Korea and the high value of the Yen, cutlery in Tsubame remains vibrant and export-oriented.

In Tsubame, the creation, growth and adaption of cutlery manufacture is intimately involved with an evolving social division of labour. Initially, the existing social division of labour used existing forging, sawing, hammering, filing and engraving expertise to learn by a slow (part-time) process of trial and error how to make cutlery. One estimate is that it took a household six years to learn how to make spoons (Patchell and Hayter 1992: 206). Once established, cutlery manufacture was organized around an increasingly highly diverse group of specialist firms integrated by a dense network of input–output linkages, common labour pools and information sharing (Figure 13.8). The largest firms have occasionaly employed up to 500–600 workers, but usually less. Typically, close marketing and purchasing linkages have not been formalized as actual contracts and both capacity and specialized subcontracting, involving activities such as grinding, polishing, plating, moulding and sharpening, were extensively developed. Linkages were often maintained on a daily basis and during the 1950s were facilitated by motorized vehicles.

In Tsubame, the social division of labour was originally controlled by wholesalers and featured exclusive supply links between contractors and subcontractors. Gradually, a group utilization system emerged whereby subcontractors supply

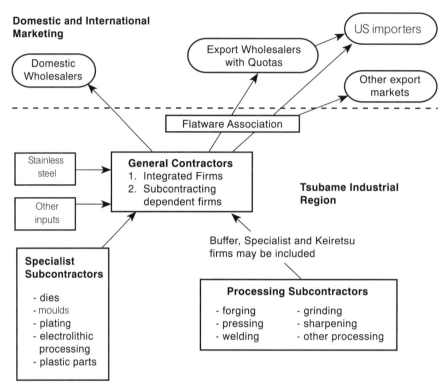

Figure 13.8 Tsubame Cutlery Industry: Subcontracting and Distribution Patterns. Source: Patchell and Hayter (1992: 208)

several contractors, and in 1935 an Export Association was created; since the 1950s this Association has been primarily responsible for organizing exports to the US, which has imposed various voluntary export quotas. In addition, the rationale for the utilization of the social division of labour changed from an emphasis on its buffer role and small-lot, high-variety capacity (for a limited product range) to a focus on the rapid mobilization for high-quality production characterized by both variety and increasing product diversity. By the 1980s (Figure 13.8), the social division of labour in Tsubame remained intact. In 1988, for example, there were over 10 000 workers employed by over 2000 firms, down modestly from the situation in the 1970s (Ika 1988: 11–15). The largest of these firms ('general contractors') employed 150–300 workers and either relied extensively on subcontracting or partially on subcontractors.

In Tsubame, cutlery manufacturers have consistently and comprehensively invested in skill formation in the labour force, new equipment and R&D in order to move up the value-added ladder within cutlery and related products. The record of the cutlery industry is one of continuous learning and innovation since its rapid adoption of western methods of production in the 1920s when cutlery was one of Japan's first light industries to be mechanized. In the rebuilding of the industry in the 1950s, firms constantly sought information on new techniques including by regularly visiting other cutlery districts and then rapidly investing in new equipment such as automatic grinding and polishing machines. This behaviour continued in the 1980s when firms adopted hydraulic presses, hard-steel presses, new dies and resistance welding involving use of lasers, ion beams and other advanced technologies (Patchell 1991).

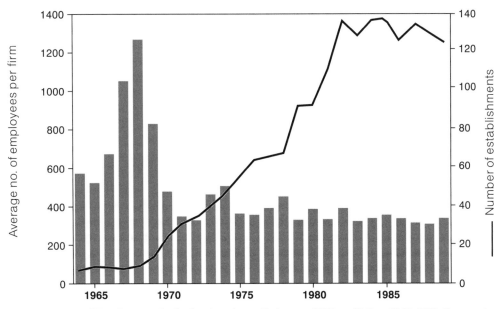

Figure 13.9   Firms and Employment in the Semiconductor Industry of Silicon Valley, 1964–1989. Source: Angel (1994)

In Tsubame, as in Solingen, automation is associated with a shift towards high-waged design-intensive production. In contrast to Solingen, where firms argue that an internal division of labour is more appropriate to quality control and to justify investments in machinery, R&D and labour training, in Tsubame, the shift towards value-added production still relies on the social division of labour. Moreover, there are signs that in the 1990s, Tsubame firms are using the specialist expertise available in the social division of labour to diversify into new kinds of products (e.g. golf clubs, road-side mirrors, tools).

Silicon Valley: a contested ideal

Arguably the most dynamic and innovative industrial district in the world is Silicon Valley (Figure 13.9). The district grew rapidly after 1950, and between 1968 and 1978 27 000 jobs were created in electronics components in Santa Clara County. Initially, the region's structure was dominated by large firms and their R&D and manufacturing activities, including those of IBM, ITT, Hewlett Packard and Fairchild (Saxenian 1983, 1994; Angel 1994). During the 1970s, there was a rapid growth of

SMEs and as a result the average size of establishment in Silicon Valley in the 1990s is much less than in the early 1960s (Figure 13.9). For national and local governments around the world, including Japan, Silicon Valley is a high-tech ideal which represents the appropriate form of leading-edge industrialization at the end of the 20th century. Attempts to clone Silicon Valley are many, although an equivalent rival is yet to appear. Yet, Silicon Valley, in its own terms, is a contested ideal.

On the one hand, the agglomeration of electronic firms that have concentrated there since the early 1950s are anchored by a few large firms, notably Hewlett Packard and Fairchild (Saxenian 1994). These large firms provided employment opportunities, markets for smaller firms and spin-off opportunities. Saxenian (1994: xx), for example, argues that Silicon Valley features 'splinter entrepreneurialism' in which new firms are continually created by employees leaving existing firms, especially the large ones. "A poster of the Fairchild family tree, showing the corporate genealogy of the scores of Fairchild spin-offs, hangs on the walls of many Silicon Valley firms . . . the tree traces the common ancestry of the region's semiconductor industry and reminds engineers of the personal ties that enable people, technology, and money to recombine rapidly into new ventures" (Saxenian 1994: 31). During the 1960s alone, 31 spin-off semiconductor firms were created mainly from Fairchild and mainly to provide inputs to established firms in the park. A common labour pool, largely based on the supply of engineering graduates from Stanford University, is a vital feature of Silicon Valley. Moreover, in addition to spin-off firms there is frequent movement of employees among firms in Silicon Valley (Angel 1991).

On the other hand, questions have been raised from the perspective of local development about how the forces of competition and co-operation are mediated in Silicon Valley. It is generally agreed that competition is driven by continuous innovation and that "The new semiconductor firms that chose to manufacture pioneered the use of low-cost, low volume flexible 'mini-fabricators' that could quickly process short runs of different designs on a single line" (Saxenian 1994: 20). For Florida and Kenney (1990a,b), however, Silicon Valley's emphasis on continuous innovation and 'short-term' manufacturing is motivated by a desire for 'super-profits' available in the early stages of product life cycles and a lack of willingness to fully pursue all manufacturing and marketing opportunities through entire life cycles, thus widening job opportunities, even if profits are less as products mature. Florida and Kenney (1990a) contrast this attitude of 'hyper competition' with the more co-operative approaches evident in Japanese production systems which seek flexibility while also achieving mass production.

Saxenian suggests that a co-operative culture has at least developed within the context of continuous innovation and relatively short product life cycles: "Loyalty grew out of reciprocal decisions to honor unwritten obligations as well as contracts and not to take advantage of one another when market conditions changed. Some firms even supported suppliers through tough times by extending credit, providing technical assistance, equipment or manpower, or helping them find new custo-

mers" (Saxenian 1994: 147). At the same time, to offset suppliers (or consumers) becoming too dependent (and vulnerable) on one customer (or supplier) several firms have developed a '20% rule': they have prevented their business from exceeding 20% of the sales of a supplier or 20% of the purchases of a consumer. In more general terms, the interdependencies that have developed among specialized, high-tech firms in Silicon Valley reflect, in Scott's terms, highly specialized, uncertain relationships in which transaction costs can be reduced by relying on the social division of labour. Clearly, the maintenance of such difficult transactions is facilitated by personal contact and geographic proximity. In turn, geographic proximity allows the realization of other forms of external economy, notably access to appropriately skilled labour, the expertise of nearby universities and so on (Angel 1991, 1994).

Florida and Kenney's (1990a) criticism is that co-operative tendencies and norms are continually threatened in Silicon Valley (to a greater degree than in Japan) because of strident competition among firms to become the first in the marketplace and by short-term attitudes, which undermines potentials to extend possibilities for manufacturing. Consequently, in their view, the mediation of the forces of competition and co-operation within Silicon Valley does not extend the benefits of technological leadership throughout society, particularly to US-based manufacturing workers. Within Silicon Valley, for example, firms have effectively opposed the establishment of unions, and mass-produced components have often been relocated in low-wage locations. At the same time, Silicon Valley has emerged as a flexibly specialized industrial district that has developed significant localization economies of scale and continues to thrive at a time when other high-tech complexes, such as Boston 128 (Norton 1992), have experienced considerable problems.

Conclusion: diverse trends in production systems

Piore and Sabel's (1994) Second Industrial Divide is based on the revival of industrial districts and especially of SME-dominated industrial districts. Many examples of contemporary industrial districts have been examined, and knowledge of such districts in California (Scott 1992, 1996; Angel 1994; Saxenian 1994) and northern Italy (Brusco 1986) has been supplemented by a better awareness of the particular role of industrial districts in Japan's industrialization (Patchell 1993a,b). A few studies have also linked vertical disintegration with a stronger role for small firms and a tendency to spatially agglomerate (Storper and Christopherson 1987). Yet, trends towards SMEs, vertical disintegration and industrial districts are not unequivocally linked. Leung (1993), for example, demonstrates how the development of subcontracting linkages by Hong Kong firms is associated with geographical dispersal of facilities within China. In addition, some established industrial districts, such as cutlery in Sheffield and Solingen, have shifted from a social division of labour towards an internal division of labour (Hayter and Patchell 1993). Even in Third Italy, Harrison (1993) argues that the intrusion of corporate hierarchies by take-over and merger is beginning to displace co-operative behaviour patterns developed by SMEs over long periods of time.

Moreover, firms that exist in vibrant industrial districts have chosen to disperse some elements of production. The most celebrated case is Silicon Valley, whose firms have established factories in Asia in the late stages of the product life cycle.

Indeed, firms in Japanese industrial districts, ranging from the small-scale case of Tsubame to the world-scale case of Toyota Town, have experimented with dispersing some aspects of production to lower cost locations. In still other cases, firms in 'rival' industrial districts are linked by joint ventures, cross-licensing and strategic alliances. While strategic alliances are typically thought of as arrangements made by already-large international firms in research-intensive industries, SMEs have also participated (Ahern 1993b). The strategic alliance between Madge Networks of the UK (Exhibit 9.1) and the Silicon Valley-based Cisco Systems of the US is an example.

In other words, industrial districts and globally dispersed production systems are in constant tension. How this tension is played out has important but contentious implications for local development. In Piore and Sabel's view, flexible specialized industrial districts are a desirable form of development, being democratic, progressive and oriented to continual improvement in working conditions as well as profitablity.

Others reject this view. An important question is this regard, concerns the direction in which the district is moving; for example, whether towards greater or lesser use of low-cost low-skilled labour or higher-cost higher-skilled labour. The latter may require greater investment and ingenuity but is socially more sustaining (Christopherson and Storper 1989).

14

CORE-FIRM-DOMINATED INDUSTRIAL DISTRICTS AND THE JAPANESE AUTO INDUSTRY

This chapter focuses on core-firm-dominated industrial districts with particular reference to the Japanese automobile industry, in which Toyota City provides the leading geographic and organizational exemplar. The industry offers a significant illustration of the transition from the fordist to the ICT techno-economic paradigm and throughout the chapter comparative attention is drawn to the US industry. According to Womack et al (1990: 11), auto manufacture is the 'industry of industries in transition' in the 20th century. During fordism, US auto firms pioneered the shift to mass production and established the US as the dominant global producer. More recently, as an expression of flexible specialization, Japanese auto firms have pioneered 'lean' production, with its remarkably efficient use of materials, space, energy, supplier organizations and labour. Moreover, lean production and its reliance on close and dense subcontracting patterns is intimately related to geographic concentration, and Toyota City, and environs, had emerged by the early 1980s as the most efficient industrial district in the global auto sector (Womack et al 1990; Williams et al 1994; see also Miyakawa 1981, 1991). Lean production, especially as illustrated by Toyota City, had become the world 'benchmark' against which auto firms globally have been forced to adjust (Cusumano 1985, 1988).

For Freeman (1988), the emergence of Japan as a global technology leader (Table 2.3; Figure 2.4) is rooted in a distinctive 'national innovation system'. Freeman summarizes the main features of this system as:

- a commitment to 'reverse engineering' to comprehensively learn best-practice technology as a basis for further innovation;
- a commitment by managers, engineers and workers to think in an integrated way about the entire production system, including with respect to product and process design;
- the use of factories as laboratories which constantly bring together R&D professionals with production engineers and workers in a manner that facilitates collective learning processes;
- an emphasis on increasing quality by constantly seeking to correct defects and up-grading products and processes; and
- the involvement of suppliers and consumers in close and stable relationships, at the core of which is the exchange of technological expertise.

Each of these characteristics is notably evident in the auto industry.

The chapter is in three main parts. The first part provides an historical overview of production trends in the industry, particularly with respect to Japan and the US, and some general differences in the auto industry of these two countries are noted. The second part examines the organizational structure of Japanese auto production systems as of the mid-1980s, paying special attention to the social division of labour and firm–supplier relationships (Sheard 1983; Asanuma 1985a,b, 1989). Finally, questions related to the nature and extent to which the principles and practices of lean production are transferred elsewhere, notably to the US, are addressed (Womack et al 1990; Rubenstein 1991; Williams et al 1994). For related discussions of automobile manufacturing in Europe, see Hudson and Schamp (1995), while Law's (1991) and Morales' (1994) studies are global in scope.

Automobile production: US Fordism challenged by Japanese flexible specialization

In 1900, the auto industry was still a craft industry and production was relatively small (Table 14.1). At that time, the US accounted for about 44% of global production and Germany about 24%. With the introduction of mass production (Womack et al 1990; Chapter 2), production levels exploded and the US quickly became the dominant national producer. Thus by 1925, production levels were over 400% of 1900 levels and the US accounted for fully 87% of global production, the UK then having emerged as the number two producer. The US remained pre-eminent until well into the 1960s, although after 1950 the share of global production accounted for by Europe and especially Japan increased rapidly. By 1980, Japan was the largest national producer of autos, and in 1990 Japan accounted for almost 31% of global output. If, in the transition from the fordist to the IC techno-economic paradigm, the geographic balance in auto production has become more equal among the US, Europe and Japan, "Japan is clearly the driving force" (Morales 1994: 15).

Historical perspective on Japan's auto industry

The Japanese and US auto industries developed independently around the beginning of the 20th

Table 14.1   Total Motor Vehicle Production in Selected Countries, 1900–1990

	1900	1925	1950	1975	1990
North America					
US	4 192	4 265 830	8 005 859	8 986 513	9 888 036
Mexico	0	0	21 575	360 678	820 558
Western Europe					
(West) Germany	2 312	62 753	306 064	3 186 208	4 660 657
Italy	0	49 400	127 847	1 458 629	1 874 672
UK	0	167 000	783 672	1 648 399	1 295 611
Sweden	0	—	17 553	366 753	335 853
Asia-Pacific					
Japan	0	—	31 597	6 941 591	13 486 796
South Korea	0	0	0	36 264	1 321 630
Eastern Europe					
Yugoslavia	0	0	0	205 567	342 727
South America					
Brazil	0	0	0	930 235	914 576
World	9 504	4 900 730	10 577 426	32 998 363	44 165 033

Source: Morales (1994: 16).

century. In the US, led by the Ford Company, the predecessor companies of GM, Chrysler and others, the industry grew rapidly, and by the 1920s and 1930s it had become the largest in the world with production increasingly concentrated in Michigan, especially the Detroit area (Boas 1961; Kroos 1974: 438–449; Rubenstein 1992: 25–46; Table 2.4). In contrast, Japanese developments were sporadic and, by the 1930s, dependent on US know-how.

Origins In Japan, the roots of the industry can be traced back to the end of the 19th century. In 1897 the first imported vehicle arrived in Yokohama, where the first importing agency was established in 1901 (Miyakawa 1991: 89). Around this time, motor vehicle manufacturing began, initially in Tokyo where Japan's machinery firms and related government-owned firms were located and where "pioneering engineers gathered to copy modern western industrial products and technologies" (Miyakawa 1991: 89). According to Takeuchi (1990: 167), auto manufacturing began in Tokyo when several individuals, using techniques from rickshaw construction and ironworking, built autos for themselves by copying European models. Another early, perhaps the first, model was financed by a bicycle shop owner in Tokyo who had visited the US in 1902 and had brought back two gasoline engines. He then hired an engineer, who had worked in a Russian factory where there was an auto which he had learnt how to operate, to develop a car. In 1911 the first auto factory was built in Japan. In these early years about 43 different models were manufactured by 13 firms in Tokyo. In the early 1900s, Osaka also became a vehicle manufacturing centre, particularly after 1911 and the growth in the production of army trucks. In Kobe, Mitsubishi Shipbuilding began manufacturing autos based on Fiat models.

Local production remained relatively small scale, however, and was unable to compete with imports led by Ford's Model T. Japan did not have a high-quality machinery industry and its limited home market was increasingly served by imports, especially from the US (Table 14.2). By

Table 14.2 Supply of Trucks and Cars in Japan 1916–1935

Year	Local production	CBU Imports	CKD Assembly
1916	294	218	
1917	250	860	
1918	195	1712	
1919	60	1579	
1920	45	1745	
1921	4	1074	
1922	—	752	
1923	—	1938	
1924	—	4063	
1925	—	1765	3437
1926	245	2381	8677
1927	302	3895	12668
1928	347	7883	24341
1929	437	5018	29338
1930	458	2591	19678
1931	436	1887	20199
1932	880	997	14087
1933	1681	491	15082
1934	2787	896	33458
1935	5089	931	30787

Source: Odaka et al (1988: 34).
CBU: Completely built units. CKD: Completely knocked down.
Note: Ford plant opened in 1925; Chevrolet plant opened in 1927; import tariff raised in 1932.

1917, imports were greater than local production, which by the early 1920s appears to have virtually disappeared. Meanwhile, car, truck and bus imports satisfied local demand and were stimulated by the Kanto earthquake of 1923 which destroyed Tokyo's railway and streetcar network. In fact, Japan removed the tariff on auto imports for a time, and Ford and Chevrolet (GM) established branch plants in Yokohama and Osaka in 1925 and 1927 respectively (Miyakawa 1991: 91). Originally, these plants were 'complete knock down' assembly operations and only assembled the parts, which were entirely imported. However, Ford (but not GM) made an effort to develop a local supply capability which ultimately provided the basis for Nissan's development in the early 1930s with the production of the Datsun in 1931 (Takeuchi 1990: 168). In fact, Toyota, then a large-scale manufacturer of textile machinery, had begun to develop an interest in auto assembly at this time, and it too sent engineers to work with Ford in Yokohama to learn production know-how (Miyakawa 1991: 94).

As reliance on US producers and imports increased, Japan's military, stimulated by the World War I and China campaigns, spearheaded a powerful lobby for a domestic auto industry. This lobbying culminated in the Automobile Industry Act of 1936 which "was enacted to assist automobile production . . . This law was intended to prevent foreign enterprises from gaining control of Japanese industries and to encourage or aid new or ailing Japanese enterprises" (Takeuchi 1990: 169). Thus, this Act required that Japanese gain majority control of the auto industry; provided local 'registered' firms with special incentives such as five-year income tax holidays and tariff exemptions on special equipment; encouraged registered firms to buy locally by making such purchases a criterion for registration; and, after 1936, Ford and GM were not allowed to expand and were more or less forced out. In addition, import tariffs on imported motor vehicles were increased from 50 to 70% and on engines from 35 to 60%, while the Yen was devalued against the dollar. Within the context of such heavy protectionism, GM and Ford sought co-operative arrangements with Nissan but, "because of heavy pressure from the Army" (Miyakawa 1991: 94), these efforts failed. In 1939, a further Act required GM and Ford to close their factories. Indeed, GM left Japan while Ford did maintain a legal presence. By 1939, Nissan, Toyota and Isuzu were the dominant three companies in Japan. Nissan and Isuzu were located in Tokyo, close to Japan's established machinery manufacturers; Toyota created its own company town, Toyota City near Nagoya, in 1938, where it had the task of developing a local supply network.

Post World War II trends   After World War II, when the Japanese auto industry was largely destroyed, the industry remained depressed until the early 1950s. In the late 1940s, technological capability was poor, there was a severe economic depression and the Labour Union Law of 1945 encouraged the formation of unions; this was soon followed by labour strife among Nissan, Isuzu and Toyota. However, following considerable debate, in 1952 the Ministry of Industry and Trade (MITI) designated the industry as a high priority in its economic restructuring programme (along with synthetic fibres, petrochemicals and electronics). At that time, the debate within Japan was about whether industrial strategy should emphasize the principle of comparative advantage and focus on traditional industries such as textiles which would take advantage of Japan's large, hard-working but low-cost labour supplies, or seek to encourage diversification into the industries of the future which required more capital, quality components and labour skill. While such important institutions as the Bank of Japan favoured the former strategy, the MITI with a minority of western advice chose the latter.

Consequently, the auto industry was protected by high tariffs and a law was passed that required supplier firms (part producers) to be officially registered; registrations were awarded to those firms who were considered the most likely to ensure quality of production. The domestication of the parts industry was further encouraged by gradual reductions in foreign exchange allocations to buy imported parts. In 1951, the Amended Law for Introduction of Foreign Capital allowed involvement by foreign auto firms in order for Japanese firms to acquire expertise. In practice, Toyota chose not to link itself with foreign firms, while Nissan entered a joint venture with Austin (UK), Isuzu with Rootes (UK), and Hino with Renault (France). These agreements, however, did not involve US firms and were temporary. Isuzu cancelled its technological co-operation agreement in 1965, and in 1969 Nissan stopped manufacturing Austin cars and started to develop its own models. In the meantime, Toyota had manufactured its own passengers since 1952. By the late 1950s, when the Japanese industry still emphasized buses and trucks, Japanese auto manufacturers had already established exports to other Asian countries; in 1957 and 1958 respectively,

Table 14.3 Production of Autos, Trucks and Buses 1950–1985 (in Thousands of Units)

	Cars	Total four-wheelers	Exports	Export ratio (%)
1950	2	32	—	
1955	20	69	1	1.8
1960	165	482	39	8.1
1965	696	1 876	194	10.3
1970	3 179	5 289	1 087	20.6
1975	4 568	6 942	2 677	38.6
1980	7 038	11 043	5 967	54.0
1985	7 647	12 271	6 730	54.8

Source: Odaka et al (1988: 41).

Nissan and Toyota began exporting to the US. The core Japanese companies also engaged in direct competition with one another, initially largely for domestic markets and subsequently for export markets.

Thus, the growth of the Japanese auto industry has largely occurred since 1950, and in the 1960s Japan began to produce cars that approached the quality of US and European models. Increases in the quantity and quality of output have been relentless (Table 14.3). In quantity terms, for example, Japan successively surpassed the output of France (1964), the UK (1966), West Germany (1967) and, benefiting from the increased demand for small cars, finally overtook the United States in 1980 to become the world's largest automobile-producing country (Sheard 1983: 52). In the major market regions of the world, Japanese firms have since gained important shares of all market segments, including large and luxury segments (Bloomfield 1991).

In 1980, Japan produced over 11 million four-wheelers, including over 7 million cars, and over half of the total production was exported. In fact, reflecting its growing global competitiveness, the export:sales ratio of Japanese autos increased consistently until 1985. With respect to direct foreign investment (DFI), Japanese firms had established branch plants in developing countries, such as Pakistan, Thailand and Chile, in the early 1960s. Since 1980, Japanese auto firms have invested massively in North America (Table 14.6)

and Europe (i.e. in their major overseas markets) – a strategy stimulated by host governments seeking to reduce growing visible balance of payments deficits with Japan. While Japan's export:sales ratio is unlikely to increase further, the global market share of Japanese-based MNCs may well increase. It also might be noted that since the early 1970s, Japan has liberalized its own policies to DFI, and GM, Ford and Chrysler have re-established themselves in Japan, although typically on a smaller scale than Japanese firms in the US.

Toyota City

By the early 1980s, Japanese firms enjoyed an enormous cost advantage over US rivals, *even in North American markets* (Figure 5.9). Among Japanese firms, both Womack et al (1990) and Williams et al (1994) cite Toyota as the most efficient auto manufacturer, it having achieved the greatest success in increasing labour productivity (Table 5.4). For these authors, by the 1980s Toyota's factories, notably its Takaoka plant, had become the world's benchmark and had achieved the leading-edge status that Ford's Highland Park facility had in 1916. Indeed, the world's biggest manufacturing firm and biggest auto manufacturer, GM, felt it advisable in 1980 to enter into a joint venture with Toyota (NUMMI) in the US in order to gain insights into the latter's competitive strengths. Meanwhile, Toyota City has become a mecca for visitors from around the world who have interests in the auto industry.

Certainly, within the auto industry Toyota epitomizes the 'Japanese' approach to development, as summarized by Freeman (1988) (see the beginning of this chapter). Toyota's plan to become an auto manufacturer was rooted in a conviction to learn from best practices elsewhere, notably in the US, and to reverse-engineer technology as a basis for further innovation. In this regard, Toyota did not even use joint ventures with foreign manufacturers to access expertise. Moreover, Toyota has strongly encouraged white and blue collar workers to think in an

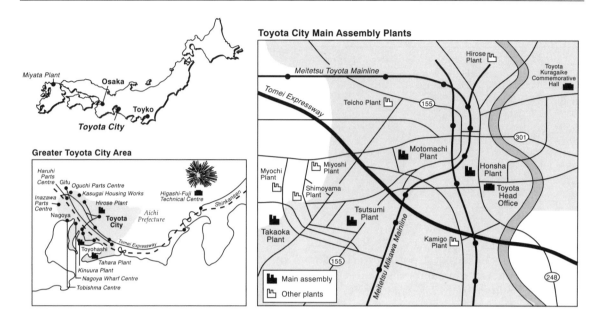

Figure 14.1   Toyota: Distribution of Major Facilities in Japan, 1994

integrated way about design; R&D and production activities are conceived as loopy, constantly interrelated processes; it pioneered functionally flexible labour relations; and it has developed close, stable relationships with its suppliers based on a commitment to technological innovation. Moreover, to a greater degree than Nissan, Honda and other Japanese auto manufacturers, Toyota emphasized the geographic concentration of its activities and that of its suppliers. From its inception, Toyota planned to create a 'company town' and simultaneously an industrial district.

Geographically, from 1937 to 1980 Toyota chose to concentrate its auto activities in and around the old fortress town of Koromo, which became Toyota City in 1959 (Miyakawa 1980, 1981, 1991). The idea for the Toyota Motor Company, which began operations in 1937, originated with Sakichi Toyoda who had visited the US in 1911 to investigate new industries, especially the auto industry (Miyakawa 1980: 41). While the family's existing business involved making textile machinery, he subsequently built a motor engine and provided funds for his son to develop the motor car business.   In developing appropriate know-how, the Toyodas were particularly helped by Ford who had operations in the Tokyo area. The Toyodas chose Koromo, however, as the location for the Toyota Motor Company. As a small isolated town whose silk industry was in depression, Koromo provided cheap land, a supportive local government, a supply of workers and an opportunity for the firm to create "an ideal industrial community" (Miyakawa 1980: 42). Nearby Nagoya offered metropolitan services and manufacturing firms. Even so, to access technological know-how, Toyota established an R&D centre in Tokyo in the 1940s, and its initial supplier association was also largely based there.

This area, which in the 1930s was remote from established centres of industry, including the fledgling auto industry in Tokyo and Osaka, now anchors the Nagoya–Hamamatsu metropolitan area, the largest centre of the Japanese auto industry which includes major Honda and Suzuki operations as well as those of Toyota (Figure 14.1; Miyakawa 1991: 101). In 1960, Toyota also relocated and substantially enlarged its R&D facilities in Nagoya.

In addition to its own assembly operations, Toyota attracted suppliers of parts to the area which were strongly, often exclusively tied to Toyota (Miyakawa 1980: 48–52). A key initiative in this regard was the formation in 1939 of the Association of Co-Development with Toyota Motor, the *Kyoho Kai*, an association of about 25 'first-tier' parts suppliers largely based in Tokyo. After World War II, this association was expanded and divided into three branches located in the Kanto region (centred on Tokyo), the Tokai region (centred on Nagoya and Toyota City) and the Kansai region (centred on Osaka). By the 1950s, the largest was the Tokai *Kyoho Kai*: in the late 1970s, for example, this association comprised 136 members, compared to the 25 in the Kansai *Kyoho Kai* and 65 in the Kanto *Kyoho Kai* (Figure 14.1). In fact, the members of these associations comprising Toyota's first-tier suppliers changed little since at least the late 1960s and until at least the late 1980s (Odaka et al 1988: 316: see also Morales 1994: 102). Moreover, the majority of the members of the Tokai *Kyho Kai* are located within Toyota City, within minutes driving time of Toyota's four assembly and several other plants in the city, and almost half sold at least 50% of their output to Toyota (Miyakawa 1980: 49). In addition to the *Kyoho Kai*, an association of component suppliers, Toyota created two other associations: the *Seiho Kai* and the *Eiho Kai* (Odaka et al 1988: 256; Miyakawa 1980: 49). The *Seiho Kai* has typically comprised 21 members which manufacture machine tools, dies and gauges; the *Eiho Kai* typically has 36 members, supplying various kinds of services related to the construction and building of facilities. Members of both these associations are largely located in Toyota City and nearby Nagoya (Miyakawa 1980: 49).

Each of the members of these associations, in turn, have nurtured the development of their own suppliers. By 1980, the 11 Toyota facilities (four assembly plants) subcontracted with 220 first-tier suppliers, who subcontracted with some 30 000 'lower' tier firms (Sheard 1983; Morales 1994:

108). Toyota City remained a company with 83% of its employment in auto-related activities (Fujita and Hill 1987). From being a rural silk-growing area in 1937, within 30 years Toyota City had become a major industrial district specializing in autos. In this district, the cause and effect relations between geographic concentration and flexible mass production is most acutely defined.

The social division of labour in the Japanese auto industry

The production system of the Japanese and US auto industries evolved according to different organizational and geographical principles until the early 1980s; since then the differences have lessened, but not disappeared, especially as aspects of Japanese production systems have been incorporated within the North American industry.

Within both Japan and the US, auto assembly and parts production has historically been concentrated in core industrial regions. In the US, however, once established in the Mid-West, the dominant corporations pursued strategies of branch plant dispersal from the 1920s until the 1980s (Rubenstein 1991: 130). In Japan, on the other hand, with the main exception of Mazda in Hiroshima (Takeuchi 1990), the leading firms have favoured geographic concentration in the Tokyo–Nagoya regions. It has only been within the last decade that the leading firms have considered locating branches in remote regions. Toyota and Nissan, for example, have built new state-of-the-art plants in the Northern Kyushu–Yamaguchi area in southern Japan.

In important respects, the Japanese industry is more competitive than its US counterpart. Thus, there are a greater number of core firms in Japan than in the US. While the US has been dominated by the 'Big Three' (GM, Ford and Chrysler), in Japan there were seven core firms in 1950, 14 in 1959, 12 in 1968 and 11 in 1987 (Nissan, Toyota, Honda, Mazda, Mitsubishi, Fuji, Isuzu, Subaru, Hino, Daihatsu and Suzuki) and while a couple of these are relatively

Table 14.4 'Make' or 'Buy' in Automobile Manufacturing: An International Comparison[a]

| Year | Purchased inputs to net sales (%) | | | | Goods and services purchased/production cost Japan[e] |
	USA[b]	West Germany[c]	France[d]	Japan[e]	
1955	58	n.a.	n.a.	72	87
1960	47	n.a.	n.a.	78	83
1965	53	63	53	73	82
1970	57	61	n.a.	77	81
1975	61	59	n.a.	80	81

Source: Odaka et al (1988: 54).
[a]Straight international comparisons are subject to some qualifications and should be adjusted for different tax systems, production subsidies, and so on. In the present table, no such adjustments have been attempted.
[b]1955–65: weighted averages of GM, Ford, Chrysler and American Motors (the 1955 figure is a geometric average of the four corporations, making use of a 1953 figure for AMC); 1970–75: weighted averages of GM, Ford and Chrysler.
[c]Weighted average of Volkswagen and Benz, except for 1970 where the Volkswagen figure is not available.
[d]Figure for Renault.
[e]Weighted averages of 7–11 corporations. Note that the numerator includes materials cost.

specialized and affiliated to a larger company, each has promoted the development of an intricate social division of labour (Odaka et al 1988: 1). That is, to a far greater degree than in the US, Japanese core companies rely on 'buying' rather than on 'making' inputs (Table 14.4). Compared to the US, West Germany or France, Japanese auto firms have consistently exhibited a stronger preference to 'buy'; in 1980, for example, Japanese firms, on average, bought 80% of their inputs. This difference remains in evidence. In 1989, Toyota's sales of US $60 billion were about half that of GM's US $127 billion, but its workforce of 91 790 was 11.9% of GM's employment, even though GM had spent the previous decade reducing jobs. Moreover, the organization of the social division of labour in the Japanese auto industry and its North American counterpart has evolved in fundamentally different ways. In particular, while Japanese companies have developed strongly hierarchical production systems, US companies have preferred lateral systems featuring many more direct contacts with subcontractors or suppliers (Figure 13.2).

The structure of subcontracting

The key feature of contemporary Japanese auto production systems is a closely delineated set of 'subcontracting layers' (Sheard 1983) or 'inter-firm relationships' (Asanuma 1989) which are arranged in a hierarchical manner around the core firms (Figure 14.2). According to Sheard (1983), "The defining criteria of a subcontracting layer is transactional distance from the automaker". Thus the core firm deals directly with the first layer of subcontractors, or the first tier of suppliers, who in turn deal with the second layer of subcontractors, or the second tier of suppliers, who in turn deal with a third layer, and so on. The exact number of suppliers is hard to state conclusively but in mid-1980s, if the main assemblers were using approximately 100–300 first-tier suppliers, the number of second- and third-tier suppliers was probably in the range of 1000–4000 (Sheard 1983; Odaka et al 1988: 2).

Although the social division of labour is extensive, this hierarchical system means that the core firms only deal directly with very few suppliers. Toyota, for example, deals directly with about 220 component suppliers, while Takeuchi (1990: 175–176) notes that Nissan subcontracts more than 50% of its direct purchases from the 105 members of its main co-operative association, Takara *Kai*. In addition, core firms purchase 'basic material supplies' (e.g. steel, plastic, rubber) from a few other core firms in other industries. In the US (and Canada), the

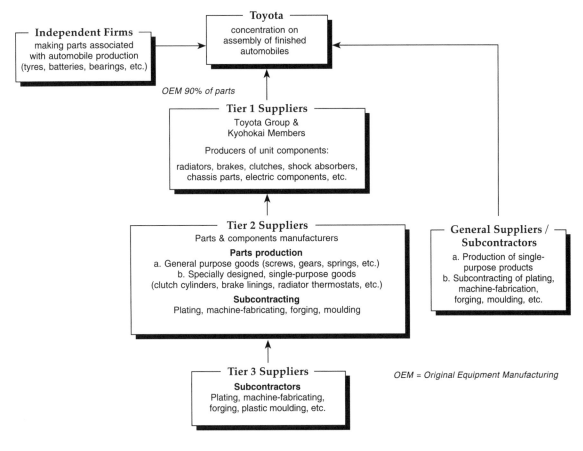

Figure 14.2   The Toyota Production System. Source: Patchell (1992: 142)

Big Three have traditionally chosen to make a much bigger share of component parts internally. Thus, in 1986, GM obtained 80% of its materials by direct purchases from 5500 suppliers. If suppliers of basic raw materials, machine tools, services and marginal inputs are included, GM dealt with around 35 000 suppliers (Asanuma 1988). At the plant level, in 1986 a new GM plant typically dealt with approximately 300 suppliers, while an old GM plant obtained supplies from over 1200 firms. On the other hand, Toyota dealt directly with about 125 suppliers in a plant that had twice the capacity of a GM plant. In other words, GM deals directly with many more suppliers, many are marginal and many only serve one plant.

The refinement of the social division of labour inherently implies specialized roles for participating members. Core firms such as Toyota and Nissan have traditionally been more specialized than their North American counterparts. Toyota, for example, specializes in engine and final vehicle assembly of components and parts (Figure 14.1). In addition, the vast majority of the firms which comprise the social division of labour are small and specialized. Even the first-tier firms comprise few firms that are truly large. According to Sheard, about 8% of major parts suppliers employ 3000 workers or more, and even these tend to be specialized, such as the tyre manufacturers. Most second- and third-tier suppliers employ less than 20 workers.

Within the social division of labour underlying the core assemblers there are two general principles organizing firm specialisms. First, the first-tier suppliers are largely product manufacturers, for example, of steering systems, radiators, head lamps. Lower-tier suppliers are more likely to perform processes, such as stamping, machining, moulding and forging (Odaka et al 1988: 258–260). Second, between the layers or tiers of suppliers specialization also occurs as, in general, labour-intensive, small-batch and low-value-added operations are transferred down to the second and third layers. Sheard (1983: 59) provides an example:

> "A large stamping firm (first layer subcontractor) might undertake to supply a side-door sub-assembly to a certain auto-maker. The firm would use its large transfer presses to make the large stampings but would farm out all smaller stampings to a smaller firm (second layer subcontractor) which would use smaller and less expensive presses. Even this firm would find it less costly to engage a local workshop (third layer subcontractor) to provide some of the component stampings. Thus, through specialization between different-sized firms with varying productive capacities, capital intensity ratios and labour costs, there results a production chain linking workshop to automaker."

The evolution of the Japanese auto industry to a large-scale producer of high-quality, reliable autos required the development of a sophisticated machine tools industry and constant improvement among all firms within the social division of labour. In this regard the core firms played a 'core' role; they established foreign markets, co-ordinated the production system and helped develop the abilities of the supplier firms by providing financing, sometimes on an equity basis but not necessarily so, by creating stable and growing demands, and by providing expertise through informal contact, demonstrations, joint projects and loans of workers and engineers. The core firms vary in aspects of their subcontracting relations in terms of the number of first-tier suppliers, the degree of control over them and the degree of exclusivity. Nissan and

Toyota developed closely related 'associations' in which exclusive relations and parent equity investment were, and remain, important features. However, even Nissan and Toyota do not have that many first-tier suppliers. Moreover, to an important degree the other companies 'tapped' into their supplier base in creating their own production systems. Consequently, one emerging trend is towards the interlocking of production systems, although this is a gradual process. The core firms still retain their own families of suppliers, and supply relationships are strong.

The kanban system

In the Japanese auto industry (and others) the development of a specialized, highly interlinked production system has been associated with the development of just-in-time (JIT) delivery systems, sometimes also called kanban systems. In fact, 'kanban' means card or visible record and is a specific type of JIT system developed by Toyota. The terms are often used interchangeably, however.

In brief, the idea of JIT is simple: "Produce and deliver finished goods just in time to be sold, subassemblies just in time to be assembled into finished goods, fabricated parts just in time to be transformed into fabricated goods" (Schonberger 1982: 16; see also Linge 1991). Conventionally, the main justification for JIT is that it reduces the need for inventory and for inventory control, thereby reducing costs. JIT reduces costs because materials do not have to be stored; stored materials represent tied up capital (they have to be paid for) and also involve an expenditure in buildings, space and inventory management. These advantages are important but there are other benefits. The whole practice of JIT typically means that deliveries are made in frequent but small lots; that is, JIT reduces lot size. In turn, small lots allow for better quality control as defects are more easily recognized at either the point of origin or destination. In other words, it is easier to check small lots and maintain higher quality standards. Indeed, to be successful, JIT requires that quality standards are strictly met since too many and frequent rejects

Table 14.5 Parts Manufactured According to Specifications Provided by the Core Firm ('ordered goods')

Parts manufactured according to drawings provided by the core firm			Parts manufactured according to drawings provided by the supplier			Parts offered by catalogue ('purchased goods')
A	B	C	D	E	F	G
The core firm provides minute instructions for the manufacturing process	The supplier designs the manufacturing process based on blueprints of products provided by the core firm	The core firm provides only rough drawings and their completion is entrusted to the supplier	The core firm provides specifications and has substantial knowledge of the manufacturing process	Intermediate region between D and F	Though the core firm issues specifications, it has only limited knowledge concerning the process	The core firm selects from catalogue offered by the supplier
Small parts assembled by firms offering assembly service	Small outer parts manufactured by firms offering stamping service	Small plastic parts used in dashboard	Seat	Brakes, bearings, tyres	Radio, electronic fuel injection system, battery	

Classification of parts and suppliers according to the degree of initiative in design of the product and the process.
Source: Asanuma (1986: 32); see also Asanuma (1985a,b).

would defeat its purpose. Moreover, JIT provides for more opportunities for greater involvement by labour in terms of, for example, the exchange of information regarding the timing and size of deliveries, and minor variations, as well as in quality control.

As Sheard points out, JIT was first introduced in the US in the form of the conveyor belt which moved a product 'just in time' to the next worker to perform his/her task. In Japan, JIT principles have been extended to the social division of labour to create a 'factory without walls' (Sheard 1983). The limited JIT system of the conveyor belt, however, was always associated with large inventories reflecting a so-called 'just in case' mentality. That is, large inventories require less co-operation among firms but provide the core firm with insurance against disruption of supply. Yet, in Japan, such disruptions have been noticeable by their absence. Bearing in mind that the Japanese road system is not particularly well developed, it is true that in recent years traffic congestion and accidents have created problems in the operation of JIT, which is becoming

increasingly computerized so that information about road conditions and accidents involving the JIT trucks can be transmitted quickly and remedial measures taken. In other respects, the operation of JIT has been utterly reliable. To get a better understanding of this reliability we need to place the kanban system within the context of inter-firm relations in the auto industry in general.

Ties that bind to mutual advantage:
the relation-specific skill

As an expression of the quality multiplier process, Asanuma's (1989) concept of the relation-specific skill provides the analytical basis for understanding both the dynamism and stability of Japanese auto production (see also Patchell 1993a,b). According to Asanuma, the relation-specific skill defines how suppliers serve the specific needs of core firms. In particular, Asanuma makes a fundamental distinction between design-approved (DA) and design-supplied (DS) firms or suppliers (Table 14.5). Thus, DS parts are parts manufactured according

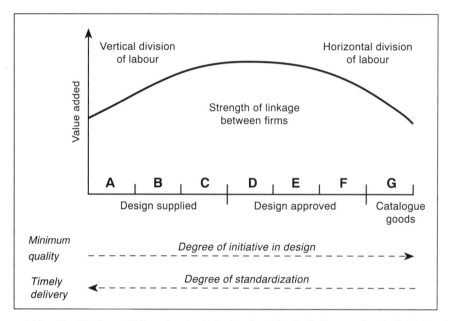

Figure 14.3   The Binding of Firms by the Relation-Specific Skill. Source: Patchell (1993a: 811)

to drawings supplied by the core firm, i.e. the core firm provides or dominates the technological expertise. DA parts, on the other hand, are parts manufactured according to designs developed by the supplier, i.e. the supplier provides or dominates the technological expertise. Depending on the degree of technological initiative undertaken by the supplier in relation to the core firm, DA and DS parts can be subdivided further. In addition, there is the special case of catalogue goods which are bought 'off the shelf' and have been researched, developed and designed exclusively by the supplier. That is, there is no exchange of 'know-how' and these firms are outside of the interactive supplier network.

The actual degree of co-operation, or what Patchell (1993a) terms 'technological bonding', between supplier and core firm in the development of a product or process is greatest in the DA and DS categories which are adjacent to one another (Figure 14.3). As Patchell notes, the strength of technological bonding declines both towards less sophisticated DS suppliers and increasingly sophisticated DA suppliers (and is least of all with respect to catalogue goods suppliers). In this

model, suppliers may develop expertise which is even greater than that held by the core company.

The relation-specific skill is the key to understanding why there is both a high level of stability and a high level of dynamism or competitiveness. Long-term stable relationships among firms in Japan's auto production systems do exist, while at the same time there is a constant striving for lower costs, better quality and new products. A key strategy in developing behaviour among suppliers which is both competitive and co-operative is the provision of incentives to them, in the form of more business from the core firm and greater profits, to develop their technological expertise. In other words, there are rewards from shifting from a DS supplier to a DA supplier. A more detailed look into the nature of inter-firm relationships provides an understanding of how this strategy works.

The basic contract   In the Japanese auto industry, as Asanuma notes, there is no single written contract between core firm and supplier which prescribes specifications of the items to be delivered, their quantities, price, etc. Rather, there are a set of contracts, related documents and established practices. The foundation of inter-firm relationships in the Japanese auto industry is provided by the so-called 'basic contract' which is exchanged when a core firm enters into a business relation with a supplier. The duration of this contract is usually one year and it is automatically renewed unless either side raises an objection. This contract defines only *general* obligations of both parties, but provides the basis for overall contractual agreements. For example, basic contracts require monthly schedules be regarded as contracts once suppliers have agreed to them. A basic contract will also typically state that kanbans are fine tunings which the core firm can introduce if the supplier agrees to the introduction of the kanban system. Typically, basic contracts require that price negotiations will occur at regular intervals, usually six months.

The specific details of inter-firm relationships are then set out in a delivery or quantity contract and in a price contract.

Quantity contract The length of a quantity contract is normally the life of a model, which Asanuma found to be typically four years, with minor model changes introduced after two years. Once a core firm selects a supplier from among rivals, the core firm rarely switches to another supplier during the life of the model or component. In addition to this 'non-switching' policy, core firms pursue a 'two vendor' policy which applies to broad groups of parts, such as head-lamps or brakes. In practice, the core firm will use at least two firms within such a category of parts as insurance against sudden disruption in supplies from one supplier, and to put pressure on all suppliers for a co-operative attitude regarding prices and quality. Then, with respect to a particular part, for example, a headlamp or brake system for a particular model of car ('sub-category of part'), the core firm will typically use one supplier whose status is stabilized during the life of the model. Sometimes an identical part may be split among two suppliers if demand is big enough; in this case, alternate suppliers might

get a fixed share of the business or the business may be split by market area, say between domestic and foreign markets.

Once supplier arrangements are established and in operation, quantities are adjusted each month depending on demand. In this regard, it might be noted that it is not widespread practice to use suppliers as capacity buffers as is the case in the West (Taylor and Thrift 1983). Both core firm and suppliers suffer during a recession for a given model.

Of course, with the introduction of a new model, formal competition among suppliers resumes. In this regard, the 'two vendor' policy ensures that there will always be a few firms capable of vigorously competing for new business because they will have the relevant expertise. This competition is not of the 'invisible hand' type characterized by an infinite number of suppliers (and consumers) but, as Itoh (1989) said, it is "the face you can see competition" (translated and quoted in Patchell 1991: 169). Everybody knows the competition. The core firm allocates new business among competing suppliers, based on several factors: its ratings of suppliers which is influenced by past performance; its policy concerning the allocation of business among suppliers; and its assessment of specific proposals. It might also be noted in this regard that core firms typically stagger model changes; this allows the core firm to keep its suppliers in constant competition for new business.

On the other hand, as already noted, a remarkable aspect of the Japanese auto industry is the stability of suppliers and the fewness of suppliers, especially within a core firm's association. For example, Asanuma (1989) notes that in 1984, there were 171 firms in Toyota's supplier association, of which 153 have been continuous members since 1973, and there had been only three exits. Such stability "has contributed to nurture a feeling among suppliers that they are in the same boat as the core firm and share a common fate, despite the fact that they have to face ratings from the core firm and engage in renegotiations individually on business terms with the core firm at regular intervals" (Asanuma

1989: 6). Asanuma's comments neatly summarize the combination of trust and paternalism with fierce competition, which are such features of the Japanese auto production system.

Price contract Prices are initially set by negotiation between core firm and supplier, taking into consideration such factors as cost, relative contributions to design and acceptable levels of profit. After six months, however, prices are renegotiated. At this time, the core firm may well ask for a price reduction of 2–10% on the basis of the argument that the supplier will have had the opportunity to move up the learning curve which defines the productivity improvements a labour force can achieve as it becomes familiar with how to manufacture a new product and as they produce more volume.

For DA suppliers who have undertaken considerable initiative to research and develop a part or process, the core firm may waive the price reduction for a year as a reward for innovativeness. If design has been shared equally by core firm and supplier the price waiver may be six months. Clearly, the size of this reward can vary from a complete price reduction waiver over varying time periods to varying percentages in the price reduction requested. Essentially, this behaviour allows the core firm to reward suppliers according to the degree of innovation, so suppliers are continually encouraged to improve. Such rewards clearly help the core firm by improving the quality of its products and by enabling it to produce more efficiently. In addition, as suppliers become more innovative, their ratings increase and they will receive more business and more challenging tasks from the core firm which will widen opportunities to be innovative.

Japanese and western auto industrial districts: summary comments In North America and Europe, as well as Japan, the auto industry's evolution was closely associated with a geographic concentration of production in large-scale industrial districts. In the US, the geographic focus centred on Michigan, in Canada on southern Ontario, and in the UK on the West Midlands and London. All of these districts created external economies of scale. The anatomy of these districts varied, however, in terms of the organization of entrepreneurial and labour-market networks and innovation (Figure 13.6).

Typically, Japanese core firms have relied more on subcontracting than their US counterparts: in the early 1980s the former typically subcontracted out about 75% of vehicle production, while the latter contracted out about 52% (Shimokawa 1993: 8–9; Yaginuma 1993: 24). US industrial districts feature higher levels of ownership integration (Figure 13.6). On the other hand, in Japanese auto industrial districts such as Toyota City, core firms organize high levels of co-operation with their suppliers and achieve greater stability than western counterparts; that is, the Japanese industrial districts feature higher levels of co-ordination (Figure 13.6). The lateral system of subcontracting developed by the US core firms involves them in a large number of direct relationships with suppliers based largely on cost. In the US, the general rule of competition favours suppliers offering a better price, and the stability of subcontracting relations is constantly threatened by the search by core firms for lower prices. In Japan, the hierarchical system of subcontracting implies fewer direct contacts for core firms who, in contrast to their US rivals, place emphasis on the stability of relations in order to help suppliers develop high-quality products and processes. In Japan, and less so in the US, core firms provide better incentives for suppliers to improve quality as well as to reduce prices.

Traditionally, highly stable firm–supplier relationships in Japan have featured many 'exclusive' relationships in which suppliers rely on one core firm as a market and other transactions. There is, however, a gradual but nevertheless clear trend for such exclusive links to decline in importance in the auto industry (Yaginuma 1993). This trend is of course a matter of degree and key relationships remain fundamentally stable.

The higher level of co-ordination achieved in the Japanese industrial districts is institutionally underpinned by keiretsu relationships, which include the supplier–firm co-operatives, and the JIT system. In this system, strong incentives to innovate exist throughout the social division of labour. The organization of internal R&D also evolved differently among US and Japanese auto firms, along the lines suggested by Freeman (1988). In particular, in Japan, firms emphasize reverse engineering and a 'loopy' approach in which there are close, interactive relations between R&D groups, production and marketing throughout the innovation process. In contrast, the US auto makers historically favoured a 'linear' approach which emphasized, extending the principles of scientific management, the formal separation of specialized functions in an innovation process, conceptualized as a linear sequence from research to marketing.

The Japanese auto industry is considered a quintessential example of flexible mass production or lean production in which high levels of efficiency are combined with high-quality production, and economies of scale and scope are combined with responsiveness to highly differentiated demands. In this latter regard, during the 1970s, Japanese auto makers rapidly escalated the number of variations of individual types of autos based on different combinations of body type, engine type, transmission type, degree of luxury, optional parts, colours and other features. Asanuma (1989), for example, reports that Crown Toyota was sold in the form of 322 variations in 1966, but that by 1978 it was offered in 101 088 variations. While Japanese auto firms have recently cut back on these variations for cost reasons, by the 1980s the industry had established itself as the world benchmark for flexible mass production in which labour productivity and product quality are mutually reinforcing trends.

One geographical consequence of the different traditions between North American (and UK) and Japanese auto makers is the greater willingness of the former to seek out lower-cost supplies from geographically dispersed sources. The greater tendency to demarcate functions, the greater reliance on the internal division of labour

and a greater preoccupation with costs combine to facilitate and motivate this search. During the 1960s and 1970s, in both North America and Europe, auto manufacturers were increasingly seeking to locationally diversify purchases as well as the auto assembly itself, in part to reduce labour costs, in part to access government support programmes, and in part to reduce risks associated with geographic concentration, such as reliance on one particular pool of labour. One example is provided by the supplier network of British Leyland's Cowley plant in the late 1970s (Figure 14.4).

Thus, in the same period that Toyota had geographically concentrated its facilities and supplier network in Toyota City and environs to its maximum extent, the Cowley plant in Oxford was relying on supplies from three metropolitan areas in the UK (in order of importance, Birmingham, London and Manchester), as well as many other places, while also importing components from the US and seven European countries.

The world car strategy Even before the post-1950 rise of the Japanese industry, the Big Three, especially Ford and GM, were already MNCs with extensive investments outside of North America, especially in Europe. Moreover, foreign operations had evolved according to distinct, independent strategies and structures, reflecting the belief that particular types of cars were needed for particular markets. Traditionally, for example, Ford and GM saw Europe as more of a small-car market than North America, because of high energy costs and narrow streets. Thus, in the 1970s, faced with a growing competitive challenge from Japan, attempts to restructure by the Big Three necessarily had to take into account the fact that their existing operations were already global and significantly decentralized.

For the Big Three, restructuring meant a growing commitment to the integration of production sytems on continental and indeed on global scales (Bloomfield 1991). This integration was heralded by the North American Auto

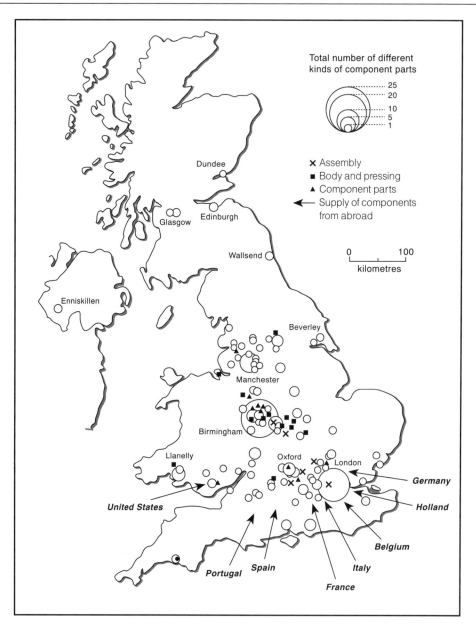

Figure 14.4 British Leyland's Factories and the Supplier Network for the Cowley Assembly Plant, 1978. Source: Miyakawa (1980: 30)

Pact of 1965, and in the 1970s underlay the so-called 'world car concept' (Bloomfield 1991: 44). Ford, in particular, sought to integrate its large North American and European operations with its more dispersed facilities in Brazil, Argentia,

Mexico, Australia and South Africa in order to produce a world car. From a marketing perspective, the world car meant a car that could be sold around the globe or at least in many parts of the globe. From a production perspective, the world

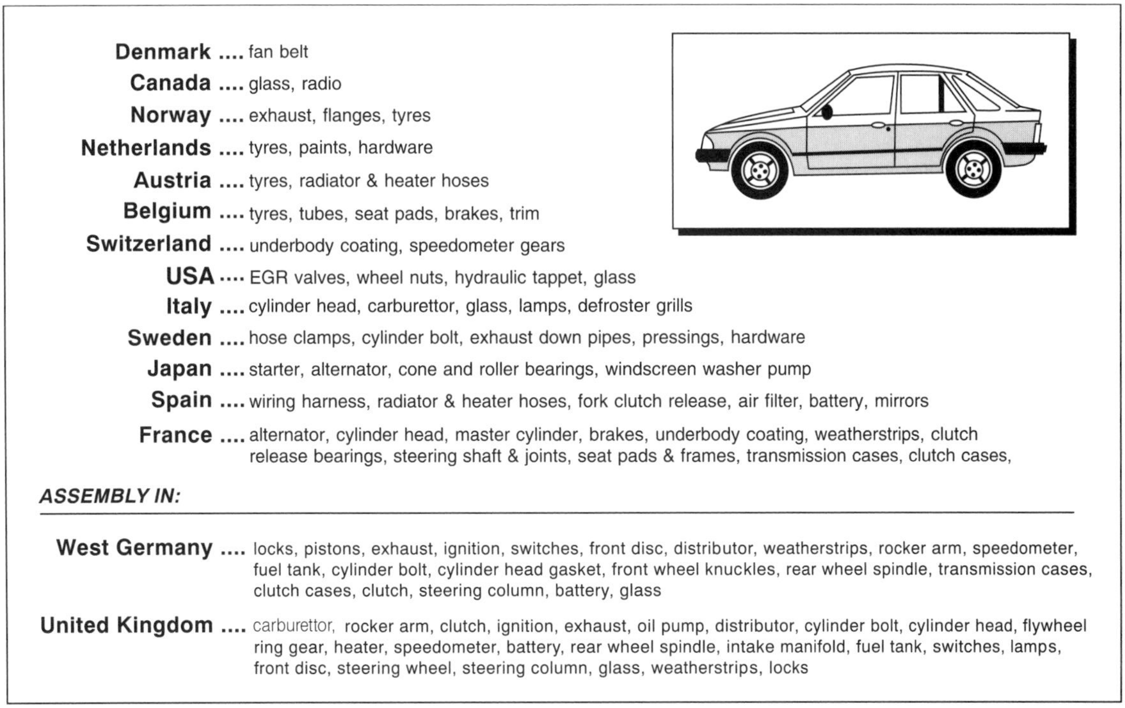

Denmark .... fan belt

Canada .... glass, radio

Norway .... exhaust, flanges, tyres

Netherlands .... tyres, paints, hardware

Austria .... tyres, radiator & heater hoses

Belgium .... tyres, tubes, seat pads, brakes, trim

Switzerland .... underbody coating, speedometer gears

USA .... EGR valves, wheel nuts, hydraulic tappet, glass

Italy .... cylinder head, carburettor, glass, lamps, defroster grills

Sweden .... hose clamps, cylinder bolt, exhaust down pipes, pressings, hardware

Japan .... starter, alternator, cone and roller bearings, windscreen washer pump

Spain .... wiring harness, radiator & heater hoses, fork clutch release, air filter, battery, mirrors

France .... alternator, cylinder head, master cylinder, brakes, underbody coating, weatherstrips, clutch release bearings, steering shaft & joints, seat pads & frames, transmission cases, clutch cases,

ASSEMBLY IN:

West Germany .... locks, pistons, exhaust, ignition, switches, front disc, distributor, weatherstrips, rocker arm, speedometer, fuel tank, cylinder bolt, cylinder head gasket, front wheel knuckles, rear wheel spindle, transmission cases, clutch cases, clutch, steering column, battery, glass

United Kingdom .... carburettor, rocker arm, clutch, ignition, exhaust, oil pump, distributor, cylinder bolt, cylinder head, flywheel ring gear, heater, speedometer, battery, rear wheel spindle, intake manifold, fuel tank, switches, lamps, front disc, steering wheel, steering column, glass, weatherstrips, locks

Figure 14.5 Ford Escort in Europe: Component Sourcing Network. Source: US Department of Transportation (1981) The US Automobile Industry, 1980, Washington DC: USGPO, 57 (see also Dicken 1992: 304)

car meant integrating research and development, servicing and producing operations to jointly design, develop and manufacture a particular car. Ford's initial attempt at a world car was the Fiesta, different parts of which were researched in the UK (the engine), Germany (drive train and brakes) and the US (interior). Assembly operations were assigned to the UK, Spain and Germany, with each plant drawing on a common set of suppliers located throughout Europe, including 45% from Ford's own branch plants (Dicken 1992: 300). Plans to build the car in Brazil were shelved, however. Ford's subsequent development of the Escort developed global integration to an even further degree (Figure 14.5). Thus, the Escort is assembled in just two locations in Europe and draws on suppliers located throughout Europe, the US, Canada and Japan. These inputs include wheel nuts and glass from the US, alternators from France, cylinder heads from Italy, mirrors from Spain,

radiators from Austria, hose clamps from Sweden and exhaust flanges from Norway. Even so, in the case of the Escort, Ford failed to produce a car manufactured on one continent and sold on others with only minor modifications; the US Escort was manufactured separately from the European Escort, and with little integration (Barnet and Cavanagh 1994: 270–271).

Ford, at least, has not given up on the world car strategy which, with its emphasis on a global supply base from affiliated operations, contrasts with the Toyota model which emphasizes a local supply base from a social division of labour. At the same time, as the Japanese companies have internationalized, their own production systems are being constantly modified. In particular, Japanese assemblers in foreign countries have relied on (long distance) inputs from established domestic suppliers, a tendency which in turn is modified by foreign investments by the suppliers themselves. In addition, Japanese firms, including Toyota, have experimented with the sourcing of supplies in low-wage locations. Indeed, for all producers, as global competition has intensified, pressures to seek out lower-cost sites remain. Lean production should not be regarded as an end state or without internal conflicts.

After lean production

The lean production system was pioneered by Japanese auto makers, especially Toyota, after the 1950s and took years to fully develop (Womack et al 1990). It continues to change (Benders 1996). At the core of lean production is still the assembly line, which not only requires fast-paced, short-rhythm work but also constant attention to quality control. Consequently, Japanese workers, as elsewhere, have experienced problems of stress and fatigue. Indeed, by the early 1990s Japanese auto makers were experiencing some problems of labour recruitment, and core companies, including Toyota, began to re-think the principles of lean production from the perspective of work organization. Thus visits were undertaken to Saturn and Uddevalla (see Chapter 12) and Toyota, for example, chose to build a new plant in Miyata, on Kyushu Island, to experiment with new ideas of work organization. The Miyata plant, it might be noted, is Toyota's first major manufacturing plant in Japan not built in its home region (Figure 14.1).

According to Benders (1996), a major theme of Toyota's plant at Miyata is to improve the quality of working life (QWL) through ergonomic changes, creating more interesting jobs and changing the reward structure. In terms of ergonomics, Toyota has reduced noise levels (e.g. by using anti-noise covers in the press shop), reduced heavy work by automation, introduced high adjustable platforms on the assembly line, and improved lighting in the assembly area. With respect to work organization, Toyota has divided the assembly line into 11 mini lines which are separated by (time) buffers of approximately five minutes to allow workers to observe the beginning and end of their part of the work in the hope

of providing for a greater sense of accomplishment. Within the mini lines, tasks have been further rearranged to enhance meaning (in contrast to established system in which tasks are fragmented with and without logical relationships between them). The *kaizen* system has also been changed. In the established system at Toyota, workers are expected to provide a minimum number of suggestions, which can be stressful and leads to trivial comments. At Miyata, there are no obligations except that the suggestions made have to be meaningful. In addition, in a general way, at Miyata Toyota has given more attention to introducing automation in a way that does not isolate workers and which can be improved, maintained and repaired on the shop floor. Finally, Toyota has changed its wage structure at Miyata so that 60% is fixed and 40% is age-based, and there is no bonus for direct productivity change (although *kaizen* still functions in this manner).

Toyota is seeking the kinds of changes in working conditions that have been achieved at Saturn and were attempted at Uddevalla for similar reasons – to ensure workers are more interested, comfortable and productive. As at Saturn and Uddevalla, Toyota has also preferred to experiment at a new location with the hope that successful behaviour can be transferred to its existing plants. The next section examines Japanese branch plant operations in North America and Europe, and raises the question of the international transfer of Japanese practices. Three prefatory points can be re-stated. First, transfer of know-how is a two-way process; for example, if Japanese auto companies are now leading innovators, they continue to observe and learn from others and to practise reverse engineering. Second, Japanese practice itself is dynamic. Third, practices in one location may not be duplicated elsewhere easily.

Japanese firms in North America and Europe

In the 30 years following 1950, the Japanese auto production system transformed itself following the destruction of World War II, to a producer of cheap trucks and cars with foreign help, to an indigenous producer of cheap cars, to a producer of reliable, high-quality small cars, to a high-value producer of a full range of cars. By the early 1970s, Japanese firms had developed technology that was transferred to the US; for example, anti-pollution equipment innovated by Honda was first sold to Toyota, and in 1973 to Ford, Chrysler and GM (Miyakawa 1991: 102). Moreover, a consistent and high-level commitment to R&D has helped enable Japanese firms to go well beyond reverse engineering and establish technological leadership, as well as organizational leadership. Indeed, it is probable that Japan's massive auto trade surpluses with North America and Europe would have been even greater in the absence of voluntary export limits and if Japanese firms had not partially replaced visible exports with direct foreign investment (Mair 1994).

Whether or not Japanese production systems can be duplicated in other economies is a matter of debate. There is a view that, in Japan, business organization, including with respect to inter-firm relations, is intimately related to cultural attitudes, notably deeply held Confucian values which underlay co-operative behaviour, respect for leadership and an unusually strong work ethic. Other differences may relate to differences in entrepreneurial attitudes, values and capabilities. For different, frankly ambiguous reasons, Williams et al (1994) claim that there is little chance of effectively transferring Japanese industrial success elsewhere, particularly that associated with Toyota whose competitive strengths are considered unique. Womack et al (1990), on the other hand, argue that lessons can be learned from Japanese models and incorporated within western economies.

Beyond a Pacific Rim context (Edgington 1990), a major wave of foreign investments by Japanese auto makers in the 1980s focused on the US (Figure 6.4) and Canada. These investments occurred as wholly owned branch plants and joint ventures with the Big Three (Table 14.6). In Europe, the Japanese auto makers invested later

and so far on a smaller scale; by the mid-1990s, Honda, Nissan and Toyota had established assembly operations in UK (Dicken 1987; Peck 1990; Jones and North 1991; Garrahan and Stewart 1992; Foley et al 1996). The motives of the Japanese companies in pursuing strategies of horizontal (and vertical) integration are primarily to expand or maintain their market share in distant markets threatened by tariff barriers, import quotas and competitive responses by local firms. For the host countries, Japanese investment promises to transfer leading-edge know-how as well as jobs. In the joint ventures, a major motive of the Big Three is to gain access to this know-how.

In the context of the transfer of Japanese production systems to North America and Europe, debate has focused especially on labour relations and subcontracting relations. With respect to labour relations, Japanese auto firms, whether in wholly owned branch plants or as part of joint ventures, have introduced at least some aspects of Japanese-style management and work practices, notably flexible work rules and broad job classifications. Mixed results have been reported as to the effectiveness and appropriateness of these practices. NUMMI, the joint venture between GM and Toyota, and managed by the latter, has performed extremely well. Toyota, it might be noted, did not re-hire any of the plant's previous managers, and employees were flown to its Takaoka plant for three weeks on-the-job training (Rubenstein 1991: 140). The plant is recognized as one of the most efficient in the US and its products have been well received. Nissan has received more criticism at its Smyrna plant in Tennessee and its plant in Washington, near Sunderland, where job flexibility is judged more as job intensification than functional flexibility (Garrahan and Stewart 1992).

Local subcontracting

From the perspective of local development, subcontracting linkages are a critical issue since they substantially define the extent of quantity and quality multiplier processes. In addition to

Table 14.6 Japanese Transplant Production Facilities in North America 1989

Firm	Location	1989 Production	Announced capacity
Assembly Plants			
Honda	Marysville, Ohio	351 670	360 000
	East Liberty, Ohio		150 000
	Alliston, Ontario	86 447	100 000
NUMMI[a]	Fremont, California	192 235	340 000
Toyota	Georgetown, Kentucky	151 150	240 000
	Cambridge, Ontario	20 859	50 000
Nissan	Smyrna, Tennessee	238 640	480 000
Mazda	Flat Rock, Michigan	216 200	240 000
Diamond Star[b]	Bloomington, Illinois	91 839	240 000
CAMI[c]	Ingersoll, Ontario		200 000
SIA[d]	Lafayette, Indiana		120 000
Assembly total		1 349 000	2 520 000
Engine plants			
Honda	Anna, Ohio		
Nissan	Smyrna, Tennessee		
Toyota	Georgetown, Kentucky		

[a]Commenced operations in 1989.
[b]General Motors/Toyota joint venture.
[c]Chrysler/Mitsubishi joint venture.
[d]Subaru/Isuzu joint venture.
Source: Womack et al (1990: 202).

the extent of local subcontracting, important questions relate to whether or not Japanese firms operating in European and North American contexts can create efficient supplier networks based on JIT principles and with a mutual commitment to enhancing quality as implied by the relation-specific skill. In practice, the local content debate has been controversial in both the US and Europe, although the latter's policy response has been stronger (Jones and North 1991).

In the US case, as of 1987, Japanese branch plants and joint ventures relied on parts from US sources for about 50–60% of their needs, or at least anticipated this level of sourcing from within the US (Table 14.7). In practice, the level of US sourcing of parts for these plants has increased since their start-up. A large share of US-source parts to Japanese branch plants or joint ventures is provided by Japanese first-tier suppliers who have followed their core company to the US. The number of Japanese suppliers increased extremely rapidly in the US (and Canada) during the 1980s,

and most of their output is destined for the Japanese core companies. Japanese firms obtain some of their parts from US suppliers but these inputs are typically bulky, low-value products, such as carpets, glass and tyres (Rubenstein 1991: 126). Japanese suppliers in the US or Japan still retain the high-value components, such as suspension systems, brakes and engines. Compared to the US-owned assemblers, the Japanese transplants rely on a significantly higher level of imported parts.

According to Rubenstein (1991: 125), Japanese firms in the US are sensitive to increasing domestic content, and Honda, for example, announced in 1989 its intention of sourcing 75% of its parts from US sources. Subsidies from state and local governments and the rising value of the Yen have encouraged such a trend. By the early 1990s, however, the Japanese production systems had not duplicated the same level of localization as that found in Japan. Morris (1989) notes, on the basis of Canadian experience, that JIT systems in Ontario are based on one-per-day

Table 14.7   Domestic Content at US Assembly Plants, 1987

	Percentage of parts produced in US
US-owned assembly plants	
Chrysler	91–88[a]
Ford	86–99[b]
GM	94–99[c]
Japanese-owned assembly plants	
Honda	60
Mazda	50
Nissan	
Cars	63
Trucks	56
Subaru/Isuzu	50[b]
Toyota	60[b]
Joint ventures	
NUMMI	
General Motors	60
Toyota	50
Diamond-Star (Chrysler/Mitsubishi)	55[b]

[a]Varies by model.
[b]Anticipated.
Source: Rubenstein (1991: 127).

deliveries (rather than multiple daily trips) and that the scale of operations is not yet sufficient to justify dense localized networks. Within North America there is also the tension that exists between the advantages of concentrating supplier networks and the advantages of a low-cost labour location in Mexico. Japanese firms in electronics and autos have already invested in maquiladoras (Kenney and Florida 1994). Mazda, as part of its arrangement with Ford, for example, produces a car, the Tracer, in Hermosillo, Mexico, and it purchases components from Mazda affiliates in South Korea as well as Japan (Rubinstein 1991: 123). In other words, if Japanese transplants are seeking to duplicate supplier networks in North America along Japanese lines, there are also signs of a willingness to disperse supply sources along North American lines.

Nissan's supplier network in the UK   Nissan's plant in Washington, near Sunderland, opened in July 1986 as the first of the Japanese European plants, with a direct employment of 1300, which

has since expanded to 3500. Since then, within the UK, Honda has established an assembly plant in Swindon, and Toyota at Burnaston in Derbyshire. However, it was Nissan's plant in Sunderland that sparked the European debate about the subcontracting practices of Japanese firms especially in relation to local content (Peck 1990). Local content is defined by the EC as the "ex-works price of a car less than the value of imports from outside the European community" (Peck 1990: 354). At start-up, Nissan's plant was a knock-down assembly plant with limited local content. By 1988, Nissan had achieved 60% local content when the first exports to the continent occurred. Indeed, Nissan saw the UK primarily as a staging point for the EC as a whole (Jones and North 1991). The French and Italian governments, however, concerned with protecting national producers, refused to allow tariff-free imports from the Sunderland plant unless local content was 80%, a level Nissan reached in 1991 (Peck 1990). In practice, Nissan's interest in creating a JIT system also reinforced its commitment to local content.

The supplier network Nissan has developed in the EC is relatively dispersed and draws from three European countries and Japan, in addition to the UK (Figure 14.6). However, while the north-east region of England did not have an established supplier network, nine of Nissan's first-tier suppliers have established their own branch plants adjacent to the assembly plant and on land owned by Nissan. These suppliers are high-quality manufacturers which perform their own R&D, largely oriented to Nissan's needs. So, Nissan has achieved very close proximity to many of its high-value components. It still imports important components from its supplier base in Japan. Lower-value components are sourced from West Germany, Belgium and Italy, and from a variety of suppliers (70 by 1989) in the UK. The latter are mainly found in the traditional auto-producing regions of the West Midlands and the South-east but are located in several other areas as well. Each potential supplier is checked by Nissan's purchasing department and then by its engineering group

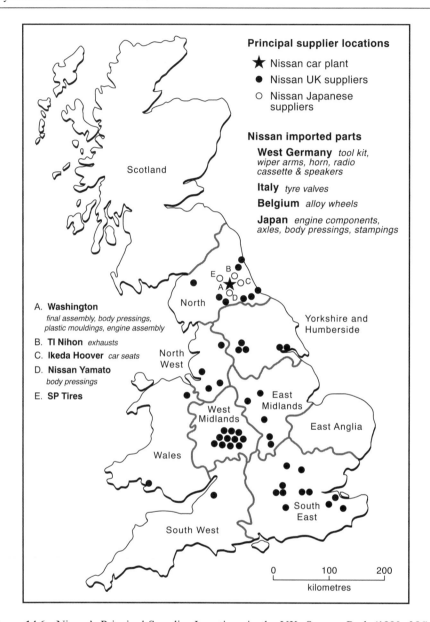

Figure 14.6   Nissan's Principal Supplier Locations in the UK. Source: Peck (1990: 356)

to ensure appropriate quality and design capabilities and potentials. If these checks reach mutual approval, prices and delivery schedules are arranged. It might also be noted that Nissan's supplier network for its Sunderland plant, as of 1989, did not include low-wage locations in developing countries.

In both North America and the UK, investments by Japanese companies have exerted profound direct effects on the geography and organization of the auto industry. The strength of the Japanese challenge has also encouraged adaptation by North American and European firms.

Adaptations by North American firms

Over the past decade or so, US auto producers have been forced to respond to the competitive threat posed by the Japanese. In terms of corporate strategy, the integration of production systems on a continental and global basis, including through the world car concept, are attempts to promote competitive advantage by promoting efficiency. The Big Three have also sought to learn directly about the nature of Japanese production systems by entering into joint ventures and strategic alliances (Table 14.6). The joint ventures in the US have generally been based on the assumption that while costs would be shared and US partners would facilitate entry of Japanese partners in the US, the latter would provide the management and organization, including many parts from Japan, in order to manufacture higher-quality, more-fuel-efficient cars. The Ford–Madza partnership is of a more comprehensive nature and stems from Ford's (friendly) acquisition of a quarter share of Mazda in 1979. This 'strategic alliance' was important to Mazda not simply to help access North American markets but as a source of finance, while Ford got control over the production of profitable cars at a time when it was experiencing great difficulty in North America. Subsequently, Mazda helped Ford design the Escort and re-started manufacturing at Ford's idle Flat Rock, Michigan, factory. The Escort was both successful (many were sold) and unprofitable. But according to Barnet and Cavanagh (1994: 273–274), Ford has learnt much from this project and alliance with Mazda: "Thanks to Mazda's strict scheduling, the new Escort has about 60 percent fewer last minute design changes . . . executives learned to pay more attention to what the workers thought about improving production. Top management learnt that up-front investment to retrain workers, the organization of a 'downsized' work force into 'quality of work' teams, and the modernization of plant and machinery would pay off in future profits". In addition, Ford has greatly increased the number of robots it uses, while many old plants have been closed and jobs lost.

Indeed, among the Big Three a substantial effort has been made to learn Japanese practices and adapt at least some of them to North American conditions, notably with respect to work practices (Holmes 1991), reconfiguration of the R&D–production–market interface, and to restructuring supplier relations. With respect to the last of these at least, Asanuma (1985a,b) sees no reason why US firms should not be successful. He argues that the reliability of Japanese suppliers in maintaining delivery schedules and quality requirements is not based on some distinctive "general moral attitude" of the Japanese. Rather, he notes the economic incentives encouraging suppliers to be reliable. These incentives begin with the 'basic contract' which allows core firms to claim financial compensation if losses occur owing to a delivery delay from the supplier. Likewise, core firms can obtain compensation if, within a specified period after receipt, defective parts are found in the plant of the core firm or after shipment of the product to the market.

In addition, in the US, until recently, contractual obligations have meant that once a part has been purchased by a core firm, then that part becomes the responsibility of the core firm as regards any malfunction. Such a tradition reflects the tendency, to borrow Asanuma's terms, for parts suppliers in the US to be DS suppliers. That is, in the US, the core firm has frequently provided blueprints, specified materials, specified manufacturing methods and the ways to test performance. As such, it is natural for core firms to assume responsibility for product liability. In Japan, supplier responsibility extends to the final consumer market so that a defective part is the joint responsibility of core firm and supplier. Again, this tradition is an incentive to suppliers to be quality conscious and is potentially a practice that can be adopted elsewhere. Similarly, in the US the length of 'contract' has traditionally been a lot shorter than in Japan and, with exceptions, typically for just one year. Consequently, among US automakers there is a tradition of many contract terminations, which reflects a continuing emphasis on finding the cheapest sources even

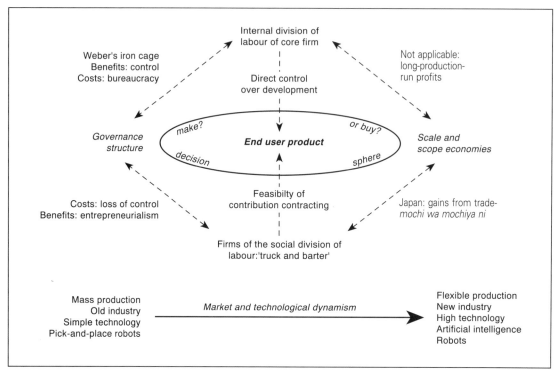

Figure 14.7   From Production Systems to Learning Systems. Source: Patchell (1993a: 812)

when parts are unchanged. Such uncertainty, however, constitutes a disincentive to undertake R&D. It is also a practice that can be changed.

US firms are attempting to adapt the key features of Japanese production systems, notably by the development of just-in-time systems, hierarchies of suppliers, and longer and closer relationships with suppliers, including with respect to stronger technological liaisons. US firms have also increased the priority assigned to quality, they are purchasing more modular components from suppliers, and they are reducing inventories (Rubenstein 1991: 128–129). In addition, US firms have begun to reduce their reliance on affiliated companies in favour of the social division of labour, and in some cases US firms have signed contracts with Japanese suppliers who have located in the US. At the same time, US firms have been active in establishing assembly and parts supplies in Mexico (Glasmeier and McCluskey 1987).

Conclusion: Japan and flexible specialization

Japan, and Japanese industry, has been at the forefront of the transition from the fordist to the IC techno-economic paradigm (Table 2.3). By the 1980s, Japan was providing global benchmarks across a wide cross-section of industries, its competitive strengths being based on principles of flexible specialization. Moreover, while Piore and Sabel (1984) conceived of flexible specialization in the context of SMEs, in Japan flexible specialization has been developed in large-scale industries (including the auto industry), organized by core firms. At the core of flexibly specialized industries is an insistent emphasis on the social division of labour. As Patchell (1993a: 813) notes, the revelation of the social division of labour is "to conceive of the evolution from production systems to learning systems" (Figure 14.7). Historically, the strategy of 'reverse engineering' placed the commitment to learning at the centre of Japan's approach to industrialization, while within-firm learning underlies *kaizen* (continuous improvement), quality circles, on-the-job training, polyvalence, and loopy R&D. Among firms, the relation-specific skill defines a process of mutual learning.

All production systems involve debates over 'make or buy', or the employment of either an internal or social division of labour (Figure 14.7). As Patchell (1993a) notes, the relative importance of scale and scope economies provide some guidance as to the resolution of this choice. Unfortunately, in practice, decisions are made under uncertainty which contributes to the arbitrariness of make/buy choices; there are, for example, no mechanical rules for comparing the advantages of control over production, permitted by the internal division of labour, with the creative potential of the social division of labour. Also, "skill differences among firms can cloud this issue substantially" (Patchell 1993a: 813). In the West, the prevailing (fordist) philosophy has placed emphasis on the internal division of labour and cost reduction. In Japan, emphasis is placed on the social division of labour and value maximization. The relation-specific skill reflects the ingenuity of Japanese firms to exploit the potential of the social division of labour more fully, without sacrificing production stability or market power. These benefits are particularly acute in situations characterized by market and technological dynamism.

Japan's pursuit of flexible specialization reflects what Marshall and Tucker (1992) label the high performance model (Figure 11.4). It is a strategy that puts priority on increasing wage levels through innovation, and the fine balancing of the forces of co-operation and competition evident in the relation-specific skill. Whether or not Japan will be able to retain its present highly coherent production systems or whether other economies can and should seek duplication remains to be seen. The social division of labour in Japan is dynamic and is subject to the intensifying forces of globalization. A degree of 'hollowing out' of the Japanese economy has occurred (Edgington 1995). At the same time, the integrity of core industries within Japan, such as the auto industry, remain intact. Toyota City still provides powerful testimony to the potential of geographic concentration.

15

PRODUCTION SYSTEMS IN HOME AND HOST ECONOMIES

There is considerable controversy regarding the role of direct foreign investment (DFI) on the nature of local production systems and in local development. Traditionally, this controversy has been debated mainly within the context of host ('branch plant') economies where, in general, the debate centres on long-term structural effects (Watts 1981). Proponents argue that foreign firms act as catalysts for desirable changes in local economic structures, via quantitative and qualitative multiplier processes (Safarian 1966). The critics suggest that in the long term, DFI, especially when associated with high levels of foreign control of industry, 'truncates' (undermines) desired structural changes (Firn 1975; Britton and Gilmour 1978; Britton 1980, 1994; Hayter 1982). Historically, this controversy is underlined by the radically different approaches adopted by different countries regarding their openness to DFI. Within the last two decades, a debate has also emerged regarding the impacts of DFI on home ('metropolitan') economies. Traditionally, these impacts have been judged to be positive, but in association with the new international division of labour (see Chapter 8) DFI has been interpreted as the cause of the 'hollowing out' and de-industrialization of core industrial regions within donor economies (see Chapter 16).

This chapter addresses these two debates but, reflecting the bias in the literature, gives particular attention to the controversy over truncation in host economies. The chapter is in four parts. First, home economies are identified and the impacts of DFI on their economic structures are briefly reviewed. Second, host economies are defined and the arguments about the catalytic/truncating role of DFI are summarized. The third part focuses on Canada, an industrialized country which has experienced unusually high and sustained levels of DFI. A final section notes some policy options for host economies in dealing with MNCs.

While the chapter is primarily pitched at a national scale, home and host economy characteristics vary within countries and regions. As Watts (1981) emphasizes with respect to host or branch plant economies, for example, at regional or local scales, 'outside' sources of investment include other domestic regions as well as foreign sources, so that for individual regions external control is typically higher than foreign control (Firn 1975). Even in countries such as Sweden, where the foreign ownership of industry is limited, during the 20th century peripheral regions have become increasingly dominated by branch plants controlled by corporations whose head-offices are located in core regions

(Fredriksson and Lindmark 1979). In branch plant dominated economies, whether defined in terms of external or foreign control, long-term growth becomes largely a question of the allocation of investment among a set of competing and geographically dispersed branch plants by corporate decision-makers located in distant head-offices. Indeed, foreign firms are of concern to host economies because of their distinctive structural characteristics related to size and entry advantages, and because of the implications for national economic and political sovereignty, i.e. for the aspiration and abilities of governments to resolve economic and social problems.

It might be noted that, apart from different ideological portrayals of MNCs as 'efficient beauty' or 'exploitative beast' (Dicken 1992), debates about DFI in home and host economies are controversial because of the underlying *counterfactual* nature of the arguments. Thus, the arguments of both proponents and critics of DFI ultimately imply that if DFI had not occurred, alternative (less or more preferable) production systems *would* have developed. Since we can never know for sure what would have happened, the interpretation of 'actual' evidence involves judgement and values. While many studies of the local impacts of DFI have favoured one approach to research designs in which this counterfactual issue is left implicit, other approaches are possible (Hayter 1982).

Home (donor) economies

Home (donor) economies, also known as metropolitan powers, are the nations, regions and cities that are the head-office homes for MNCs which make decisions about the allocation of investment around the globe. Indeed, international firms have long provided symbols of national industrial prowess and the defining characteristics of the economic basis for the concept of 'metropolitan power', whether metropolis is interpreted in terms of nations (Innis 1930: 385; Lower 1973) or cities (Hall 1966; Friedmann

1986; Wheeler 1987; Sassen 1991). That is, in an economic sense, metropolitan status and power is fundamentally defined by the global reach of head-offices and related control functions. Decisions made within metropoles, especially with regard to the nature, timing and location of investment and technological change, shape economic livelihoods elsewhere. In contrast, the economic fortunes of branch plant economies or hinterlands depend on decision-making and technology based elsewhere. Home economies and metropoles are therefore 'controlling economies' and the most important are the location of the head-offices of the largest MNCs which exercise the greatest global reach. In this context, control refers to autonomy over corporate strategies and structures (see Chapter 11). Host economies may be thought as the mirror images of home economies; they are dominated by branch plant operations which are controlled by head-offices based in other jurisdictions, i.e. home economies. In practice, economies may take on the characteristics of primarily home or host economies, or some mixture of the two.

Home economies are not distributed randomly or evenly but reveal high levels of concentration, within nations (Wheeler 1987, 1988) as well as globally (Hymer 1960). Home economies are most simply defined by reference to the 'origins', really the locus of decision-making control, of DFI. Historically, on a global scale, the most significant home economies are well known and relatively few (Table 11.2). From the late 19th century until the 1970s, MNCs based in the US and UK provided the lion's share of DFI, although some other industrialized European countries are noteworthy. As noted in Chapter 11, for example, Dunning (1993) estimates that in 1914, MNCs based in the UK and US controlled almost two-thirds of outwards DFI, and that France and Germany accounted for over 22% (Table 11.2). In 1938, the UK and US remained dominant with the latter's importance rapidly increasing. In the 1950s, the expansion of US MNCs was particularly spectacular and, according to Dunning, by 1960 US giants accounted for over 49% of the world's total

DFI. The UK remained in second place and MNCs based in the Netherlands had also risen to prominence. Since 1960, the dominance of US-based MNCs has declined, while those based in several European countries (notably Germany) and in Japan have rapidly become more important. Even so, Dunning estimates that in 1988 the US accounted for over 30% of world DFI, almost two times greater than the next most important country, which was still the UK. Germany's role has been steadily increasing since the late 1950s, while Japan's rise is even more recent, particularly since the late 1960s when the Japanese government lifted controls on outwards DFI. Canadian, French, Dutch and Swiss MNCs also provide noteworthy levels of DFI. Controlling sources of DFI among developing countries has been increasing but remains a tiny fraction of global totals.

Dunning's figures (Table 11.2) are broadly confirmed by the UN's estimates of the distribution of the sources of DFI between 1960 and 1985 (Table 15.1). According to the latter, in

1985 the US and the UK accounted for about half of global DFI, while the former indicates that in 1988 these two countries account for about 47% of DFI. Clearly, the sourcing of DFI has become more diverse; many countries, including developing countries, have firms that control DFI and this trend is likely to continue. Nevertheless, the principal home economies of the ICT techno-economic paradigm are the US, several west European countries and Japan. While in each of these regions there has been inwards DFI, in each case outward DFI exceeds inwards DFI (Table 15.2).

In the US case, its predominant role as supplier of DFI, and as the world's principal home economy in the second half of the 20th century, has been balanced by increasingly substantial inwards flows of DFI in recent years. Indeed, between 1975 and 1983, the ratio of outwards to inwards DFI decreased considerably, and in 1988, inwards DFI actually exceeded outwards DFI. However, domestic control of the US economy is at high levels so that the US clearly remains a home economy. Although there are national variations and trends vary, in the 1980s the major industrial powers in Europe had ratios of outwards to inwards DFI in excess of 1.0. Both Australia and Canada are noteworthy by their much greater relative reliance on inwards DFI compared to sourcing outwards DFI. Japan's outwards DFI, on the other hand, far outstripped inwards DFI during the 1970s and

Table 15.1 Foreign Direct Investment in the World Economy: The Changing Relative Importance of Leading Source Nations, 1960–1985

Country of origin	Percentage of world total outward direct investment		
	1960	1975	1985
US	47.1	44.0	35.1
UK	18.3	13.1	14.7
Japan	0.7	5.7	11.7
West Germany	1.2	6.5	8.4
Switzerland	3.4	8.0	6.4
Netherlands	10.3	7.1	6.1
Canada	3.7	3.7	5.1
France	6.1	3.8	3.0
Italy	1.6	1.2	1.7
Sweden	0.6	1.7	1.3
Developed market economies	99.0	97.7	97.2
Developing market economies	1.0	2.3	2.7
World total	100.0	100.0	100.0

Source: based on UNCTC (1988: Table 1.2).

Table 15.2 The Changing Balance of Inward and Outward Direct Investment

	Foreign direct investment ratios[a]	
	1975	1983
Western Europe	1.20	1.56
United States	4.48	1.66
Japan	10.65	12.32

[a]FDI ratio = $\dfrac{\text{Outward stock}}{\text{Inward stock}}$

Source: Dicken (1992: 55).

1980s, and its economy remains overwhelmingly under domestic control. By the 1980s at least, Japan constituted a classic home economy in a way that the US had done for most of the 20th century, and the UK had done until World War II.

The impacts of (outwards) DFI on the structure of home economies

Traditionally, DFI was unequivocally seen as a source of metropolitan wealth, stability, diversity and control. Moreover, this control was typically seen as exercised in the interests of the metropole itself. Recently, this view has been complicated by the new international division of labour thesis (Fröbel et al 1980) which suggests that MNCs are closing down factories in home economies and relocating them to lower-wage hinterlands. Thus, the impacts of contemporary patterns of DFI on home economies in terms of economic structure, trade balance and employment opportunities is potentially complicated, imposing both benefits and costs. To assess these impacts, Hawkins (1972) distinguishes several types of impacts: home-office effects, supporting-firm effects, export-stimulating effects and production-displacement effects. The first three of these effects contribute positively to home economies and the production-displacement effect negatively.

The home-office effect

The home-office effect is the employment and income generated in the principal control functions, notably head-offices and R&D centres, of MNCs in return for services provided to branch plants. Indeed, an immediate corollary of DFI is the expansion of control functions (and jobs) in home economies, most obviously manifested by increasingly large head-offices and research and development (R&D) laboratories. Similarly, direct foreign disinvestment results in head-office downsizing (Wheeler 1987). Overwhelmingly, but not completely, MNCs have concentrated their head-offices in home economies and within these home economies in relatively few metropolitan areas. Thus, globally, major concentrations of head-offices are located in the US, Europe and Japan. According

to Friedmann (1986), for example, among OECD countries, the 'primary' metropolitan areas are New York, Chicago, Los Angeles, London, Paris, Rotterdam, Frankfurt, Zurich and Tokyo. Within these countries there are 'secondary' centres such as Miami, Houston, San Francisco, Toronto, Brussels, Milan, Vienna and Sydney, while Friedman (tentatively) labels Singapore and Hong Kong as examples of primary and secondary cities in semi-peripheral countries. Lower down the urban hierarchy, the head-offices of parent firms that are international in scope tend to be smaller and fewer and more of the head-office functions tend to be subsidiary head-offices controlled by parent companies in the higher-tier centres. The autonomy of subsidiary companies, including with respect to investment planning and regional spheres of influence, is accordingly constrained according to the mandates set by parent companies (see Chapter 11).

Corporate R&D facilities, although relatively more dispersed than head-offices, are geographically concentrated among industrialized countries and within industrial countries in particular regions and cities (Malecki 1979, 1991; Healey and Watts 1987). Globally, corporate R&D is overwhelmingly concentrated among rich countries, especially the OECD countries (Malecki 1991: 136–143). As Malecki (1991: 136) points out, poor countries simply cannot afford more than a limited amount of research. Using scientists and engineers as an index of research activity, the absolute and relative (per capita) disparities between developed and developing countries are huge (Table 15.3). Among the OECD countries, the US, Japan and Germany are the three largest in terms of GNP and the largest in terms of R&D investments. In terms of gross domestic expenditure on R&D (GERD) as a percentage of GNP, in 1990 the US spent 2.77%, Germany 2.73% and Japan 2.88% (Hayter 1996). These figures include military R&D. If limited to civilian R&D, the respective GERDs are about 1.75%, 2.40% and 2.86%, so that in absolute terms civilian R&D in the US and Japan is about the same.

Table 15.3 Scientists and Engineers: Global Distribution 1985

	Scientists and engineers	
	Number (000s)	Engaged in R&D (1980 per million population)
Africa	1 623	91
Asia	32 670	272
Latin America	4 746	252
Europe	37 369	1 732
North America	33 247	2 678
Oceania	1 105	1 483
Developed countries	81 247	2 984
Developing countries	29 513	127

Source: Unesco (1988: Tables 5.1, 5.2) and Malecki (1991: 138).

Overwhelmingly, industry R&D has been dominated by international firms who have largely concentrated their R&D centres in home economies (Mansfield et al 1979). Such a geographic concentration of R&D in home economies allows for greater security over R&D processes, facilitates communication with head-offices, provides a mechanism to integrate and control dispersed corporate facilities, and reflects corporate desires to access related research institutions and pools of educated labour. To use the terminology of Howells and Wood (1992), most R&D is either 'home-market' based, i.e. exclusively concentrated in home economies, or 'host-country' based, whereby firms still concentrate their main R&D efforts in home economies but establish smaller-scale programmes in host countries to adapt technology created in parent laboratories for local markets and conditions (Britton and Gilmour 1978; Thwaites 1978). Howells and Wood also recognize 'world' based R&D in which MNCs locate R&D in several countries to supply technology to international markets. For example, IBM performs its long-term basic R&D in Zurich as well as Ruschlikon, New York (its headquarters location) and Silicon Valley (Kelly and Keeble 1990). Other examples of world-based R&D are the German-based MNCs that have relocated

genetic R&D to the US to escape strict German regulations, as well as to access US scientific expertise (Blau 1994).

There is evidence of a modest trend towards the internationalization of R&D. Howells and Wood (1992: 23) note that US-based MNCs conducted just over 9% of their R&D efforts in foreign countries in 1990 (a slight increase compared to the 1960s), while foreign-owned R&D accounted for over 11% of industry R&D in the US in 1988 (up from almost 5% in 1977). The latter trend notably includes Japanese MNCs in the electronic and automobile industries and German companies in the pharmaceutical industry, although firms from a variety of countries are represented (Angel and Savage 1996; Florida and Kenney 1994). Indeed, most foreign R&D occurs among the major home economies themselves and in most cases remains focused on the adaptation of products for local markets (Howells and Wood 1992; Pearce and Singh 1992).

Within countries, control functions are geographically concentrated, for the most part in metropolitan cities. In this context, the US case has been particularly well documented over the years with respect to both head-offices (Ullman 1958; Pred 1974; Borchert 1978) and R&D activities (Malecki 1979, 1980; see Figures 8.9 and 8.10). Moreover, if the geographic concentration of control activities is an important feature of the US economy, in other OECD countries concentration is even greater. Thus, London in the UK, Paris in France, Rotterdam in the Netherlands, Tokyo in Japan, and Toronto in Canada are control centres with no close national rival. The local impacts of such concentrations should not be underestimated. In employment terms, corporate head-offices and R&D laboratories can be very large. In the 1980s, for example, GM had a number of R&D centres, several of which employed over 1000 scientists and engineers, while IBM's basic R&D centre in New York employed over 4000 professionals.

The head-office jobs of these firms also numbered in the thousands. These jobs are typically high income and while the recessions of the early 1980s and 1990s, essentially for the first time, witnessed permanent lay-offs of managers and R&D professionals, within restructured corporations, these jobs are once again perceived as high income and stable. In addition to direct impacts, head-offices and R&D centres are deeply embedded in networks of information exchange with a variety of service-sector organizations (Wheeler 1986).

The supporting firm effect   Supporting firm effects are the jobs and sales created by firms and organizations in the home economy as a result of supplying services and information to head-offices and R&D centres in support of foreign-based branch plants. Head-offices and R&D centres purchase many service inputs from the private and public sectors, including universities, in home economies which are stimulated by the needs generated by DFI. The most significant are from so-called producer services which are intermediate-demand inputs to the production of goods or other services and which primarily comprise high-order office activities such as computing, legal, accounting, design, advertising, architectural, consulting and engineering services (Daniels 1987; Marshall et al 1988; Beyers 1989). Indeed, producer services are the most rapidly growing sector in most advanced economies in terms of employment, and they offer considerable export potential (Beyers and Alvine 1985; Coffee and Polèse 1987, 1989; Coffee 1996).

Geographically, high-level producer services are closely tied to the head-offices they serve (Semple 1985; Coffee 1996: 347). The supply of producer services to head-offices primarily depends on close, frequent personal contacts, which are facilitated by agglomeration. Concentration in metropolitan areas also provides access to skilled sources of labour and to complementary services (Coffee and Polèse 1987). In theory, rapid developments in telecommunications threaten to sever the close spatial links between head-offices and producer services. In practice, face-to-face contact is deemed to allow forms of information exchange not possible by long-distance communication so that geographic

concentration is likely to remain important. Similarly, liaisons between R&D centres and private and public sector organizations are advantaged by proximity.

Export stimulus and production displacement effects   The export stimulus effect involves the creation of jobs in resource and manufacturing activities which would not have occurred in the absence of DFI. This effect includes employment and sales associated with component parts manufactured by affiliated plants or by independent suppliers in the home economy for export to branch plants, and the machinery (capital goods) that is purchased from suppliers in the home economy to establish branch plant operations. Over time, DFI is also a source of learning and market know-how for parent companies and suppliers in the home economy, thus widening opportunities for export.

On the other hand, the production-displacement effect occurs when DFI replaces visible exports and may even increase visible imports if branch plants export back to the home economy. Nakagjo (1980), writing from a Japanese perspective, disaggregates the export stimulus and production displacement effects on visible exports and imports (Figure 15.1). According to this scheme, the home economy benefits from export-inducing, export-opportunity and import-opportunity loss effects and losses from export-substitution and reverse-import-promoting effects. Since the onset of de-industrialization in Europe and North America in the 1970s, a growing number of studies conclude or at least imply that production-displacement effects by MNCs based in these regions far outweigh export-stimulus effects (Frank and Freeman 1978; Owens 1980; Bluestone and Harrison 1982; Taylor and Thrift 1981; Donaghu and Barff 1990). These studies express particular concern for the loss of blue collar jobs due to relocation through the mechanism of DFI. Owens (1980), for example, noted that about 40% of the factories established in Eire by UK firms between 1958 and 1978 contributed directly to job loss in the UK. The implications for

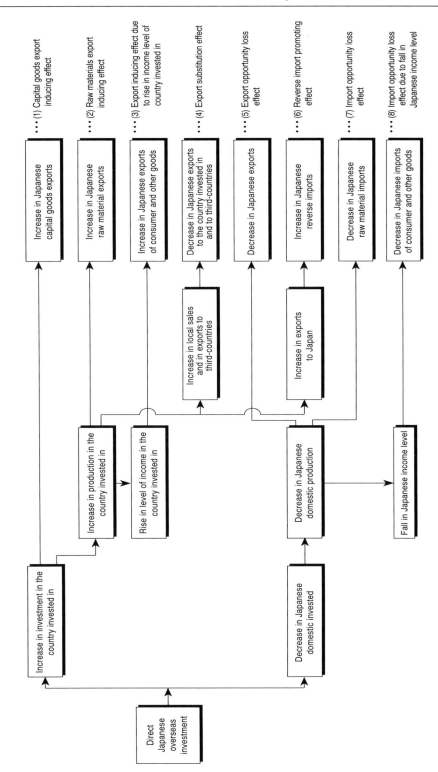

Figure 15.1 A Japanese Assessment of the Effect of Direct Foreign Investment on Visible Imports and Exports in the Donor Economy. Source: Nakagio (1980: 472)

component and machinery suppliers, as well as for head-office and related services, however, are not assessed.

Empirically, it is not easy to estimate 'aggregate' home-office, supporting-firm, export-stimulus and production-displacement effects. These effects are direct and indirect, they involve goods and services within and outside the firm, and they need to be traced across international boundaries. The implications of market learning and access are particularly difficult to quantify and it needs to be recognized that firms may lose exports (or resources) if DFI does not occur. Employment levels and skills are also profoundly affected by technological change. New technology may be seen as both displacing jobs, maintaining the profitability of surviving jobs, and creating new jobs within the firm and for suppliers. Within particular corporations and time periods, all these effects may occur simultaneously. GM, for example, has closed down auto plants in the US Midwest where it has significantly reduced jobs, but has opened new factories in Mexico. At the same time, its home office and R&D facilities in the midwest remain massive, it has extensively modernized facilities in the US, built new facilities, such as the Saturn plant, and sought to extend its subcontracting system in the Midwest. Philips, the Dutch giant, is another case in point (Figure 8.5). During the 1980s, its employment did decline in the Netherlands relative to overseas where it has built new factories, including in low-wage locations (de Smidt 1990). Yet Philips has also extensively modernized its Dutch base, especially its core region where R&D and innovative manufacturing processes are concentrated.

In the past, several studies which sought to take into account all the various effects of DFI on home economies concluded that these effects are predominantly favourable. Nakagjo's (1980) assessment with respect to Japan, Britton and Gilmour's (1978) review of American studies, and Morgan's (1979) analysis of British experience all emphasize the job-enhancing, export stimulus and head-office effects and the supporting-firm export effect. In terms of the latter, Nakagjo gives particular emphasis to exports of Japanese-made machinery to Japanese branch plants, while Britton and Gilmour and Morgan stress the benefits derived from invisible exports of business services. Possibly, contemporary 'aggregate' assessments of DFI impacts on home economies would be more equivocal.

Global cities as the apogee of home economies

Geographically, control functions and supporting activities are highly concentrated in major metropolitan areas where the advantages of localization and urbanization economies of scale are strong. According to Sassen (1991), while other cities are globally important (Hall 1966; Friedmann 1986), New York, London and Tokyo are the world's most important global cities from which firms in a variety of sectors organize the world economy (see also Wheeler and Mitchelson 1989; Mitchelson and Wheeler 1994). New York, London and Tokyo are the metropoles which historically have most fully expressed the structure of home economies and are above all centres where the control functions of leading industrial, financial, trading and business service corporations, especially MNCs, are located. As Mitchelson and Wheeler (1994: 87) note, the world cities of New York, London and Tokyo, to which Soja (1991) would add Los Angeles, are places which "provide information that other places need. And through the economies of localization, they sustain specialties that are demanded across continents if not the globe." According to Soja, the four metropoles of New York, London, Tokyo and Los Angeles are the dominant nodes which are strategically located to ensure round-the-clock servicing of the financial and informational needs of increasingly globally integrated MNCs. As MNCs have continued to globalize their operations during the IC techno-economic paradigm, however, Mitchelson and Wheeler (1994) argue that New York has strengthened its leading position, at least in the US and probably globally, as a command and control centre of the information on which MNCs rely. For them, Los Angeles plays a more localized role.

The global reach of these world metropoles is ultimately rooted in the external economies of scale and scope provided by dense networks of information and goods exchange within a diversity of industrial and business service districts. Thus, within each of these cities, agglomerations of the head-offices of industrial MNCs and of financial districts, comprising banks, insurance companies and stock exchanges, are supplemented by concentrations of wide-ranging producer services, wholesaling and distribution activities, speciality retail centres, R&D activities and a plethora of manufacturing activities serving domestic and export markets. These metropoles are extremely dynamic, however, and if their position as command and control centres appears to be strengthening, the survival of industrial districts is more problematical. In the cases of New York and London, de-industrialization has occurred and unemployment is a significant problem. In Tokyo, during the 1980s, rapidly escalating land costs, the increasing value of the Yen, congestion and the impacts of globalization have also created problems for manufacturing industry.

Tokyo's evolving industrial districts From the Meiji Restoration of the 1860s, Tokyo emerged as Japan's dominant industrial centre in terms of level of output and technological capability. By the 1970s, a wide range of industrial districts had evolved, largely involving a social division of labour. Within Tokyo's 'inner zone', the southern district was occupied by auto and electronic assembly and suppliers; the central district was dominated by a highly differentiated printing and publishing activities; and the eastern district was occupied by a diversity of consumer goods industries manufacturing over 50 different types of products, including dolls, shoes, knit wear, stockings, binoculars, toys, neckties, handbags, pencils and many foodstuffs and beverages (Ide and Takeuchi 1980; Takeuchi 1994: 20). In addition, the northern part of the inner zone housed a mixed range of industries, and examples of heavy industry could also be found. By the 1970s, however, the forces of suburbanization and dispersal were already well established, and since then they have increased rapidly. According to Takeuchi (1994, 1995), these gathering forces of dispersal are not simply associated with the decline of inner Tokyo as a manufacturing centre in favour of the intermediate and outer zones. Rather, led by the electronics and auto industries, a new intra-metropolitan division of labour has been created within a 'high tech corridor' in which the research-intensive activities of inner Tokyo, especially in the southern district, give way to R&D centres, innovative manufacturing and finally to large-scale assembly operations in the outer zone (Figure 15.2). Nissan and Isuzu (autos), Sony and Toshiba (electronics), Canon (cameras and copiers) and NEC (computers) are examples of core-firm-dominated production systems at the forefront of these changes.

In the innermost zone, and towards the south, Takeuchi identifies the emergence of a new 'hard centre' which includes a new industrial park for specialized machine processors near Tokyo (Haneda) airport, a small R&D centre which specializes in the production of pilot models, and so-called 'system houses' which include technically sophisticated software producers and consultants to aid in the design and development of new products and processes. The next zone, focusing on the Kawasaki and Yokohama areas, has become a major R&D centre as high-tech firms such as Fujitso, Hitachi, NEC and Toshiba have established laboratories there, along with other research institutes and innovation centres. Further out, component manufacturers of increasing size are found, while in the outer zone large-scale assembly factories have been established which are in turn attracting component suppliers. Thus, while the level of manufacturing activity in inner Tokyo has decreased, particularly with respect to activities based on low labour costs, there has been a growth of technically sophisticated activities, and in the outer zone manufacturing output has increased. It might also be noted the consumer goods industries remain within the inner and intermediate zones of Tokyo. While routinized

(a) *c.* **1980**

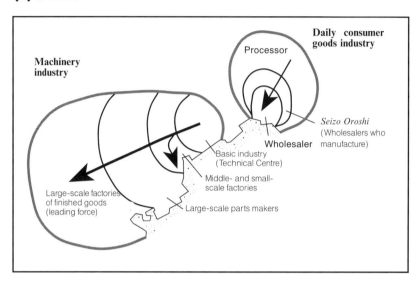

(b) *c.* **1990: The High-tech Corridor**

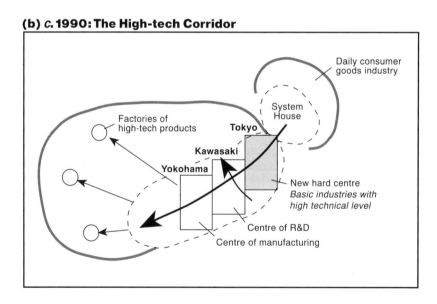

Figure 15.2   The Evolving Production System of Tokyo. Source: Takeuchi (1994: 204, 207)

operations have located in low-wage countries, efforts are being made to renew industrial districts by the adoption of electronics engineering and processing, and the creation of more complex products (Takeuchi 1994: 208). Meanwhile, printing and publishing has survived, increasingly in the form of a graphic arts community.

An example of contemporary location dynamics of manufacturing in Tokyo is provided by the Canon corporation, especially in regard to its camera production (Figure 15.3). Established in the inner part of southern Tokyo in 1933 as a camera manufacturer, Canon established its head-office and main manufacturing plant in the 1950s at nearby Ota-ku, Shinmaruko, where it developed an extensive system of subcontractors. Canon began mass production in 1961 at its Toride plant in the outer zone of Tokyo, where there was cheap land and available labour, while retaining high-quality production at Shinmaruko (Takeuchi 1994: 25–30). In 1963, another plant in southern Tokyo in Kawasaki was built to manufacture intermediate-quality cameras. In response to technological change and rising land and labour costs in Tokyo, Canon radically re-configured its camera production system while diversifying into new products, notably calculators, copiers and office equipment. Thus, a compact-camera factory was located in Taiwan in 1970 and while this factory now produces single-lens-reflex cameras, factories in Malaysia and China in 1990 manufacture zoom and non-zoom-type compact cameras respectively. Within Japan, camera production at Toride was relocated to a more distant plant at Fukushima in 1973, and to the even more remote location of Ohita on Kyushu in 1982. In 1990, Canon ceased to make compact cameras in Japan and high-quality camera production was consolidated on Kyushu. Meanwhile, the Toride plant shifted to the production of office equipment, notably colour copiers. The Fukushima plant now manufactures bubble jet printers, and a plant at Otsunomiya in the outer zone of Tokyo that once manufactured lenses now makes video cameras. Other factories established near to the main Ota plant that originally manufactured high-quality cameras have shifted to other products, while the main Ota plant itself has become a development centre. In addition, Canon has established new R&D laboratories in inner Tokyo, Tsukuba (Japan's major planned science park) and Osaka.

While maintaining R&D for cameras in the Tokyo region and high-quality camera production on Kyushu, lower-cost assembly has been shifted to Asia. Cameras, however, now account for less than one-third of Canon's sales, and various office equipment products such as copiers, calculators and printers are manufactured in Japan and in several foreign countries including Germany, the US, France, the UK, South Korea and Singapore. In some of these countries Canon also has R&D centres. While it may be argued that to some extent Canon has displaced Japanese production, especially in Tokyo, with foreign factories, a Tokyo location may not have been practical in many cases. In addition, Canon's DFI gains access to markets and skilled human resources in Europe and North America as well as to cheap labour in Asia. Moreover, Canon has continued to rely on about 300 subcontractors within the Tokyo region where its head-office and central research laboratories, as well as innovative manufacturing, remain important.

Host economies

It is from the control and command centres of the world, notably New York, London and Tokyo, that most DFI is directed and organized. In national terms, the distribution of DFI has been somewhat more diversified than its origins (Table 11.3). Developing countries, for example, have been important recipients of DFI throughout the 20th century, although it is interesting to note that in relative terms, developing countries have been less important in recent decades than in the early decades of the century. Indeed, since World War II the majority of DFI has occurred among advanced industrial countries. In this

(a) Canon's Tokyo Region Locations

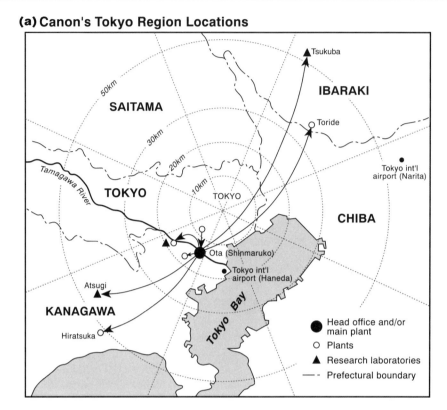

(b) Canon's Camera Production Elsewhere

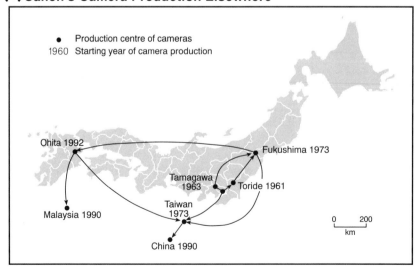

Figure 15.3   Features of Canon's Evolving Camera Production System. Source: Takeuchi (1994)

period, west European countries have consistently been major recipients of DFI, as has Canada, while in the 1980s DFI has expanded substantially in the US. Japan, on the other hand, has accounted for a relatively small amount of total DFI and is clearly not a 'host economy'.

Host economies are nations, regions and cities whose industrial base is characterized by high levels of DFI and foreign control. For foreign-owned firms, ultimate determinations over strategies and structures are held by parent companies' head-offices located in foreign political jurisdictions. At regional and city scales, external control defines activities that are foreign controlled or are controlled by domestic companies whose parent head-offices are elsewhere in the country. In general terms, host economies are economies dominated by foreign (and externally) controlled branch plants.

Conventionally, branch plants and subsidiary companies are defined as foreign controlled if more than 50% of equity shares are held by companies whose head-offices are located in a foreign jurisdiction. That is, effective control is equated with the location of head-offices, and not particularly according to where individual shareholders reside. It is often argued that companies achieve control with less than 50% of equity in subsidiaries so that the conventional measure of foreign control is likely to be conservative of effective control (Watts 1981; Hayter 1985). Joint ventures involving a 50:50 split in equity between a local and a foreign-owned company would also be classified as domestic according to the conventional definition. Yet, such 50:50 joint ventures normally have extremely limited decision-making discretion and changes in their operations require permission from both parents. In this case again, the conventional definition of foreign control is too conservative. In practice, however, it is difficult to simply relate control to some statistical level of equity and the particular way control is exercised is influenced by a variety of factors (Dicken 1976; see Chapter 11).

The major host economies

Given the caveats as to definition and meaning of control, there are substantial variations in the importance of foreign control to national economies, no matter what variables are used. In terms of employment in 1984, in such economically important countries as Canada and Spain, for example, foreign affiliates account for over 40% of national totals, while in Mauritius, foreign affiliates account for over 65% of employment (Table 15.4). In contrast, in Japan, the relative contributions of foreign firms to employment are virtually negligible, in the US the share is small, and Germany, the world's third largest industrial power, has less than 10% of its employment in foreign affiliates. Similar remarks are pertinent for employment and production in the manufacturing sector, both with respect to developed and developing countries (Table 15.5). Among developed countries in 1980, for example, whereas a remarkable 56.6% of Canada's manufacturing production was foreign controlled, in Japan just 4.2% was under foreign control.

In addition to employment and production (the conventionally used criteria), other variables can be used to measure foreign ownership and define host economies. Wilkins (1994), for example, suggests the share of inward DFI stock as a percentage of gross domestic product. According to this measure, in Canada's case, DFI stock accounted for over one-third of the economy in 1970, although it has declined considerably since (Table 15.6). In addition, relative measures of foreign ownership can be usefully supplemented by rankings of host economies in terms of absolute levels of inward DFI (Table 15.7). In 1914 and 1982, for example, the US was the leading host country according to this criterion. In fact, since the 1970s, absolute levels of DFI in the US have been significant and considerably higher than for any other country (McConnell 1980). From this perspective, Canada, with an economy one-tenth the size of that of the US, also stands out as a historically significant host economy: Canada is ranked third in 1914, first in 1929 and 1975, and second in

Table 15.4   Foreign Affiliates' Share of Employment for Selected Host Countries, 1976–1988

Host country	(Year)	Employment in foreign affiliates (%)	Host country	(Year)	Employment in foreign affiliates (%)
Mauritius	(1984)	65.4	Austria	(1985)	13.5
Spain	(1977)	46.6	UK	(1987)	13.0
Canada	(1976)	41.0	France	(1984)	12.0
Malaysia	(1988)	32.2	Italy	(1985)	11.8
Singapore	(1988)	32.0	West Germany	(1982)	8.3
Argentina	(1984)	26.8	Portugal	(1984)	8.2
Fiji	(1985)	25.0	Peru	(1988)	6.5
Belgium	(1985)	18.0	US	(1987)	3.7
Brazil	(1987)	16.2	Korea	(1986)	2.7
Chile	(1979)	15.0	Japan	(1986)	0.4

Source: Wilkins (1994: 33).

Table 15.5 The Share of Foreign Firms in the Manufacturing Industry of Individual Nations *c.* 1980

Nation	Foreign percentage share of national	
	Manufacturing employment	Manufacturing production
Developed Countries		
Australia	21.8	22.7
Canada	44.3	56.6
France	19.0	27.8
West Germany	16.8	21.7
Italy	18.3	23.8
Japan	1.8	4.2
United Kingdom	13.9	21.2
United States	3.0	—
Sweden	5.7	7.3
Developing Countries		
Brazil	30.0	44.0
India	13.0	13.0
South Korea	10.0	11.0
Malaysia	34.0	44.0
Mexico	21.0	39.0
Singapore	58.0	83.0

Sources: OECD (1981: Table 1); UNCTNC (1985: 100).

Table 15.6 Canada: Inward Foreign Direct Investment Stock as a Percentage of Gross Domestic Product, 1970, 1980 and 1989

Year	Investment stock (% of GDP)
1970	34.0% (1)
1980	24.1% (2)
1989	20.5% (13)

Source: adapted from Wilkins (1994: 30).

1982. While several developing countries were ranked in the top five in 1914 and 1929, only Brazil constituted a leading recipient of DFI in 1982.

Since there is no consensus as to how to simply define a host economy in terms of some minimum proportionate level of foreign ownership or some

absolute level of DFI, caution needs to be exercised in relying on one measure in one particular time period. Clearly, while the US has been the most significant host economy for DFI over the past decade, the US is not a branch-plant-dominated economy. Despite massive recent inflows of DFI, overall levels of foreign ownership of the US economy remain low and the US, as the world's principal industrial power, has been the world's principal home economy this century. Similarly, the emergence of Japan and Germany as industrial powers is associated with their role primarily as home economies (a characteristic shared by Sweden and some other countries, albeit on a smaller scale). According to OECD data, while Germany has received significantly higher levels of DFI than Japan, foreign ownership of manufacturing employment is about 17%, a little less than in Italy or France and a little more than

Table 15.7 The Top Five Host Economies in Terms of Foreign Direct Investment 1914–1982

1914	1929	1975	1982[a]	
US	Canada	Canada	US	(124.7)
Russia	US	US	Canada	(56.3)
Canada	India	UK	UK	(51.3)
Argentina	Cuba	Germany	Germany	(32.1)
Brazil	Mexico	Netherlands	Brazil	(21.6)

Source: adapted from Wilkins (1994: 20–23).

[a]Figures in parentheses refer to the inward stock of foreign direct investment in billions of US dollars.

in the UK (Table 15.5). According to Wilkins' estimates of the share of total employment in foreign affiliates, the relative level of foreign control in Germany is substantially reduced and is well below the level in the UK (Table 15.4). In the case of the UK, inflows of DFI have grown considerably in importance in recent decades (Table 15.5), and in some regions, such as Scotland (Firn 1975), levels of foreign (and external) control are high (Watts 1981). In addition, the industrial base of the UK has been shrinking.

Evidently, variations in the relative importance of foreign control are not simply a function of the size of national economies or their stage of economic development. When compared with Canada, for example, foreign firms play a far less significant role in smaller, high-income countries such as Belgium and Sweden and in bigger, high-income countries such as the US, Japan and Germany. The same point can be made about bigger and smaller developing countries. The spectacular economic growth of South Korea, for example, although substantially helped by imports of US machinery and Japanese components, has been achieved without much loss of domestic control of industry. Brazil and Malaysia show different tendencies. National variations in the relative importance of foreign control must be seen in terms of the differential choices made by different countries, or at least by different governments.

In fact, international flows of DFI exhibit significant interdependencies, especially among advanced countries which act as both host and home economies. Predominantly, however, host economies are those in which foreign-owned branch plants play a dominating role. In this regard, the available measures confirm that among advanced countries, Canada is *the* example of a host economy. Bearing in mind the size of the Canadian economy, absolute levels of DFI have been significant, and in turn these inflows have created high relative levels of foreign control in the key manufacturing and resource producing sectors of the economy. Indeed, according to both OECD data (Table 15.5) or Wilkins' estimates (Table 15.4), foreign ownership is quantum leaps greater in Canada than virtually all other advanced countries, and high levels of foreign ownership have been significant for a long period of time. Perhaps not surprisingly, the debate over the pros and cons of foreign ownership has a long history in Canada (Marshall et al 1936; see also Watkins 1963; Safarian 1966; Watkins Report 1968; Levitt 1970).

The paradox of DFI for host economies

There is a fundamental paradox between the operations of international firms and local development which ultimately derives from the fact that DFI imposes both benefits and costs on host economies. The paradox is apparent within the theory of location advantage of international firms (see Chapter 11). According to this theory, firms that grow internationally need to have some kind of entry or competitive advantage(s) in order to overcome spatial barriers to entry. In addition, the process of internationalization means, to some extent at least, that market-based transactions are replaced by flows of goods and services (administered transactions) within the international operations of the firm.

From the perspective of the host economy, the entry advantages of foreign firms equate to an established bundle of production, technological and marketing know-how, along with financial resources, which can readily be invested in branch plants. International firms can potentially speed up growth rates in host economies by applying their proven abilities to build and operate factories, the security of which is typically enhanced by links to affiliated markets, distribution systems, input sources and a wide variety of already established corporate services and networks. That is, DFI can act as a *catalyst* for local development.

For host economies, the alternative strategy of encouraging domestic firms to copy and re-develop the entry advantages of foreign firms may face significant costs and uncertainties, as well as requiring longer time horizons. Foreign firms offer to host economies the powerful and seductive advantages of faster, more secure growth. Foreign firms may also be locationally flexible within host economies and willing to locate in designated regions, where domestic firms are weak.

At the same time, branch plants are part of an internal and international set of administered transactions and these affiliated international linkages replace potential local linkages with domestic (non-affiliated) firms. Yet quantity and quality multiplier effects depend upon the extent of these local linkages (see Chapter 13). Admittedly, the extent to which branch plants are integrated internationally within parent companies in relation to integration with the local economy, does vary. The differential performance of Ford and GM in Japan in the 1920s testifies to this variation, even among firms in the same industry and host economy (Takeuchi 1990). Thus, the local impacts of individual branch plants depend upon corporate policy and how these plants function within corporate systems (Dicken 1976). Notwithstanding individual corporate variations, MNCs are internally

integrated systems in which the chosen level and forms of integration provide institutionalized constraints within which individual branch plants operate.

The economic costs imposed by DFI on host economies relate to the extent to which integration of foreign-owned subsidiaries and branch plants within the international structures of parent companies replaces or pre-empts local linkage development. From this perspective, local linkages are *truncated* and host economies are truncated economies (Britton and Gilmour 1978: 96; Hayter 1982). In contrast to independent firms who control the investment decision process and can choose the extent to which they wish to depend on other firms, truncated firms do not have full responsibility for the entire manufacturing process of a product line because parent companies supply functions and services as well as authority over the nature, level, location and timing of investment decisions. Moreover, the nature and forms of integration are structured to reflect the priorities of the MNC rather than those of particular nations.

If truncation results from the integration of branch plants within parent companies, from the latters' perspective such integration is rational. Thus, the duplication of services and functions already provided in the corporate system in foreign-based plants may be inefficient and undermine possibilities for firm-level economies of scale. For the parent company, integration is also essential to preserve and control the entry advantages that it has developed over time. Indeed, MNCs are hesitant to voluntarily transfer their core expertise beyond the boundaries of the firm. Consequently, and paradoxically, the need for MNCs to internalize and integrate internationally reduces their local impacts in generating structural changes within host economies. Such impacts are best illustrated in economies experiencing high levels of foreign control. Canada is the notable example.

Canada as a truncated production system

Countries have varied in their openness to DFI. In general, those countries extolling the virtues of economic liberalism, notably the US, the UK, Canada and Australia, have been relatively open to DFI. According to this philosophy, the 'free' flow of DFI and 'free' trade stimulate competition and allow for the most efficient allocation of resources. Even so, as inwards DFI has increased in the US and the UK, so has the criticism of it. On the other hand, Canada and Australia have long been host economies and for many decades significant shares of industrial activity have been under foreign control. Admittedly, in Canada's case, exports of DFI have been greater than imports since 1975, although as MacPherson (1996: 72) points out, if the use of retained earnings by foreign companies is considered, Canada continues to be a net importer of capital. In effect, foreign companies have used profits generated in Canada as investment sources to expand operations and the extent of foreign control. Moreover, DFI in Canada, especially by US-based MNCs, has frequently been accomplished by acquisition rather than by the building of greenfield plants and this tendency has been an important feature of recent patterns (McNaughton 1992a).

In Canada and Australia, the structural impacts of DFI continue to be debated (Crough and Wheelwright 1982; Grant 1983; Morris 1983; Hayter 1988; Clements and Williams 1989; McLoughlin and Cannon 1990; Corvari and Wisner 1993; Britton 1994; MacLachlin 1996). In both countries, initially strong British interest was superseded by US MNCs, and in recent years by growing DFI from elsewhere but especially by Japanese MNCs (Edgington 1990). During the 20th century, however, the dominating foreign influence has been that of US MNCs, and this influence has been greater in Canada than in more distant and isolated Australia (Britton 1990).

The criticisms of DFI in Canada (and Australia) are that branch plants create truncated production systems. In Canada, the Gray Report (1972), which formed the basis for some modest changes in public policy towards DFI, coined the term 'truncation' when it defined a truncated firm as a subsidiary company which does not carry out all the functions involved in the development of a product; rather, it relies on the parent company for one or all of these functions, such as marketing or R&D. The problem arises when a lot of firms in a regional economy are truncated and they collectively rely on foreign-based parent companies for various corporate functions. Such functions then become 'underdeveloped' or 'underrepresented' in the host economy. Thus, a truncated economy or production system is defined as one in which the existence of foreign-controlled branch plants replace or pre-empt *economically viable* indigenous development (Hayter 1982: 277). That is, branch plant reliance on parent companies for key functions serves to transfer demands for these same functions outside of the host economy even though the host economy has the location conditions to support such functions.

Characteristics of truncated production systems within Canada

In Canada, the idea of truncation can be traced at least as far back as Marshall et al (1936), who concluded that 'industrial science' in Canada's private sector (what is now called in-house R&D) would have been greater with lower levels of American ownership of Canadian industry. This study recognized that US-based firms who had already invested in R&D operations in the US rarely established similar facilities in Canada. In effect, US-owned branch plants transferred demands for R&D back to their head-office. In the opinion of Marshall et al (1936), if Canadian firms had been more important, greater levels of in-house R&D in Canada would have occurred. Within Canada, those authors that have developed this line of thinking to argue that the Canadian economy is truncated comprise a 'nationalist' argument (Britton and Gilmour 1978; Clement and Williams 1989).

Historically, two broad categories of DFI in Canada may be discerned, both of which have contributed to truncating effects and to the country's core–periphery structure (Figure 15.4). On the one hand, DFI has established branch plants manufacturing mature products in the core primarily to serve Canadian markets, which are also largely in the core (Britton 1976, 1980, 1994; Britton and Gilmour 1978). On the other hand, a major motive of DFI in Canada's peripheral regions is to extract resources in order to export commodities and primary manufactures, primarily to home economies (Hayter 1981; 85, 88). In practice, the motives for DFI are less clear-cut, and especially with the growing trend towards free trade between Canada and the US (and now Mexico), motives have become more varied (MacPherson 1996). At the same time, any assessment of the implications of free trade for Canada must consider how branch plants will respond (MacLachlan 1996).

7For Canada, according to the nationalist argument, the high levels of DFI in both primary and secondary manufacturing have resulted in production systems with strong international links (especially with the US) but fragmented and weak domestic links. In particular, truncation for Canada implies production systems in which decision-making, R&D, local multiplier effects and exports of high-value-added manufactures are under-represented, while imports of high-value goods, services and technology and commodity exports are over-represented. Using the US as the home economy, these characteristics can be outlined (Figure 15.4; see Britton and Gilmour 1978: 95).

1. Many of Canada's largest head-offices are subsidiaries of US-based MNCs whose lines of authority are with the US parent rather than across in Canada. To a noteworthy degree, US firms have favoured geographically proximate Canadian regions for investment; thus firms in the Midwest and north-east have typically located in Ontario and Quebec, and firms in California have located in British Columbia (Ray 1971; MacNaughton 1992b). When DFI is spread across Canada, subsidiaries are not necessarily integrated but frequently report directly to the US head-office. The size, location and scope, and any restructuring, of these head-offices is determined by parent companies.

2. Overwhelmingly, US-based MNCs in Canada have preferred to concentrate R&D operations in the US. In many cases, MNCs have acquired Canadian subsidiaries with established R&D operations which have then been closed down. In other cases, DFI pre-empts R&D investments which the MNC has already established in the US (or some other home country). With a few exceptions, the R&D that does exist in US branch plants is usually small in scale and meant to adapt innovations from elsewhere to Canadian circumstances.

3. Branch plants in secondary manufacturing rarely have the mandate to export except under highly prescribed conditions (Williams 1983). Rather, branch plants manufacture mature products to serve domestic markets, and sales to other regions are the responsibility of the parent company or other subsidiaries. Thus branch plants have neither the authority nor the marketing capability to export secondary manufactured products. Moreover, given that such plants typically do not have much independent technological capability, interest and ability to create innovative products with export potentials is weak. It might also be noted that historically many US branch plants were established in Canada behind a high tariff wall, in some cases operating as 'miniature replicas', i.e. producing the full product range of parent company operations. As technological change permitted increasing economies of scale, these plants became increasingly inefficient.

4. In the resource sector, branch plants are primarily concerned with exporting commodities (staples) to the US, typically to affiliated operations and sometimes to affiliated operations in foreign countries. Value added then occurs in the US (or overseas) rather than in Canada. In Watkins' terminology, this role represents a 'staple trap'. Product diversification of staples requires R&D but R&D is concentrated in home countries where desires for product diversification are more likely to be realized.

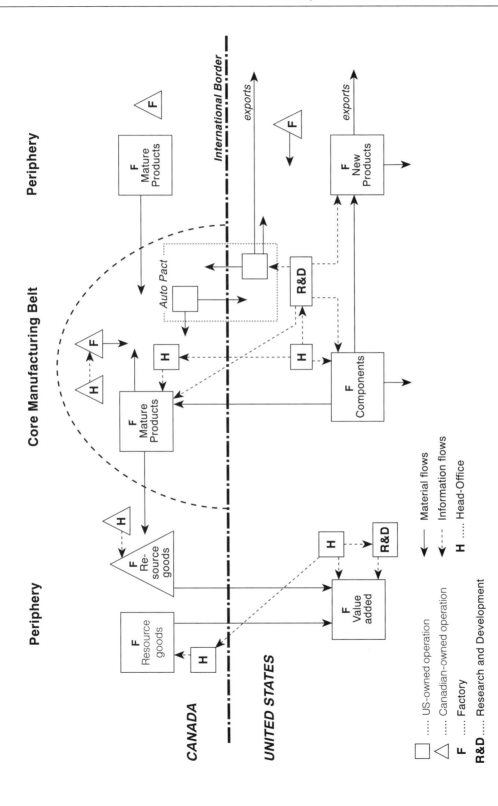

Figure 15.4 Canada's Truncated Production System in Relation to the US. Source: based on Britton and Gilmour (1978)

5. The local multiplier effects of US-owned branch plants are dampened by reliance on components from affiliated factories in the US and on services and technological expertise from parent head-offices and R&D laboratories. As a result, opportunities for spin-offs, liaisons and market niches for SMEs in manufacturing, business services and relationships with universities in Canada are lost. Within secondary manufacturing in Canada, the evidence indicates that foreign-owned firms do have higher import propensities to import than domestic firms and are more likely to import than to export (Britton 1994; MacLachlan 1996).

6. Finally, it might be noted that the high level of foreign ownership of Canadian industries not only directly limits R&D, valued-added activity, and exports of innovative products among branch plants, but does so indirectly by limiting opportunities for growth by domestic firms (Britton and Gilmour 1978). These indirect effects also have implications for how truncation is evaluated (Hayter 1982).

In practice, many caveats can be offered to the truncation model. US branch plants in secondary manufacturing, for example, export on a substantial scale to certain export markets, as permitted by parent companies. In the past, US branch plants in Canada exported to Commonwealth markets when Commonwealth preference existed, and occasionally exports, such as oil exploration equipment to Libya, which are politically sensitive from a US base, are made from a Canadian branch plant. The most significant manufactured exports from Canada, however, are in the automobile industry. The Canadian production of cars is entirely foreign-owned and until the arrival of Japanese branch plants in the 1980s, production was entirely under the control of the Big Three US companies. Following the Auto Pact of 1965, the industry was rationalized on continental lines so that Canadian (and US) branches began to produce particular models for the combined (tariff-free) US/Canadian markets instead of just for the smaller Canadian market (Holmes 1983, 1992). The 'fixed' (or 'managed') trade deal has certainly enhanced operating efficiencies and stabilized a substantial part of North America car assembly production in Canada more or less in accordance with Canada's market size. Moreover, in recent years, Canadian unions have been more effective than their US counterparts in negotiating favourable settlements (Holmes and Rusonick 1991). At the same time, the Auto Pact confirmed the concentration of head-offices and R&D in the US Midwest, subcontracting relations have tended to favour US suppliers, and if Canadian branch plants export, they do so only to the US with respect to the allocated model types. In addition, if the Big Three, and now Japanese companies, have invested in Canada there are no Canadian companies that can do likewise in the US and Japan.

Trade and technological truncation Canada's truncated production systems are certainly high-wage systems closely integrated with the US economy. At the same time, Canada's truncated production systems are fragmented, specialized and reveal high-income and employment leakages. According to Britton (1980, 1994), these structural characteristics are ultimately reflected in Canada's trade (and employment) patterns. In terms of trade, for example, from 1950 to the late 1970s, Britton (1980) noted that Canada generally had large balance of payments surpluses on visible trade (Britton and Gilmour 1978). However, exports were overwhelmingly dominated by a narrow range of commodities and primary manufactures such as lumber, newsprint, pulp, copper and wheat; the main exception to this orientation is autos after 1965, but, by agreement, the autos (and components) exported were roughly offset by cars imported. Imports, on the other hand, comprise a diversity of manufactured goods, and high-value manufactured imports in particular gained increasing and dominating shares of Canada's domestic markets. With respect to invisible trade, Canada invariably experienced large balance of payments deficits and one consistently important item constituted payments for business services, such as head-office services and R&D services from parent companies.

As Britton (1994) has recently indicated, despite the upheavals of the 1980s, Canada's trade structure has scarcely changed in an aggregate sense (Table 15.8). Thus in 1991 a surplus on visible trade was more than offset by a deficit on invisibles. Moreover, the visible surplus primarily dependent on staples (raw materials and primary manufactures) and, apart from autos, the deficit in end products, especially high-tech end products, was considerable. Patterns of trade and technology are intimately related. On the one hand, reliance on staple exports and foreign ownership has not been conducive to aggressive investment in R&D. On the other hand, failure to fully invest in R&D limits innovation and export diversification.

The extent of Canada's under-representation in R&D, especially industry R&D, is well known and documented. On a per capita basis, Canada's R&D efforts are relatively small when compared to other OECD countries (Table 15.9). Thus, in terms of R&D as a percentage of GDP, the intensity of Canada's performance is less than half that of the US, Japan and Sweden, and among long-established high-income countries,

Canada is comparable only to Australia. Moreover, Canada's (and Australia's) performance is even worse if only R&D performed by industry is considered (Table 15.9). In Canada's case, in 1986, half of its R&D was funded by industry, a level higher only than that for Australia. Weaknesses in private sector R&D, especially in-house R&D, is particularly important because of the distinctive role such R&D plays in creating firm-specific advantages in a way that other forms of R&D, such as government and university R&D, cannot duplicate (Hayter 1988, 1996). An additional implication is that the various kinds of technological liaisons generated by in-house R&D are only weakly developed in Canada in comparison to most other industrialized countries.

In general, the argument that foreign ownership has technologically truncated the Canadian economy has been debated in the context of secondary manufacturing (Safarian 1966; Britton 1976). It is also relevant to the staple industries. In the case of the forest product production system, for example, Canada has a massive supply base with 17% of the world's forest

Table 15.8   Canada: International Trade, 1991

Trade components	Current C $billion		
	Exports	Imports	Trade balance
Merchandise	141 728.3	135 947.9	5 780.4
Raw and processed	75 150.8	41 204.4	33 946.4
End products	64 015.6	90 455.1	(26 439.5)
Auto	32 590.7	31 137.1	1 453.6
Other	31 424.9	59 318.0	(27 893.1)
Technology intensive	14 238.0	22 471.0	(8 233.0)
Non-merchandise	38 239.008	72 264.0	(34 025.0)
Services	24 024.0	36 069.0	(12 045.0)
Total	179 967.3	208 211.9	(28 244.6)

Sources: Statistics Canada (1991a,b); Industry Canada (1994).

Table 15.9   R&D performance: Selected Countries, 1986

	Total R&D/GDP (%)	Industrial R&D/GDP (%)	Industrial R&D Total R&D (%)
US	2.89	2.07	71.9
Japan	2.81	1.88	66.8
Sweden	2.79	1.97	70.7
Germany	2.66	1.92	72.2
France	2.38	1.40	58.7
UK	2.33	1.47	63.1
Switzerland	2.28	1.69	74.3
Netherlands	2.06	1.16	56.5
Italy	1.47	0.85	56.9
Canada	*1.35*	*0.68*	*50.8*
Australia	1.14	0.39	33.9
Spain	0.48	0.28	57.5

Source: Britton (1990: 282).

resource, and Canada is the world's largest exporter of forest products. Traditionally, however, these exports emphasized a narrow range of low-value commodities, and indigenous R&D to create innovative products has lagged behind other countries (Hayter 1982, 1985, 1988). In the pulp and paper industry, for example, R&D levels in absolute terms are no greater than those of Finland (and lower if paper-machinery manufacturers are counted); in per capita terms, as well as absolutely, R&D levels in Canada are much lower than in the US, Sweden, Germany and Japan (Table 15.10). By relying on imported technology, the potential of the forest industry to create jobs in R&D, value added and machinery manufacturing and to diversify exports is reduced.

Branch plant restructuring Free traders in Canada have long argued that to reduce the inefficiencies associated with miniature replicas and other protected branch plant operations, tariffs on imported manufactured goods should be eliminated. The counter, 'nationalist' argument is that removal of tariff protection undermines the rationale for branch plants in secondary manufacturing (Britton 1977). In this view, free trade encourages parent companies to close branches in Canada in favour of concentration in the US where costs are lower and markets larger. According to the nationalists, only innovative manufacturing can be sustained in the high-cost Canadian economy (Britton and Gilmour 1978;

Table 15.10 Corporate R&D Expenditures in the Pulp and Paper Industry, 1986

	R&D expenditure (US$m)	R&D as % of sales
US	1224	1.25
Japan	230	0.75
Sweden	86	1.35
Finland	48	0.75
Canada	50	0.30

Source: Science Council of Canada (1991).

Britton 1994). The technological changes and globalization associated with the IC paradigm give further credence to this view.

In practice, tariff barriers protecting Canada's secondary manufacturing sector have been progressively lowered since the 1950s, culminating in the Free Trade Agreement of 1989 (FTA) with the US and its extension as the North American Free Trade Agreement (NAFTA) to include Mexico. In this regard, it might be noted that the Auto Pact of the 1960s provided a *fixed* or managed trade deal in which US MNCs got tariff-free access to the Canadian–American market and Canada received substantial production commitments (but no R&D). In other sectors, free trade has not implied any similar commitment and restructuring has frequently meant branch plant closure in Canada. MacLachlan (1996) provides a case study of McGraw-Edison of Canada, a diversified manufacturer of electrical goods, a subsidiary of McGraw-Edison of Chicago. By 1970, the Canadian subsidiary comprised several product groups and at least 14 plants, although these plants were more closely integrated with US operations than with each other. As sales and profits declined, however, Canadian capacity was gradually reduced, a Canadian head-office eliminated in 1977 and the last plant closed in 1983 as Canadian markets were increasingly served from US locations. An example from the resource sector is provided by the Spruce Falls newsprint mill of Spruce Falls, Ontario (Hayter 1993). This mill, owned by the *New York Times* and Kimberly Clark of Philadelphia was built in the 1930s to supply newsprint to the former. The mill was a classic truncated branch plant, manufacturing large volumes of a low-value commodity for a parent firm and without any independent marketing or R&D capability. In late 1991, however, the US partners sold the mill (to former employees) thus ending a 60-year relationship. The *New York Times* had discovered that a Finnish firm produced a better quality paper and had decided to enter a joint venture with this firm to build a plant in the US, thus negating the need for Canadian supplies. The new owners of Spruce

Falls had production know-how but no marketing connections or the finance to modernize the plant and product-mix.

Ultimately, the nationalists in Canada argue that for a country of its size and wealth, Canada could have been expected to develop more diversified, locally coherent production systems and that greater levels of domestic control would have provided more options for the path of industrialization. The reasons for this situation are varied but high levels of foreign ownership and an emphasis on commodity and mature products have discouraged R&D investments which in turn has inhibited innovation and diversification.

Host country policies towards DFI

Historically, different approaches to DFI are illustrated by Canada and Sweden which, while differing in language, history and geography, are two northern, resource-rich 'late developer' countries that industrialized in the late 19th century (Laxer 1989). In both countries, industrialization was predicated on the establishment of national tariffs on imported manufactured goods, and in both countries the expansion of agricultural settlement and railroad building provided domestic markets for manufactured goods. In Canada, however, an 'open door' policy towards DFI in the resource and manufacturing sectors was adopted, whereas in Sweden retention of domestic control of the critical manufacturing sector was a central plank of development policy. Stimulated by the tariff, branch plants grew rapidly in Canada and levels of foreign ownership soon reached unprecedented levels. By 1914, there were already 450 American branch plants and subsidiaries in Canada (Field 1914: 29–32); by 1926, 30% of Canadian manufacturing and 32% of Canadian mining and smelting was under American control (Laxer 1989: 15–16). Moreover, Canadian policy, apart from the financial sector, offered no restrictions on the entry of foreign firms, including if entry was by acquisition, and did not discriminate in any way against foreign firms; for

example, with respect to taxes on profits, limits to profit repatriation or limits on growth within Canada. In addition, policy was not concerned with the transfer of know-how from foreign firms to domestic manufacturers. According to Laxer, dependency on foreign investment offered Canadian policy-makers the 'easy' route of industrial development for a late follower.

In Sweden, on the other hand, industrialization strategies were explicitly predicated on Litz's model of autarkic nationalism rather than Anglo-Saxon liberalism (Laxer 1989: 99–113). Foreign companies, especially British and German companies, were certainly interested in investing in Sweden, and in Sweden's mining sector, foreign ownership amounted to 20% in 1900 (although of minor importance in other sectors). Yet there was little tolerance for foreign ownership in Sweden and, indeed, Sweden took measures to reverse foreign ownership in key economic sectors. In 1916, for example, the government passed a law preventing foreign ownership of Swedish resources and the strategic iron and steel industries. In the 1930s, Krupp, the giant German steel and armaments company, was forced to relinquish control of Bofors, while another German company was forced to close its Swedish airplane factory. These policies also set the stage for the future. Thus, in the 1980s, foreign subsidiaries in Sweden accounted for only 4% of the total economy and 6% of manufacturing (Laxer 1989: 103). These figures contrast sharply with those for Canada (Tables 15.4 and 15.5).

Different attitudes towards DFI are not just historically interesting. These choices leave important legacies which inevitably impact on contemporary choices of industrial policy. There are other countries, such as Australia, and many regions, such as Scotland and northern England, which have branch plant structures and which share Canada's problem of technological truncation (Firn 1975; Thwaites 1978). Countries such as Sweden, on the other hand, are now facing more insistent pressure to pursue open-door policies to DFI than they have in the past. Then there are the countries of eastern Europe and

Russia which are seeking to radically restructure their industrial base (Hamilton 1995; Schamp and Berentsen 1995), and poor countries such as China and Mexico which are trying to develop through industrialization (Auty 1982; Wang and Bradbury 1990). All of these countries are faced with the problem of how best to bargain with MNCs; the paradoxical impacts of DFI on local development are well known, even if the precise policy solutions remain uncertain. In order to increase the host-country benefits of DFI, while mitigating the costs, various policy suggestions have been offered related to minimum levels of local procurement, local decision-making participation, world product mandates, local training schemes, restrictions on methods of entry and special taxation levels. Without reviewing these and related policies in detail, two general observations can be made. First, many developing countries have sought to spatially concentrate DFI in 'special economic zones' or 'export processing zones'. Although rarely discussed as such, these policies are a variant of what used to be called growth centre strategies. Second, many countries emphasize the attraction of DFI in the form of joint ventures with local partners.

Export processing zones

For several decades, especially since the early 1970s, many developing countries have sought to spatially channel DFI towards so-called export processing zones (EPZs), which are enclaves deliberately created to host foreign-owned branch plants. EPZs have become an increasingly significant policy tool in Africa, Asia and Latin America (Figure 12.4). One report indicates that by 1984, 70 EPZs had been located in 35 Third World countries (Wang and Bradbury 1990: 309), while a 1986 report identifies 116 EPZs and notes that 74 developing countries had established an EPZ or were planning to do so (Dicken 1992: 183). Moreover, these EPZs have been established in poor countries, such as China (Sit 1986; Leung 1990; Wang and Bradbury 1990) and Mexico (Sklair 1989; South 1990; Wilson 1992), and if Mexico's political economy is now

officially and wholeheartedly capitalistic, China is seeking, so far, to embrace MNCs while maintaining principles of communism (Sit 1986).

According to UNIDO (1980: 6), an EPZ is defined as:

"a relatively small, geographically separated area within a country, the purpose of which is to attract export-oriented industries, by offering them especially favourable investment and trade conditions as compared with the remainder of the host country. In particular, the EPZs provide for the importation of goods to be used in the production of exports on a bonded duty free basis."

In practice, EPZs are designed to be dominated by DFI (although domestic firms can be important). By the mid-1980s, the sizes of some EPZs was impressive (ILO 1988). In 1986 in Asia, for example, major EPZs were well developed in Hong Kong (89 000 workers in several EPZs), Singapore (217 000 workers in several EPZs), Taiwan (over 80 000 workers in four EPZs), Malaysia (almost 82 000 workers in 11 EPZs), South Korea (140 000 in three EPZs) and the Philippines (39 000 workers in three EPZs). Mexico, however, following the introduction of its Border Industrialization Programme in 1965, has created the world's largest EPZ which effectively runs along its border with the US and which has attracted a large number of *macquiladoras* (assembly plants). In 1965, for example, 12 macquilas employed less than 4000 workers. By 1988 almost 1400 factories employed over 300 000 workers, predominantly in such centres as Ciudad Juarez, Tijuana, Matamoros, Nogales, Reynosa and Mexicali. As with the Asian EPZs (Table 12.2), electrical product assembly plants dominate and the workforce predominantly comprises young women.

In China, a key feature of *kaifang zhengce*, the open policy introduced in the late 1970s, is the creation of special economic zones (SEZs). The Shenzhen SEZ, located just outside Hong Kong, is perhaps the best known. Total foreign capital invested in Shenzhen in 1980 was US $26.57 million, and reached US $585.52 million in 1992

Table 15.11 Distribution of Foreign Investment among Sectors in Shenzhen, 1979–1992 (US$10 000)

	1979	1984	1986	1990	1992
Industrial	645	8 235	31 874	32 306	44 056
Commerce, restaurant and service	200	5 251	974	858	807
Transportation and communications	31	881	416	440	300
Properties	553	4 412	2 965	5 576	9 503
Agriculture and fishery	30	57	138	28	55
Construction			13	6	122
Finance and insurance			8 948	5 913	14 008
Others	78	4 177	3 605	6 730	2 688

Source: Shenzhen Tongjijuu (1993).

(Table 15.11). DFI accounted for 87.8% in 1980 and 54.9% in 1992. In terms of sectoral distribution, the industrial sector has been the most important for DFI since 1979, when it accounted for 42.1% of the total. By 1992, 61.6% of DFI was in the industrial sector, principally to establish relatively low-skilled and low-value factories producing consumer goods primarily for export (Sit 1985; Eng forthcoming). In theory, the SEZs are expected to absorb the know-how and resources of foreign firms and then transmit these characteristics to indigenous Chinese enterprises located in 'inland' regions through *Nei Lian* (inland linkages). The presence of DFI, however, is no guarantee of technology transfer and the extent to which DFI is generating industrial linkages throughout China has been questioned (Lai 1985; Auty 1992; Fan 1992).

Joint ventures

Even for new plants, the paradox of DFI in host economies means that there is nothing automatic about the extent to which MNCs will disseminate best practice know-how in terms of production, engineering, marketing, management, labour relations and procurement. Indeed, MNCs may resist the transfer of know-how which defines the core of its entry advantage. In advanced countries, levels of development ensure that some transfer occurs, and in several countries, includ-

ing the US, UK and Canada, the dissemination of know-how is largely left to local business practice or individual initiative by MNCs and domestic firms. In countries with less modern industrial experience and lower levels of education and training, however, there is a substantial gap between the know-how of MNCs and levels of local technological capability. In this regard, industrializing countries have frequently relied on a policy of joint venture.

In contemporary times (e.g. since *glasnost and perestroika*) throughout the former centrally planned economies of eastern Europe, there has a been a rapid growth of joint ventures in a variety of industries (Michalak 1993). East Germany differs somewhat only in that there is an existing pool of domestic MNCs already available from the former West Germany. In China, within the SEZs, foreign firms can only participate if they provide at least 25% of the capital in any particular venture. While it is possible for 100% foreign-owned branch plants to exist, Chinese policy has given particular preference to joint ventures of various kinds. In the past, Japanese policy, for example in the auto industry in the 1950s (see Chapter 14), permitted domestic firms to engage in joint ventures with foreign firms. A few other developing countries have also sought to promote joint ventures in a variety of industries.

For host economies, the attraction of joint ventures lies in the promise of accessing and diffusing the know-how of MNCs without forfeiting control of the economy. Thus, local participation in plants utilizing best-practice technology and management systems provides a learning experience for local managers and engineers as well as for workers. In turn, these managers, engineers and workers potentially can be an important source of know-how to supplying firms and/or in the setting up of rival companies. Moreover, as Japanese experience has revealed, once local learning from the foreign partner has gone as far as it can, joint venture arrangements can be terminated in favour of local control. In this regard, it will be interesting to see how joint ventures evolve in China, where presumably a communist government would assign a high priority to domestic control, and in eastern European countries where principles of socialism are well established.

In the meantime, it is important to recognize that joint ventures themselves do not inevitably enhance know-how within local economies. Learning processes are themselves problematical and, as already noted, MNCs may wish to limit the extent of transfer of know-how. Joint venture plants, like wholly owned branch plants, are typically integrated with parent company operations world-wide, and this company-wide know-how is beyond the reach of particular host economies. Moreover, joint venture companies are typically established with highly restricted mandates to manufacture a specific product in specific companies for specific markets, as agreed to by the partners. The management of joint ventures are typically limited to operational matters as directed by specific mandates. Joint ventures, in other words, are not inherently dynamic organizations. Any change in their structure requires agreement by parent companies. Also, if local partners can eventually acquire full control, in many countries so can foreign partners. Indeed, for MNCs, joint ventures provide access to local know-how and a way of evading foreign-ownership criticisms. Even in joint ventures,

advantages may progressively accumulate in favour of MNCs (see Chapter 11).

Conclusion

The controversy over the local impacts of DFI have been forcefully expressed in the context of developing countries. Thus, Frank's (1967) argument that MNCs actively underdevelop poorer countries by undermining local self-sufficiency in favour of 'cash' exports, extracting surplus profit, distorting infrastructure provision and imposing First World priorities on Third World conditions has been criticized for its crudity but remains at the heart of dependency theory formulations. Others have argued that the greatest benefits of DFI in host economies can be seen in a developing country context. Jacoby (1972: 30), for example, argues:

"The economic, political, technological, and cultural effects of multinational corporate investment are most striking when the host country is less developed than when it is relatively advanced ...[In central America, for example] the American corporation played an innovating and catalytic role, founding new industries, transmitting technological and managerial skills as well as capital, and in many cases creating entire social infrastructures of schools, housing, health facilities, and transportation in order to conduct its business."

Not withstanding these different views, DFI continues to explode around the globe, albeit in a spatially selective manner. Moreover, the essential rationale for host countries for seeking DFI as a source of development, is the potential of DFI to act as catalyst. To be (overly) brief, the rationale for open-door policies to DFI is that branch plants can provide locally scarce resources or expertise to the host economy in the form of finance, production and technological know-how, access to markets and related advantages. Branch plants thus provide a quicker way to industrialize than learning by doing. Branch plants, it is suggested, can also provide a 'demonstration' of various kinds of 'best practices' which local firms can imitate. Models of truncation offer different visions of the impacts of DFI. The explosion of DFI around the globe will doubtless remain controversial.

16

DE-INDUSTRIALIZED REGIONS: RESTRUCTURED AND REJUVENATING?

Production systems are dynamic and even the longest established, most powerful industrial districts are not immune to the gales of creative destruction. Not that long ago, the American Manufacturing Belt, the UK's Axial Belt and Germany's Ruhr were considered the epitome of self-sustaining industrial agglomerations. Indeed, their eternity and essence seemed to be neatly captured by Myrdal's (1957) concept of circular and cumulative causation, a model widely disseminated in economic geography textbooks (Hurst 1974: 328). Yet, in the 1970s and 1980s, these same industrial districts experienced profound de-industrialization, offering the defining examples of the processes of circular cumulative causation in rapid reverse. By the mid-1980s, this decline was increasingly seen as irreversible (Scott 1988a; Crandall 1993). These regions, however, have not been passive to their unanticipated predicament and there are signs of economic rejuvenation which features a strong role for a restructuring manufacturing sector (Stohr 1990a,b,c; Grabher 1991; Florida 1994).

This chapter examines de-industrialized regions and pursues three main themes. First, de-industrialization is defined and its local impacts reviewed. Second, the complexity of de-industrialization is emphasized by briefly reviewing various theories that have been put forward to explain the phenomenon. Third, the kinds of initiatives that have been undertaken to help rejuvenate de-industrialized regions are summarized. The discussion focuses particularly on experiences within the American Manufacturing Belt, the UK Axial Belt and the Ruhr. A witty supplement to this discussion is Michael Moore's (1989) film, *Roger and Me,* which explores the human impact of de-industrialization, the schism between corporate and community priorities, the growing disparities between the 'haves and have nots', a cause of de-industrialization and the community responses to de-industrialization with specific reference to Flint, Michigan, a single-industry auto town in the American manufacturing belt, and, in 1937, the founding home of the United Auto Workers of America.

The dimensions of de-industrialization

It is well known that the manufacturing sector among industrialized economies, in relation to the service sector, has been in *relative* decline throughout the 20th century, especially when measured in terms of direct employment (Table 3.14). However, the term 'de-industrialization', only began to be widely used in the 1970s, notably in relation to the UK economy (Singh 1977) where the manufacturing sector was experiencing *abso-*

lute decline. Indeed, according to Thirlwell (1982), de-industrialization is best defined as absolute declines in manufacturing activity, particularly as measured by employment levels in a nation (or region) over a long period of time. At the same time, it needs to be emphasized that there are no commonly accepted yardsticks for *precisely* defining de-industrialization in terms of minimum levels of (net) job loss, minimum time period or even with respect to the most appropriate geographic boundaries. In practice, de-industrialization is used rather loosely in reference to varying degrees of job loss, time periods and geographic scales. Moreover, employment change does not necessarily imply that investment and production changes in the same direction; job loss may be associated with modernization.

Given these caveats, the UK provides the defining case of national de-industrialization (Singh 1977; Blackaby 1979: Thirlwell 1982). Thus the UK began to lose manufacturing jobs in an absolute sense before other industrial countries and, at least until the 1990s, on a relative scale far

greater than elsewhere. According to OECD data, between 1966 and 1994 about half of the industrial jobs in the country disappeared (Table 3.3). Other countries have experienced de-industrialization, if not to the same degree. In particular, Germany and France have recorded substantial job losses and the oldest industrial regions, such as the Ruhrgebiet, are widely judged to have de-industrialized. According to the OECD, industrial jobs in the US as a whole in 1994 are significantly greater than in 1966, although there has been little net change since 1980 (Table 3.3). A more precise (and narrow) definition of the manufacturing sector based on US government data, however, reveals significant job losses which have been especially severe in the Manufacturing Belt (Figure 3.9). The general consensus is that this region has experienced de-industrialization (Bluestone and Harrison 1982; Crandall 1993).

However defined, de-industrialization should be recognized as a process, specifically as a process of cumulative causation in reverse (Figure 16.1). In this view, industrial decline

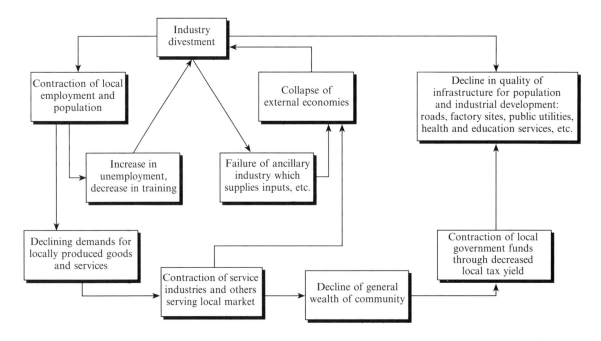

Figure 16.1 Myrdal's Process of Cumulative Causation in Reverse. Source: adaptations based on Keeble's (1967) Illustration of the Cumulative Causation Model

gathers its own momentum as declines in sales and job losses in one industry translate into declining purchases from other industries, who in turn cut back jobs and sales. As the process deepens, the ability of firms to find capital for investment and innovation becomes more difficult. Yet, if modernization is not attempted, equipment inevitably becomes increasingly obsolete and less efficient. At the same time, attempts at innovation can fail, potentially entrenching conservative attitudes and preoccupation with cost-cutting rather than market development. Indeed, a preoccupation with cost-cutting without reference to market roles tends to encourage firms to focus solely on paths of rationalization and closure. Such a viscious, interrelated, cost-cutting spiral of declining market shares, jobs and investments are important themes in the depiction of de-industrialization in the UK and the US (Bluestone and Harrison 1982; Eatwell 1982). Yet, if these core regions, which were once thought to be the very definitions of self-sustaining growth, could within a generation represent self-sustaining decline, presumably yet another 'U-turn' is possible.

Recessions

De-industrialization is a long-term process in which there are permanent reductions in jobs and industrial capacity. In practice, job losses and divestments have typically occurred during particular economic downturns or recessions. Recessions, however, do not automatically imply de-industrialization. Indeed, during the 1950s and 1960s, recessions were typically associated with temporary job reductions and temporary increases in rates of unemployment. During this period, in the UK and much of Europe, full employment typically meant unemployment rates of 2–3%; recessionary unemployment rates of 5–6% were regarded as severe, and politically unacceptable. In North America at this time, corresponding rates were modestly higher but the general pattern was similar.

During the 1970s, however, recessions became increasingly severe and the recession which began in the late 1970s and early 1980s was marked by downturns in jobs, profits and capacity not seen among western capitalist economies, most notably the UK, France, Italy and the US, since the 1930s (Townsend 1983: 29). Bearing in mind that the particular timing of recessions has varied among countries and among regions within countries (Norcliffe 1987; Green, et al 1994), it was this recession that confirmed that western economies were in the throes of fundamental change. Unemployment rates well above 10% occurred widely, and many regions, including in the American Manufacturing Belt, the industrial heartland of the UK and the Ruhr (Table 16.1), recorded rates well in excess of this level. By 1982, for example, unemployment in Flint, Michigan, was over 27%. Moreover, many of these job losses were permanent and did not 'bounce back' on the next up-swing. In 1987, communities in the Ruhr were still experiencing unemployment levels in excess of 15%. It should be noted that in all these regions, job losses typically involved high-income union jobs and white collar jobs, including those of management.

Thus, recessions which in the 1950s and 1960s meant temporary change, have heralded permanent changes in the 1980s and 1990s. As part of this shift, there has been a dramatic shift in what is considered socially and politically acceptable levels of unemployment. In countries such as the UK, US and Canada, for example, unemployment rates of 8–10% are now widely considered 'normal'. Indeed, 'recovery' from recessions has frequently not reduced unemployment rates significantly and the idea of 'jobless growth' has gained currency. If the unemployment rate has become a less sensitive indicator of the business cycle than it used to be, however, the consequences of unemployment for individuals are an ongoing and increasing problem.

Unemployment: the social consequences

Unemployment is the immediate, most significant problem created by recession, and when

Table 16.1   Unemployment Rates in the Recessionary Crisis in the Early 1980s: Selected Localities in Germany, the US and the UK

Germany	1980	1982	1984	1987
Duisberg – Oberhausen, NRW	6.2	10.2	15.6	15.9
Dortmund – Unna, NRW	6.0	11.4	15.4	15.6
Germany	3.8	7.5	9.1	8.3
United States	*1978*	*1980*	*1982*	*1984*
Detroit, MI	8.3	18.5	20.3	12.2
Flint, MI	8.8	18.2	27.1	13.4
Lansing, MI	6.5	9.2	14.2	8.7
United Kingdom	*1978*	*1980*	*1982*	*1984*
Coventry, West Mid	6.3	8.2	13.9	
Sheffield, S. Yorks	4.3–5.5	7.9–12.9	13.3–14.6	13.6–15.1
United Kingdom	5.6–6.6	8.4–12.4	12.3–13.3	12.4–13.4

Sources: Aring et al (1989: 172); Clark (1986: 130); Healey and Clark (1984: 306); and South Yorkshire Statistics (1979, 1982, 1984).
Notes: NRW refers to North Rhine Westphalia, MI to Michigan, West Mid to West Midlands and S. Yorks to South Yorkshire. The unemployment data for Sheffield and the UK are ranges for highest/lowest monthly rates. In 1982 and 1984 the range is for 1982/3 and 1983/4.

combined with de-industrialization it becomes particularly serious. In the UK's Axial Belt, the American Manufacturing Belt and the German Ruhr, soaring unemployment meant alternative sources of employment were not locally available, at least of the same income. Moreover, as unemployment spread nationally, migration became less of an effective response for individuals, even if they were in a position to move. Even in Germany, with its record of employment creation based around comprehensive and well-articulated training schemes, there has been no easy solution to rapidly rising unemployment. Since Germany's unemployment rate rose to over 9% by 1984, declining local employment opportunities for laid-off workers in the Ruhr were not particularly ameliorated in the rest of the country where there was also an increasing number of people looking for work. In fact, unemployment in the Ruhr (and Germany) remained high during the late 1980s. In March of 1988, for example, the labour exchange region of Dortmund recorded unemployment rates of 18.5%, while the other major labour exchanges in the Ruhr were all higher than 16% at this

time (Aring et al 1989: 65). By March 1989, rates of unemployment had declined from these peaks but the average for the Ruhr was still 14.7% and for Germany as a whole 8.4%.

Within the context of de-industrialization, therefore, the lay-off of workers contributes immediately to unemployment. One study, which interviewed 350 of 600 workers laid off by the closure of a steelworks in Sheffield in the 1980s found that 41% registered for unemployment and fully 68% were not employed in the immediate aftermath of redundancy (Table 16.2). These rates varied among groups of workers, with the most 'favoured' group being males under 40 years of age, which still experienced an unemployment rate of 29% (and a non-employment rate of 34%). This study also found that lay-offs and bleak job prospects did not serve to radicalize people or to shift established political beliefs; in contemplating future prospects the question of 'betraying class' was simply irrelevant (Hudson and Sadler 1986). A more comprehensive study in the Ruhr found unemployment to be concentrated among the unskilled although by no means limited to those 'without qualifications'. In addition, significant shares of the unemployed were out of work for at

Table 16.2 Immediate Post-redundancy Situation of Laid-off Workers at a Sheffield Steel Works, early 1980s

	% Unemployed	% Non-employed	(No.)
Males			
Under 40	29	34	(63)
40–55	52	56	(64)
55 or over	42	84	(179)
TOTAL	42	67	(306)
Females			
Under 40	32	58	(19)
Over 40	32	72	(25)
TOTAL	32	64	(44)
Skilled manual	45	59	(106)
Semi- and unskilled	51	81	(126)
Routine non-manual	30	66	(45)
Other non-manual	29	60	(73)
ALL	41	68	(350)

Source: Noble (1987: 280).
Note: The first column of data comprise individuals officially registered as unemployed.

Table 16.3 Ruhr Region: Selected Characteristics of Unemployed, 1986

Labour exchange region	Without qualifications (%)	Length of unemployment (%)		Age of unemployed (%)	
		1–2 years	>2 years	45–54	55–64
Duisberg	64.5	20.3	22.7	16.0	21.8
Oberhausen	57.6	20.4	17.8	18.0	18.1
Essen	63.5	19.3	24.4	20.6	14.1
Bochum	67.3	19.7	23.6	16.6	16.4
Gelsenkirchen	60.7	18.8	22.9	19.0	16.7
Recklinghausen	56.4	18.8	21.6	17.3	14.6
Dortmund	61.4	18.6	25.6	17.4	18.8
North Rhine Westphalia	61.6	18.0	18.9	18.1	13.6

Source: Kommission Montanregionen (1989: 174). See also Aring et al (1989: 45).

least one year and were at least 45 years of age (Table 16.3). Notwithstanding statistical characteristics, unemployment in old industrial districts has bitten deeply, affecting skilled and unskilled, young and old, those with families of their own and those without, while the loss of jobs by older workers has not presented younger workers with opportunities.

Unemployment of this magnitude imposes considerable problems for individuals, families and communities. Individual workers who are permanently laid-off often suffer shock, profound uncertainty as to what to do and a profound loss of personal esteem and dignity. In a de-industrializing society, alternative work typically does not exist and in the initial rounds of job loss in the 1970s and 1980s policy infrastructure in the form of counselling and training of the kind needed by middle-aged people just did not exist. There is now at least a much better appreciation of the mental and physical health problems of sickness, stress and anomie, as well as such social problems as drug addiction, drunkenness and family abuse, suffered by long-term workers who suddenly find themselves out of a job. These problems are by no means restricted to the unemployed (or to 'workers') and unemployment does not inevitably cause these problems. Rather, groups of unemployed workers (and managers) tend to experience higher rates of personal and family problems than similar groups of employed workers and managers. Moreover, help is not necessarily available beyond a relatively short period.

In communities which now have an established history of de-industrialization, young people who have never worked, at least on a full-time basis, are an important group. For young people who have yet to discover 'the dignity of a job', not having a job may not constitute a loss of dignity. Indeed, there may well be a need to stimulate 'traditional' beliefs in the worthiness of work itself as well as to actually create job opportunities for young people with little or no skill and often little education. It was the decline of these regions that signalled the beginning of the long-term problem of job creation now facing most industrialized (as well as developing) societies. Indeed, in the early 1990s, in many long-established de-industrialized communities, generational unemployment in which children become youths and then parents without any history of waged work has become an important trend.

The issue of minority employment, and affirmative action, is also seriously complicated by de-industrialization. In the US in particular, but also in the UK, minorities tend to experience higher rates of unemployment. A similar trend exists in Germany, where minorities are often immigrants or guest workers. Yet, if black and brown minorities are 'over-represented' among the unemployed, white majorities have also been laid-off in massive numbers. In de-industrialized regions where unemployment rates are generally high and affect all groups, albeit some more than others, affirmative action necessarily becomes more controversial. In many de-industrialized regions the jobs lost have been primarily high-paid jobs involving males, while the jobs that have been created have tended to be targeted towards lower-paying occupations involving females. In de-industrialized regions this trend also imposes difficult social problems as within families traditional gender roles are often reversed and family income declines. More generally, women are more likely to be employed, but at low wages. For single-parent families, which usually, involves mothers as the main or only parent, working for low wages brings its own stresses to child care. These social consequences of de-industrialization give urgency to the challenge of rejuvenation.

Local examples of de-industrialization

Just as industrialization is a geographically uneven process, so is de-industrialization. Within countries, de-industrialization is a spatially selective process in particular time periods (see Chapter 3, especially Figures 3.8–3.10). Similarly, de-industrialization affects some industries more than others at particular times. In the case of (West) Germany, from the mid-1970s to the mid-1980s one of the hardest hit regions was North Rhine Westphalia, where the average decline in manufacturing jobs was almost 14%. Within this region, the biggest losses occurred in the Ruhr-gebeit (the industrial region of the Ruhr) where there were also substantial local variations in the rate of manufacturing decline, including a 34% job loss in the Gelsenkirchen labour market region and over 26% in the Duisberg–Oberhausen labour market region. In terms of industrial composition, job losses in the Ruhr were led by the iron/steel and coal industries, and sometimes both (Table 16.4).

Throughout Europe and the US, many iron, steel and coal industries that were established in the 19th century in peripheral regions have been

Table 16.4 Industrial Employment in the Ruhrgebiet 1976–1986

Labour market region	Industry employment			Iron, steel and coal[a]		
	1976	1986	% change	1976	1986	% change
Duisberg-Oberhausen	142 311	105 136	−26.1	86 702	60 074	
Essen-Mulheim	88 595	68 813	−22.3	10 196	7 329	−27.9
Bochum	89 881	74 374	−17.3	29 351	18 717	
Gelsenkirchen	50 669	33 480	−33.9	17 337	12 408	−28.4
Dortmund-Unna	137 365	103 374	−24.7	66 749	43 525	
Recklinghausen	110 620	89 584	−19.0	47 341	36 473	−23.0
North Rhine Westphalia[b]	2 265 205	1 953 441	−13.8	400 564	293 592	

Source: Kommission Montanregionen (1989: 155). See also Aring et al (1989: 40).
[a]Essen, Mulheim and Gelsenkirchen and Recklinghausen only produced coal. The figures for the other labour market regions ('Arbeitsmark region') are mainly for iron and steel.
[b]North Rhine Westphalia is the federal state within which the Ruhr is located.

in decline for a considerable period of time (Martin 1988). In recent decades, iron and steel and coal making regions and communities within industrial heartlands have also experienced massive disruption (Hudson and Sadler 1983, 1986, 1989; Webber 1986). De-industrialization, however, has not been confined to the iron, steel and coal industries. Thus, in the US and especially the UK, the 1970s and 1980s witnessed the decline of a wide range of end-product industries, not least of which was the auto industry. In 1979 and 1980, for example, General Motors (GM), Ford and Chrysler closed down 20 facilities employing 50 000 workers in the US which as a result of 'knock-on' effects on suppliers, led to the closure of another 100 plants and a further 80 000 job losses (Bluestone and Harrison 1982: 36). In Michigan alone, specialized auto cities lost over 100 000 workers between 1978 and 1982 (Table 16.5). For cities such as Flint, these job losses escalated further during the 1980s.

Indeed, de-industrialization has a cumulative and circular nature which entangles networks of places and industries in a spiral of decline. Declines of final product industries which assemble complex products such as ships, autos, motor bikes, trucks, and various consumer electronics have particularly widespread implications for specialist component and processing

Table 16.5 Job Losses in the Transportation Equipment Industry: Michigan Auto Cities, 1978 and 1982

City	1978	1982	% change
Ann Arbor	21 800	16 100	−26.1
Battle Creek	2 400	2 200	−8.3
Bay City	4 400	2 600	−40.9
Detroit	260 400	175 400	−32.6
Flint	62 200	45 700	−26.5
Grand Rapids	4 800	3 500	−27.1
Jackson	4 900	2 600	−46.9
Kalamazoo	3 600	2 800	−22.2
Lansing	27 400	24 400	−10.9
Saginaw	15 600	10 700	−31.4
Michigan	432 700	303 200	−29.9

Source: Clark (1986: 130).

manufacturers as well as for basic material suppliers, such as iron and steel firms. Thus de-industrialization is a process which fragments industrial districts directly and indirectly. Two examples are provided by Coventry and Sheffield, located in the northern spokes of the UK's Axial Belt (Figure 3.4).

The examples of Coventry and Sheffield Prior to the Industrial Revolution, Sheffield was a small, isolated community whose major manufacturing activity was cutlery. It was here, however, that

Benjamin Huntsman, a clockmaker from nearby Doncaster, pioneered his crucible steel making process in the 1740s, thus establishing Sheffield as a founding spoke of the northern part of the Axial Belt. Sheffield's growth in the 19th century was rapid and specialized, fuelled largely by investments in iron and steel and tool making and to some extent by cutlery; by 1900 it housed a population of almost half a million people. Coventry's growth as a city of over 300 000 people occurred principally in the 20th century. During the fordist techno-economic paradigm, both cities enjoyed considerable prosperity and neither was included within the framework of UK's traditional regional policies of the 1950s and 1960s. In fact, Coventry's growth, as part of the West Midlands, was formally restricted. Between the late 1970s and mid-1980s, however, the decline of these two cities was extraordinarily fast.

Declines in manufacturing employment in Sheffield began in the late 1960s or early 1970s and increased rapidly after 1978. Indeed, by 1984 every other Sheffield job that existed in 1978 was lost (Table 16.6). Moreover, the decline was strongest in the three dominant, most export-oriented industries, notably metal manufacturing, mechanical engineering and engineers' small tools. Within this same period, Sheffield lost its distinctiveness as a manufacturing city in the UK: in 1978 the city's dependence on manufacturing was 12% above the national average, yet in 1984 it was just 3% above (Watts 1991a: 41). Data for South Yorkshire, in which Sheffield comprises about half the population, reveal that job losses between 1979 and 1985 were concentrated among relatively large plants (Table 16.7). In fact, 98% of job losses occurred in plants employing at least 100 employees in 1979 (Watts 1991a: 44). Survey evidence of large plant

Table 16.6   Sheffield: Manufacturing Employment Change 1978–1984

Industry	1978	1984	Job loss	%
Metal manufacturing	59 200	22 300	36 900	−62
Mechanical engineering	13 200	7 700	5 500	−42
Engineers small tools	10 000	5 100	4 900	−49
Food and drink	9 000	6 500	2 500	−28
Hand tools, implements	6 800	4 300	2 500	−37
Non-metallic minerals	6 100	4 100	2 000	−33
Other manufacturing	17 200	13 000	4 200	−24
MFG Total	121 500	63 000	59 100	−48

Source: Watts (1991a: 43).

Table 16.7   Job Losses From Plant Closures, South Yorkshire, 1979–1985

Plant size (no. of employees)	Employment		Job losses	
	1979	1985	Number	Percentage
11–99 (1979) } 10–99 (1985) }	30 784	26 807	−3 977	−13
100–199	21 994	15 219	−6 775	−31
200–499	35 655	25 202	−10 463	−29
500–999	40 109	17 092	−23 017	−57
1000 and over	58 236	24 198	−34 038	−58

Source: Watts (1991a: 44).

closures within Sheffield indicates that most (70%) were externally controlled and that the context for closure decisions varied (Watts 1991b). Thus some plants were closed to facilitate consolidation at another existing site within Sheffield or at an existing site outside of Sheffield, while still others meant decisions to drop a production line entirely. The motivations for closure also vary (see Table 4.8).

Similar to Sheffield, Coventry's economy has traditionally specialized in three manufacturing industries: motor vehicles, mechanical engineering and electrical engineering (Table 16.8). Of these, motor vehicles is easily the most important, and it experienced its greatest absolute and relative decline between 1974 and 1982. Indeed, in this period, Coventry lost almost half of its manufacturing jobs that existed in 1974; Coventry's rate of decline in three quite different industries approached that noted for Sheffield in an overlapping time period. In fact, most of Coventry's job losses also occurred after 1978 and the onset of recession (Healey and Clark

1985: 1364). Moreover, externally controlled large companies and plants dominated Coventry's job loss, as they did in Sheffield. Indeed, one company, British Leyland accounted for a quarter of the jobs lost (Healey and Clark 1984). Also, older plants were more likely to be closed than newer ones (Healey and Clark 1985: 1358).

To summarize employment change in Coventry, Healey and Clark (1984, 1985) use a 'components of change' analysis in which the overall net change in employment is disaggregated into categories representing different contexts of change which, statistically at least, are comprehensive and mutually exclusive in the time period under investigation (Lloyd and Mason 1978; Healey 1983; Watts 1987: 142–144). In Coventry's case, Healey and Clark distinguish between births, branch plant openings, transfer openings and closings, closures and *in situ* change between 1974 and 1982 (Table 16.9). There were components of growth but these were overwhelmed by the components of

Table 16.8 Manufacturing Employment Change in Coventry, by Industry 1974–1982

| Industry | Employment | | Employment change |
	1974 (%)	1982 (%)	Total (%)
Vehicles	59 995 (52)	26 626 (43)	−33 369 (−56)
Mechanical engineering	22 423 (19)	12 492 (20)	−9 931 (−44)
Electrical engineering	14 441 (13)	11 327 (18)	−3 114 (−22)
Other	18 458 (16)	11 964 (19)	−6 494 (−35)
Total	115 317	62 409	−52 908 (−46)

Source: Healey and Clark (1985: 1354).

Table 16.9 Components of Employment Change in Manufacturing, Coventry Metropolitan District, 1974–1982

Job gain components		Job loss components	
Births	1 040		
Branch plants	818	Closures	13 506
Transfer openings	3 689	Transfer closures	2 891
In situ expansion	3 720	*In situ* decline	45 778
	+9 267		−62 175

Total employment in 1974: 115 317; in 1982: 62 409.
Source: based on Healey and Clark (1984: 307).

decline. Predominantly, job loss resulted from *in situ* contractions. Thus, out of a net loss of 53 000 manufacturing jobs, 46 000 jobs were lost as plants contracted in size and a further 13 000 jobs were lost in complete closures.

There are a variety of factory contexts in which job loss occurs. Factory closure, for example, may mean a complete loss of jobs or it may involve a full or partial transfer to alternative, existing or new locations within the region or elsewhere. As Massey and Meegan (1982) point out, *in situ* job loss can also occur through *rationalization* in which factory capacity is reduced; *modernization*, which comprises investment in new technology which replaces (direct) jobs, even if capacity is unchanged or increased; and *intensification*, in which existing job loads are spread over fewer employees or the same employees are assigned more tasks. Moreover, these contexts are not mutually exclusive: rationalization, modernization and intensification can all occur in the same factory in the same time period. At the same time, the different production contexts have different implications for local development. It makes a considerable difference to local community development if, for example, job loss occurs as part of investment in new capacity and technology or as part of the rationalization and closure of factories.

One approach to statistically interpreting employment change over time in particular places is shift-share analysis. Shift share basically distinguishes an 'industry mix effect' and a 'regional effect' (Fothergill and Gudgin 1982; Watts 1987: 231–237; Appendix 1). The industry mix effect assesses the extent to which employment change in a region results from the extent to which its specialisms were nationally fast-growing or slow-growing. The regional effect assesses the extent to which employment trends relate to factors internal to the region (and which can either reinforce or offset industry mix effects). In Coventry's case, for example, the industry mix effect largely accounts for its rapid growth in the 1950s and 1960s and its decline in the 1970s, in tandem with the national fortunes of the auto industry (Healey and Clark 1984: 308). Unfortu-

nately, in addition to other problems, the conventional shift-share technique does not say anything about the national fortunes of industry (but see Norcliffe and Featherstone 1990). Consequently, an analysis of Coventry's (or Sheffield's) performance in terms of national trends does not recognize the factors affecting the de-industrialization of the country as whole. Indeed, the intense, concentrated job losses experienced in de-industrialized regions of the size and complexity of the Axial Belt, the Manufacturing Belt and the Ruhr suggest the existence of broader, longer-term forces of de-industrialization.

The explanation of de-industrialization

The proponents of a 'post-industrial' economy interpret de-industrialization as a particularly expressive indicator that modern economies are experiencing long-term structural change in which manufacturing activities are inevitably declining in significance. In this view, apart from the people directly involved, de-industrialization is not a cause for concern. They may well ask, as Singh (1977) observes, since agricultural decline ('deruralization') did not provoke long-term concerns for employment and economic growth, why should de-industrialization be any different? Singh's answer to this question suggests that modern economies need to remain concerned about the vitality of manufacturing for two main reasons. First, manufacturing exports and imports can make massive contributions to balance of payments surpluses or deficits. Second, the manufacturing sector remains a primary source of technological progress and productivity increases which provide major contributions to economic growth. In the US, for example, between 1929 and 1968 productivity increases in the manufacturing sector were twice those in the service sector.

In the context of the balance of payments, the service sector has been a major positive influence for some time in both the US and the UK. Whether or not the role of the service sector can be expanded in this regard to compensate for increasing de-industrialization is debatable. As Singh (1977) notes, a substantial part of the UK's invisible surplus stems from returns to capital outflows, including DFI, from the UK in previous time periods. The same point can be made regarding the US. Yet, the capital outflows in the first place reflect industrial strength and the existence of large corporations with the resources and expertise necessary for international expansion. The problem is that de-industrialization is by definition an erosion of industrial strength and it is the more industrially powerful economies of Japan and Germany that are spawning an increasing share of capital outflows which ultimately become the basis for invisible earnings. With respect to tourism, another important item of invisible trade (and services), Singh suggests it will be hard for the UK to increase its present role as an exporter as influxes of foreign tourists to Britain are offset by outflows of British tourists. Singh believes that the balance between the inflows and outflows is not likely to change much. London does export financial services but whether or not the export of such services can be increased or even maintained as de-industrialization proceeds, is debatable. For Singh (1977), de-industrialization constitutes a significant economic problem for the UK. Moreover, the points he made in the mid-1970s are still relevant in the mid-1990s and are not inappropriate to the situation facing the US. For policy-makers, however, the problem of de-industrialization is a complex one.

The complexity of de-industrialization

Various theories of de-industrialization have been put forward. These theories differ in ideological perspective, the national and regional context in which they are developed, the extent to which they are formally presented and empirically supported, and in the explanatory emphasis placed on international trade, government bureaucracies, labour unions, investment, management, historical and cultural factors, economic philosophy and the organization of

innovation. A summary 'listing' of the better known theories in terms of their primary explanatory perspective underlines the contested and complex nature of the de-industrialization process.

Trade perspectives First, there are theories related to international trade (Singh 1977). Essentially these theories argue that de-industrialization occurs in a region as a result of foreign firms overpowering local firms in domestic and international markets. Trade-related explanations of de-industrialization can be classified into two types. The first approach emphasizes that it is imports from low-wage developing countries that are undermining industry in high-wage economies. A second, contrary version of the trade argument emphasizes that de-industrialization of some high-wage economies is caused by imports from some other high-wage economies who provide products that consumers prefer for non-price-related as well as price-related reasons. These non-price characteristics relate to quality, servicing, reliability, delivery and marketing. A particularly important, contemporary interpretation of this second version, primarily associated with the US, is that massive Japanese exports are undermining US industries (and those elsewhere), especially as these exports far exceed Japanese imports. In fact, Japanese trade policies have been accused for some time of 'export targeting', in which the industries of other countries are supposedly systematically undermined while high levels of visible and invisible trade barriers are maintained around Japan's domestic markets.

Until relatively recently at least, actual trade performance suggests that imports from poor countries have not been a widespread source of de-industrialization in established industrial regions. While imports from the NICs, and developing countries in general, are definitely becoming more important, during the 1970s and the critical recession of the early 1980s these imports were still overwhelmed by trade among industrialized countries (Tables 3.4 and 3.5). The main exception to this observation relates to textiles. In the case of the clothing industry, in which wages are the main cost, exports from developing countries have been particularly important. For example, between 1970 and 1975, they increased from 38 to 53% of the world total (Steed 1981: 270). In those segments of the clothing industry where design and fashion are not important, these exports constituted a major threat to industrial production in advanced countries, including in the New England region of US and the cotton textile centres within the UK's Axial Belt. In some instances, claims that job losses are driven by cheap imports do not bear scrutiny. The Sheffield-based cutlery industry, for example, constantly complained about 'cheap imports' and 'unfair competition' from Asian countries, which in the 1950s and early 1960s focused on Japan and subsequently shifted to Hong Kong and Korea (Hayter and Patchell 1993). Yet import penetration has been a relentless trend since the 1950s and has been dominated by imports from other advanced countries. In this regard, the UK's entry into the EU simply served to speed up this process and to shift the source of manufactured imports to western Europe, especially Germany where wages in cutlery have long been higher (Hayter 1986a).

In general, it seems that the de-industrialization of the Manufacturing Belt, Axial Belt and the Ruhr in the 1970s and early 1980s is not related to such imports. Rather, the main trade-related problems faced by industries in these regions have originated in other industrialized countries. All three countries have pursued liberal import policies for decades and in all three countries manufactured imports are substantial; in 1990, for example, the US accounted for over 15% of the world's imports of manufactured goods, while Germany, in second place, accounted for over 8% and the UK (in fourth place) accounted for another 6.2% (Table 3.4). Moreover, in the US and the UK in the trade of manufactured goods, manufactured imports have consistently exceeded exports and, especially in the UK, import penetration is extensive (Table 3.7). In the case of the UK,

Singh's (1977) interpretation of de-industrialization in terms of declining competitiveness and an increasing inability to meet foreign competition from other advanced countries in domestic as well as foreign markets, has considerable credence (Freeman 1979).

At the same time, questions may be raised as to why British manufacturing has failed to meet export competition in so many instances. That is, trade failure is a proximate cause of industrial decline. In the case of Germany, de-industrialization has occurred without a similar loss of export competitiveness.

Investment perspectives A second set of theories of de-industrialization primarily focus on investment. The best known, and most widely cited in the geographical literature, is the capital mobility hypothesis which emphasizes the relocation of capital from established industrial regions, including the American Manufacturing Belt, the UK Axial Belt and the German Ruhr, to new industrial spaces. Other investment-oriented theories paradoxically emphasize either the failure of industry to modernize or the success of industry in modernizing and replacing labour with capital.

Bluestone and Harrison's (1982) famous study argued that the de-industrialization of the American Manufacturing Belt stemmed from the tendency of major corporations in the region to invest in new facilities in the American South or low-wage countries such as Mexico. In this view, job loss is directly linked with outwards foreign investment. For example, "During the 1970s, General Electric expanded its worldwide payroll by 5,000 but did so by adding 30,000 foreign jobs and reducing its US employment by 25,000. RCA followed the same strategy . . . General Motors has given up its plans to build a new . . . plant in Kansas City . . . and instead has shifted its capital to one of its facilities in Spain" (Bluestone and Harrison 1982: 6–7). Similar arguments have been made in UK and German contexts (Fröbel et al 1980; Massey 1984; Taylor and Thrift 1981). In the US, Bluestone and Harrison (1982) further suggest that profits are too often used to fund corporate growth by acquisition rather than internally. They were particularly critical of conglomerate growth based on facilitating the movement of capital between industries on the basis of short-term financial thinking.

While the capital mobility hypothesis is compelling, questions can be raised as to the extent to which it can account for de-industrialization, particularly in the early 1980s. In a detailed study of plant closures in the UK, for example, Townsend (1983: 41) found some examples of "the transfer of production to Third World countries . . . but, by and large the evidence is lacking on this point". Indeed, in aggregate terms, direct foreign investment (DFI) has remained concentrated within the industrialized countries (Tables 11.2 and 11.3). DFI also generates benefits for donor economies (Figure 15.3). Foreign investment tendencies also need to be related to the levels and motivation of domestic investments.

The relationship of investment with de-industrialization is itself controversial. On the one hand, investment in technological change which modernizes existing plants is frequently associated with job loss (Massey and Meegan 1982). A well-known model of industry evolution also features the tendency of investment to be job-replacing over time as firms shift their emphasis from developing products to increasing the efficiency of processes (Abernathy and Utterbach 1978; Chapman 1992). Yet, as this model notes, investment in new technology can be job-enhancing, while technology itself has to be manufactured although not necessarily in the same region as it is adopted. Moreover, in the absence of changes in investment and technology, i.e. the failure to modernize, the viability of industry will almost certainly deteriorate. Indeed, specifically in the British case, Eatwell (1982) has emphasized industry's widespread failure to invest and adopt new technology as a primary reason for de-industrialization. As an alternative to a commitment to long-term modernization, Eatwell emphasizes the importance attached to 'degenerate productivity growth', i.e. the more intensive

use of old equipment, which at best provides a short-term solution to the long-term problem of re-investment. In this view, it is the lack of investment and technical change, rather than too much technology, that has more important consequences for de-industrialization.

Neoconservative perspectives Neoconservative or 'right wing' views are rooted in the belief that the economy works best when individuals are free to choose and to pursue their self-interest in the context of market forces which are regulated by competition. Competition, within and between consumers and firms, is the great regulator which ensures fair prices, wages and profits and an efficient allocation of resources. In this view, restrictions on competition limit the pursuit of efficiency and may unfairly subsidize the inefficient.

Neoconservatives have essentially interpreted de-industrialization in the US and the UK as a result of restrictions on the 'rights' of individuals, as firms, workers and consumers, to choose and therefore to compete effectively. It should be noted that the emphasis in this view is on individual freedoms and competition rather than on co-operation and collective rights. In the context of industrial decline, neoconservatives have been particularly concerned with restrictions on competition that result from (1) the growth of the public sector; (2) the strength of trade unions; and (3) high income taxes. In the context of the UK, for example, Bacon and Eltis (1976) argue that government bureaucracies divert resources from sectors regulated by competition to one where non-market considerations predominate. In their view, the growth of public sector 'crowds out' private sector initiative by absorbing labour resources, making them expensive, introducing excessive 'red tape' and by diverting capital from growing to declining industries for political reasons. They argue that governments have considerable powers of taxation which allows governments to fund bureaucracies and monopolies without facing the discipline of the market-place. In the conservative view, however, high taxes are a double disincentive to the effective operation of the economy: they reduce returns on investments by firms and dissuade individuals from looking for work by providing welfare. Moreover, in the conservative view, unions similarly constrain the economic freedoms of firms by preventing downwards movements in wages, constantly bargaining for increases in wages and non-wage benefits, and limiting managerial control.

Because of its close association with Reagonism and Thatcherism, i.e. with the economic philosophies of the national governments of the US and UK during the 1980s, as well as in other places, the neoconservative view has been of considerable importance. Whether it provides the most appropriate interpretation for de-industrialization is another matter. Government bureaucracy, for example, may well be a problem. However, it is not particularly clear why excess government is a particular problem in the UK and the US. The share of national employment accounted for by the public sector in these countries during the 1960s and 1970s is broadly consistent with other advanced countries. It is also clear that for some time labour shortages have not been a particular problem in the UK or the US, so the private sector has scarcely been 'crowded out'. A similar point can be made as regards taxes (Kuttner 1984: 187). Some countries with extremely high rates of growth have imposed relatively heavy tax burdens to finance strongly redistributive social policies, and capital taxes have been much higher in Germany and Japan than in either the US or the UK.

De-industrialization in the UK and the US (and Germany) has frequently featured unionized industries and communities where labour relations are well established and extremely difficult to change (Clark 1986). Indeed, radical arguments, such as those offered by Bluestone and Harrison (1982), mesh with neoconservative views that a major motivation underlying the relocation of capital is to escape embedded labour relations. Even so, de-industrialization has not been limited to union industries and the *in situ* restructuring of unionized industries has

occurred. In addition, in the US and especially the UK, industrial unions have been in decline for some time and at least in the UK union decline has not led to an employment turnaround. There is also evidence that any productivity problems in the UK that arose from work practices were more than offset by lower real wages than other industrialized countries (Williams et al 1989). More generally, after four years of Thatcherism, the UK's export–import balance in manufactured goods became negative for the first time in 1983 and subsequently deteriorated "alarmingly" (Williams et al 1989: 73). For some observers, neoconservative solutions have become part of the problem (Eatwell 1982). In the case of the UK, for example, monetary policy which allowed 'markets' to maintain an overvalued pound in the 1980s was particularly devastating to export competitiveness in manufacturing.

Management perspectives Apart from the conservative view which believes management to be unfairly shackled, most theories of de-industrialization at least implicitly raise questions about managerial choices and attitudes. Other studies, notably by Williams et al (1983, 1989) in a UK context, have brought these questions to the forefront. Williams et al (1983, 1989) centre their argument on the concept of 'enterprise calculation', which in a broad sense refers to "the exercise of managerial discretion within external and institutional constraints" (Williams et al 1989: 81). In other words, enterprise calculation refers to how firms develop long-range planning or strategies (Ansoff 1965). Williams et al (1989) focus particularly on the financial, production and market calculations or plans of major corporations in several important UK industries during the 1970s and 1980s, including the auto industry, and have concluded that these calculations led to failure because of their internal inconsistency. In the case of Austin Rover, for example, Williams et al (1987, 1989) suggest that massive corporate miscalculations over market size and share within the UK, while simultaneously withdrawing from export markets, undermined a potentially effective restructuring of production and labour relations and aggravated financial problems.

It also might be noted that labour relations systems are agreements between management and labour, and in the collective bargains developed in the UK and the US, management, as well as labour, preferred an adversarial system which left management in full control over long-term planning and the timing, scale and location of investment. Moreover, as noted previously, union demands for job demarcation reinforced management's established commitment to scientific management or Taylorism (Marshall and Tucker 1992).

A more general, historically based critique of British management traditions is provided by Wiener (1981), who argues that British elites never embraced industrialization and urbanization. Rather, even as industrialization became such a dominating trend in the UK, the landed aristocracy maintained cultural hegemony and shaped the aspirations of the new industrial class to reflect their own 'bucolic' interests. Industry became a means to an end rather than an end in itself. Moreover, in the UK, public (i.e. private) schools and the Oxbridge axis have provided a remarkably narrow funnel for the education of leaders in government, finance, industry and aristocracy, and one which has not highly valued industry.

Institutional perspectives There are explanations of de-industrialization that incorporate some of the arguments already mentioned within broader frameworks which stress the role of domestic institutions and domestic industrial cultures. There are at least two important expressions of this approach, one emphasizing national variations in industrial culture and economic philosophy (Eatwell 1982; Dyson 1983), and the other emphasizing variations in national innovation systems (Freeman 1982).

In the first version – in which industrial culture is interpreted as a distinct set of beliefs, government–business relations and microeconomic practices – a distinction is made between economies, notably the US and the UK, in which economic liberalism prevails, from economies in which economic philosophies of statism and corporatism prevail, such as in Japan and Germany respectively (Dyson 1983). In the former economies, perhaps especially the UK, it is argued that liberalism has led to a preoccupation with cost-cutting, destructive forms of competition based on narrowly defined concepts of self-interest, and schisms in the financial and industrial sectors of the economy in which the priorities and preferences of the former are given precedence over the latter. Economic liberalism, in particular, is accused of encouraging a preoccupation with 'short-term' financial performance at the expense of longer-term industrial viability. Governments are also criticized for their 'short-termism'. Pollard (1982), for example, is critical of British 'stop–go' policies in the 1950s and 1960s, particularly of the regular attempts by successive governments to inhibit investment, justified on the short-term basis of reducing inflation during business up-turns despite the evident need for much of British industry to modernize. In contrast, in statist and corporatist industrial cultures, competitive relations are ameliorated by stronger tendencies towards co-operation, community relations and a willingness to adopt long-term perspectives towards industrial change.

In a related institutional perspective, Freeman (1982) and others (Freeman and Perez 1988; Nelson 1988) argue that the nature of innovation systems not only varies between advanced capitalist countries and both developing and centrally planned economies but also within the former. Again, a particular distinction is made between the innovation systems developed in Japan and Germany and those in the US and the UK. National innovation systems are defined in terms of the organization of all those processes underlying innovation, specifically research and development, marketing and production activities occurring in business, and more generally in terms of education, training, research and public sector social and industrial policies which provide the infrastructure underlying innovation.

According to Freeman (1988), Japan has a highly coherent innovation system which closely links in a highly flexible, 'loopy' way manufacturing, R&D and marketing and gives priority to reverse engineering and incremental innovation as well as to more radical technical change (see Figure 2.9). In the US and UK, on the other hand, innovation systems were often organized in a highly structured, linear way and have often not been complemented by appropriate investments in manufacturing processes (Florida and Kenney 1990a,b). For the past half-century, the US and the UK have also given much greater attention to military technology than either Japan or Germany. From this perspective, de-industrialization is at least partly explained with respect to differential abilities to innovate products and processes and to fully exploit their potentials, specifically with respect to manufacturing.

Traditionally, there have been substantial variations in national industrial cultures and innovation systems, and the industrial supremacy of countries that exposed economic liberalism has been successfully challenged by rival industrial cultures. Yet no industrial culture has proven immune to de-industrialization and the lines between rival industrial cultures are becoming increasingly blurred (Katzenstein 1985). Such distinctions reflect part of the legacy of industrialization and add to the appreciation of the complexity of de-industrialization.

The various theories of de-industrialization not only provide different explanations but also different policy prescriptions. In neoconservatism, for example, self-interested competition and individualism are the foundations of wealth and progress. Thus, conservatives recommend reducing the power of unions, deregulation with respect to rules governing business, the privatization of the public sector, and lower taxes in order to increase individual economic freedoms as much as possible. Critics of management, on the other hand, may be more skeptical of giving managers more freedom! Those who argue that innovation is the key to industrial rejuvenation are more likely to recommend industrial policies which involve more rather than less government

and stronger ties of co-operation among the main institutions of the economy (Britton and Gilmour 1978; Freeman 1982). In practice, choice of policy is perhaps as much ideological as it is a considered response to the testing of theories.

Towards rejuvenation

Within the industrial core regions of the American Manufacturing Belt, the UK's Axial Belt and the German Ruhr, de-industrialization was an unanticipated phenomenon. During the fordist techno-economic paradigm these regions were certainly not the fastest-growing regions within their respective countries (see Chapter 3). Yet they continued to experience some growth, unemployment rates were low and they remained industrial powerhouses of global significance. The UK, the US and Germany all introduced regional policies in the 1950s and 1960s but these policies largely targeted regions that were geographically and economically marginal, i.e. outside the industrial cores. Indeed, *within* the regions, cities and communities comprising the industrial cores, economic development was generally not considered to be a policy problem. Experience indicated that economic development could be left to the private sector. Consequently, particularly in the UK and the US, when de-industrialization started to savage local industrial structures, there were few or no economic development agencies and programmes in place, at national or local levels, to plan for economic diversification, and limited capacity for dealing with the social and psychological implications of large-scale lay-offs or policies specifically dealing with plant closure.

In other words, the regions, cities and communities within the industrial core regions were not politically or socially prepared for the economic crisis of de-industrialization that dramatically gathered steam in the late 1970s and early 1980s. Moreover, as de-industrialization became apparent, national governments, especially in the UK and the US, adopted a brand of conservatism which disparaged government interference in the form of subsidies, including regional policy. In any case, the increasingly 'national' nature of economic problems rendered regional policy less meaningful. In addition, the kinds of 'top-down' regional policies introduced and administered by national governments in the 1960s were increasingly criticized from within the regional planning and development literature (Friedmann and Weaver 1979; Trist 1979; Sharpe 1991). From this perspective, top-down regional policy was too remote, inflexible and non-democratic to respond to the myriad local variations in problems, opportunities and values. Rather, local development required local initiative – 'bottom-up' planning in which local participation and locally generated ideas play crucial roles. For the industrial core regions experiencing de-industrialization for the first time in the 1970s, policies of rejuvenation have typically represented some form of locally inspired 'bottom-up' development.

Such a trend does not mean that higher levels of government are absent from local development planning. Higher levels of government remain important sources of funds, co-ordination and policies that affect local development in a variety of ways. Much funding of local development projects still comes from national and regional governments but is often negotiated on an *ad hoc* basis, involving a variety of departments, and for projects which have been generated locally. Moreover, many policies can only be effectively introduced by national or regional governments to ensure uniformity and to reduce the risk of businesses avoiding communities with such a policy (Watts 1991b). At the local level, Harvey (1989) has characterized trends in planning as a shift from 'managerialism' to 'entrepreneurialism', as cities and communities once solely concerned with service provision and tax collection ('managerialism') now actively promote economic development ('entrepreneurialism'). Cox and Mair (1988, 1991) interpret this trend as a revival of boosterism designed to re-establish capital's hegemony over communities which are reduced to competing with one another. In practice, it is difficult to so categorize

the diversity of local development initiatives (Wild and Jones 1994).

The diversity of local responses

In practice, local development is highly diverse in terms of process, involving a wide variety of organizations representing a wide variety of community interests, and highly diverse in terms of outcomes (Cooke 1989). Trist (1979), for example, documents several cases from the 1970s of locally inspired development from the UK, the US and Canada, each of which is highly distinctive in terms of local agency, financing arrangements and sectoral emphasis. For example, Sudbury, a remote mining town in Ontario, Canada, responded to job loss in its economic base by developing the idea of an international-scale winter games facility, while Philadelphia's response to burgeoning unemployment was to build on its existing expertise in pharmaceuticals and medicine by developing a world-wide institute for the ethical testing of new drugs. Both ideas were developed from within the community, involved local funds and attracted the development of senior governments. Since the recession of the early 1980s, the diversity of local responses has escalated.

Throughout the old industrial core regions the diversity of responses to de-industrialization prompted Herrschel (1995) to argue that the national context of these responses has become unimportant, or at least less important. One of the cities cited by Herrschel is Dortmund in the Ruhr, and its response to de-industrialization is instructive. Between 1970 and 1985, Dortmund lost 42% of its productive employment, and total employment in the city declined by 40 000 – relative losses much greater than for Germany as a whole (Table 16.10). In 1987, unemployment was still around 16% (Table 16.1). In response to these problems, and particularly after major job losses implemented by Hoesch, its dominant employer, Dortmund has introduced at least 10 economic policy initiatives which have been organized by a remarkable range of organizations at local and state levels (Table 16.11). A

Table 16.10 Employment Development 1970–1985 in Dortmund, the Ruhrgebiet and the Federal Republic of Germany as a Whole

| | 1970 | | 1985 | | Variation 1970–1985 |
	Absolute	%	Absolute	%	%
Dortmund					
Agriculture, forestry	2 270	0.8	820	0.3	−64.0
Producing sector	139 880	50.3	81 496	34.2	−41.0
Services	136 150	48.9	156 512	65.5	+14.6
Total	278 300	100.0	238 828	100.0	−14.2
Ruhrgebiet					
Agriculture, forestry	39 700	1.8	21 063	1.0	−46.9
Producing sector	1 222 800	55.4	837 745	41.8	−31.5
Services	943 600	42.8	1 143 088	57.1	+21.1
Total	2 206 100	100.0	2 001 896	100.0	−9.3
Federal Republic of Germany					
Agriculture, forestry	2 262 000	8.5	1 390 000	5.4	−38.6
Producing sector	12 987 000	48.9	10 461 000	41.0	−19.5
Services	11 311 000	42.6	13 680 000	53.6	+20.9
Total	26 560 000	100.0	25 531 000	100.0	−3.9

Source: Hennings and Kunzmann (1990: 203).

predominant thrust of these policies is to encourage innovation; for example, by providing support to new technology-oriented firms, aiding existing firms to innovate, providing special help to Hoesch to modernize its iron and steel works, helping the establishment of public research facilities, the building of a technology centre (completed in 1985) and by creating a science and technology park. In addition, the recycling of derelict land provides space for development while enhancing environmental values, an important location factor for high-tech industry. Similar observations may be made regarding policies to improve old public infrastructure originally built to meet the needs of the iron, steel and coal complex. In practice, a strong thrust of Dortmund's economic policies are to support small and medium-sized enterprises. In addition, it is worth noting that in each of these 10 policies at least five, usually more, different organizations played 'lead' roles or were 'highly involved' (see also Foley and Watts 1996a,b).

It remains to be seen whether or not these initiatives will be successful. Henning and Kunzmann's (1990: 221) interim assessment is positive with respect to impacts on both the local industrial base and the local labour market. In the UK's Axial Belt and the American Manufacturing Belt, other communities are having some success in adjusting ravaged economies (Healey and Clark 1985; Marshall 1990; Roberts et al 1990; Foley and Watts 1996a,b). At the same time, there has been criticism of local development and even if judged from their own objectives there have been failures. *Roger and Me,* for example, painted a rather dismal picture of the manifold attempts to promote development by Flint, attempts which included the promotion of small firms (e.g. a firm which manufactured lint rollers); tourism, including the construction of an industrial museum (auto world) and a convention hotel; the retail sector; construction of a new prison; and visits by Ronald Reagan and an evangelist, among others, to provide personal advice on coming to grips with de-industrialization. Flint's efforts,

Table 16.11 Actors at Local and State Levels involved in Local Development Policies in Dortmund

Economic development programme/policy	Local government				Politicians and political committees			Public and semi-public institutions					Regional		State ministries				Others	
	LED	UPD	MPD	UPU	KP	CED	CUP	CIC	CT	TSA	LDC	LO	DA	ARM	MSI	LS	CT	LTU	HW	U
1 Locating technology-oriented new enterprises	■	●	●	–	●	●	○	●	–	○	–	○	○	–	●	–	○	–	–	○
2 Helping existing enterprises innovate	●	–	–	–	○	○	–	■	●	●	–	●	–	–	●	○	○	○	○	■
3 Locating public research facilities	○	●	●	–	■	○	○	●	–	–	–	–	○	–	○	○	○	○	–	■
4 Technology centre (linking university research to private enterprises)	■	–	–	–	●	●	○	■	–	–	–	○	–	–	●	–	–	–	–	■
5 Science and technology park	■	●	○	–	●	●	●	●	–	–	–	–	○	–	○	●	●	–	–	●
6 Aiding start ups	●	○	○	–	○	○	○	■	■	○	–	○	–	–	●	–	–	–	–	○
7 Supporting voluntary employment initiatives	○	○	○	○	■	●	○	–	–	○	–	●	–	–	○	●	●	●	○	○
8 Providing industrial land:																				
Developing virgin land	■	●	●	○	●	●	●	○	○	○	–	–	●	○	○	–	●	–	●	–
Recycling derelict land	■	●	●	○	●	●	●	○	○	○	●	–	○	○	○	–	●	–	●	–
9 Improving public infrastructure roads, freeways, canal and harbour	●	●	●	○	●	●	●	■	○	○	–	–	●	○	●	●	●	●	●	–
10 Providing subsidies for steel industry to aid the process of modernization	–	○	○	–	●	●	○	●	–	–	–	●	–	–	●	●	●	●	■	–

Note: ■, lead; ●, highly involved; ○, supportive or involved; –, not involved.
LED = Department of Local Economic Development; UPD = Urban Planning Department; MPD = Municipal Property Department; UPU = Urban Policy Unit; KP = Key politicians; CED = Committee for Economic Development; CUPD = Committee for Urban Planning and Development; CIC = Chamber of Industry and Commerce; CT = Chamber of Trades; TSA = Trade Supervisory Authority; LDC = Land Development Corporation; LO = Labour Office; DA = District Authority; ARM = Intercommunal Association of Ruhr Municipalities; MSI = Ministry of Economy, Medium Scale Industries and Transportation; LS = Ministry of Labour and Social Affairs; CT = Ministry of Country and Town Development; LTU = Local trade unions; HW = Hoesch-Werke AG; U = University.
Source: Hennings and Kunzmann (1990: 221).

however, seem to have largely failed; the most significant investments, i.e. the museum and the convention hotel, have both closed.

Attempts to rejuvenate de-industrialized regions are by no means limited to manufacturing. De-industrialization has been on such a scale that cities such as Dortmund, Sheffield and Pittsburgh are much less specialized on manufacturing, and other sectors have inevitably come to the fore. There have been particularly widespread attempts to develop tourism, revitalize retailing and promote services as, in general, to use Zukin's (1991) term, these cities have sought to become centres of consumption rather than production. In a German context, Soyez (1986) has noted the potential of 'industrial tourism', one aspect of which is the potential for regions to exploit their industrial heritage in the form of museums, working models and tours of existing operations. The same trend exists elsewhere. In Sheffield's case, attempts to replace lost industrial jobs have involved industrial tourism, the development of international class facilities for a variety of sports, and the creation of a large new shopping complex (Meadowhall) designed to attract shoppers from outside the city region. The sports complex, developed originally for the University games, by the way, was actively opposed by the central government and was very much a 'local' initiative. It might also be noted that the Meadowhall Shopping Centre and the sports complex are on sites formerly occupied by some of Sheffield's best-known steel firms. Needless to say, the replacement of one of the founding hearths of the Industrial Revolution with shops, and the rapidity with which this change has happened, is broadly felt within the community (Exhibit 16.1).

Possibilities for flexible specialization

One common theme of most local policy initiatives throughout de-industrialized core regions (and beyond) concerns the promotion of new and small firms. However, most of these policies offer help on an individual basis, often across sectors, and without a sense of a broader strategy. Yet there is a growing appreciation for

Exhibit 16.1
Meadowhall Ghosts

Bernadette Hayter
Robert Road
Sheffield, 1995

I sit and watch
The streams teem by
And catch the eye
Of men of steel.

Now set in bronze
Their skills once proud
Encased by shroud
Of glist and glint.

Our conscience salved
We've helped scars heal.
How must they feel?
A slow sad death.

New pristine place
Once cast and mould
The men of gold
From sweat and toil.

Theirs once the stream
Of labouring men
Forever stemmed
Redundant now.

This deafening tramp
On polished floor
Now drops no more
Hammer on steel.

Hypnotic trudge
A shopper's dream
The swelling sea
Of consuming greed.

God Midas dwells
'Neath mighty dome
This is the home
Of purchased joy.

From young and old
Brass now is tossed
In melting pot
Of High Street names

Times once so hard
Seems no-one cares
Nothing compares
This Meadowhell.

Two old steel pals
Share bench and gaze
Across the maze
They once knew well.

Over the scene
Their wry eyes scan
No less than sham
This plastic world.

"Who gives a sod
For us now mate?
's like rag man's feight
For t'last balloon."

Note: Meadowhall Shopping Centre is built on the sites of former steel works. With respect to the last two lines, a 'ragman' refers to 'rag and bone men' who plied the street with horse and cart making a living by collecting unwanted items from households (clothes, pots, etc.) which were recycled for cash. In exchange for these items, the ragmen gave balloons to children. 'Feight' is Sheffield dialect for 'fight'.

the development benefits of flexible specialized industrial districts which compromise interacting populations of small firms, including those organized around core firms (Piore and Sabel 1984; see Chapter 13). These benefits are particularly evident when interactions are based on mutually rewarding innovative behaviour. An interesting and important question concerns whether or not the old industrial core regions can develop, or re-develop, such systems. In the Axial Belt, Ruhr and American Manufacturing Belt, recent studies suggest that such a possibility is realistic (Grabher 1991; Florida 1994; Healey and Dunham 1994).

In the case of the Ruhr, Grabher (1991) argues that the traditional vertically integrated iron, steel and coal complexes, "cathedrals in the desert", have been dismantled and are being replaced by networks of firms specializing in various facets of environmental technology. The traditionally dominant firms ('cathedrals') have themselves diversified into higher-value production, and some have become general contractors organizing 'loose coalitions' of SMEs whose initial specialisms in environmental technology focused on solving existing environmental problems in the region. Grabher (1991) suggests that these firms are now shifting their emphasis from 'end of pipe' environmental technology to preventive ('front of pipe') technology, a shift which has considerable implications for export. In this way, the Ruhr is developing new specialisms linked to the old specialisms, and the new skills are being learnt and developed within a newly evolving social division of labour. In the Axial Belt of the UK, there is also evidence that Coventry achieved a modest revival in the late 1980s, at least partially on the basis of a 'new regime' of subcontracting, especially within the auto industry, which features relationships which are based more on quality, stability and the specialization of suppliers than had previously been the case (Rawlinson 1991; Healey and Dunham 1994: 1296).

An even larger-scale trend towards the re-organization and re-vitalization of the manufacturing sector has been raised in the context of the American Manufacturing Belt (Florida 1994). The conventional view is that the American Manufacturing Belt has been extensively de-industrialized and the region is continuing to lose jobs (Crandell 1993). Indeed, according to Crandell (1993: 18), between 1977 and 1987 the Manufacturing Belt lost more jobs than any other region in the US and this loss is 'more than expected', taking into account industrial structure. Job losses in manufacturing in the region were also substantial between 1980 and 1993 (Figure 2.9).

Yet, if the powerful advantages incorporated within industrial core regions on the scale of the American Manufacturing Belt can be reversed within a decade, then the possibility that the relentless downward spiral of de-industrialization can be arrested should be recognized. Indeed, Florida (1994) has argued that just such a reversal is under way in the American Manufacturing Belt, particularly with respect to the core Midwest states of Ohio, Indiana, Illinois, Michigan, Wisconsin and Minnesota, plus Pennsylvania and New York. He notes that within the wider processes of creative destruction, rejuvenation and new forms of production organization are not limited to 'new industrial spaces' but are possible in old industrial districts.

Florida notes that in 1990, the Industrial Midwest produced US$250 billion worth of manufacturing goods, about one-quarter of the national total (Table 16.12); if New York and Pennsylvania are included, manufactured output in 1990 amounted to US$365 billion, or 36% of the national total. Moreover, after the massive real decline in manufactured output between 1977 and 1982 of over −29%, between 1982 and 1987 real manufacturing output grew at almost 16% a year. Real growth in manufacturing value added was even faster (and output for all sectors in the economy was greater in 1990 than in 1977). In addition, real investment in 1990 in the Midwest was similar to 1977 levels. On the other hand, in real dollar values, manufactured output in 1990 was still only 82% of 1977 levels, and employment levels in the 1990s are

Table 16.12 Trends in Key Economic Indicators for the Industrial Midwest, 1970–1990

	1977	1982	1987	1990	% Change		
					1977–1982	1982–1987	1987–1990
Output ($ billions)	754.6	648.3	780.0	819.0	−14.1	20.3	5.0
	425.0	*581.5*	*823.4*	*994.8*	36.8	41.6	20.8
Manufacturing output	248.0	175.8	203.9	204.5	−29.1	15.9	0.3
($ billions)	*139.7*	*157.7*	*215.3*	*248.4*	12.9	36.5	15.4
Manufacturing value-	301.5	−26.00	274.1	281.0	−26.0	22.8	2.5
added ($ billions)	*169.8*	*200.1*	*289.4*	*323.8*	17.9	44.6	11.9
Manufacturing value-	56.8	49.7	60.1	58.4	−12.5	20.9	−2.9
added per employee	*32.0*	*44.6*	*63.5*	*70.9*	39.3	42.3	11.7
($ thousands)							
Capital expenditures	23.1	17.9	20.0	20.6	−22.6	11.7	3.0
($ millions)	*13.0*	*16.1*	*21.1*	*25.0*	23.3	31.5	118.5

Source: Florida (1994).
Notes: Figures in the top row for each variable are in constant 1985 dollars. Figures in italics represent nominal dollars.

substantially lower than in 1980 (Figure 2.9). Growth rates in manufacturing between 1987 and 1990 were also considerably reduced and the early 1990s was another recessionary period.

The basis for Florida's (1994) optimistic scenario is that underlying the statistical record of improvement (even if this improvement slowed in the late 1980s and is not measured by jobs) there is a "deeper transformation of the organization of production" occurring in the Midwest. According to Florida, this transformation involves a shift away from Taylorist work principles towards 'high performance organizations' based on a higher-skilled, more flexible workforce organized in teams and committed to high-quality, innovative production (Figure 11.4; Hayter forthcoming). In this view, DFI from Japan, as well as from some major US firms such as Xerox, is leading the transformation by establishing new plants which are establishing new labour bargains based on team work, continuous training and a commitment to innovation and product quality. In addition, Florida notes that high-performance organizations in the region have sought to develop suppliers which are similarly committed to high-performance principles. In autos and electronics products in particular there are signs of flexibly specialized production systems. Florida further notes the increasing export commitment among firms in the Midwest, including small and medium-sized firms.

According to this analysis, rejuvenation in the Midwest depends on the acceptance of the high-performance model. Even if it is a management goal, such acceptance is likely to take many years, particularly when the workforce is unionized, as the Powell River experience in British Columbia reveals (see Chapter 12). Any shift towards high performance, or other models for that matter, is likely to be geographically uneven within the Midwest (Pollard and Storper 1996).

Conclusion

According to the theory of cumulative advantage, established industrial regions, historically best evidenced by the UK's Axial Belt, the German Ruhr and the American Manufacturing Belt, have enormous advantages over more peripheral and newer centres of industrialization. According to the theory of new industrial spaces, more peripheral and newer centres of industry offer powerful attractions to new investment over the established regions. The historical record shows that the core industrial regions of the UK, Germany and the US have often experienced the ups and downs of business cycles and the ferocity of full-blown depression, such as that of the 1930s. Following the depression of the 1930s, however, the industrial base of the core industrial regions emerged intact and began to grow, much stimulated by World War II and the subsequent long boom. In contrast, the crisis of de-industrialization that they have experienced since the 1970s during the IC techno-economic paradigm represents a fundamental turning point for these regions. Their industrial power has been challenged, part of their industrial base has been permanently removed, and future industrial vitality requires significant changes in organization and perhaps in industry mix as well.

The UK's Axial Belt, the German Ruhr and the American Manufacturing Belt now know that their industrial strengths are not inviolate or a non-extinguishable heritage. Indeed, such is the destruction in the UK's Axial Belt that the ability to generate new skills out of the old skills has been seriously compromised. Throughout the old industrial regions, communities have looked to non-manufacturing sectors to replace lost manufacturing jobs. There are limits to the potentials of these strategies, however, and the rejuvenation of manufacturing is an important and legitimate task of local development. From this perspective, there is a growing weight of opinion that the success of plans for industrial rejuvenation depends on how well factories, firms and production systems develop as mutually interactive learning systems and how well governments can introduce policies which are supportive of learning and innovation (Storper 1993).

17

INDUSTRIAL TRANSFORMATION AND JOBS: CONTEMPORARY DILEMMAS

From a long-term perspective, industrial geographies are constantly shaped by the complex, dynamic forces rooted in Schumpeter's (1939, 1943, 1947) gales of creative destruction. Within and among regions, firms and people co-operate and compete in relationships which are continually enhanced and threatened by the forces of demand and supply and by technological and organizational innovation. One of the implications of processes of creative destruction is that there are no conclusions in the sense of some 'end state(s).' From this perspective, too-literal claims for the 'end of history' or the 'end of geography' fundamentally misrepresent the dynamics of change, and the problem of industrial transformation. In this brief, concluding chapter a few comments are offered on contemporary dilemmas underlying the problem of industrial transformation (see preface to Part I). These comments primarily centre on the issue of jobs (Barnet and Cavanagh 1994; Rifkin 1995; Eatwell 1996a).

The forces of creative destruction unleashed by the Industrial Revolution created new technologies, institutions and geographies, forcing existing systems to adapt or diminish and perhaps disappear. Since the late 18th century, on an ever-increasing scale, specialized industrial communities of varying size and complexity have achieved great wealth by exporting to other regions and nations while simultaneously becoming more open and vulnerable to outside or 'global' forces of change. Globalization, in other words, is not a new phenomenon. The nature and extent of industrialization, and of globalization, however, has changed significantly, both quantitatively and qualitatively (Table 2.3 and Figure 2.4). In recent decades, for example, globalization has become 'deeper' in terms of geographic scope and the diversity of ways in which production is configured and factories and communities linked to one another. Moreover, in this current phase of industrialization, whether labelled the ICT techno-economic paradigm and the fifth Kondratieff, or the post-Second Industrial Divide or the Third Industrial Revolution, at global and local scales, there are distinctly new features and dilemmas concerning employment and the nature of work.

At the core of the jobs problem is the ability of the global economy to produce vast quantities of products (and indeed services) with fewer employees. Significant increases in production no longer apparently imply increasing employment opportunities; they may even imply the reverse. In addition to the spectre of jobless growth, there is a particular concern about the supply of 'good' jobs and growing income inequalities. In turn, these concerns inevitably raise questions about the efficacy of personal, community and corporate investments in education and training. But potential solutions also raise another significant dilemma. In particular, assuming (good) jobs can be found, there is the matter of whether or not the environment can sustain increases in the demands associated with increases in the incomes of the more or less five billion people which populate the earth, most of whom are indeed likely to want to consume more. These various dilemmas vary between developed and developed economies, among nations within these (and related) categories and among localities within nations.

The jobs challenge

The Industrial Revolution created a massive jobs problem in large part by making redundant many workers in agriculture, which in 1800 was everywhere the predominant economic activity. Industrializing countries, however, were able to absorb these surplus workers, plus many more generated by growing birth rates, within the rapidly expanding manufacturing sector. Subsequently, among industrialized countries, as employment increases in manufacturing declined the service sector absorbed a growing share of the workforce. Yet it is myopic to think that services are some kind of 'natural' solution to job-creation problems, for reasons already given (see Chapter 3). Manufacturing is a source of productivity, wealth and bargaining power, and is closely integrated with services as well as the primary sector. In some countries, at least, de-industrialization is such a social problem because services are not providing ready-made solutions to employment problems. In fact, there are real concerns about employment potentials within the service sector (Rifkin 1995: 141–164).

The scope of the problem

In recent decades the global workforce has grown rapidly (Johnston 1991). Thus ('official') counts by the OECD and related organizations calculated that the global workforce expanded from almost 1597 million people in 1970 to over 2164 million in 1985 (Table 17.1). These same studies estimated that the global workforce will expand to about 2753 million by 2000, of which 14.6% would be in the OECD countries, down from 19.2% in 1970. The workforce in the OECD is growing, but much less rapidly than in developing countries. While between 1985 and 2000 the workforces in OECD countries is estimated to grow at about 0.5% per year, in developing regions the rate of growth is estimated at 2.1% (Table 17.1). In absolute numbers, between 1970 and 1985 developing countries added almost half a billion workers, and it is estimated that they will add over half a billion workers between 1985 and 2000. Even in the OECD, where the workforce has almost stopped growing, between 1985 and 2000 a further 29 million workers are anticipated.

In aggregate terms, expansion of the global workforce is impressive. In gender terms, among developed regions in 1987 about 40% of the workforce was female, and in developing countries this figure was almost 35% by 1985 (Table 17.1). In addition, the migration of workers among developing countries, and between developing countries and OECD countries, is occurring on an increasing scale. Barnet and Cavanagh (1994: 296), for example, report that by the early 1990s, around 75 million people were leaving poor countries each year, largely in search of work. Indeed, they suggest that in Jordan and Bangladesh the foreign exchange generated by workers in foreign countries constituted their principal export earnings. Moreover, there has been a widespread tendency for developing countries to invest in education and create an increasingly larger pool of workers who perceive better and more appropriate job prospects among OECD countries (Johnston 1991: 121–122). In the case of the US, many immigrant workers originate from the more rapidly growing NICs (Barnet and Cavanagh 1994: 299).

If there are more jobs globally now than ever before, however, there is also more unemploy-

Table 17.1 Distribution of the World's Workforce, 1970–2000

Country of region	Labour force 1970 (million)	Labour force 1985 (million)	Female share (%)	Labour force 2000 (million)	Labour force annual growth rate 1985–2000 (%)
World[a]	1 596.8	2 163.6	(36.5)	2 752.5	1.6
OECD[a]	307.0	372.4	(40.9)	401.3	0.5
US	84.9	122.1	(44.1)	141.1	1.0
Japan	51.5	59.6	(39.9)	64.3	0.5
Germany	35.5	38.9	(39.3)	37.2	−0.3
UK	25.3	28.2	(41.4)	29.1	0.2
France	21.4	23.9	(42.5)	25.8	0.5
Italy	20.9	23.5	(36.5)	24.2	0.2
Spain	13.0	14.0	(32.6)	15.7	0.8
Canada	8.5	12.7	(43.2)	14.6	0.9
Australia	5.6	7.4	(39.7)	8.9	1.3
Sweden	3.9	4.4	(48.0)	4.6	0.3
Developing regions[a]	1 119.9	1 595.8	(34.7)	2 137.7	2.1
China	428.3	617.9	(43.2)	761.2	1.4
India	223.9	293.2	(26.2)	383.2	1.8
Indonesia	45.6	63.4	(31.3)	87.7	2.2
Brazil	31.5	49.6	(27.2)	67.8	2.1
Pakistan	19.3	29.8	(11.4)	45.2	2.8
Thailand	17.9	26.7	(45.9)	34.5	1.7
Mexico	14.5	26.1	(27.0)	40.6	3.0
Turkey	16.1	21.4	(34.0)	28.8	2.0
Philippines	13.7	19.9	(32.0)	28.6	2.4
South Korea	11.4	16.8	(34.0)	22.3	1.9
USSR	117.2	143.3	(48.3)	155.0	0.5

[a] Totals include some countries not listed in table. With respect to the female share of the total workforce, for OECD 1985 data used, and for developing regions 1987 data used.
Source: Johnston (1991: 117, 119).

ment. According to estimates provided by the International Labour Organization (ILO) for the early 1990s, more than 800 million people across the globe are unemployed or underemployed (Rifkin 1995: xv). Of this number, around 700 million are estimated to be in developing countries (Barnet and Cavanagh 1994: 294) where the growth in the working population is particularly rapid, along with its feminization as female participation rates continue to grow (Johnston 1991: 116; Barnet and Cavanagh 1994: 294).

Among the major OECD countries it is clear that the scope of the jobs challenge has increased since 1970 (Table 17.2). Thus, all of the countries comprising the G7 witnessed significant increases in unemployment levels from 1983 to 1992 compared to two decades earlier (Eatwell 1996b: 3–5). The lowest rate of increase occurred in the US where unemployment rates in the latter period were 1.5 times greater than in the earlier period. The biggest increases occurred in France and Germany. As a result, the prevailing tendency for unemployment rates to be lower in Europe than in either the US or Canada in the 1950s and 1960s has been reversed since the early 1980s. Apart from Japan, however, unemployment rates are significantly higher than they were in the 1950s and 1960s. Even in Japan, unemployment has increased.

Table 17.2 Unemployment in the G7 Countries, 1964–1973 and 1983–1992

	1964–1973	1983–1992
US	4.46	6.69
Canada	4.23	9.64
UK	2.94	9.79
West Germany	0.79	6.03
France	2.23	9.70
Italy	5.48	10.13
Japan	1.22	2.71

Source: OECD.

Moreover, there are few indications that unemployment, especially in Europe, will decline in the near future. Indeed, in Europe the situation in the 1990s seems to be worsening (Rifkin 1995: 199). Many countries have experienced double-digit unemployment recently, and in Germany unemployment in the early 1990s was around 4 million. In Spain, unemployment of close to 20% has been recorded. The US has managed to increase its overall workforce faster than other OECD countries (Table 17.1), while reducing unemployment. In Canada, on the other hand, 1996 unemployment rates continue to hover around 10%.

Throughout the OECD, including the US, there is concern for the quality of jobs being created and for the uncertain nature of changing work patterns (Christopherson 1989). Barnet and Cavanagh (1994: 293), for example, cite studies which suggest that half of the jobs created in the US between 1980 and 1987 were part-time or contracted out on a part-time basis, often involving women paid at lower wage levels. Kenney and Florida (1994: 271) estimate that part-time workers in general earn between 20 and 40% less than permanent employees for comparable work. Part-time workers are also much less likely to have pension plans. Thus, even in the US where the aggregate size of the workforce is increasing, there is concern that a growing share of jobs are relatively low paid and unstable. Similar fears have been expressed for Europe (Standing 1992). In addition, peripheral labour segments, traditionally dominated by minorities

and women, are increasingly supplemented by white males, especially those with limited education and skill. Indeed, several authors have noted that income polarization is occurring in North America and Europe through various kinds of contracting-out strategies in which core workers are replaced by financially flexible workers.

The scope of the job problems is in fact much greater outside of the OECD where job supply problems are compounded by low incomes and meagre support networks. In attempts to broaden opportunities and increase productivity, many developing countries have expanded education and training. As in the case of the OECD, especially given the cost and length of time involved, questions can be asked regarding the purpose of training and education. Moreover, education and training typically increases the mobility of individuals and poor countries have to be particularly careful that investments which increase the supply of skilled human resources are matched with appropriate job and income prospects. Several east European countries face similar dilemmas.

Yet knowledge is vital to the realization of human potential and, even if personal (and community) choices have become more complicated, there is a generally positive relationship between education, income levels and economic growth. Moreover, globally mobile labour can enhance productivity and help both sending and receiving countries (Johnston 1991); the search for employment flexibility need not be deleterious to worker and community welfare (Standing

1992; see Chapter 12). Needless to say, working out socially desirable employment policies will not be easy. One important reason relates to prevailing high levels of unemployment – especially when it is realized that there is no consensus as to why these high levels exist.

Alternative explanations for high unemployment According to the theory of techno-economic paradigms, and related Kondratieff cycle models, periods of deep recession and high unemployment facilitate the structural (i.e. institutional and technological) changes necessary for employment-expanding growth, especially for industrialized countries. As noted, globally, jobs have been created (Table 17.1). In addition, among OECD countries, the US has been a leader in this regard and Japan's unemployment rates of the 1990s are less than half the rates of the US during the long boom of the 1950s and as low as European rates were then. Moreover, in contrast to previous techno-economic paradigms, industrialization is occurring on a much broader geographic scale.

At the same time, the structural nature of the unemployment problem (and the fragility of the growth that has occurred) raises questions as to why. The theory of techno-economic paradigms itself is organized around technological and institutional innovation and these two perspectives form the basis for competing theories of high levels of unemployment. In the best-known argument, technology is seen as the prime culprit (Barnet and Cavanagh 1994; Rifkin 1995). From this perspective, the relationship between technology and employment in the present techno-economic paradigm is different from previous ones. Rather than creating new jobs, technology is seen as replacing the need for work. Rifkin (1995), for example, paints a relentless trend towards automation in which the future is captured by the workerless factory and machines that think. Example after example is provided of machines replacing people, including in developing countries. Moreover, for Rifkin (1995: 141), the loss of industrial jobs will soon be followed by the 'last service worker' as the new

information technologies replace the need for middle management and low-skilled employees in a wide range of service occupations.

Eatwell (1996b) offers a different explanation for high unemployment, especially among the OECD countries. He argues that "Whatever technological changes may have done to the *composition* of employment, there is no evidence that the speed of technological change is behind the growth in unemployment throughout the G7" (Eatwell 1996b: 5). He reasons that if technology is replacing workers, then productivity should be increasing. In fact, among all the G7 countries overall productivity growth and manufacturing productivity growth has been much less in the 1980s and 1990s than in the 1960s (Eatwell 1996b: 6). The smallest declines have occurred in the US where manufacturing productivity was 3.1 per annum between 1964 and 1973, and 2.8 per annum between 1983 and 1992. In the same time periods, the decline for West Germany was from 4.0 to 2.4 and that for Japan from 9.6 to 5.7. In both periods, overall productivity growth is less than for manufacturing.

Moreover, in Eatwell's analysis, unemployment has not simply been caused by investment in developing countries: so far the G7 has tended to enjoy a trade balance with the faster-growing NICs. Rather, Eatwell (1996b: 9) argues that the primary culprit for high unemployment lies in the slow down of demand. Eatwell notes that for all G7 countries the rate of growth of real GDP is less in the 1980s and 1990s than in the 1960s. In the US, for example, the annual rate of growth declined from 4.0 to 2.9 between the 1964–1973 and 1983–1992 periods. For Germany and Japan, the respective declines were from 4.5 to 2.9 and from 9.6 to 4.0. Real GDP growth also declined among developing countries.

In turn, Eatwell suggests that the persistence of the decline in demand into the 1990s stems from changes in the structure of international finance and related impacts on domestic policies. In particular, since the early 1970s, the elimination of fixed exchange rates and the deregulation of financial markets has provided the basis for an explosion of short-term capital movements which are purely speculative. Unfortunately, these movements have undermined the abilities of national governments to regulate the economy as financial speculators have more money to move around than most national governments have in reserve. Consequently, the preferences of the financial community have exercised an increasingly important influence on government policy and this interest has been more concerned with reducing government expenditures and controlling inflation than with maintaining full employment.

These differing interpretations of the unemployment problem lead to differing policy prescriptions. Those who stress the relentless march of technology stress ways of re-distributing available work and income among more people in more equitable ways. Eatwell and others, first and foremost, want to regulate the global financial system, notably by reducing what appear to be huge rewards for speculation; for example, by fixing exchange rates or taxing international financial transactions. The two sets of solutions are not necessarily incompatible. Stabilizing the international financial system would permit long-term thinking in which investments in physical assets would be given greater priority and allow national governments somewhat more leeway in formulating distinctive polices, including with respect to how to cope with technological innovation and the implications of technology for employment.

Industry and the environment If policies can be introduced to stimulate global GDP and productivity increases, and/or if global demands can be increased by more equitable work and income distribution, then so would industrial output. Whether or not the manufacturing sector's direct employment share remains the same or declines, manufacturing processes need material inputs and energy to create material outputs. Industry has always made great demands on the environment and in the long-term processes of creative destruction, environmental values have tended to be destroyed first and if possible re-created later. Indeed, environmental demands are closely associated with income levels and the pattern of industrialization. On a global basis, it is the rich industrial countries, rather than the poorer

countries, that on a per capita (and absolute) basis consume vastly more energy and resources, and discharge more unwanted pollutants into the air and waters of the world.

The destruction of environmental values by industry has led to major attempts to reduce environmental damage and to restore environmental resources and amenity. These attempts have generally been more successful in the OECD countries. In contrast, the environmental record of industrialized (former) communist countries has generally been appalling. Moreover, in OECD countries, shifts towards more environmentally conscious manufacturing processes are occurring; in the US, for example, Florida (1996) provides systematic evidence in support of a positive relationship between 'high performance' organizations (see Chapter 12) and their innovative strategies with respect to work organization, R&D and product development, and innovative strategies to pre-empt environmental problems. In this view, environment and economy are not inevitably part of a zero sum game but can be mutually reinforcing. Indeed, environmental industries themselves have become large-scale job generators in some countries.

On the other hand, many developing countries, including the population giants of China and India, are seeking modern industry and modern technology, and environmental impacts are often either ignored or dismissed as the hypocrisy of rich countries. Yet it is not clear just how far industrialization can proceed without violating environmental limits, unless specific measures are undertaken. Industrialization and improvements in the standard of living, for example, need vast supplies of water. In India and China, which contain large populations in semi-arid areas, where is this water going to come from? Resolutions to these problems will require technological and institutional innovations to ensure environmental values are properly incorporated in financial accounting.

Industry and local development: turning the kaleidoscope?

Globalization, led by MNCs, incorporates powerful forces of standardization, long felt in

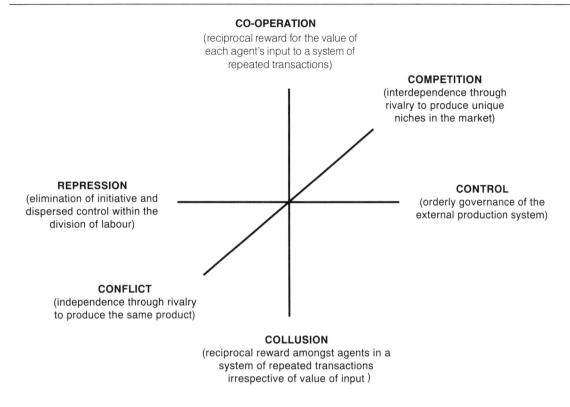

CO-OPERATION
(reciprocal reward for the value of
each agent's input to a system of
repeated transactions)

COMPETITION
(interdependence through
rivalry to produce unique
niches in the market)

REPRESSION
(elimination of initiative and
dispersed control within the
division of labour)

CONTROL
(orderly governance of the
external production system)

CONFLICT
(independence through rivalry
to produce the same product)

COLLUSION
(reciprocal reward amongst agents in a
system of repeated transactions
irrespective of value of input)

Note: Each axis represents relationships among firms.

Figure 17.1   The Processes of Regional Industrial Evolution. Source: Patchell (1996: 484)

branch plant economies (Levitt 1970), but in fact extending over the globe. Yet the forces of localism are powerful too, as the lives of the great majority of individuals for the great majority of time are spatially highly circumscribed around home and work (Hägerstrand 1970). Moreover, each locality approaches development problems and dilemmas from a different situation and with a different legacy. Patchell (1996) defines an institutional model of regional industrial change which explicitly incorporates the different ways regions around the world have organized themselves to resolve the problem of industrial transformation.

In Patchell's (1996) framework, region's regenerate economically through human relationships which comprise processes of co-operation,

competition and control (Figure 17.1). He notes that a great variety of strategies have been developed in different parts of the world to allow people within regions to mediate the tensions between co-operation, competition and control (and their opposites, respectively collusion, conflict and repression) in the creation of specialized, interdependent and hopefully efficient economies. People within regions need to develop forms of co-operation (or collusion), competition (or conflict) and control (or repression). Patchell uses the metaphor of the kaleidoscope to express the various choices regions make combining these three processes.

Patchell notes two important 'extreme' types of regional industrial evolution. The evolutionary ideal occurs when 'co-operation promotes control and competition'. Control ensures that complex production systems can be co-ordinated, competition encourages innovation to penetrate markets or reduce costs, and co-operation constrains both hyper-competition (which is destructive of long-term thinking and stability) and excessive control which dampens innovation. In contrast, an 'involutionary economy' occurs when firms collude, a dominant firm enforces control, or firms insist on extreme forms of independence, thus reinforcing the status quo. In summary, the "involutionary economy is a trap where short-range self interest cuts off firms and entrepreneurs from the information they need to develop . . . New and differentiated values are not created; rather, production of old products or services is simply parceled out to more hands" (Patchell 1996: 496).

In reality, regional economies can be located somewhere between these extremes. Patchell himself discusses several alternatives located in Europe, North America and Asia, each of which provides a different turn of the kaleidoscope. Interesting questions relates to how, if at all, people and organizations in particular regions can actually turn the kaleidoscope themselves. Can historically hyper-competitive regions become fundamentally more co-operative? Can regions historically experiencing high levels of control become more competitive? How deep are the forces of institutional inertia? Such changes are difficult, but people and organizations can learn, and learning increases the propensity for innovation. The idea of regional change needs to be given more attention (Ahern 1993a; Patchell 1993a; Storper 1993).

Conclusion – but not for industrial geography

The challenge for industrial geography is to understand the industrial transformation problem as it evolves over time and space. The extraordinary dynamic nature of this problem ensures the vitality and importance of industrial geography. For the late 1990s, Patchell's (1996) kaleidoscope provides an appropriate metaphor for analysing regional industrial choice. It is a metaphor that emphasizes differentiation and choice across space as well as time. Moreover, it attaches central importance to people themselves and how they compete, control and co-operate to ensure a livelihood. From this perspective, the distinctiveness of local industrial geographies is intimately related to the goals of industrial transformation and the underlying attitudes and values of local populations; institutional arrangements, for example with respect to industrial organization, labour markets, trade, technology and environmental matters; and the various ways globalizing forces are absorbed, mediated and manipulated (Glasmeier et al 1993; Peck 1996; Taylor 1996). Looking down the kaleidoscope, an emerging theme of crucial importance is the idea of regions (and therefore employees, families, firms, other organizations and production systems) as learning systems and how learning systems vary from place to place. Simply put, the places that comprehensively organize themselves for learning and innovation will increase their chances of meeting the goals of industrial transformation (Cooke and Morgan 1990; Patchell 1993a; Storper 1993; Florida 1994; Hayter 1996).

Appendix 1

INDEXES OF LOCALIZATION, DIVERSIFICATION AND SHIFT-SHARE TRENDS

In manufacturing geography, and related fields such as regional economics and regional planning, there are several indexes of localization and diversification, and a method of analysis known as shift-share, which have been widely used to descriptively summarize aggregate statistical trends. Using data from British Columbia, and Canada in general, these methods are briefly summarized in this appendix.

By way of introduction, it might be noted that manufacturing activities may be measured according to several variables. The most common of these variables are simple counts of factories and firms, employment, sales, value added and physical quantity measures. In land-use studies, floor-space size is another variable that is used. Each variable has its particular advantages and disadvantages. Measures of physical quantity (e.g. tonnes of steel, board feet or cubic metres of lumber, dozens of knives and forks) are readily understood and available data which can be used to map the size of an industry over space and time. Such measures, however, are hard to use in inter-industry analysis and are complicated by qualitative considerations, such as high- and low-quality steel. Monetary measures, such as value added, get around the standardization problem and facilitate comparisons across industries (cross-sectional analysis). Value added is defined

as the value of products sold by a factory minus the value of material inputs; that is, value is added primarily by labour (and management) and by machines. While value-added data are frequently provided in government censuses, quite a lot of information needs to be collected to derive the value-added figure.

In practice, the most commonly used variable within manufacturing geography to measure the size, trend and evolution of industries is employment. Generally speaking, employment data are easy to obtain, there is no problem with inflation and data are intrinsically interesting and meaningful. Because of its wide usage, some limitations associated with employment as a measure of size might be given particular note. First, not all employees are equal; they vary by skill, income and efficiency. Second, for the firm, labour is a factor of production, similar to capital, material and land. Consequently, labour can be replaced or substituted to some degree by these other factors of production (and vice versa). Thus, for capital-intensive industries, the number of employees may underestimate the industry's importance within an economy in terms of exports and income generated. Similarly, jobs may overestimate the importance of a labour-intensive industry. Moreover, because of substitution, direct employment change may or may not

Table A1.1   Employment in Manufacturing Industries of British Columbia and Canada, 1978 and 1993

	British Columbia				Canada			
	1978		1993		1978		1993	
Industry	Employment	%	Employment	%	Employment	%	Employment	%
Food & beverages	19 671	12.8	20 234	13.6	229 906	12.84	216 101	13.1
Tobacco	0	0.0	114	0.1	8 778	0.5	4 778	0.3
Rubber & plastics	1 456	1.0	4 530	3.1	60 455	3.4	73 374	4.5
Leather & allied products	274	0.2	191	0.1	24 415	1.4	12 818	0.8
Primary textiles	118	0.1	140	0.1	67 808	3.8	18 346	1.1
Textile products	1 253	0.8	1 483	1.0	20 003	1.1	27 646	1.7
Clothing	2 400	1.6	3 976	2.7	99 517	5.6	82 737	5.0
Wood	50 955	33.3	41 939	28.2	119 004	6.6	109 961	6.7
Furniture & fixtures	1 752	1.1	2 230	1.5	46 613	2.6	44 654	2.7
Paper & associated products	20 655	13.5	17 469	11.8	126 783	7.1	101 926	6.2
Printing & publishing	7 422	4.8	11 688	7.9	98 037	5.5	124 867	7.6
Primary metals	8 941	5.8	4 711	3.2	121 996	6.8	84 416	5.1
Fabricated metals	10 088	6.6	10 585	7.1	156 665	8.7	132 606	8.0
Machinery	5 753	3.8	5 505	3.7	92 113	5.1	74 379	4.5
Transportation equipment	8 523	5.6	5 675	3.8	178 636	10.0	209 879	12.7
Electrics & electronic products	3 210	2.1	5 390	3.6	114 279	6.4	118 629	7.2
Non-metal mineral products	4 172	2.7	4 636	3.1	55 843	3.1	42 661	2.6
Petroleum & coal products	1 146	0.7	894	0.6	20 383	1.1	14 084	0.9
Chemicals	2 751	1.8	3 196	2.2	84 786	4.7	90 490	5.5
Other manufacturing	2 644	1.7	3 898	2.6	64 829	3.6	63 050	3.8
Total	153 184	100.0	148 484	100.0	1 790 849	100.0	1 647 402	100.0

Source: *Statistics Canada*, cat. 31–203.

accurately reflect the prospects of an industry in a region. Massey and Meegan (1982), for example, documented the employment and output change of several industries in the UK between 1968 and 1972. In the case of locomotives and synthetic fibres, both industries declined by about 20% in terms of jobs. However, while the output of the locomotive industry declined by a similar amount in the same period, synthetic fibres recorded a big output gain. Clearly, change in the two industries raises very different types of implications for the communities in which they are located.

It is also important to emphasize that the following indexes and shift-share analysis are not theories, nor is there any question of (formal) statistical significance. They are what might be called 'quick and dirty' (or crude) methods for statistically summarizing employment data in a simple way. Their statistical interpretation is based on common sense. These methods, however, are useful in summarily describing tables of information and providing insights into the nature of regional industry composition and change over time, the wider significance of which depends on the theoretical argument presented. Bearing in mind these caveats, employment data are used in the following calculation. In particular, the examples draw from employment for 20 manufacturing industries for British Columbia (BC) and for Canada in 1978 and 1993 (Table A1.1)

The location quotient

The simplest measure of determining the (statistical) extent to which an industry is concentrated in an area relative to some larger benchmark region is provided by the location quotient (LQ):

LQ of industry i =

$$\frac{\dfrac{\text{employment of industry in study region}}{\text{employment of all industries in study region}}}{\dfrac{\text{employment of industry } i \text{ in benchmark region}}{\text{total industrial employment in benchmark region}}}$$

The rationale underlying the location quotient index is that if the LQ of an industry is greater than 1, the industry is 'over-represented' in the case study region in comparison to the benchmark region. Such an activity is likely to be relatively specialized within the region and a net contributor to exports. If the LQ is less than 1, the activity is 'under-represented' in the region and a net importer.

For example, the LQ of the wood industries in BC in 1993, with Canada as benchmark is

$$\text{LQ} = \frac{41\,939}{148\,484} \bigg/ \frac{109\,961}{1\,647\,402} = \frac{0.282}{0.067} \text{ or } \frac{28.2\%}{6.70\%} = 4.2$$

In the case of paper and associated products, the LQ for BC in 1993 is

$$\frac{11.8\%}{6.2\%} = 1.9$$

In the case of transportation equipment, the LQ for BC in 1993 is

$$\frac{3.8\%}{12.7\%} = 0.30$$

According to our assumptions, the wood and paper industries are important specialisms within BC which are significant exporters, while transportation equipment is relatively unimportant in comparison to other parts of Canada. Again, it should be noted that differences in the size of LQs does not involve any question of statistical significance. One convention is to consider LQs above 1.5 as clearly denoting specialisms. LQs assume equal efficiency and consumption patterns throughout the benchmark region, which is often, but not always, the nation. The nation, however, is a meaningful political unit and one within which data classification schemes are typically standardized.

Coefficient of specialization

The simplest measure for determining the (statistical) extent to which a region is diversified among a set of industries compared to a larger benchmark region is the coefficient of specialization (S_c). The S_c is calculated by subtracting, for each industry, the percentage share of an industry in the benchmark region from the percentage share of the same industry in the case study region.

S_c varies from 0 (perfect diversification) to 1 (perfect specialization). Note that S_c tells you the amount by which employment in a region would have to be redistributed to get the same distribution as the nation as a whole. For example, the coefficient of specialization for BC in 1978, using Canada as a benchmark region, is 0.33 (Table A1.2). The coefficient for 1993 is 0.28, indicating that BC's manufacturing has become slightly more specialized between 1978 and 1993. It should be noted that specialization and diversification are defined only in terms of employment composition across a set of industrial categories. If other variables, or a different set of categories, are used the index may change.

The *Lorenz curve* is a graphical method of portraying a region's industrial diversification in relation to the nation. One way of deriving the Lorenz curve is to (1) rank each industry for both the case study region and the benchmark region on the basis of number employed; (2) derive a frequency distribution; and (3) plot on a graph the cumulative share of each industry for both case study and benchmark regions against the ranks of industrial categories in order of numbers employed. The greater the difference between the curves, the greater the degree of specialization of the case study region.

As an example, the Lorenz curve for BC and Canada in 1993 is shown in Figure A1.1.

Shift-share analysis

Shift-share analysis classifies employment change over time in a region in comparison to a

Table A1.2 Coefficient of Specialization – An Analysis of Manufacturing Industry in British Columbia (Canada as Benchmark)

Industry	1978			1993		
	% Employed in British Columbia	% Employed in Canada	% Difference	% Employed in British Columbia	% Employed in Canada	% Difference
Food & beverages	12.8	12.8	0.0	13.6	13.1	0.5
Tobacco	0.0	0.5	-0.5	0.1	0.3	-0.2
Rubber & plastics	1.0	3.4	-2.4	3.1	4.5	-1.4
Leather & allied products	0.2	1.4	-1.2	0.1	0.8	-0.6
Primary textiles	0.1	3.8	-3.7	0.1	1.1	-1.0
Textile products	0.8	1.1	-0.3	1.0	1.7	-0.7
Clothing	1.6	5.6	-4.0	2.7	5.0	-2.3
Wood	33.3	6.6	26.6	28.2	6.7	21.6
Furniture & fixtures	1.1	2.6	-1.5	1.5	2.7	-1.2
Paper & associated products	13.5	7.1	6.4	11.8	6.2	5.6
Printing & publishing	4.8	5.5	-0.6	7.9	7.6	0.3
Primary metals	5.8	6.8	-1.0	3.2	5.1	-2.0
Fabricated metals	6.6	8.7	-2.2	7.1	8.0	-0.9
Machinery	3.8	5.1	-1.4	3.7	4.5	-0.8
Transportation equipment	5.6	10.0	-4.4	3.8	12.7	-8.9
Electrics & electronic products	2.1	6.4	-4.3	3.6	7.2	-3.6
Non-metal mineral products	2.7	3.1	-0.4	3.1	2.6	0.5
Petroleum & coal products	0.7	1.1	-0.4	0.6	0.9	-0.3
Chemicals	1.8	4.7	-2.9	2.2	5.5	-3.3
Other manufacturing	1.7	3.6	-1.9	2.6	3.8	-1.2
Total % difference			33.0			28.4
% Difference/100 = coefficient of specialization			0.33			0.28

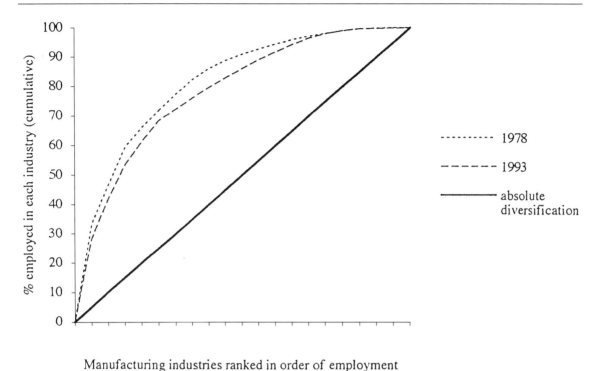

Figure A1.1   Lorenz Curve showing Degree of Manufacturing Diversification in British Columbia, 1978 and 1993

benchmark region into three categories or components: national share (N_j), structural shift (P_j) and differential shift (D_j). The national-share component represents the amount by which regional employment would have changed if it had changed at exactly the national rate for all industries over the study period. If the region grows at a different rate to the nation, there is a (positive or negative) employment *shift*. This total shift comprises the structural shift, also known as the proportionality shift and the industry mix effect, which is the amount of regional employment change attributable to the type of industries in the region (whether they are nationally fast- or slow-growing). Second, the net shift comprises the differential shift, also known as the regional effect, which is the amount of regional employment change attributable to location factors within the region.

In algebraic notation, these components can be expressed as:

$$G_j = E_{jt} - E_{jo}$$
$$= (N_j + P_j + D_j)$$
$$N_j = E_{jo}(E_t/E_o) - E_{jo}$$
$$(P + D)_j = E_{jt} - (E_t/E_o)E_{jo}$$
$$= (G_j - N_j)$$
$$P_j = \Sigma_i[(E_{it}/E_{io}) - (E_t/E_o)]E_{ijo}$$
$$D_j = \Sigma_i[E_{ijt} - (E_{it}/E_{io})E_{ijo}]$$
$$= (P + D)_j - (P_j)$$

where E_j is the total employment in case study region j; E is the total benchmark region employment; o and t are the initial and terminal time periods; and i is the industry subscript.

A shift-share analysis of BC between 1978 and 1993, using Canada as the benchmark region,

Table A1.3 Shift and Share Analysis of Manufacturing Industry in British Columbia, 1978–1993

Industry	BC % 1978	Canada % 1978	LQ 1978	National change %	Employment in BC		BC change %	Expected employment	Total shift	Expected employment	Diff. shift	Comp. shift
					1978	1993						
Food & beverages	12.8	13.6	1.0	-6.0	19671	20234	2.9	18097	2137	18490	1744	393
Tobacco	0.0	0.1	0.0	-45.6	0	114	—	0	114	0	114	0
Rubber & plastics	1.0	3.1	0.3	21.4	1456	4530	211.1	1340	3190	1767	2763	428
Leather & allied products	0.2	0.1	0.1	-47.5	274	191	-30.3	252	-61	144	47	-108
Primary textiles	0.1	0.1	0.0	-72.9	118	140	18.6	109	31	32	108	-77
Textile products	0.8	1.0	0.7	38.2	1253	1483	18.4	1153	330	1732	-249	579
Clothing	1.6	2.7	0.3	-16.9	2400	3976	65.7	2208	1768	1995	1981	-213
Wood	33.3	28.2	5.1	-7.6	50955	41939	-17.7	46879	-4940	47083	-5144	204
Furniture & fixtures	1.1	1.5	0.4	-4.2	1752	2230	27.3	1612	618	1678	552	67
Paper & associated	13.5	11.8	1.9	-19.6	20655	17469	-15.4	19003	-1534	16605	864	-2397
Printing & publishing	4.8	7.9	0.9	27.4	7422	11688	57.5	6828	4860	9453	2235	2625
Primary metals	5.8	3.2	0.9	-30.8	8941	4711	-47.3	8226	-3515	6187	-1476	-2039
Fabricated metals	6.6	7.1	0.8	-15.4	10088	10585	4.9	9281	1304	8539	2046	-742
Machinery	3.8	3.7	0.7	-19.3	5753	5505	-4.3	5293	212	4645	860	-647
Transportation equipment	5.6	3.8	0.6	17.5	8523	5675	-33.4	7841	-2166	10014	-4339	2172
Electrics & electronics	2.1	3.6	0.3	3.8	3210	5390	67.9	2953	2437	3332	2058	379
Non-metal minerals	2.7	3.1	0.9	-23.6	4172	4636	11.1	3838	798	3187	1449	-651
Petroleum & coal	0.7	0.6	0.6	-30.9	1146	894	-22.0	1054	-160	792	102	-262
Chemicals	1.8	2.2	0.4	6.7	2751	3196	16.2	2531	665	2936	260	405
Other manufacturing	1.7	2.6	0.5	-2.7	2644	3898	47.4	2432	1466	2571	1327	139
Total	100.0	100.0	1.0	-8.0	153184	148484	-3.1	140929	7555	141183	7301	254

illustrates how the technique works in practice (Table A1.3). Thus, in Canada, between 1978 and 1993, manufacturing as a whole declined by -8.0%. If BC's industries had also changed at this rate, its total 'expected' employment in 1993 would have been 140 929. Thus number defines the regional share effect. In fact, BC's manufacturing employment in 1993 was 148 484 so that there was a net (positive) shift of 7 555 workers. The net shift is positive because BC's (-3.1%) decline in manufacturing employment was less rapid than Canada's (-8.0%). Most of this net shift is accounted for by the differential shift. Basically, the differential shift is the difference between the employment expected if growth had occurred at the national rate and the employment levels actually reached. In the case of food and beverages, for example, in Canada the industry declined by -6.0%. If this industry had declined at the same rate in BC, the employment level in 1993 would have been 18 490. In fact, the level was 20 234, so the differential shift in this case is 1744. The composition effect of the food and beverage industry can then be found by simply subtracting the differential shift (1744) from the total shift (2137), the result of which is 393. The other industries can be figured out in the same way.

In this case, according to shift-share analysis, BC's net positive effect is largely accounted for by favourable regional factors. Apparently, during the time period in question, BC's industry mix did not exert noticeable effects. The differential and industry-mix effects, however, varied considerably by individual industry. Thus, substantial negative differential effects were experienced by the wood industry and by transportation equipment. Shift-share analysis thus provides insights into aggregate level change and industry level change. The reasons for employment change and their implications for local development, however, cannot be simply inferred from shift-share. Thus, nationally, the wood industry declined and employment in BC's wood industry declined particularly rapidly. In part, BC's decline is accounted for timber supply limitations and rising costs; but there was also massive investment and technological change in the industry which has become increasingly capital-intensive. In this period the wood industries remained a vibrant part of BC's economy. In contrast, BC did not share in the national growth in the transportation industry. The auto and related industries which experienced substantial national growth are overwhelmingly concentrated in Ontario and Quebec, and while shipbuilding has struggled nationally, federal government support has been limited to Quebec and Atlantic Canada and the industry has virtually disappeared from BC.

Clearly, great care should be exercised in the use of shift-share analysis. The technique provides statistical results which have to be carefully interpreted as many factors affect employment change which are not always easily classified into the two shift components, while the technique is sensitive to industry aggregation and the time periods used. It can, however, be used to raise questions about employment change and it has often been helpful in isolating the effects of industry structure.

Appendix 2

INPUT–OUTPUT ANALYSIS: AN INTRODUCTION

In a large, relatively complex region a new plant may buy a variety of inputs from firms, organizations and households within the region. The most sophisticated technique for estimating multipliers which take into account direct and indirect transactions is provided by input–output (I/O) analysis. The data used in the following example are taken from Hoover (1971: 224–234). An excellent introduction to I/O analysis is Miernyk (1965).

An input–output table is derived in three basic steps. First, a gross flows or transaction table is calculated on the basis of all the actual flows of goods and services, usually measured in terms of sales that occur in a regional economy in a period of time, usually a year. This table shows 'actual' data; for example, data collected by a survey of all firms and organizations in a region. The second step involves the creation of a technical coefficients table which standardizes the gross flows table in terms of $1 worth of output. The third, and most complicated step, involves calculating the inverse matrix, or Leontief matrix, which shows in tabular form all the direct and indirect and induced effects which occur when the (export) sales of an industry increase by $1. Once this table is derived, various kinds of analyses can be performed, including the calculation of income multipliers.

This appendix briefly, and without mathematics (see Miernyk 1965), illustrates these steps and calculates one type of income multiplier. The discussion assumes that local income is generated by exports (Hoover (1971) also discusses supply-generated local income).

The gross flows or transactions table

The first step in I/O analysis is to derive a gross flows table which is simply an inter-sectoral accounting framework (Table A2.1). Firms are surveyed to determine which industries and sectors they purchase their inputs from, and which industries and sectors they sell their outputs to. Usually an I/O table is classified into (a) an intermediate sector, (b) a payments sector, and (c) a final demands sector.

This table simply tells us the dollar value of inputs and outputs for sectors within a (hypothetical) region during a specific time period. For example, firms in industry A produce $4300 of output, including $300 worth of goods that firms in industry A sold to other firms in industry A, $400 worth of goods to B, $100 to C, $500 to D, and $3000 to *other* final demands. For the purposes of this illustration, final demands can be regarded as exports. Reading down the column you can see the dollar distribution of inputs to produce the above outputs. The payments sector includes all imports from other regions.

It should be noted that in practice, I/O tables typically comprise many more rows and columns and that the payments and final demands sectors include more than just exports and imports. In

Table A2.1 Hypothetical Gross Flows Table (Actual $)

From/to	Intermediate sector by industry				Final demands	Output totals
	A	B	C	D		
Intermediate sector						
A	300	400	100	500	3000	4300
B	50	200	1000	300	1300	2850
C	1000	200	100	700	1100	3100
D	0	800	200	500	1100	2600
Payments sector	2950	1250	1700	600		
Input totals	4300	2850	3100	2600		

Source: Hoover (1971: 226).

principle, all transactions are included in an I/O table, including payments and receipts to and from governments (e.g. taxes and sales) inside and outside the region and purchases from households (primarily for labour) and sales to households. Indeed, since it is households in a region which ultimately receive the income generated, the household sector is vital to the calculation of multipliers. There are other transactions (e.g. inventory depletion and addition, capital depreciation and expenditure) which are also normally included but which are ignored in this discussion.

It should also be noted that it is the intermediate sector that is the focus for analysis, and researchers can make choices as to what to include within this sector or classify as part of the payments and final demands sectors. In some analyses, researchers may prefer to place households and local government in the latter sectors; in other cases, researchers may wish to include households and local governments within the intermediate sector (this is known as 'closing' the table with respect to households and local governments). In this example, households have been left in the payments and final demands sectors for now.

The input–coefficient table

The next step in I/O analysis is to 'standardize' the above gross flows data by calculating a set of input coefficients (often called technical coefficients) which indicate the dollar value of inputs

Table A2.2 Input Coefficients

Purchases (in dollars) from:	Per dollars worth of gross output in:			
	A	B	C	D
Intermediate sector				
A	0.070	0.140	0.032	0.192
B	0.012	0.070	0.323	0.115
C	0.233	0.070	0.032	0.269
D	0	0.281	0.065	0.192
Payments sectors	0.687	0.438	0.549	0.230
Totals	1.000	1.000	1.000	1.000

Source: Hoover (171: 228).

Table A2.3 Direct and Indirect Effects of a $1 Increase in Final Demand

Total added $ sales by intermediate sector	Per dollar of increased sales to final demand by:			
	A	B	C	D
A	1.118	0.289	0.157	0.359
B	0.126	1.234	0.439	0.352
C	0.297	0.204	1.171	0.501
D	0.068	0.452	0.247	1.400
Total	1.609	2.259	2.014	2.612
Payments sector	1.003	0.999	1.000	0.751

Source: Hoover (1971: 229).

to produce $1 worth of output for each intermediate sector (Table A2.2).

By reading down each column we can find the input requirements from each industry to produce $1 worth of output of each industry. This table indicates, for example, that for firms in industry A to produce $1 worth of final demands, these firms buy 7¢ of inputs from other firms in industry A, 1.2¢ from B; 23.3¢ from C and 68.7¢ from the payments sector. These input coefficients are derived from the data in Table A2.1. Thus to produce $4300 worth of output requires firms in industry A to purchase $300 worth of inputs from other firms in industry A (Table A2.1). Therefore, to produce $1 worth of output, firms in industry A need to buy $300/$4300 = $0.07.

Leontief's inverse matrix

The previous table is useful in that it shows the direct effects of any increase in the final demands for a given sector. For example, if final demands for industry A increase by $250, then this will have certain *direct* impacts on supplying sectors (i.e. $250 \times 0.07 + 250 \times 0.012$, etc.). However, if sector D is to expand to meet A's requirements, B will stimulate input requirements of its own, etc. These *indirect* effects can be calculated iteratively (but this would be exhausting!). Fortunately, Leontief showed how matrix algebra could be applied to derive all direct and indirect effects (see Table A2.3).

From this table we can estimate total direct and indirect impacts (but not induced impacts) as a result of the increase in the final demands for any given sector. Thus if A's final demands increase by $1, the total (direct and indirect) increase in the size of regional shipments is $1 \times 1.118 + 1 \times 0.126 + 1 \times 0.297 + 1 \times 0.068 = 1.609$. Variations in the size of this shipment or linkage multiplier are an indication of the strength of the local linkage ties each industry has with the local

Table A2.4 Input Coefficients: I/O Table Closed with Respect to Households

	A	B	C	D	H
A	0.070	0.140	0.032	0.192	0.308
B	0.012	0.070	0.323	0.115	0.019
C	0.233	0.070	0.032	0.269	0.019
D	0	0.281	0.065	0.192	0.135
H	0.442	0.105	0.323	0.154	0.038
Total	1.000	1.000	1.000	1.000	1.000

Source: Hoover (1971: 247).

Table A2.5 Direct, Indirect and Induced Effects of a $1 Increase in Final Demand

	A	B	C	D	H
A	1.482	0.537	0.473	0.699	0.593
B	0.233	1.307	0.532	0.452	0.174
C	0.466	0.400	1.318	0.659	0.276
D	0.270	0.590	0.422	1.589	0.329
H	0.906	0.618	0.786	0.846	1.476

Source: Hoover (1971: 248).

economy. In this economy, for example, industry D generates the most inter-industry shipments.

Income multipliers

In order to use the I/O table to derive income multipliers, specific reference must be made to households. If we assume that households will behave similar to industries and use additional revenue to purchase goods, then households can be considered just like any other industry within the Intermediate Sector. Thus assuming in our gross flows table (a) payments from households to A, B, C, D and H were 1900, 300, 1000, 400 and 200, respectively, and (b) purchases by households from A, B, C, D and H were 1600, 200, 200, 700 and 200, respectively, a revised input coefficient table can be presented (Table A2.4).

Reading down column A, the table indicates that to produce $1 worth of final output, A purchases inputs (labour) $0.442 from households. This defines the direct income effect of each $1 worth of output. Industry A generates the highest direct income effects and industry B the lowest.

From this, a table of direct, indirect and induced impacts may be obtained by inverting the matrix (Table A2.5).

In Table A2.5, induced effects, which are defined as the income generated from the expenditures by households on goods and services, are included. In the case of industry A, by closing the table with respect to households, total household (direct, indirect and induced) income generated by a $1 increase in final demand is $0.906. A so-called Type II multiplier can be defined as follows:

$$\text{Type II multiplier} = \frac{\text{direct, indirect and induced income change}}{\text{direct income change}}$$

For industry A, the Type II multiplier is 0.906/ 0.442 = 2.25. Type II multipliers can readily be calculated for industries B, C and D.

In calculating multipliers of this sort, important assumptions are made about behaviour. For example, it is assumed that the pattern of purchasing by industries (and households) will remain constant in the manner defined by the input coefficient table. In the short term, bottlenecks and changing preferences may modify such an assumption, while in the long term, technological change, economies of scale and changing demands need to be considered (Miernyk 1966).

REFERENCES

Abernathy WJ and Utterbach JM 1978 Patterns of industrial innovation. *Technology Policy* **81**: 40–7

Abler R, Adams RS and Gould P 1971 *Spatial organization: the geographer's view of the world* Englewood Cliffs, NJ: Prentice-Hall

Aguillar FJ 1967 *Scanning the business environment* New York: Macmillan

Ahern R 1993a The role of strategic alliances in the international organization of industry. *Environment and Planning A* **25**: 1229–46

Ahern 1993b Implications of strategic alliances for small R&D intensive firms. *Environment and Planning A* **25**: 1511–26

Aitken HGH 1961 *American capital and Canadian resources* Cambridge, MA: Harvard University Press.

Alchian AA 1950 Uncertainty, evolution and economic theory. *Journal of Political Economy* **58**: 211–21

Alden J and Morgan R 1974 *Regional planning: a comprehensive view* New York: John Wiley

Alexandersson G 1967 *Geography of manufacturing* Englewood Cliffs, NJ: Prentice-Hall

Alonso W 1964 Location theory. In J Friedmann and W Alonso (eds) *Regional development and planning: a reader* Cambridge, MA: MIT Press

Alvstam C and Ellegard K 1990 Volvo: the organization of work: a determinant of the future location of manufacturing enterprises. In M de Smidt and E Wever (eds) *The corporate firm in a changing world economy* London: Routledge, pp. 183–206

Amin A and Robins K 1990 The re-emergence of regional economies? The mythical geography of flexible production. *Environment and Planning D* **8**: 7–34

Amirahmadi H and Saff G 1993 Science parks: a critical assessment. *Journal of Planning Literature* **8**: 107–23

Amirahmadi H and Wallace C 1995 Information technology, the organization of production, regional development. *Environment and Planning A* **27**: 1745–76

Anderson M and Holmes J 1995 High-skill, low wage manufacturing in North America: a case study from the automotive parts industry. *Regional Studies* **29**: 655–71

Angel D 1991 High technology agglomeration and the labor market. *Environment and Planning A* **23**: 1501–16

Angel DP 1994 *Restructuring for innovation: the remaking of the US semiconductor industry* New York: Guildford

Angel DP and Savage LA 1996 Global localization? Japanese research and development laboratories in the USA. *Environment and Planning A* **28**: 819–34

Ansoff HI 1965 *Corporate strategy* London: Penguin

Aring J, Butzin B, Danielzyk R and Helbrecht I 1989 *The Ruhr region in crisis*? Oldenburg: University of Oldenburg (in German)

Armstrong H and Taylor J 1978 *Regional economic policy and analysis* Oxford: Philip Allan

Arrighi G and Drangel J 1986 The stratification of the world-economy: an exploration of the semi-peripheral zone. *Review* **10**: 9–74

Asanuma B 1985a The organization of parts purchases in the Japanese automotive industry. *Japanese Economic Studies* Summer: 32–53

Asanuma B 1985b The contractural framework for parts supply in the Japanese automotive industry. *Japanese Economic Studies* Summer: 54–78

Asanuma B 1986 Transactional structure of parts supply in the Japanese and electric machinery industries: a comparative analysis. *Technical Report No. 3* Faculty of Economics, Kyoto University

Asanuma B 1988 Japanese manufacturer–supplier relationships in international perspective: the automobile case. Paper presented to an Australia–

Japan Research Centre Conference on *Japanese Corporate Organisation and Adjustment* September, Canberra: Australian National University

Asanuma B 1989 Manufacturer–supplier relationships in Japan and the concept of the relation specific skill. *Journal of the Japanese and International Economies* 3: 1–30

Atkinson J 1985 The changed corporation. In D Clutterback (ed.) *New patterns of work* Aldershot, Hants: Gower, pp. 13–34

Atkinson J 1987 Flexibility or fragmentation? The United Kingdom labour market in the eighties. *Labour and Society* 12: 87–105

Averitt RT 1968 *The dual economy: the dynamics of American industry* New York: Norton

Auty RM 1975 Scale economies and plant vintage: toward a factory classification. *Economic Geography* 51: 150–62

Auty RM 1982 Industrial policy reform in China: structural and regional imbalances. *Transactions of the Institute of British Geographers* 17: 481–94

Auty R 1985 Multinational corporations and regional revenue retention in a vertically integrated industry. *Regional Studies* 17: 3–17

Auty RM 1993 *Sustaining development in mineral economies: the resource curse thesis* London: Routledge

Bacon R and Eltis WA 1976 *Britain's economic problem: too few producers* London: Macmillan

Bagnasco A and Sabel CF (eds) 1995 *Small and medium-sized enterprises* London: Pinter

Bain JS 1954 Economies of scale, concentration and the condition of entry in twenty manufacturing industries. *American Economic Review* 64: 15–39

Bain JS 1959 *Industrial Organization* London: John Wiley

Bairoch P 1982 International industrialization levels from 1750 to 1980. *Journal of European Economic History* 11: 269–333

Bairoch P 1993 *Economics and world history: myths and paradoxes* Hemel Hempstead: Harvester Wheatsheaf

Bale J 1976 *The location of manufacturing industry* Edinburgh: Oliver and Boyd

Bale JR 1977 Industrial estate development and location in post-war Britain. *Geography* 62: 87–92

Baranson J 1969 *Automotive industries in developing countries* Washington: International Bank for Reconstruction and Development

Barber RJ 1970 *The American Corporation* New York: Dutton

Barff RA and Austen J 1993 'It's gotta be da shoes': domestic manufacturing, internation sub-contracting, and the production of athletic footwear. *Environment and Planning A* 25: 1103–14

Barff RA and Knight PL 1988 The role of federal military spending in the timing of New England's employment turnaround. *Papers of the Regional Science Association* 65: 151–66

Barnes T 1987 Homo economicus, physical metaphors, and universal models in economic geography. *The Canadian Geographer*. 31: 299–308

Barnes T 1988 A reply to 'Lets keep economics and geography in economic geography'. *The Canadian Geographer* 32: 347–50

Barnes T 1990 Analytical political economy: a geographical introduction. *Environment and Planning A* 22: 993–1006

Barnes T and Hayter R 1992 'The: little town that did.' Flexible accumulation and community response in Chemainus, British Columbia. *Regional Studies* 26: 647–63

Barnes T and Hayter R 1994 Economic restructuring, local development and resource towns: forest communities in coastal British Columbia. *Canadian Journal of Regional Studies* 17: 289–310

Barnes T, Hayter R and Grass E 1990 MacMillan Bloedel: corporate restructuring and employment change. In M De Smidt and E Wever (eds) *The corporate firm in a changing world economy: case studies in the geography of enterprise* London: Routledge, pp. 145–65

Barnet RJ 1993 The end of jobs. *Harper's Magazine* September: 47–52

Barnet RJ and Cavanagh J 1994 *Global dreams: imperial corporations and the new world order* New York: Simon and Schuster

Barnet RJ and Muller RE 1975 *Global reach: the power of multinational corporations* New York: Simon and Schuster

Barr BM 1983 Industrial parks as locational environments. In FEI Hamilton and GJR Linge (eds) *Spatial analysis, industry and the industrial environment. Vol 3. Regional economies and industrial systems* London: John Wiley, pp. 423–40

Barr BM, Waters NM and Matthews AS 1984 Firms and industrial parks: insights from principal coordinates analysis. In BM Barr and NM Waters (eds) *Regional diversification and structural change* BC Geographical Series, Vancouver: Tantalus, pp. 121–37

Bartlett W 1993 The evolution of workers' cooperatives in southern Europe: a comparative perspective. In C Karlsson, B Johannisson and D Storey (eds) *Small business dynamics* London: Routledge, pp. 57–77

Bates T 1995 Analysis of survival rates among franchise and independent small business start-ups. *Journal of Small Business Management* 33: 26–36

Bathelt H and Hecht A 1990 Key technology industries in the Waterloo region: Canada's technology triangle (CTT). *The Canadian Geographer* 34: 225–34

Baumol WJ 1959 *Business behaviour, value and growth* New York: MacMillan

BC Business 1995 High technology entrepreneur: John MacDonald. *BC Business* October: 41

Beattie E and Watts HD 1983 Some relationships between manufacturing and the urban system *Geoforum* **14**: 125–32

Bechter DM and Chmura C 1990 The competitiveness of rural country manufacturing during a period of dollar appreciation *Regional Science Perspectives* **20**: 54–88

Behrman JN 1969 *Some patterns in the rise of the multinational enterprise* Chapel Hill: University of North Carolina

Bell D 1974 *The coming of the post-industrial society: a venture in social forecasting* London: Heinemann

Benders J 1996 Leaving lean? Recent changes in the production organization of some Japanese carplants. *Economic and Industrial Democracy* **17**: 9–38

Berg M 1985 *The Age of Manufactures 1700–1820* London: Fontana

Berggren C and Rehder RR 1993 Uddevalla and Saturn: the global quest for more competitive and humanistic manufacturing organizations. *Proceedings of the Fourth Symposium on Cross Cultural Consumer and Business Studies* December: 193–7

Berle AA and Means G 1932 *The modern corporation and private property* New York

Berry BJL 1992 *Long wave rhythms in economic development and political behavior* Baltimore: John Hopkins University Press

Berry BJL, Conkling EC and Ray DM 1993 *The global economy: resource use, locational choice, and international trade* Englewood Cliffs, NJ: Prentice Hall

Best MH 1990 *The new competition* Cambridge, MA: Harvard University Press

Beyers WB 1972 On the stability of regional interindustry models: the Washington data for 1963 and 1967. *Journal of Regional Science* **12**: 363–74

Beyers WB 1989 *The producer services and economic development in the United States: the last decade* Washington DC: Economic Development Administration

Beyers WB and Alvine MJ 1985 Export services in a postindustrial society. *Papers of the Regional Science Association* **57**: 33–45

Beyers W, Bourque P, Seyfried W and Weeks E 1970 *Input–output tables for the Washington economy, 1967* University of Washington: Graduate School of Business Administration

Bianchi A 1992 Tax systems holds hidden – possible scary – surprises. *The Financial Post* October 5: U3

Bilkley WJ 1978 An attempted integration of the literature on the export behaviour of firms. *Journal of International Business Studies* **9**: 133–46

Birch DL 1979 *The job generation process* In MIT program on neighbourhood and regional change. Cambridge, MA: MIT Press

Birley S 1989 The start-up. In P Burns and J Dewhurst (eds) *Small business and entrepreneurship* London: Macmillan Education, pp. 8–31

Birley S, Cromie S and Myers A 1991 Entrepreneurial networks: their emergence in Ireland and overseas. *International Small Business Journal* **9**: 56–74

Blackaby FT 1979 *De-industrialisation* London: Heinemann

Blackbourn A 1974 The spatial behaviour of American firms in Western Europe. In FEI Hamilton (ed.) *Spatial perspectives on industrial organization and decision making* London: John Wiley, pp. 245–64

Blair JM 1972 *Economic concentration: structure, behavior and public policy* New York: Harcourt Brace Jovanovich

Blau J 1994 Germany hopes new law will boost gene research. *Research-Technology Management* **37**: 3–4

Bloomfield GT 1991 The world automotive industry in transition. In CM Law (ed.) *Restructuring the global automobile industry* London: Routledge, pp. 19–60

Bloomquist LE 1988 Performance of the rural manufacturing sector. In DL Brown, N Reid, H Bluestone, DA McGranahan and S Mazie (eds) *Rural economic development in the future: prospects for the 1980s* Rural development research report 69. Washington, DC: US Department of Agriculture Economic Research Service

Bluestone B and Harrison B 1982 *The deindustrialization of America* New York: Basic Books

Boas CW 1961 Locational patterns of American automobile assembly plants, 1895–1958. *Economic Geography* **37**: 218–30

Bolton J 1971 *Small firms: report of the commissioner of inquiry on small firms* London: HMSO

Borchert JR 1978 Major control points in American economic geography. *Annals of the Association of American Geographers* **68**: 214–32

Bourque P 1969 Income multipliers for the Washington economy. *University of Washington Business Review* **28**, Winter: 5–15

Bourque 1971 An input–output analysis of economic change. *University of Washington Business Review*, **30**, Summer: 5–22

Bradbury JH 1979 Toward an alternative theory of resource-based town development in Canada. *Economic Geography* **55**: 147–66

Bramham D 1996 Children at labor. *The Vancouver Sun* May 25: D3–4

Braverman H 1974 *Labor and monopoly capital* New York: Monthly Review Press

Britton JNH 1976 The influence of corporate organization and ownership on the linkages of industrial plants: a Canadian inquiry. *Economic Geography* **52**: 311–24

Britton JNH 1980 Industrial dependence and technological underdevelopment: Canadian consequences of direct foreign investment. *Regional Studies* **14**: 181–99

Britton JNH 1981 Industrial impacts of foreign enterprise: a Canadian technological perspective. *Regional Studies* **14**: 181–200

Britton JNH 1989a Innovation policies for small firms. *Regional Studies* **23**: 167–73

Britton JNH 1989b A policy perspective on incremental innovation in small and medium sized enterprises. *Journal of Entrepreneurship and Regional Development* **1**: 179–90

Britton JNH 1990 Industrial policy in Canada and Australia: technology change and support for small firms. In R Hayter and PD Wilde (eds) *Industrial transformation and challenge in Australia and Canada* Ottawa: Carleton University Press, pp. 277–98

Britton JNH 1991 Reconsidering innovation policy for small and medium sized enterprises: the Canadian case. *Environment and Planning C* **8**: 189–206

Britton JNH 1994 The new competition and the foreign ownership question in Canada revisited. Paper presented at a conference in honour of HA Innis *Regions, Institutions and Technology: Reorganizing Economic Geography in Canada and the Anglo-American World* Toronto: University of Toronto

Britton JNH and Gilmour JM 1978 *The weakest link: a technological perspective on Canadian industrial underdevelopment* Background Study 43, Ottawa: Science Council of Canada

Brohman J 1996 Postwar development in the Asian NICs: does the neoliberal model really fit? *Economic Geography* **72**: 107–30

Brusco S 1986 Small firms and industrial districts: the experience of Italy. In D Keeble and E Wever (eds) *New firms and regional development in Europe* London: Croom Helm

Brusco S 1990 The idea of the industrial district: its genesis. In F Pike, G Becattini and W Sengenberger (eds) *Industrial districts and inter-firm co-operation in Italy* Geneva: International Institute for Labour Studies, pp. 10–19

Brusco S and Righi E 1989 Local government, industrial policy and social consensus: the case of Modena. *Economy and Society* **18**: 405–24

Burns P 1989 Strategies for success and routes to failure. In P Burns and J Dewhurst (eds) *Small business and entrepreneurship* London: Macmillan Education, pp. 32–67

Burns P and Dewhurst J 1989 *Small business and entrepreneurship*. London: Macmillan Education

Business Week 1993 The little giants. *Business Week* 6 September: 40–6

Butler C 1995 Male bonding. *Sales and marketing management* July: 66–72

Cannon JB 1975 Government impact on industrial location. In L Collins and DF Walker (eds) *Locational dynamics of manufacturing activity* London: John Wiley, pp. 109–34

Cannon JB 1980 The impact of incentives on manufacturing change: the Georgian Bay region of Ontario. *The Canadian Geographer* **32**: 248–57

Caves RE 1971 International corporations: the industrial economics of foreign investment. *Economica* **38**: 1–27

Caves RE 1974 Industrial organization. In JH Dunning (ed.) *Economic analysis and the multinational enterprise* London: Allen Unwin, pp. 115–46

Caves RE 1980 Industrial organization, corporate strategy and structure. *Journal of Economic Literature* **43**: 64–92

Chambers EJ and Percy MB 1992 *Western Canada in the international economy* Edmonton: The University of Alberta Press

Chandler AD 1962 *Strategy and structure: chapters in the history of American industrial enterprise* New York: Doubleday

Chapman K 1980 Environmental policy and industrial location. *Area* **12**: 209–16

Chapman K 1981 Issues in environmental impact assessment. *Progress in Human Geography* **5**: 190–210

Chapman K 1982 Environmental policy and industrial location in the United States. In R Flowerdew (ed.) *Institutions and Geographical Patterns* London: Croom Helm, pp. 141–68

Chapman K 1985 Raw material costs and the development of the petrochemical industry in Alberta. *Transactions of the Institute of British Geographers* **10**: 138–48

Chapman K 1992 Continuity and contingency in the spatial evolution of industries: the case of petrochemicals. *Transactions of the Institute of British Geographers* **17**: 47–64

Chapman K and Humphrys G (eds) 1987 *Technical change and industrial policy* Oxford: Basil Blackwell

Chapman K and Walker D 1987 *Industrial location* Oxford: Basil Blackwell

Chapman SD 1972 *The cotton industry in the industrial revolution* London: Macmillan

Chinitz B 1961 Contrasts in agglomeration: New York and Pittsburgh. *American Economic Review, Papers and Proceedings Supplement* **51**: 279–89

Chisholm M 1990 *Regions in recession, restructuring and resurgence?* London: Unwin Hyman

Christaller W 1966 *Central places in southern Germany* (translated by CW Baskins) Englewood Cliffs, NJ: Prentice-Hall

Christopherson S 1989 Flexibility in the US service economy and the emerging division of labour. *Transactions, the Institute of British Geographers* **14**: 131–43

Christopherson S and Storper M 1989 The effects of flexible specialization on industry politics and the labour market: the motion picture industry. *Industrial and Labour Relations Review* **42**: 331–47

Christy CV and Ironside RG 1987 Promoting 'high technology' industry: location factors and public policy. In K Chapman and G Humphrys (eds) *Technical change and industrial policy* Oxford: Basil Blackwell, pp. 233–52

Clark C 1940 *The conditions of economic progress* London: Macmillan

Clark C 1966 Industrial location and economic potential. *Lloyd's Bank Review* **82**: 1–17

Clark GL 1981 The employment relation and the spatial division of employment. *Annals of the Association of American Geographers* **71**: 412–24

Clark GL 1986 The crisis of the midwest auto industry. In AJ Scott and M Storper *Production, work and territory: the geographical anatomy of industrial capitalism* London: Allen and Unwin, pp. 127–48

Clark GL 1989 *Unions and communities under seige* Cambridge University Press

Clark GL 1991 Limits of statutory responses to corporate restructuring illustrated with reference to plant closing legislation. *Economic Geography* **61**: 22–41

Clark GL 1993 Global interdependence and regional development: business linkages and corporate governance in a world of financial risk. *Transactions of the Institute of British Geographers* **18**: 309–25

Clark GL 1994 Strategy and structure: corporate restructuring and the scope and characteristics of sunk costs. *Environment and Planning A* **25**: 5–26

Clark GL and Wrigley N 1995 Sunk costs: a framework for economic geography. *Transactions of the Institute of British Geographers* **20**: 204–23

Clarke IM 1982 The changing international division of labour within ICI. In M Taylor and N Thrift (eds) *The geography of multinationals* London: Croom Helm, pp. 90–116

Clarke TA 1991 Capital constraints on non-metropolitan accumulation: rural progress in the United States of America since the 1960s. *Regional Studies* **7**: 169–90

Clements W and Williams G (eds) *The new Canadian political economy* Montreal: McGill-Queen's University Press

Coase RH 1937 The nature of the firm. *Economica* **4** 386–405

Coffee WJ 1996 The role and location of service activities in the Canadian space economy. In JNH Britton (ed.) *Canada and the global economy* Montreal-Kingston: McGill-Queen's Press, pp. 335–51

Coffee WJ and Polèse M 1987 Trade and the location of producer services: a Canadian perspective. *Environment and Planning A* **19**: 597–611

Coffee WJ and Polèse M 1989 Producer services and regional development: a policy oriented perspective. *Papers of the Regional Science Association* **67**: 13–67

Cohen AJ 1984 Technological change as historical process: the case of the US pulp and paper industry. *Journal of Economic History* **44**: 755–99

Collins L and Walker DF (eds) *Locational dynamics of manufacturing activity* London: Wiley

Cooke P (ed.) 1989 *The changing face of urban Britain* London: Unwin Hyman

Cooke P and Morgan K 1990 *Learning through networking: regional innovation and the lessons of Baden-Worttemburg* Cardiff: University of Wales

Cooke P and Morgan K (in press) *The collective entrepreneur: strategies for corporate and regional renewal* Oxford: Blackwell

Cooper MJM 1975 The industrial location decision making process. *Occasional Paper 34* Centre for Urban and Regional Studies, University of Birmingham

Corvari R and Wisner R 1993 *Foreign multinationals and Canada's international competitiveness* Working Paper 16. Ottawa: Investment Canada

Cox KR and Mair A 1988 Locality and community in the politics of local development. *Annals of the Association of American Geographers* **78**: 307–25

Cox KR and Mair A 1991 From localized social structures to localities as agents. *Environment and Planning A* **23**: 197–213

Crandall R 1993 *Manufacturing on the Move* Washington, DC: Brookings Institute

Cromley RG and Leinbach TR 1981 The pattern and impact of the filter-down process in nonmetropolitan Kentucky. *Economic Geography* **57**: 208–74

Cross M 1981 *New firm formation and regional development* Aldershot: Gower

Crough G and Wheelwright T 1982 *Australia: a client state* Ringwood, Victoria: Penguin

Crow B and Thomas A 1985 *Third world atlas* Milton Keynes: Open University Press

Curran J and Blackburn RA (eds) 1991 *Paths of enterprise: the future of the small business* London: Routledge

Curran J and Blackburn RA 1994 *Small firms and local economic networks: the death of the local economy* London: Paul Chapman

Cusumano MA 1985 *The Japanese automobile industry: technology and management Nissan and*

Toyota The Council on East Asian Studies: Harvard University Press

Cusumano MA 1988 Manufacturing innovation: lessons from the Japanese auto industry. *Sloan Management Review* **30**: 29–40

Cyert RM and March JG 1963 *A behavioral theory of the firm* Englewood Cliffs, NJ: Prentice-Hall

Daly MT and Stimson RJ 1994 Dependency in the modern global economy: Australia and the changing face of Asian finance. *Environment and Planning A* **26**: 415–34

Daniels PW 1987 Producer-services research: a lengthening agenda. *Environment and Planning A* **19**: 569–71

Davidsson P, Lindmark L and Olofsson C 1994 New firm formation and regional development in Sweden. *Regional Studies* **28**: 395–410

Davie MF 1984 *Paper machine evolution since 1920* Pointe Claire: Paprican

De Smidt M 1990 Philips; a global electronics firm restructures its home base. In M de Smidt and E Wever (eds) *The corporate firm in a changing world economy* London: Routledge, pp. 55–76

De Smidt M and Wever E (eds) *The corporate firm in a changing world economy* London: Routledge

Denison EF 1968 *Why growth rates differ: postwar experience in nine western countries* Washington, DC: Brookings Institution

Dewhurst J 1989 The entrepreneur. In P Burns and J Dewhurst (eds) *Small business and entrepreneurship.* London: MacMillan Education, pp. 68–93

Dicken P 1971 Some aspects of the decision-making behaviour of business organizations. *Economic Geography* **47**: 426–37

Dicken P 1976 The multiplant business enterprise and geographical space: some issues in the study of external control and regional development. *Regional Studies* **10**: 401–12

Dicken P 1977 A note on location theory and the large business enterprise. *Area* **9**: 138–43

Dicken P 1983 Japanese manufacturing investment in the United Kingdom. *Area* **15**: 273–84

Dicken P 1987 Japanese penetration of the European automobile industry: the arrival of Nissan in the United Kingdom. *Tijdschrifte voor Economische en Sociale Geografie* **79**: 94–107

Dicken P 1988 The changing geography of Japanese foreign investment in manufacturing industry: a global perspective. *Environment and Planning A* **20**: 633–53

Dicken P 1992 *Global shift: the internationalization of economic activity* New York: Guilford Press

Dicken P 1994 Global–local tensions: firms and states in global space-economy. *Economic Geography* **70**: 101–28

Dicken P and Lloyd PE 1980 Patterns and processes of change in the spatial distribution of foreign-controlled manufacturing employment in the United Kingdom, 1963–75. *Environment and Planning A* **12**: 1401–26

Dicken P and Thrift N 1992 The organization of production and the production of organization: why business enterprises matter in the study of geographical industrialization. *Transactions of the Institute of British Geographers* **17**: 279–92

Doeringer P and Piore M 1971 *Internal labour markets and manpower analysis* Lexington: DC Heath

Donaghu MT and Barff R 1990 Nike just did it: subcontracting and flexibility in athletic footwear production. *Regional Studies* **24**: 537–52

Duncan JS and Ley D 1982 Structural marxism and human geography: a critical assessment. *Annals of the Association of American Geographers* **72**: 30–59

Dunning JH 1958 *American investment in British manufacturing industry* London: George Allen and Unwin

Dunning JH 1973 The determinants of international production. *Oxford Economic Papers, New Series* **25**: 289–336

Dunning JH 1977 Trade, location of economic activity and the MNE: a search for an eclectic approach. In B Ohlin, PO Hessleborn and PM Wijkman (eds) *The international allocation of economic activity* London: MacMillan

Dunning JNH 1980 Towards an eclectic theory of international production: some empirical tests. *Journal of International Business Studies* **11**: 9–31

Dunning JH 1988 *Explaining international production* London: Unwin Hyman

Dunning JH 1993 *Multinational enterprises and the global economy* Wokingham: Addison-Wesley

Dunning JH and Norman G 1983 The theory of the multinational enterprise: an application to multinational office location. *Environment and Planning A* **15**: 675–92

Dyck I 1989 Integrating home and wage workplace: women's daily lives in a Canadian suburb. *The Canadian Geographer* **33**: 329–41

Dyson K 1983 The cultural, idealogical and structural context. In K Dyson and S Wilks (eds) *A comparative study of state and industry* Oxford: Oxford University Press, pp. 26–66

Eatwell J 1982 *Whatever happened to Britain?* London: Duckworth

Eatwell J (ed.) 1996a *Global unemployment* New York: ME Sharpe

Eatwell J 1996b Unemployment on a world scale. In J Eatwell (ed.) *Global unemployment* New York: ME Sharpe, pp. 3–20

Edgington DW 1987 Influences on the location and behaviour of transnational corporations: some examples taken from Japanese investment in Australia. *Geoforum* **18**: 343–59

Edgington DW 1990 *Japanese business down under: patterns of Japanese investment in Australia* London: Routledge

Edgington DW 1994 The geography of endaka: industrial transformation and regional employment changes in Japan, 1986–1991. *Regional Studies* **28**: 521–35

Edgington DW 1995 Locational preferences of Japanese real estate investors in North America. *Urban Geography* **16**: 373–96

Edgington DF and Hayter R (in press) International trade, production chains and corporate strategies: Japan's timber trade with British Columbia. *Regional Studies* **31**

Edwards RC 1979 *Contested terrain: the transformation of the workplace in the twentieth century* New York: Basic Books

Eichner AS 1969 *The emergence of oligopoly* Baltimore: John Hopkins

Elson D and Pearson R 1981 Nimble fingers make cheap workers: an analysis of women's employment in Third World export manufacturing. *Feminist Review* Spring: 87–107

Encarnation DJ 1992 *Rivals beyond trade: America versus Japan in global competition* Oxford: Blackwell

Eng I (forthcoming) The rise of manufacturing towns: urban development and externally driven industrialization in the Pearl River Delta of China. *International Journal of Urban and Regional Restructuring*

Eng I and Lin Y 1996 Seeking competitive advantage in an emerging open economy: foreign direct investment in Chinese industry. *Environment and Planning A* **28**: 1113–38

England KV 1993 Suburban pink collar ghettos: the spatial entrapment of women? *Annals of the Association of American Geographers* **83**: 225–42

Erickson RA 1974 The regional impact of growth firms: the case of Boeing 1963–68. *Land Economics* **50**: 127–36

Erickson RA 1976 The filtering-down process: industrial location in a non-metropolitan area. *Professional Geographer* **28**: 254–60

Erickson RA 1989 Export performance and state industrial growth. *Economic Geography* **65**: 280–92

Erickson RA and Hayward D 1991 The international flows of industrial exports from US regions. *Annals of the Associations of American Geographers* **81**: 371–90

Erickson RA and Leinbach TR 1979 Characteristics of branch plants attracted to nonmetropolitan areas. In RE Lonsdale and HL Seyler *Nonmetropolitan industrialization* Washington, DC: VH Winston, pp. 57–78

Erickson RA and Wasylenko M 1980 Firm location and site selection in suburban communities. *Journal of Urban Economics* **8**: 68–85

Estall RC and Buchanan RO 1980 *Industrial activity and economic geography* London: Hutchinson (first edition 1963)

Ettlinger N 1990 Worker displacement and corporate restructuring: a policy conscious appraisal. *Economic Geography* **66**: 67–82

Ewers HJ and Wettmann RW 1980 Innovation-oriented regional policy. *Regional Studies* **14**: 161–79

Fan CC 1992 Regional impacts of foreign change in China, 1984–89. *Growth and Change* **23**: 129–59

Fellner W 1949 *Competition among the few* New York: Angustus M Kelley

Fernandez-Kelley MP 1989 Broadening the scope: gender and international economic development. *Sociological Forum* **4**: 611–35

Field FW 1914 *Capital investments in Canada* Montreal: Monetary Times of Canada

Firn JR 1975 External control and regional development. *Environment and Planning A* **7**: 393–414

Firn JR and Swales JK 1978 The formation of new manufacturing establishments in central Clydeside and the West Midlands. *Regional Studies* **12**: 199–213

Fisher AGB 1935 *The clash of progress and society* London: Macmillan

Fleming DK 1967 Coastal steelworks in the Common Market. *Geographical review* **57**: 48–72

Fleming DK and Krumme G 1968 The Royal Hoesch Union: case analysis of adjustment patterns in the European steel industry. *Tijdschrift voor Economische en Sociale Geografie* **58**: 177–99

Florida R 1994 The economic transformation of the industrial midwest, paper presented at the University of Toronto, Conference held in honour of HA Innis: *Regions, institutions and technology: reorganizing economic geography in Canada and the Anglo-American world*

Florida R 1996 Lean and green: the move to environmentally conscious manufacturing. *California Management Review* **39**: 80–105

Florida R and Kenney M 1988a Venture capital, high technology and regional development. *Regional Studies* **22**: 33–48

Florida R and Kenney M 1988b Venture capital-financed innovation and technological change in the USA. *Research Policy* **17**: 119–37

Florida R and Kenney M 1990a *The breakthrough illusion* New York: Basic Books

Florida R and Kenney M 1990b High technology restructuring in the USA and Japan. *Environment and Planning A* **22**: 233–52

Florida R and Kenney M 1994 The globalization of Japanese R&D investment in the United States. *Economic Geography* **70**: 344–69

Florida R and Smith DF 1993 Venture capital formation, investment, and regional industrialization. *Annals, Association of American Geographers* **83**: 434–51

Foley PD and Watts HD 1996a New process technology and the regeneration of the manufacturing sector of an urban economy. *Urban Studies* **33**: 445–57

Foley P and Watts HD 1996b Production site R&D in a mature industrial region. *Tijdschrift voor Economische en Sociale Geografie* **87**: 136–45

Foley P, Hutchinson J, Herbane B and Tait G 1996 The impact of Toyota on Derbyshire's local economy and labour market. *Tijdschrift voor Economishe en Sociale Geografie* **87**: 19–31

Forest Sector Advisory Council 1991 *Newsprint cost study 1986–90* Vancouver: Price Waterhouse

Fothergill S and Gudgin G 1982 *Unequal growth: urban and regional employment change in the UK* London: Heinemann Educational

Fothergill S, Kitson M and Monk S 1985 Rural industrialization: trends and causes. In MJ Healey and B Ilberry (eds) *Industrialization of the countryside* Norwich: Geobooks, pp. 214–37

Frank AG 1967 *Capitalism and underdevelopment in Latin America* New York: Monthly Review Press

Frank RH and Freeman RT 1978 *The distributional consequences of direct foreign investment* New York: Academic Press

Fredriksson CG and Lindmark LG 1979 From firms to systems of firms: a study of interregional interdependence in a dynamic society. In FEI Hamilton and GJR Linge (eds) *Spatial analysis, industry and the industrial environment 1*, New York: Wiley, pp. 155–86

Freeman C 1979 Technical innovation and British trade performance. In F Blackeby (ed.) *Deindustrialization* London: Heinemann, pp. 56–77

Freeman C 1982 *The economics of industrial innovation* 2nd edition London: Francis Pinter

Freeman C 1987 *Technology policy and economic performance: lessons from Japan* London: Pinter

Freeman C 1988 Japan: a new national system of innovation? In G Dosi, C Freeman, R Belson and G Silverberg (eds) *Technical change and economic theory* London: Pinter, pp. 330–48

Freeman C and Perez C 1988 Structural crises of adjustment, business cycles and investment behaviour. In G Dosi, C Freeman, R Nelson, R Silverberg and L Soete (eds) *Technical Change and Economic Theory* London: Pinter, pp. 38–66

Freeman C, Clark J and Soete L 1982 *Unemployment and Technical Innovation: A Study of Long Waves and Economic Development* London: Frances Pinter

Friedmann D 1988 *The misunderstood miracle: industrial development and political change in Japan* Ithaca: Cornell University Press

Friedmann J 1986 The world city hypothesis. *Development and Change* **17**: 36–53

Friedmann J and Weaver C 1979 *Territory and function: the evolution of regional planning* London: Edward Arnold

Fröbel F, Heinrichs J and Kreye O 1980 *The new international division of labour* Cambridge: Cambridge University Press

Fruin MW 1992 *The Japanese enterprise system* Oxford: Clarendon Press

Fuchs M and Schamp E 1990 Standard Electrik Lorenz: introducing CAD into a telecommunications firm: its impact on labour. In M de Smidt and E Wever (eds) *The corporate firm in a changing world economy* London: Routledge, pp. 77–99

Fuentes A and Ehrenreich B 1987 Women in the global factory. In R Peet (ed.) *International capitalism and industrial restructuring* Boston: Allen and Unwin, pp. 201–15

Fujita K 1991 Women workers and flexible specialization: the case of Tokyo. *Economy and Society* **20**: 260–82

Fujita K and Hill RC 1987 *Toyota city: corporation and community in Japan* East Lansing: Department of Sociology, Michigan State University

Fusi DS 1991 Major new sites for science continue to spring up around the world. *Site Selection* **36**: 610–15

Galbraith JK 1952 *American capitalism* Boston: Houghton Mifflin

Galbraith JK 1967 *The new industrial state* London: Hamish Hamilton

Galbraith JK 1983 *The anatomy of power* Boston: Houghton Mifflin

Gann D and Senker P 1993 Construction robotics: technological change and work organization. *New Technology, Work and Employment* **8**: 3–9

Garrahan P and Stewart P 1992 *The Nissan enigma* London: Mansell

Gates B 1992 Learning to play the American game: rules can be different – and harder. *The Financial Post* October 5: U2

George RE 1970 *A leader and a laggard: manufacturing industry in Nova Scotia, Quebec and Ontario* Toronto: University of Toronto Press

George RE 1974 *The life and times of industrial estates limited* Halifax: Institute of Public Affairs, Dalhousie University

Gereffi G 1989 Rethinking development theory: insights from East Asia and Latin America. *Sociological Forum* **4**: 505–33

Gereffi G 1995 Global production systems and Third World development. In B Stallings (ed.) *Global change, regional response: the new international context of development* Cambridge: Cambridge University Press, pp. 100–43

Gereffi G and Korzeniewicz M 1990 Commodity chains and footwear exports in the semiperiphery.

In W G Martin (ed.) *Semiperipheral states in the world economy* New York: Greenwood, pp. 45–68

Gereffi G and Korzeniewicz M (eds) *Commodity chains and global capitalism* Westport: Greenwood Press

Gertler M 1988 The limits to flexibility: comments on the post-fordist vision of production and its geography. *Transactions of the Institute of British Geographers* **13**: 419–32

Gertler M 1992 Flexibility revisited: districts, nation states and the forces of production. *Transactions of the Institute of British Geographers* **17**: 259–78

Gibb DC (ed.) 1985 *Science parks and innovation centres: their economic and social impact* Amsterdam: Elsevier

Gibb R and Michalak W 1994 *Continental trading blocs: the growth of regionalism in the world economy* London: John Wiley

Gillespie A, Howells J, Williams H and Thwaites A 1987 Competition, internalisation and the regions: the example of information technology production industries in Europe. In M J Breheny and RW McQuaid (eds) *The development of high technology industries: an international survey* London: Croom Helm: 113–42

Gilmour JM 1974 External economies of scale, inter-industrial linkages and decision making in manufacturing. In FEI Hamilton (ed.) *Spatial perspectives on industrial organization and decision making* London: Wiley, pp. 335–62

Gilmour JM and Murricane K 1973 Structural divergence in Canada's manufacturing belt. *The Canadian Geographer* **42**: 1–18

Glasmeier A 1988 Factors governing the development of high-tech industry agglomeration formation. *Regional Studies* **22**: 287–301

Glasmeier AK and McCluskey RE 1987 US auto parts production: an analysis of the organization and location of a changing industry. *Economic Geography* **63**: 142–59

Glasmeier A, Thompson J and Kays AJ 1993 The geography of trade policy: trade regimes and location decisions in the textile and apparel complex. *Transactions of the Institute of British Geographers* **18**: 19–35

Gooch GD and Castensson R 1991 The transfer of technology from Great Britain to Sweden 1825–1850. *Geografisca Annaler* **73B**: 175–85

Goodman LW 1987 *Small nations, giant firms* New York: Holmes and Meier

Goodman E, Bamford J and Saynor P (eds) 1989 *Small firms and industrial districts in Italy* London: Routledge

Gordon MR, Edwards R and Reich D 1982 *Segmented work, divided workers: the historical transformation of labor in the United States* Cambridge: Cambridge University Press

Gould A and Keeble D 1984 New firms and rural industrialisation in East Anglia. *Regional Studies* **18**: 189–201

Gould P and White R 1974 *Mental maps* London: Penguin

Grabher G 1991 Rebuilding cathedrals in the desert: new patterns of co-operation between large and small firms in the coal, iron, and steel complex of the German Ruhr area. In EM Bergman, G Maier and F Todtling (eds) *Regions reconsidered: economic networks, innovation, and local development in industrialized countries* London: Mansell, pp. 59–78

Grant P 1983 Technological sovereignty: forgotten factor in the 'hi-tech razzamatazz'. *Prometheus* **1**: 239–70

Grass E and Hayter R 1989 Employment change during recession: the experience of forest product manufacturing plants in British Columbia, 1981–1985. *The Canadian Geographer* **33**: 240–52

Gray Report 1972 *Foreign direct investment in Canada* Ottawa: Government of Canada

Green MB and Cromley RG 1984 Merger and acquisition fields for large United States cities 1955–70. *Regional Studies* **18**: 291–301

Green AC, Owen DW and Winnett CM 1994 The changing geography of recession: analyses of local unemployment time series. *Transactions of the Institute of British Geographers* **19**: 142–62

Gregerson HM and Contreras A 1975 *US investment in the forest-based sector in Latin America* Baltimore: Johns Hopkins

Grieco JM 1982 Between dependency and autonomy: India's experience with the international computer industry. In TH Moran (ed.) *Multinational corporations: the political economy of foreign direct investment* Lexington: DC Heath

Grotz RE 1990 The demands of technological change on vocational training and retraining: the West German experience. *Tijdschrift voor Economische en Social Geografie* **81**: 170–81

Gudgin G 1978 *Industrial location processes and regional employment growth* Saxon House

Gudgin G and Fothergill S 1984 Geographical variation in the rate of formation of new manufacturing firms. *Regional Studies* **18**: 203–6

Gwynne RN 1979 Oligopolistic reaction. *Area* **11**: 315–19

Hägerstrand T 1970 What about people in regional science? *Regional Science Association* **26**: 7–24

Hakanson H 1979 Toward a theory of location and corporate growth. In FEI Hamilton and GJR Linge (eds) *Spatial analysis, industry and the industrial environment, vol 1: industrial systems* London: John Wiley, pp. 115–38

Hall P 1966 *The world cities* New York: McGraw-Hill

Hall P 1981 The geography of the fifth Kondratieff cycle. *New Society* March 26: 537–7

Hall P 1985 The geography of the fifth Kondratieff. In P Hall and A Markusen (eds) *Silicon Landscapes* London: Allen and Unwin, pp. 1–19

Hamilton FEI (ed.) 1974 *Spatial perspectives on industrial organization and decision making* London: Wiley

Hamilton FEI 1976 Multinational enterprise and the European Economic Community. *Tijdschrift voor Economische and Social Geografie* **67**: 258–78

Hamilton FEI 1995 Re-evaluating space: locational change and adjustment in central and eastern Europe. *Geographische Zeitschrift* **83**: 67–86

Hamilton FEI and Linge GJR (eds) 1979 *Spatial analysis, industry and the industrial environment. 1 Industrial systems* London: Wiley

Hamilton RT 1989 Unemployment and business formation rates: reconciling time-series and cross-sectional evidence. *Environment and Planning A* **21**: 249–55

Hanson S and Johnson I 1985 Gender differences in work-trip length: explanations and implications. *Urban Geography* **9**: 367–78

Hanson S and Pratt G 1988 Spatial dimensions of the gender division of labour in a local labour market. *Urban Geography* **9**: 193–219

Hanson S and Pratt G 1991 Job search and the occupational segregation of women. *Annals of the Association of American Geographers* **811**: 229–53

Hanson S and Pratt G 1995 *Gender, work and space* London: Routledge

Haren CC and Holling RW 1979 Industrial development in nonmetropolitan America: a locational perspective. In RE Lonsdale and HL Seyler (eds) *Nonmetropolitan industrialization* Washington DC: VH Winston, pp. 13–45

Harrington JW 1985 Corporate strategy, business strategy, and activity location. *Geoforum* **16**: 349–56

Harrington JW and Warf B 1995 *Industrial location: principles, practice and policy* London: Routledge

Harris CD 1954 The market as a factor in the localization of industry in the United States. *Annals of the Association of American Geographers* **44**: 315–48

Harrison B 1992 Industrial districts: old wine in new bottles. *Regional Studies* **26**: 469–83

Harrison B 1993 The Italian industrial districts and the crisis of the cooperative form. *European Planning Studies* **2**: 3–22

Harrison RT and Hart M 1983 Factors influencing new-business formation: a case study of northern Ireland. *Environment and Planning A* **15**: 1395–412

Hart M and Gudgin G 1994 Spatial variations in new firm formation rates. *Regional Studies* **28**: 367–80

Hart PE 1962 The size and growth of firms. *Economica* **24**: 29–39

Hartshorne R 1928 Location factors in the iron and steel industry. *Economic Geography* **4**: 241–52

Hartshorne R 1929 The iron and steel industry of the United States. *Journal of Geography* **28**: 133–53

Harvey D 1982 *The limits to capital* Chicago: University of Chicago Press

Harvey D 1988 The geographical and geopolitical consequences of the transition from Fordist to flexible accumulation. In G Sternlieb and JW Hughes (eds) *America's new market geography: nation, region and metropolis* Rutgers, NJ: Centre for Urban Policy Research, pp. 101–34

Harvey D 1989 From managerialism to entrepreneurialism: the transformation in urban governance in late capitalism. *Geografisca Annaler* **71B**: 3–17

Harvey D 1990 Between space and time: reflections on the geographical imagination. *Annals of the Association of American Geographers* **80**: 418–34

Hawkins RG 1972 Job displacement and the multinational firm. *Occasional Paper 3* Washington DC: Centre for International Studies

Hayter R 1976 Corporate strategies and industrial change in the Canadian forest product industries. *Geographical Review* **66**: 209–28

Hayter R 1978 Locational decision-making in a resource based manufacturing sector: case studies from the pulp and paper industry of British Columbia. *Professional Geographer* **30**: 240–49

Hayter R 1979 Labour supply and resource based manufacturing in isolated communities: the experience of pulp and paper mills in North Central British Columbia. *Geoforum* **10**: 163–77

Hayter R 1981 Patterns of entry and the role of foreign-controlled investments in the forest product sector of British Columbia. *Tijdschrift voor Economische en Social Geografie* **72**: 99–111

Hayter R 1982 Truncation, the international firm and regional policy. *Area* **14**: 277–82

Hayter R 1985 The evolution and structure of the Canadian forest sector: an assessment of the role of foreign ownership and control. *Fennia* **163**: 439–50

Hayter R 1986a The export dynamics of firms in traditional industries during recession. *Environment and Planning A* **18**: 729–50

Hayter R 1986b Export performance and export potentials: western Canadian exports of manufactured end products. *The Canadian Geographer* **30**: 26–39

Hayter R 1988 *Technology and the Canadian forest-product industries: a policy perspective* Science Council of Canada, Background Study 54, Ottawa: Ministry of Supply and Services

Hayter R 1992 International trade relations and regional industrial adjustment: implications of the 1980s' North American softwood lumber dispute for British Columbia. *Environment and Planning A* **24**: 153–70

Hayter R 1993 International trade and the Canadian forest industries: the paradox of the North American free trade agreements. *Zeitschrift für Kanada-Studien* **23**: 81–94

Hayter R 1996 Research and development. In PW Daniels and W Lever (eds) *The global economy in transition* London: Longman, pp. 164–92

Hayter R (in press) High performance organizations and employment flexibility: a case study of in situ change at the Powell River paper mill, 1980–1994. *The Canadian Geographer*

Hayter R and Barnes T 1990 Innis' staple theory, exports and recession: British Columbia 1980–86. *Economic Geography* **60**: 156–73

Hayter R and Barnes T 1992 Labour market segmentation, flexibility and recession: a British Columbian case study. *Environment and Planning D* **23**: 333–53

Hayter R and Edgington D F (in press) Cutting against the grain: a case study of MacMillan Bloedal's Japan strategy. *Economic Geography*

Hayter R and Holmes J 1993 Booms and busts in the Canadian paper industry: the case of the Powell River paper mill. *Discussion Paper 27*, Department of Geography, Simon Fraser University, Burnaby, BC

Hayter R and Holmes J 1994 Recession and restructuring at Powell River, 1980–94: employment and employment relations in transition. *Discussion Paper 28* Department of Geography, Simon Fraser University, Burnaby, BC

Hayter R and Ofori-Amoah B 1992 The ADIA programme and the manufacturing sector of the Okanagan Valley, British Columbia. *The Canadian Geographer* **36**: 56–65

Hayter R and Patchell J 1993 Different trajectories in the social divisions of labour: the cutlery industry in Sheffield, England, and Tusubame, Japan. *Urban Studies* **30**: 1427–45

Hayter R and Storey K 1979 Regional economic development in Atlantic Canada. *Plan Canada* **19**: 95–105

Hayter R and HD Watts 1983 The geography of enterprise: a reappraisal. *Progress in Human Geography* **7**: 157–81

Healey MJ 1981 Locational adjustments and the characteristics of manufacturing plants. *Transactions of the Institute of British Geographers* **6**: 394–412

Healey MJ 1982 Plant closures of multi-plant enterprises – the case of a declining industrial sector. *Regional Studies* **16**: 37–51

Healey MJ 1983 Components of locational change in multi-plant enterprises. *Urban Studies* **20**: 394–412

Healey MJ and Clark D 1984 Industrial decline and government response in the West Midlands: the case of Coventry. *Regional Studies* **18**: 303–18

Healey MJ and Clark D 1985 Industrial decline in a local economy: the case of Coventry, 1974–1982. *Environment and Planning A* **17**: 1351–67

Healey MJ and Dunham PJ 1994 Changing competitive advantage in a local economy: the case of Coventry, 1971–90. *Urban Studies* **31**: 1279–1301

Healey MJ and Ilbery BW 1990 *Location and change: perspectives on economic geography* Oxford: Oxford University Press

Healey MJ and Rawlinson MB 1993 Interviewing business owners and manager: a review of methods and techniques. *Geoforum* **24**: 339–55

Healey MJ and Watts HD 1987 The multi-plant enterprise. In WF Lever (ed.) *Industrial change in the United Kingdom* London: Longman, pp. 149–66

Hecht S and Cockburn A 1989 *The fate of the forest* London: Penguin

Heelas P and Morris P (eds) 1992 *The values of the enterprise culture: the moral debate* London: Routledge

Heilbroner R 1985 *The nature and logic of capitalism* New York: Norton

Heilbroner 1992 *Twenty-first century capitalism.* Concorde, Ontario: Anansi

Henning G and Kunzmann KR 1990 Priority to local economic development: industrial restructuring and local development responses in the Ruhr area – the case of Dortmund. In WB Stohr (ed.) *Global challenge and local response* London: Mansell, pp. 199–223

Herbert R 1996 What a denial Nike is – of sweatshop economics. *The Vancouver Sun* 11 June

Herrigel GB 1989 Industrial order and the politics of industrial change: mechanical engineering. In PK Katzenstein (ed.) *Industry and politics in West Germany: toward the third republic* Ithaca: Cornell University Press, pp. 185–220

Herrschel T 1995 Local policy restructuring: a comparative assessment of policy responses in England and Germany. *Area* **27**: 228–41

Hewings G 1977 *Regional industrial analysis and development* London: Methuen

Hiebert D 1990 Discontinuity and the emergence of flexible production: garment production in Toronto, 1901–1931. *Economic Geography* **66**: 229–53

Hill C 1954 Some aspects of industrial location. *Journal of Industrial Economics* **2**: 184–92

Hills RL 1988 *Papermaking in Britain 1488–1988.* London: Athlone Press

Hirsch S 1967 *Location of industry and international competitiveness.* Oxford: Clarendon Press

Hirsch S 1976 An international trade and investment theory of the firm. *Oxford Economic Papers* **28**: 258–70

Hirst P and Thompson G 1992 The problem of 'globalization': international economic relations, national economic management and the formation of trading blocs. *Economy and Society* **21**: 357–95

Hirst P and Zeitlin J 1991 Flexible specialization versus post-fordism: theory, evidence and policy implications. *Economy and Society* **20**: 1–56

Ho SPS and Heunemann RW 1984 *China's open door policy: the question for foreign technology and capital* Vancouver: University of British Colombia Press

Hoare AG 1973 The spheres of influence of industrial location factors. *Regional Studies* **7**: 301–14

Hollander S 1965 *The sources of increased efficiency: a study of Dupont rayon plants* Cambridge, MA MIT Press

Holmes J 1983 Industrial reorganization, capital restructuring and locational change: an analysis of the Canadian automobile industry in the 1960s. *Economic Geography* **59**: 25–71

Holmes J 1986 The organization and locational structure of production subcontracting. In AJ Scott and M Storper (eds) *Production, work and territory* Boston: Allen and Unwin, pp. 80–106

Holmes J 1987 Industrial reorganization, capital restructuring and locational change: an analysis of the Canadian automobile industry in the 1960s. In K Chapman and G Humphrys (eds) *Technical change and industrial policy* Oxford: Basil Blackwell, pp. 121–56

Holmes J 1992 The continental integration of the North American automobile industry: from the Auto Pact to the FTA. *Environment and Planning A* **24**: 95–119

Holmes J 1996 Restructuring in a continental production system. In JNH Britton (ed.) *Canada and the global economy* Montreal-Kingston: McGill-Queen's University Press, pp. 230–54

Holmes J and Hayter R 1994 Recent restructuring in the Canadian pulp and paper industry. *Discussion Paper No. 26* Department of Geography. Simon Fraser University

Holmes J and Rusonick A 1991 The break-up of an international labour union: uneven development in the North American automobile industry and the schism in the UAW. *Environment and Planning A* **23**: 9–35

Hoover EM 1971 *An introduction to regional economics* New York: Alfred Knopf

Howells J R 1986 Industry–academic links in research and innovation: a national and regional development perspective. *Regional Studies* **20**: 472–6

Howells J and Wood M 1992 *The globalisation of production and technology* London: Belhaven Press

Howes C and Markusen AR 1993 Trade, industry and economic development. In H Noponen, J Graham and AR Markusen (eds) *Trading industries, trading regions* New York: The Guilford Press, pp. 1–44

Hudson R 1984 Producing an industrial wasteland: capital, labour and the state in North East England. In RL Martin and B Rowthorne (eds) *Deindustrialization and the British space economy* London: Macmillan

Hudson R and Sadler D 1983 Region, class and the politics of steel closures in the European Community. *Society and Space* **1**: 405–28

Hudson R and Sadler D 1986 Contesting work closures in Western Europe's old industrial regions: defending place or betraying class? In AJ Scott and M Storper (eds) *Production, work and territory* Boston: Allen and Unwin, pp. 172–194

Hudson R and Sadler D 1989 *The international steel industry: restructuring state policies and localities* London: Routledge

Hudson R and Schamp E 1995 (eds) *Towards a new automobile map of Europe: new production concepts and spatial restructuring* New York: Springer

Huff DL 1960 A topographical model of consumer space preferences. *Papers and Proceedings of Regional Science Association* **6**: 159–73

Hunter H 1955 Innovation, competition and locational changes in the pulp and paper industry. *Land Economics* **31**: 314–27

Hurst ME 1974 *A geography of economic behaviour* North Scituate, MA: Duxbury Press

Hutton T and Ley D 1987 Location, linkages, and labor: the downtown complex of corporate activities in a medium size city, Vancouver, British Columbia. *Economic Geography* **63**: 125–40

Hymer SH 1960 The international operations of national firms: a study of direct investment. PhD dissertation, Cambridge: MIT

Hymer SH 1972 The efficiency (contradictions) of multinational corporations. In G Paquet (ed.) *The multinational firm and the nation state* New York: Macmillan, pp. 49–65

Hymer SH and Rowthorn R 1970 Multinational corporations and international oligoploy: the non-American challenge. In CP Kindleberger (ed.) *The international corporation* Cambridge: MIT Press

Ide S and Takeuchi A 1980 *Jiba sangyo*: localized industry. In T Shoin (ed.) *Geography of Japan* Tokyo: the Association of Japanese Geographers, pp. 299–319

Ika H 1988 Flying Tsubame. *Research on Tsubame City* **6**: 1–41 (in Japanese)

Innis HA 1930 *The fur trade in Canada: an introduction to Canadian economic history* Toronto: Toronto University Press

ILO (International Labour Office) 1988 *Economic and social effects of multinational enterprises in developing countries* Geneva: ILO

ILO (International Labour Office) 1993 *Yearbook of Labour Statistics* Geneva: ILO

Itoh M 1989 Interfirm relations and continuous transactions. In K Imai and R Komiya (eds). *The*

Japanese Firm Tokyo: Tokyo University (in Japanese)

Jacoby NH 1972 The multinational corporation. In A Kapoor and PD Grub (eds). *The multinational enterprise in transition* Princeton: the Darwin Press, pp. 141–55

Jenson J 1989 The talents of women, the skills of men: flexible specialization and women. In S Wood (ed.) *The transformation of work* London: Unwin, pp. 141–55

Jessop R 1992 Fordism and post-fordism: a critical re-formulation. In M Storper and AJ Scott (eds) *Pathways to industrialization and regional development* London: Routledge, pp. 46–69

Johannisson B 1986 Network strategies: management technology for entrepreneurship and change. *International Small Business Journal* **5**: 19–30

Johannisson B 1993 Designing supportive contexts for emerging enterprises. In C Karlsson, B Johannisson and D Storey (eds) *Small business dynamics* London: Routledge, pp. 117–42

Johanson J and Vahlne JE 1977 The internationalization process of the firm – a model of knowledge development and increasing foreign market commitments. *Journal of International Business Studies* Spring: 23–32

Johanson J and Widerscheim-Paul F 1975 The internationalization of the firm – a model of knowledge development and increasing foreign market commitments. *Journal of International Business Studies* **88**: 23–32

Johnson M 1985 Postwar industrial development in the southeast and the pioneer role of labor-intensive industry. *Economic Geography* **61**: 46–65

Johnson M 1991 An empirical update on the product-cycle explanation and branch plant location in the nonmetropolitan US South. *Environment and Planning A* **23**: 397–410

Johnson PS and Cathcart DG 1979 New manufacturing firms and regional development: some evidence from the Northern region. *Regional Studies* **13**: 269–80

Johnson S 1991 Small firms and the UK labour market: prospects for the 1990s. In J Curran and RA Blackburn (eds) *Paths of enterprise: the future of small business* London: Routledge, pp. 88–113

Johnston RJ 1984 The world is our oyster. *Transactions of the Institute of British Geographers* **9**: 443–59

Johnston RJ 1987 *Geography and geographers: Anglo-American human geography since 1945* London: Edward Arnold

Johnston WB 1991 Global work force 2000: the new world labor market. *Harvard Business Review* March–April: 115–27

Joint Economic Congress Survey 1982 *Location of high technology firms and regional economic development* Washington: US Government Printing Office

Jones G 1986 *British multinationals: origins, management and performance* Aldershot: Gower

Jones NK and North J 1991 Japanese motor industry transplants: the west European dimension. *Economic Geography* **67**

Karlsson J, Johannisson B and Storey D (eds) 1993 *Small business dynamics: international, national and regional perspectives* London: Routledge

Katona G and Morgan JN 1952 The quantitative study of factors determining business decisions. *Quarterly Journal of Economics* **66**: 67–90

Katzenstein PJ 1985 *Small states in world markets: industrial policy in Europe* Ithaca, New York: Cornell University Press

Keeble DE 1967 Models in economic development. In RJ Chorley and P Haggett (eds) *Models in Geography* London: Methuen, pp. 243–303

Keeble DE 1976 *Industrial location and planning in the United Kingdom* London: Methuen

Keeble DE 1980 Industrial decline, regional policy and the urban–rural manufacturing shift. *Environment and Planning A* **12**: 945–61

Keeble DE 1984 The urban–rural manufacturing shift. *Geography* **69**: 163–6

Keeble DE 1989 High technology industry and regional development in Britain: the case of the Cambridge phenomenon. *Environment and Planning C* **7**: 153–72

Keeble DE and Walker S 1994 New firms, small firms and dead firms: spatial patterns and determinants in the United Kingdom. *Regional Studies* **28**: 411–27

Keeble DE, Owens PL and Thompson D 1982 Regional accessibility and economic potential in the European community. *Regional Studies* **16**: 419–31

Kelley MR and Brooks H 1988 *The state of computerized automation in US manufacturing* Centre for Business and Government, JF Kennedy School of Government, Cambridge: Harvard University Press

Kelly T and Keeble DE 1990 IBM: the corporate chamelon. In M de Smidt and E Wever (eds) *The corporate firm in a changing world economy: case studies in the geography of enterprise* London: Routledge, pp. 21–54

Kemper NJ and De Smidt M 1980 Foreign manufacturing establishments in the Netherlands. *Tijdschrift voor Economische en Social Geografie* **71**: 21–40

Kenney M 1986 Schumpeterian innovation and entrepreneurs in capitalism: a case study of the US biotechnology industry. *Research Policy* **15**: 21–31

Kenney M and Florida R 1994 Japanese maquiladoras: production organization and global commodity chains. *World Development* **22**: 27–44

Kepner CH and Tregoe BC 1965 *The rational manager* New York: McGraw Hill

Kerr C 1954 The balkanization of labour markets. In E Bakke (ed.) *Labor mobility and economic opportunity*, Cambridge: MIT Press, pp. 92–110

Khan A and Hayter R 1984 The linkages of new manufacturing firms: an exploratory enquiry in the Vancouver metropolitan area. *Albertan Geographer* **20**: 1–13

Kilby P 1971 *Hunting the hefflelump: entrepreneurship and economic development* New York: The Free Press

Kioke K 1988 *Understanding industrial relations in modern Japan* Basingstoke: Macmillan

Kioke K and Inoke T 1990 *Skill formation in Japan and southeast Asia* Tokyo: University of Tokyo Press

Knickerbocker FT 1973 *Olipololistic reaction and multinational enterprise* Boston: Harvard University Press

Knox PL, Bartels EH, Holcomb B, Bohland JR and Johnston RJ 1988 *The United States – a contemporary human geography* Harlow, Essex: Longman Scientific Technical

Knox PN and Agnew J 1994 *The geography of the world economy* London: Routledge

Kobrin SJ 1987 Testing the bargaining hypothesis in the manufacturing sector in developing countries. *International Organization* **41**: 609–38

Kolde EJ 1972 *International Business Enterprise* Englewood Cliffs, NJ: Prentice Hall

Kondratieff ND 1978 The long waves in economic life. *Lloyds Bank Review* **129**: 41–60

Kraar L 1996a Need a friend in Asia: try the Singapore connection. *Fortune* March 4: 177–83

Kraar L 1996b Daewoo's daring drive into Europe. *Fortune* May 13: 145–52

Krebs G 1982 Regional inequalities during the process of national economic development: a critical approach. *Geoforum* **13**: 71–81

Kroos HE 1974 *American economic development: the progress of a business civilization* New York: Prentice-Hall

Krumme G 1969a Towards a geography of enterprise. *Economic Geography* **45**: 30–40

Krumme G 1969b Notes on locational adjustment patterns in industrial geography. *Geografiska Annaler* **51B**: 15–19

Krumme G 1970 The inter-regional corporation and the region. *Tijdschrift voor Economische en Sociale Geografie* **61**: 318–33

Krumme G 1981 Making it abroad: the evolution of Volkswagen's North American production plans. In FEI Hamilton and GJR Linge (eds) *Spatial analysis,* *industry and the industrial environment, vol. 2. International industrial systems* New York: Wiley, pp. 329–56

Krumme G and Hayter R 1975 Implications of corporate strategies and product cycle adjustments for regional employment changes. In L Collins and DF Walker (eds) *Locational Dynamics of Manufacturing Activity* New York: Wiley, pp. 325–56

Kudo A 1994 IG Farben in Japan: the transfer of technology and managerial skills. In G Jones (ed.) *The making of global enterprise* London: Frank Cass, pp. 159–83

Kuhn RL 1982 *Mid-sized firms: success strategies and methodologies* New York: Praeger

Kuhn RL 1985 *To flourish among giants* London: John Wiley

Kuttner R 1984 *The economic illusion: false choices between prosperity and social justice* Boston: Houghton Mifflin

Lai CF 1985 Special economic zones: the Chinese road to socialism? *Environment and Planning D: Society and Space* **3**: 63–84

Langlois RN and Robertson PL 1995 *Firms, markets and economic change: a dynamic theory of business institutions* London: Routledge

Laulajainen R 1982 Temporal hierarchy in corporate space. *Geojournal* **6**: 399–408

Laulajainen R 1995 Corporate geography relaunched. *Geographical review of Japan* **68**: 185–97

Law CM 1980 The foreign company's location investment decision and its role in British regional development. *Tijschrift voor Economische en Sociale Geografie* **71**: 15–20

Law CM 1983 The defence sector in regional development. *Geoforum* **14**: 169–84

Law CM (ed.) 1991 *Restructuring the global automobile industry: national and regional impacts* London: Routledge

Laxer G 1989 *Open for business: the roots of foreign ownership in Canada* Toronto: Oxford University Press

Lees C and Hinde S 1995 Scandal of football's child slavery. *The Sunday Times* 14 May: 6

LeHeron RB 1976 Best practice firms and productivity changes in the Pacific Northwest plywood and veneer industry, 1960–72: some regional growth implications. *Environment and Planning A* **8**: 163–72

LeHeron RB 1980 Exports and linkage development in manufacturing firms: the example of export promotion in New Zealand. *Economic Geography* **56**: 281–99

LeHeron RB 1990 Resource development and the evolution of New Zealand forestry companies. In R Hayter and PD Wilde (eds) *Industrial transformation and challenge in Australia and Canada* Ottawa: Carleton University Press, pp. 195–212

Leigh R and North DJ 1978 Regional aspects of acquisition activity in British manufacturing activity. *Regional Studies* **12**: 227–45

Leinbach TR and Amrhein C 1987 A geography of the venture capital industry in the USA. *Professional Geographer* **39**: 146–58

Lester RA 1967 *Manpower planning in a free society.* Princeton: Princeton University Press

Leung CK 1990 Locational characteristics of foreign equity joint venture investment in China, 1979–85. *Professional Geographer* **42**: 403–21

Leung CK 1993 Personal contacts, subcontracting linkages, and development in the Kong Kong–Zhujiang Delta region. *Annals of the Association of American Geographers* **83**: 272–302

Lever WF 1972 Industrial movement, spatial association and functional linkage. *Regional Studies* **6**: 371–84

Lever WF 1985 Theory and method in industrial geography. In M Pacione (ed.) *Progress in industrial geography* London: Croom Helm, pp. 10–39

Lever-Tracy C 1988 The flexibility debate: part-time work. *Labour and Industry* **1**: 210–41

Levitt K 1970 *Silent surrender: the multinational corporation in Canada* Toronto: Macmillan

Ley DF 1980 Liberal idealogy and the post-industrial city. *Annals of the Association of American Geographers* **70**: 238–58

Ley DF 1985 Downtown or the suburbs? A comparative study of two Vancouver head-offices. *The Canadian Geographer* **29**: 30–43

Ley DF 1986 Alternative explanations for inner city gentrification: a Canadian assessment. *Annals, Association of American Geographers* **76**: 521–35

Ley DF and Hutton T 1987 Vancouver's corporate complex and producer services sector; linkages and divergence within a provincial staple economy. *Regional Studies* **21**: 413–24

Ley DF and Mills C 1990 Labour markets, housing markets and changing family in Canada's service economy. In R Hayter and PD Wilde (eds) *Industrial transformation and challenge in Australia and Canada* Ottawa: Carleton University Press, pp. 93–107

Library of Congress 1968 *Paper making: art and craft* Washington: Library of Congress

Lindberg O 1953 An economic-geographical study of the Swedish paper industry. *Geografisca Annaler* **35**: 28–40

Linge GJR 1991 Just-in-time: more or less flexible. *Economic Geography* **61**: 316–332

Lipietz A 1986 New tendencies in the international division of labour: regimes of accumulation and modes of regulation. In AJ Scott and M Storper (eds) *Production, work and territory: the geographical anatomy of industrial capitalism* London: Routledge, pp. 16–40

List F 1922 *The national system of political economy* London: Longmans Green

Litvak IA and Maule CJ 1981 *The Canadian multinationals* Toronto: Butterworths

Lloyd GIH 1913 *The cutlery trades: the economics of small scale production* London: Longmans Green

Lloyd PE and Dicken P 1977 *Location in space: a theoretical approach to economic geography* 2nd edition, London: Harper and Row

Lloyd P and Mason C 1978 Manufacturing industry in the inner city. *Transactions of the Institute of British Geographers* **3**: 66–90

Lloyd P and Mason C 1984 Spatial variations in new firm formation in the United Kingdom: comparative evidence from Merseyside, Greater Manchester and South Hampshire. *Regional Studies* **18**: 207–20

Lonsdale RE and Seyler HK (eds) 1979 *Nonmetropolitan industrialization* New York: John Wiley

Lorenz EH 1992 Trust, community and cooperation: towards a theory of industrial districts. In M Storper and AJ Scott (eds) *Pathways to industrialization and regional development* London: Routledge, pp. 195–204

Loveman G and Sengenberger W 1991 The emergence of small scale production: an international comparison. *Small Business Economic* **3**: 1–38

Lovering J 1990 Fordism's unknown successor: a comment on Scott's theory of flexible accumulation and the re-emergence of regional economies. *International Journal of Urban and Regional Research* **12**: 171–85

Lower AR 1973 *Great Britain's woodyard: British America and the timber trade 1763–1867* Montreal: McGill-Queen's Press

Luostarinen R 1980 *Internationalization of the firm* Helsinki: Helsinki School of Economics

Luttrell WF 1962 *Factory location and industrial movement: a study of recent experiences in Great Britain* London: National Institute of Economic and Social Research

Lutz RA 1994 Implementing technical change with cross functional teams, *Research-Technology Management* **37**: 14–18

Lynn M 1995 Quiet tycoon runs rings round hamstrung IBM, *The Sunday Times* 7 May: 2.4

Lyons B 1993 Small subcontractors in UK engineering, *Small Business Economics* **5**: 101–9

Machlup F 1949 *The basing point system: an economic analysis of a controversial pricing practice* Philadelphia: Blakiston

Machlup F 1967 Theories of the firm: marginalist, behavioural, managerial, *American Economic Review* **57**: 1–33

Mack RP 1971 *Planning on uncertainty* New York: Wiley Interscience

Mack RS and Schaeffer PV 1993 Nonmetropolitan manufacturing in the United States and the product cycle theory: a review of the literature. *Journal of Planning Literature* **8**: 124–39

MacLachlan I 1996 Organization restructuring of US-based manufacturing subsidiaries and plant closures. In JNH Britton (ed.) *Canada and the global economy* Montreal: McGill-Queen's University Press, pp. 195–214

MacPherson AD 1987 Industrial innovation in the small business sector: empirical evidence from metropolitan Toronto. *Environment and Planning A* **20**: 953–71

MacPherson AD 1988 New product development among small Toronto manufacturers: empirical evidence on the role of new technical service linkages. *Economic Geography* **64**: 62–75

MacPherson AD 1994 The impact of industrial process innovation among small manufacturing firms: empirical evidence from western New York. *Environment and Planning A* **26**: 121–36

MacPherson AD 1996 Shifts in Canadian direct investment abroad and foreign direct investment in Canada. In JNH Britton (ed.) *Canada and the global economy* Montreal-Kingston: McGill-Queen's University Press, pp. 69–83

MacPherson AD and McConnell JE 1992 Recent Canadian direct investment in the United States: and empirical evidence from Western New York State, *Environment and Planning A* **24**: 121–36

Maddison A 1991 *Dynamic forces in capitalist development* Oxford: Oxford University Press

Mair A 1994 *Honda's global local corporation* New York: St Martin's Press

Malecki EJ 1979 Locational trends in R&D by large US corporations 1965–77. *Economic Geography* **55**: 309–23

Malecki EJ 1980 Dimensions of R&D Location in the United States, *Research Policy* **9**: 2–22

Malecki EJ 1986 Research and development and the geography of high technology complexes. In J Rees (ed.) *Technology, regions and policy* Totowa, NJ: Rowman and Littlefield, pp. 51–74

Malecki EJ 1987 The R&D location decision of the firm and 'creative' regions, *Technovation* **6**: 205–22

Malecki EJ 1991 *Technology and Economic Development* Harlow: Longman

Malecki EJ and Bradbury S 1992 R&D facilities and professional labour: labour force dynamics in high technology. *Regional Studies* **26**: 123–36

Malmgren HB 1961 Information, expectations and the theory of the firm, *Quarterly Journal of Economics* **75**: 399–421

Mandel E 1980 *Long waves of capitalist development* Cambridge: Cambridge University Press

Mansfield E, Teece D and Romeo R 1979 Overseas research and development by US-based firms, *Economica* **46**: 187–96

Mantoux P (1966) *The industrial revolution in the eighteenth century* London: Methuen

Marchak P 1995 *Logging the globe* Montreal: McGill-Queen's University Press

Marglin S 1983 The power of knowledge. In F Stephens (ed.) *Work organization* London

Markusen A 1985 *Profit cycles, oligopoly and regional development* Cambridge, MA: MIT Press

Markusen AR 1987 *Regions: the economics and politics of territory* London: Mansell

Markusen A 1994 Studying regions by studying firms. *The Professional Geographer* **46**: 477–90

Markusen A, Hall P and Glasmeier A 1986 *High Tech America* Boston: Allen and Unwin

Markusen A, Hall P and Deitrick S and Campbell S 1991 *The rise of the gunbelt* New York: Oxford University Press

Marquand D 1992 The enterprise culture: old wine in new bottles? In P Heelas and P Morris (eds) *The values of the enterprise culture: the moral debate* London: Routledge, pp. 61–72

Marris R 1968 *The economic theory of 'managerial' capitalism* New York: Basic Books

Marshall A 1922 *The principles of economics* London: Macmillan

Marshall HA, Southard FA and Taylor KW 1936 *Canadian–American industry: a study in international investment* Toronto: Ryerson

Marshall JN 1987 Industrial change, linkages and regional development. In WF Lever (ed.) *Industrial Change in the United Kingdom* Harlow: Longman, pp. 108–22

Marshall JN, Wood P, Daniels PW, McKinnon A, Bachtler J, Damesick P, Thrift N, Gillespie A, Green A and Leshon A 1988 *Services and uneven development* Oxford: Oxford University Press

Marshall M 1990 Regional alternatives to economic decline in Britain's industrial heartland: industrial restructuring and local intervention in the West Midlands conurbation. In WB Stohr (ed.) *Global challenge and local response* London: Mansell, pp. 163–98

Marshall R 1994 Organizations and learning: systems for a high-wage economy, In C Kerr and PD Staudoher (eds) *Labor economics and industrial relations: markets and institutions* Cambridge: Harvard University Press, pp. 601–45

Marshall R and Tucker M (1992) *Thinking for a living: work, skills and the future of the American economy* New York: Basic Books

Martin R 1988 The political economy of Britain's north–south divide, *Transactions, Institute of British Geographers* **13**: 389–418

Martin R and Rowthorn B (eds) 1986 *The geography of deindustrialisation* Basingstoke, Hampshire: Macmillan

Martin R, Sunley P and Wills J 1993 The geography of trade union decline: spatial dispersal or regional resilience? *Transactions, the Institute of British Geographer* **18**: 36–62

Marx K 1978 *Capital* vol 2, Harmondsworth: Penguin

Mason CM 1983 Some definitional problems in new firm research. *Area* **15**: 53–60

Mason CM 1989 Explaining recent trends in new firm formation in the UK: some evidence from South Hampshire. *Regional Studies* **23**: 331–46

Massey D 1979 A critical evaluation of industrial location theory. In FEI Hamilton and GJR Linge (eds) *Spatial analysis, industry and the industrial environment, 1 Industrial systems* London: Wiley pp. 57–72

Massey D 1984 *Spatial divisions of labour: social structures and the geography of production* London: Macmillan

Massey D 1988 Uneven development: social change and spatial divisions of labour. In D Massey and J Allen (eds) *Uneven re-development: cities and regions in transition* London: Hodder and Stoughton, pp. 250–76

Massey D and Allen J (eds) 1984 *Geography matters!* Cambridge: Cambridge University Press

Massey D and Meegan R 1982 *The anatomy of job loss* London: Methuen

Massey D, Quintas P and Wield D 1992 *High tech fantasies: science parks in society and space* London: Routledge

Mather AS 1990 *Global forest resources* London: Belhaven

McArthur R 1987 Innovation, diffusion and technical change: a case study. In K Chapman and G Humphrys (eds) *Technical change and industrial policy* Oxford: Basil Blackwell, pp. 26–50

McConnell JE 1980 Foreign direct investment in the United States. *Annals of the Association of American Geographers* **70**: 259–70

McConnell JE 1983 The international location of manufacturing investments: recent behaviour of foreign-owned corporations in the United States. In FEI Hamilton and GJR Linge (eds) *Spatial analysis, industry and the industrial environment* vol 3, London: Wiley, pp. 337–58

McConnel JE 1986 Geography of international trade. *Progress in Human Geography* **10**: 471–83

McDermott P 1973 Spatial margins and industrial location in New Zealand. *New Zealand Geographer* **29**: 64–74

McDowell L 1992 Valid games: a response to Erica Schoenberger. *The Professional Geographer* **44**: 212–15

McLafferty S and Preston V 1991 Gender, race and commuting among service sector workers. *The Professional Geographer* **43**: 1–15

McLelland DC 1961 *The achieving society* Princeton, NJ: Van Nostrand

McLoughlin P and Cannon JB 1990 Whither regional policy? Australian and Canadian perspectives. In R Hayter and PD Wilde (eds) *Industrial transformation and challenge in Australia and Canada*. Ottowa: Carleton University Press, pp. 259–276

McManus JC 1972 The theory of the international firm. In G Paquet (ed.) *The multinational firm and the nation state* New York: Macmillan, pp. 66–93

McNaughton RB 1992a Patterns of foreign direct investment in Canada 1985–89: Canadian-based foreign controlled firms versus US and overseas investors. *The Canadian Geographer* **36**: 50–6

McNaughton RB 1992b US direct investment in Canada 1985–89. *The Canadian Geographer* **36**: 181–8

McNee RB 1958 Functional geography of the firm with an illustrative case study from the petroleum industry, *Economic Geography* **34**: 321–7

McNee RB 1960 Towards a more humanistic economic geography: the geography of enterprise. *Tijdschrift voor Economische en Social Geografie* **51**: 201–5

McNee RB 1963 The spatial evolution of the Sun Oil Company. *Annals of the Association of American Geographers* **53**: 609

McNee RB 1974 A systems approach of understanding the geographic behaviour of organizations, especially large corporations. In FEI Hamilton (ed.) *Spatial perspectives on industrial organization and decision-making* London: Wiley, pp. 47–75

Mensch A 1979 *Stalemate in technology: innovation overcame the depression* Cambridge, MA: Ballinger

Meyer JR, Kain JF and Wohl M 1969 *The urban transportation problem* London: Oxford University Press

Michalak WZ 1993 Foreign direct investment and joint ventures in East-Central Europe: a geographical perspective. *Environment and Planning A* **25**: 1573–91

Miernyk WH 1965 *The elements of input–output analysis* New York: Random House

Miller EW 1962 *A geography of manufacturing* Englewood Cliffs, NJ: Prentice-Hall

Milne S 1991 Small firms, industrial reorganisation, and space: the case of the UK high-fidelity audio sector. *Environment and Planning A* **23**: 833–52

Mitchelson RL and Wheeler JO 1994 The flow of information in a global economy: the role of the American urban system in 1990. *Annals of the Association of American Geographers* **84**: 87–107

Miyakawa Y 1980 Evolution of industrial system and industrial community. *Science Reports of the Tohoku University* 7th Series (Geography) **30**: 21–64

Miyakawa Y 1981 Evolution of industrial system and industrial community. *Science Reports of the Tohoku University* 7th Series (Geography) **31**: 49–84

Miyakawa Y 1991 The transformation of the Japanese motor vehicle industry and its role in the world: industrial restructuring and technical evolution. In CM Law (ed.) *Restructuring the global automobile industry* London: Routledge, pp. 88–113

Monck CSP, Porter RB, Quintas P, Storey DJ and Wynarczyk P 1988 *Science parks and the growth of high technology firms* London: Croom Helm

Moore M 1990 *Roger and Me* Warner Bros Video

Morales R 1994 *Flexible production: restructuring of the international automobile industry* Cambridge: Polity Press

Moran TH 1975 Multinational corporations. In F Tugwell (ed.) *The politics of oil in Venezuela* Stanford: Stanford University Press

Morgan AD 1979 Foreign manufacturing by UK firms. In F Blackaby (ed.) *De-industrialization* London: Heinemann, pp. 78–101

Morgan K 1994 Reversing attrition: the auto cluster in Baden-Wurttemburg, paper presented at the University of Toronto, conference in honour of HA Innis. *Regions, institutions and technology: reorganizing economic geography in Canada and the Anglo-American world*

Morgan K and Sayer A 1988 *Microcircuits of capital: 'sunrise' industry and uneven development* Cambridge: Polity Press

Moriarty BM 1983 *Industrial location and community development* Charlotte: University of North Carolina

Morrill RL 1974 *The Spatial Organization of Society* 2nd edition, Belmont: Wadsworth

Morris J 1988 New technologies, flexible work practices and regional social spatial differentiation: some observations from the United Kingdom. *Environment and Planning D* **6**: 301–19

Morris J 1989 Japanese inward investment and the 'importation' of sub-contracting complexes: three case studies. *Area* **21**: 269–77

Morris PJ 1983 Australia's dependence on imported technology – some issues for discussion. *Prometheus* **1**: 144–59

Moses LN 1958 Location and the theory of production. *Quarterly Journal of Economics* **73**: 373–99

Murphy AB 1992 Western investment in east-central Europe: emerging patterns and implications for state stability. *Professional Geographer* **44**: 249–59

Myrdal G 1957 *Economic theory and underdeveloped regions* London: Duckworth

Nakagjo S 1980 Japanese direct investment in Asian newly developing countries and intra-firm division of labour. *The Developing Economies* **18**: 463–83

Nakamura H 1990 *New mid-sized firm theory* Tokyo: Tokyo Keizai Shinpansha (in Japanese)

Nelson J 1989 The new global sweatshop. *Canadian Forum*

Nelson RR 1988 Institutions supporting technical change in the United States. In G Dosi, C Freeman, RR Nelson, G Silverberg and L Soete (eds) *Technical change and economic theory* New York: Pinter, pp. 312–39

Newfarmer RS and Topic S 1982 Testing dependency theory: a case study of Brazil's electrical industry. In MJ Taylor and N Thrift *A geography of multinationals* London: Croom Helm, pp. 33–60

Niosi J 1985 *Canadian multinationals* Toronto: Between the Lines

Nishioka H and Krumme G 1973 Location conditions, factors and decisions: an evaluation of selected location surveys. *Land Economics* May: 195–205

Noble I 1987 Unemployment after redundancy and political attitudes: some empirical evidence. In RM Lee (ed.) *Redundancy, lay-offs and plant closures: their character, causes and consequences* London: Croom Helm, pp. 280–304

Norcliffe GB 1975 A theory of manufacturing places. In L Collins and DF Walker (eds) *Location dynamics of manufacturing activity* London: John Wiley, pp. 19–58

Norcliffe GB 1984 Nonmetropolitan industrialization and the theory of production. *Urban Geography* **5**: 25–42

Norcliffe GB 1987 Regional unemployment in Canada in the 1981–1984 recession. *The Canadian Geographer* **311**: 150–9

Norcliffe GB 1994 Regional labour market adjustments in a period of structural transformation: an assessment of the Canadian case. *The Canadian Geographer* **38**: 2–17

Norcliffe GB and Featherstone D 1990 International influences on regional unemployment patterns in Canada during the 1981–84 recession. In R Hayter and PD Wilde (eds) *Industrial Transformation and Challenge in Australia and Canada* Ottawa: Carleton University Press, pp. 71–92

Norton RD 1992 Agglomeration and competitiveness: from Marshall to Chinitz. *Urban Studies* **29**: 155–70

Norton RD and Rees J 1979 The product cycle and the spatial decentralization of American manufacturing. *Regional Studies* **13**: 141–51

Oakey RP 1981 *High technology industry and industrial location* Aldershot: Gower

Oakey RP 1984 *High technology small firms* London: Frances Pinter

Oakey RP 1985 British university science parks and high technology small firms. *International Small Business Journal* **4**: 58–67

Oakey RP 1991 Government policy towards high technology: small firms beyond the year 2000. In J Curran and RA Blackburn (eds) *Paths of enterprise: the future of the small business* London: Routledge, pp. 128–48

Oakey RP 1993 High technology small firms: a more realistic evaluation of their growth potential. In C Karlsson, B Johannisson and D Storey (eds) *Small business dynamics* London: Routledge, pp. 224–43

Oakey RP, Thwaites AT and Nash PA 1982 Technical change and regional development: some evidence on regional variations in product and process innovation. *Environment and Planning A* **14**: 1073–86

Odaka K, Ono K and Adachi F 1988 *The automobile industry in Japan: a study of ancillary firm development* Oxford: Oxford University Press

Ofori-Amoah B 1993 Technology choice and diffusion in the manufacturing sector: the case of the twin-wire in the Canadian pulp and paper industry. *Geoforum* **24**: 315–26

Ofori-Amoah B 1995 Regional impact of technological change: the evolution and development of the twin-wire paper machine from 1950 to 1988. *Environment and Planning A* **21**: 1503–20

Ofori-Amoah B and Hayter R 1989 Labour turnover characteristics at the Kitimat pulp and paper mill: a log-linear analysis. *Environment and Planning A* **21**: 1491–510

OECD 1979 *The impact of newly industrializing countries* Paris: OECD

OECD 1981 *International investment and multinational enterprises: recent international investment trends* Paris: OECD

OECD 1993 *Small and medium-sized enterprises: technology and competitiveness* Paris: OECD

O'Farrell PN 1980 Multinational enterprises and regional development: Irish evidence. *Regional Studies* **13**: 141–52

O'Farrell PN 1986 *Entrepreneurs and industrial change* Dublin: IMI

O'Farrell PN and Crouchley R 1984 An industrial and spatial analysis of new firm formation in Ireland. *Regional Studies* **18**: 221–36

O'Farrell PN and Hitchens DNMW 1988 The relative competitiveness and performance of small manufacturing firms in Scotland and the mid-west of Ireland. *Regional Studies* **22**: 399–416

Ó'hUallacháin B 1985 Spatial patterns of foreign investment in the United States. *Professional Geographer* **37**: 155–63

Ó'hUallacháin B 1986 The role of foreign direct investment in the development of regional industrial systems: current knowledge and suggestions for a future American research agenda. *Regional Studies* **20**: 151–62

Ó'hUallacháin B and Reid N 1992 Source country differences in the spatial distribution of foreign direct investment in the United States. *Professional Geographer* **44**: 272–85

Oinas P 1995a Types of enterprise and local relations. In B van der Knapp and RB LeHeron (eds) *Human resources and industrial spaces: a perspective on globalisation and localization* London: John Wiley, pp. 177–95

Oinas P 1995b Organisations and environments: linking industrial geography and organisation theory. In S Conti, EJ Malecki and P Oinas (eds) *The industrial enterprise and its environment: spatial perspectives* Aldershot: Averbury, pp. 143–67

Owens PR 1980 Direct foreign investment – some implications for the source economy. *Tijdschrift voor Economische en Social Geografie* **71**: 50–62

Palloix C 1977 The self-expansion of capital on a world scale. *Review of Radical Political Economics* **9**: 1–28

Park SO 1990 Daewoo: corporate growth and spatial organization. In M de Smidt and E Wever (eds) *The corporate firm in a changing world economy: case studies in the geography of enterprise* London: Routledge, pp. 207–33

Park SO and Markusen A 1995 Generalizing new industrial districts: a theoretical agenda and an application from a non-western economy. *Environment and Planning A* **27**: 81–104

Park SO and Wheeler JO 1983 The filtering down process in Georgia: the third stage of the product life cycle. *The Professional Geographer* **35**: 18–31

Patchell JR 1991 The creation of production systems within the social division of labour of the Japanese robotics industry: the impact of the relation specific skill. Unpublished PhD thesis, Simon Fraser University

Patchell JR 1992 *Shinchintaisha*: Japanese small business revitalization. *Business: The Contemporary World* **4**: 50–61

Patchell JR 1993a From production systems to learning systems: lessons from Japan. *Environment and Planning A* **25**: 797–815

Patchell JR 1993b Composing robot production systems: Japan as a flexible manufacturing system. *Environment and Planning A* **25**: 923–44

Patchell JR 1996 Kaleidoscope economies: the processes of cooperation, competition and control. *Annals of the Association of American Geographers* **86**: 481–506

Patchell JR and Hayter R 1992 Community industries and enterprise strategies in Japan: the case of the Tsubame cutlery industry. *Growth and Change* **23**: 199–216

Patchell JR and Hayter R 1995 Skill formation and Japanese production systems. *Tijdschrift voor Economische en Sociale Geografie* **86**: 339–56

Pavitt RM and Townsend J 1987 The size distribution of innovating firms in the UK: 1945–83. *Journal of Industrial Economics* **35**: 297–306

Pearce RD and Singh S 1992 *Globalizing research and development* London: Macmillan

Peck FW 1985 The use of matched pairs research design in industrial surveys. *Environment and Planning A* **17**: 981–9

Peck FW 1990 Nissan in the North East: the multiplier effects. *Geography* **75**: 355–7

Peck JW 1992 Labour and agglomeration: labour control and flexibility in local labour markets. *Economic Geography* **68**: 325–47

Peck J 1993 Regulating labour: the social regulation and reproduction of local labour markets. In A Amin and N Thrift (eds) *Globalization, institutions, and regional development in Europe* New York: Guildford, pp. 147–76

Peck J 1996 *Work-place: the social regulation of labour markets* New York: Guilford Publications

Peck F and Townsend AR 1984 Contrasting experience of recession and spatial restructuring: British shipbuilders, Plessey and metal box. *Regional Studies* **18**: 319–38

Peet JR 1983 Relations of production and the relocation of United States manufacturing industry since 1960. *Economic Geography* **59**: 112–43

Penrose E 1959 *The theory of the growth of the firm* New York: Wiley

Perez C 1983 Structural change and the assimilation of new technologies in the economic and social system. *Futures* **15**: 357–75

Perrucci R 1994 *Japanese auto transplants in the heartland* New York: Aldine de Gruyter

Peterson WC (ed.) 1988 *Market power and the economy* Boston: Kluwer Academic

Pfirrmann O 1995 Path analysis and regional development: factors affecting R&D in west German small and medium sized firms. *Regional Studies* **27**: 605–18

Phelps NA 1992 External economies, agglomeration and flexible accumulation. *Transactions of the Institute of British Geographers* **17**: 35–46

Pinch SP, Mason CM and Witt STG 1989 Labour flexibility and industrial restructuring in the UK 'Sunbelt': the case of Southampton. *Transactions of the Institute of British Geographers* **14**: 418–35

Pinelo AS 1973 *The multinational corporation as a force in Latin American politics: a case study of the international petroleum corporation in Peru* New York: Praeger

Piore M and Sabel C 1984 *The second industrial divide: possibilities for prosperity* New York: Basic Books

Pocock D and Hudson R 1978 *Perception of the urban environment* London: Macmillan

Polanyi K 1944 *The great transformation* New York: Rinehart

Pollard J and Storper M 1996 A tale of twelve cities: metropolitan employment change in dynamic industries in the 1980s. *Economic Geography* **66**: 1–21

Pollard S 1981 *Peaceful conquest. The industrialization of Europe 1760–1970* Oxford: Oxford University Press

Pollard S 1982 *The wasting of the British economy* London: Croom Helm

Porter ME 1990 *The competitive advantage of nations* New York: Free Press

Pounce RJ 1981 *Industrial movement in the United Kingdom, 1966–75* London: HMSO

Pratten C 1991 *The competitiveness of small firms* Cambridge: Cambridge University Press

Pred AR 1965 Industrialization, initial advantage and American metropolitan growth. *Geographical Review* **55**: 158–85

Pred AR 1967 *Behaviour and location: foundations for a geographic and dynamic location theory: Part 1* University of Lund, Lund Studies in Geography B, No. 27

Pred AR 1969 *Behaviour and location: foundations for a geographic and dynamic location theory: Part II* University of Lund, Lund Studies in Geography B, No. 28

Pred AR 1974 *Major job providing organizations and systems of cities.* Washington, DC, Association of American Geographers, Commission on College Geography, resource paper 27

Pyke F, Becattini G and Sengenberger W (eds) 1990 *Industrial districts and inter-firm cooperation in Italy* Geneva: International Institute for Labour Studies

Randall JE and Ironside RG 1996 Communities on the edge: an economic geography of resource-dependent communities in Canada. *The Canadian Geographer* **40**: 17–35

Rawlinson M 1991 Subcontracting in the motor industry: a case study in Coventry. In CM Law (ed.) *Restructuring the global automobile industry: national and regional impacts* London: Routledge, pp. 215–30

Rawstrom EM 1958 Three principles of industrial location. *Transactions of the Institute of British Geographers* **25**: 132–42

Ray DM 1971 The location of United States' manufacturing subsidiaries in Canada. *Economic Geography* **47**: 389–400

Ray MD 1965 *Market potential and economic shadow* Chicago: University of Chicago, Department of Geography, Research Paper No. 101

Rees J 1972 The industrial corporation and location decision analysis. *Area* **4**: 199–205

Rees J 1974 Decision-making, the growth of the firm and the business environment. In FEI Hamilton (ed.) *Spatial perspectives on industrial organisation and decision-making* London: John Wiley, pp. 189–211

Rees J 1978 On the spatial spread and oligopolistic behaviour of large rubber companies. *Geoforum* **9**: 319–30

Rees J 1979 Technological change and regional shifts in American manufacturing. *The Professional Geographer* **31**: 45–54

Rees J 1989 Regional development and policy. *Progress in Human Geography* **13**: 576–88

Rees J, Hewings GJD and Stafford HA (eds) 1981 *Industrial location and regional systems* London: Croom Helm

Rees J, Briggs R and Oakey R 1984 The adoption of new technology in the American machinery industry. *Regional Studies* **18**: 489–504

Rees J, Weinstein BL and Gross HT 1988 *Regional patterns of military procurement and their implications* Washington: Sunbelt Institute

Rees KG 1993 Flexible specialization and the case of the remanufacturing industry in the lower mainland of British Columbia. Unpublished MA thesis, Department of Geography, Simon Fraser University

Rees KG and Hayter R (forthcoming) Flexible specialization, uncertainty and the firm: enterprise strategies in the wood remanufacturing industry of the Vancouver metropolitan area, British Columbia. *The Canadian Geographer*

Reynolds PD and Maki W 1991 *Regional characteristics affecting business growth: assessing strategies for promoting regional economic well-being.* Project report submitted to Rural Poverty and Resource Program, The Ford Program

Reynolds PD, Miller B and Maki WR 1993 Regional characteristics affecting business volatility in the United States, 1980–84. In C Karlsson, B Johannisson and D Storey (eds) *Small business dynamics* London: Routledge, pp. 78–115

Reynolds P, Storey DJ and Westhead P 1994 Cross-national comparisons of the variation in new firm formation rates. *Regional Studies* **28**: 443–56

Ricks DM, Fu MCY and Arpan JS 1974 *International business blunders* Columbus: Grid Inc.

Rifkin J 1995 *The end of work* New York: Putnam

Rigby DL 1991 Technical change and profits in Canadian manufacturing: a regional analysis. *The Canadian Geographer* **35**: 251–63

Roberge R 1972 The timing, type, and location of adaptive inventive activity in the eastern Canadian pulp and paper industry: 1806–1940, PhD dissertation, Clark University, Ann Arbor, Michigan: University Microfilms

Roberts P, Collis C and Noon D 1990 Local economic development in England and Wales: successful adaptation of old industrial areas. In WB Stohr (ed.) *Global challenge and local response* London: Mansell, pp. 135–62

Robertson DH 1928 *The control of industry* London: Nisbet

Robinson EAG 1931 *The structure of competitive industry* Welwyn: Nisbet

Robinson K 1995 Industrial location and air pollution controls: a review of evidence from the USA. *Progress in Human Geography* **19**: 222–44

Rothwell R and Zegweld W 1982 *Innovation and the small and medium sized firm* Hingham, MA: Kluwer Nijhoff

Rowley G 1985 Urban renewal and industrial change in an inner city area: the Sheffield cutlery industry. *East Midland Geographer* **8**: 187–96

Rubenstein JM 1991 The impact of Japanese investment in the United States. In CM Law (ed.) *Restructuring in the global automobile industry.* London: Routledge, pp. 114–42

Rubenstein JM 1992 *The changing US auto industry: a geographical analysis* London: Routledge

Rubinstein S, Bennett M and Kochan T 1993 The Saturn partnership: co-management and the reinvention of the local union. In BE Kaufman and M Kleiner (eds) *Employee representation: alternatives and future directions* University of Wisconsin: Industrial Relations Research Association, pp. 339–70

Rumelt RP 1974 *Strategy, structure and economic performance* Cambridge, MA: Harvard University Press

Sabel CF 1989 Flexible specialization and the re-emergence of regional economies. In P Hirst and J Zeilin (eds) *Reversing industrial decline? Industrial structure and policy in Britain and her competitors* Oxford: Berg, pp. 17–70

Sabel CF and Zeitlin J 1985 Historical alternatives to mass production: politics, markets and technology in nineteenth century industrialization. *Past and Present* **108**: 133–76

Sabel CF, Herrigel G, Deer R and Kzis R 1989 Regional prosperities compared: Massachusetts and Badenwurttemberg in the 1980s. *Economy and Society* **18**: 374–404

Sadler D 1984 Works closure at British Steel and nature of the state, *Political Geography Quarterly* **3**: 297–311

Safarian AE 1966 *Foreign ownership of Canadian industry* Toronto: McGraw Hill

Safarian AE and Dobson W (eds) (1995) *Benchmarking the Canadian business presence in Asia* Hong Kong Bank of Canada Papers on Asia, Vol 1 Toronto: Centre for International Business

Saso M 1990 *Women in the Japanese workplace* London: Hilary Shipman

Sassen S 1991 *The global city: New York, London, Tokyo* Princeton: Princeton University Press

Sayer A 1982a Explaining manufacturing shift: a reply to Keeble. *Environment and Planning A* **14**: 119–25

Sayer A 1982b Explanation in economic geography: abstraction versus generalisation. *Progress in Human Geography* **6**: 68–89

Sayer A 1986 Industrial location and a world scale: the case of the semiconductor industry. In A J Scott and M Storper (eds) *Production, work and territory: the geographical anatomy of industrial capitalism* Boston: Allen and Unwin, pp. 107–23

Sayer A 1989 Postfordism in question. *International Journal of Urban and Regional Research* **13**: 666–95

Sayer A and Morgan K 1985 A modern industry in a declining region: links between method, theory and policy. In D Massey and R Meegan (eds) *Politics and method: contrasting studies in industrial geography* London: Methuen, pp. 144–68

Sayer A and R Walker 1992 *The new social economy: reworking the division of labour* Oxford: Blackwell

Saxenian A 1981 The genesis of Silicon Valley. *Built Environment* **9**: 7–17

Saxenian A 1994 *Regional advantage: culture and competition in Silicon Valley and Route 128* Cambridge: Harvard University Press

Schamp EW 1987 Technology parks and international competition in the Federal Republic of Germany. In B van der Knapp and E Wever (eds) *New technology and regional development* London: Croom Helm, pp. 119–35

Schamp EW and Berentson WH 1995 Institutions in the industrial transition of central and eastern Europe. *Geographsiche Zeitschrift* **83** 65–6

Scherer FM 1980 *Industrial market structure and economic performance.* Chicago: Rand McNally

Schmenner RW 1982 *Making business location decisions.* Englewood Cliffs, NJ: Prentice-Hall

Schoenberg E 1985 Foreign manufacturing investment in the United States: competitive strategies and international location. *Economic Geography* **61**: 241–59

Schoenberger E 1986 Competition, competitive strategy, and industrial change: the case of electronics components. *Economic Geography* **62**: 321–33

Schoenberger E 1987 Technological and organizational change in automobile production: spatial implications. *Regional Studies* **21**: 199–214

Schoenberger E 1988 Multinational corporations and the new international division of labour: a critical appraisal. *International Regional Science Review* **11**: 105–19

Schoenberger E 1989 Some dilemmas of automation: strategic and operational aspects of technological change in production. *Economic Geography* **65**: 232–47

Schoenberger E 1990 US manufacturing investments in Western Europe: markets, corporate strategy, and the competitive environment. *Annals of the Association of American Geographers* **80**: 379–93

Schoenberger E 1991 The corporate interview as an evidentiary strategy in economic geography. *The Professional Geographer* **44**: 180–9

Schoenberger E 1992 Self criticism and self-awareness in research: a reply to Linda McDowell. *The Professional Geographer* **44**: 215–18

Schoenberger E 1994 Corporate strategy and corporate strategists: power, identity and knowledge within the firm. *Environment and Planning A* **26**: 435–51

Schonberger R 1982 *Japanese manufacturing techniques* New York: The Free Press

Schumpeter JA 1939 *Business cycles: a theoretical, historical and statistical analysis of the capitalist process* New York: McGraw-Hill

Schumpeter JA 1943 *Capitalism, socialism and democracy* New York: Harper

Schumpeter JA 1947 The creative response in economic history. *Journal of Economic History* **7**: 149–59

Scott AJ 1982 Production system dynamics and metropolitan development. *Annals of the Association of American Geographers* **72**: 185–200

Scott AJ 1986 Industrialization and urbanization: a geographical agenda. *Annals of the Association of American Geographers* **76**: 25–37

Scott AJ 1987 The semiconductor industry in south east Asia: organization, location and the international division of labour. *Regional Studies* **21**: 143–60

Scott AJ 1988a *New industrial spaces* London: Pion

Scott AJ 1988b *Metropolis: from the division of labour to urban form* Los Angeles: University of California Press

Scott AJ 1990 The technopoles of southern California. *Environment and Planning A* **22**: 1575–1605

Scott AJ 1992 The role of large producers in industrial districts: a case study of high technology systems houses in southern California. *Regional Studies* **26**: 265–75

Scott AJ 1996 The craft, fashion, and cultural-products industries of Los Angeles: competitive dynamics and policy dilemmas in a multisectoral image-producing complex. *Annals of the Association of American Geographers* **86**: 306–23

Scott AJ and Mattingly DJ 1989 The aircraft and parts industry in Southern California: continuity and change from the inter-war years to the 1990s. *Economic Geography* **65**: 48–71

Scott AJ and Storper M (eds) 1986 *Production, work, territory: the geographical anatomy of industrial capitalism* London: Allen Unwin

Scott AJ and Storper M 1990 Regional development reconsidered. *Working Paper No. 1*, Lewis Centre for Regional Policy Studies, University of California, Los Angeles

Scott AJ and Storper M 1992 Industrialization and regional development. In M Storper and AJ Scott (eds) *Pathways to industrialization and regional development* London: Routledge, pp. 3–20

Semple RK 1985 Quaternary place theory: an introduction. *Urban Geography* 6: 285–96

Semple RK and Phipps AG 1982 The spatial evolution of corporate headquarters within an urban system. *Urban Geography* 3: 258–79

Servan-Schreiber JJ 1968 *The American challenge* London: Hamish Hamilton

Shafer M 1983 Capturing the mineral multinationals, advatange or disadvantage? *International Organization* 37: 93–120

Shaffer RE and Pulver GC 1985 Regional variations in capital structure of new small businesses: the Wisconsin case. In D J Storey (ed.) *Small firms in regional economic development* Cambridge: Cambridge University press, pp. 166–92

Sharpe R 1991 Social movements in the local economic development process. *Environments* 21: 56–8

Shaw G 1993 Ski patrol leads to better back pack. *The Vancouver Sun* 24 September: D1–2

Sheard P 1983 Auto-production systems in Japan: organization and locational features. *Australian Geographical Studies* 21: 49–68

Shenzhen Tongjijuu 1993 *Shenzhen tongji nianjian, 1993* (Annual Statistics of Shenzhen, 1993). Beijing: Zhongguo tongji, pp. 352–3

Shepherd WG 1979 *The economics of industrial organization* Englewood Cliffs, NJ: Prentice Hall

Shimokawa K 1993 Keiretsu transactions: Japan and the United States compared. *Japanese Economic Studies* 22: 49–66

Simon HA 1955 A behavioural model of rational choice. *Quarterly Journal of Economics* 69: 99–118

Simon HA 1957 *Models of man: social and rational* New York: John Wiley

Simon HA 1959 Theories of decision-making in economics and behavioral science. *American Economic Review* 49: 253–83

Simon HA 1960 *The new science of management decision* New York: Harper Row

Simon H 1992 Lessons from Germany's midsize giants. *Harvard Business Review* March–April; 115–23

Singh A 1977 UK industry and the world economy: a case of deindustrialization? In AP Jacquemin and HW De Jong (eds) *Welfare aspects of industrial markets 2* Leiden: Martinus Nijhoff Social Science Division, pp. 183–214

Sit V 1985 Special economic zones of China: a new type of export processing zone? *The Developing Economies* 23: 68–87

Sklar RL 1975 *Corporate power in an African state: political impact of multinational corporations in Zambia* Berkeley: University of California Press

Sklair L 1989 *Assembling for development: the maquila industry in Mexico and the United States* Boston: Unwin Hyman

Skúlason JB 1994 Foreign investment and the bargaining process: case study of the aluminum industry in Iceland. Unpublished MA thesis, Department of Geography, Simon Fraser University

Smith A 1986 *The wealth of nations* (reprint) Harmondsworth: Penguin

Smith DM 1966 A theoretical framework for geographical studies of industrial location. *Economic Geography* 42: 95–113

Smith DM 1971 *Industrial location: an economic geographical analysis* New York: Wiley

Soja E 1991 Poles apart: urban restructuring in New York and Los Angeles. In JH Mollenkopf and M Castells (eds) *Dual City Restructuring* New York: Russell Sage Foundation, pp. 361–76

South R 1990 Transnational 'maquiladora' location. *Annals of the Association of American Geographers* 80: 549–70

South Yorkshire County Council, 1979, 1982, 1984 *South Yorkshire Statistics* Barnsley: South Yorkshire County Council

Soyez D 1986 Industrietourismus. *Erkunde* 40: 105–11

Soyez D 1988a Stora lured abroad? A Nova Scotian case study in industrial decision-making and persistence. *The Operational Geographer* 16: 11–14

Soyez D 1988b Scandinavian silviculture in Canada: entry and performance barriers. *The Canadian Geographer* 32: 133–140

Soyez D 1991 Stora Kopparberg in Canada: negotiating an enterprise in a foreign host environment. In A Buttimer, J van Buren and N Hudson-Rodd (eds) *Land life lumber leisure: local and global concern in the human use of woodland* Ottawa

Soyez D 1996 Industrial resource use and transnational conflict patterns: geographical implications of the James Bay Hydropower schemes, Canada. In MJ Taylor (ed.) *Environmental change: industry, power, industry and policy* Aldershot: Avebury

Springate D 1978 Difficulties associated with DREE's current approach. In NH Lithwick (ed.) *Regional economic policy: the Canadian experience* Toronto: McGraw-Hill Ryerson, pp. 257–83

Stafford HA 1960 Factors in the location of the paperboard container industry. *Economic Geography* 36: 260–6

Stafford HA 1969 An industrial location decision model. *Proceedings of the Association of American Geographers* 1: 141–5

Stafford HA 1972 The geography of manufacturers. *Progress in Geography* **4**: 181–215

Stafford HA 1974 The anatomy of the location decision: content analysis of case studies. In FEI Hamilton (ed.) *Spatial perspectives on industrial organisation and decision-making* New York: Wiley, pp. 169–87

Stafford HA 1977 Environmental regulations and the location of US manufacturing: speculations. *Geoforum* **8**: 243–8

Stafford HA 1979 *Principles of industrial facility location.* Atlanta: Conway Publishing

Stafford HA 1985 Environmental protection and industrial location. *Annals, Association of American Geographers* **75**: 227–40

Stafford HA 1991 Manufacturing plant closure selection within firms. *Annals of the Association of American Geographers* **81**: 51–65

Stafford HA and Watts HD 1990 Abandoned products, abandoned places: plant closures by multi-product multi-locational firms in urban areas. *Tijdschrift voor Economische en Sociale Geografie* **81**: 161–9

Standing G 1989 European unemployment, insecurity and flexibility: a social dividend solution. *World Employment Programme Working Paper 23* Geneva: ILO

Standing G 1992 Alternative routes to labour flexibility. In M Storper and AJ Scott (eds) *Pathways to industrialization and regional development* London: Routledge, pp. 255–75

Starbuck WH 1971 Organizational metamorphoses. In WH Starbuck (ed.) *Organizational growth and development* London: Penguin, pp. 275–98

Steed GPF 1976a Centrality and locational change: printing, publishing, and clothing in Montreal and Toronto. *Economic Geography* **47**: 371–83

Steed GPF 1976b Locational factors and dynamics of Montreal's large garment complex. *Tijdschrift voor Economische en Sociale Geografie* **67**: 151–68

Steed GPF 1976c Standardization, scale, incubation, and inertia: Montreal and Toronto clothing industries. *The Canadian Geographer* **20**: 298–309

Steed GPF 1978 Global industrial systems – a case study of the clothing industry. *Geoforum* **9**: 35–47

Steed GPF 1981 International location and comparative advantage: the clothing industries in developing economies. In FEI Hamilton and GJR Linge (eds) *Spatial analysis, industry and the industrial environment* vol 2, London: Wiley, pp. 265–303

Steed GPF 1982 *Threshold firms: backing Canada's winners* Science Council of Canada, Background Study 48, Ottawa: Minister of Supply and Services

Steed GPF and DeGenova D 1983 Ottawa's technology-oriented complex. *The Canadian Geographer* **27**: 262–78

Steindl J 1947 *Random processes and the growth of firms* New York: Hafner

Stevenson HH and Sahlman WA 1989 The entrepreneurial process. In P Burns and J Dewhurst *Small business and entrepreneurship* London: Macmillan Education, pp. 94–157

Stohr WB 1990a Synthesis. In WB Stohr (ed.) *Global challenge and local economic response* London: Mansell, pp. 1–19

Stohr WB 1990b Introduction. In WB Stohr (ed.) *Global challenge and local economic response* London: Mansell, pp. 20–34

Stohr WB 1990c On the theory and practice of local development in Europe. In WB Stohr (ed.) *Global challenge and local economic response* London: Mansell, pp. 36–54

Storey DJ 1981 New firm formation, employment change and the small firm: the case of Cleveland County. *Urban Studies* **18**: 335–45

Storey DJ 1982 *Entrepreneurship and the new firm* London: Croom Helm

Storey DJ 1994 *Understanding the small business sector* London: Routledge

Storey DJ and Johnson S 1987 *Job generation and labour market change* Basingstoke, Hants: Macmillan

Storper M 1981 Towards a structural theory of industrial location. In J Rees, GJD Hewings and HA Stafford (eds) *Industrial location and regional systems* New York: Bergin, pp. 17–40

Storper M 1985 Oligopoly and the product cycle: essentialism in economic geography. *Economic Geography* **61**: 260–82

Storper M 1989 The transition to flexible specialization in the US film industry: external economies, the division of labour and the crossing of industrial divides. *Cambridge Journal of Economics* **13**: 273–305

Storper M 1992 The limits to globalization: technology districts and international trade. *Economic Geography* **68**: 60–93

Storper M 1993 Regional worlds of production: learning and innovation in the technology districts of France, Italy and the USA. *Regional Studies* **27**: 433–55

Storper M and Christopherson S 1987 Flexible specialization and regional industrial agglomerations: the case of the US motion picture industry. *Annals of the Association of American Geographers* **77**: 104–117

Storper M and Harrison B 1991 Flexibility, hierarchy and regional development: the changing structure of industrial production systems and their forms of governance. *Regional Policy* **20**: 407–22

Storper M and Scott AJ 1989 The geographical foundations and social regulation of flexible production complexes. In J Wolch and M Dear (eds) *The power of geography: how territory shapes social life* Boston: Unwin Hyman

Storper M and Scott AJ (eds) 1992 *Pathways to industrialization and regional development* London: Routledge

Storper M and Walker R 1989 *The capitalist imperative: territory, technology, and industrial growth* New York: Basil Blackwell

Streeck W 1989 Skills and the limits of neo-liberalism: the enterprise of the future as a place of learning. *Work, Employment and Society* 3: 89–104

Streeck W 1992 Training and the new industrial relations: a strategic role for unions. In M Regini (ed.) *The Future of Labour Movements* London: Sage, pp. 250–69

Struyk RJ and James FJ 1975 *Intrametropolitan industrial location: the pattern and process of change*. D C Heath

Studley V 1977 *The art and craft of handmade paper* New York: Van Nostrand Reinhold

Suarez-Villa L 1984 The manufacturing process cycle and the industrialization of the United States-Mexico borderlands. *Annals of Regional Science* 18: 1–23

Sunley P 1992 Marshallian industrial districts: the case of the Lancashire cotton industry in the inter-war years. *Transactions of the Institute of British Geographers* 17: 306–320

Suyama S 1996 Restructuring of production-distribution system in Wajima *Shikki* industry. *Science Reports of the Institute of Geoscience* University of Tsukuba, Section A: Geographical Sciences 17: 63–85

Takeuchi A 1990 Nissan Motor Company: stages of international growth, locational profile, and subcontracting in the Tokyo region. In M de Smidt and E Wever (eds) *The corporate firm in a changing world economy* London: Routledge, pp. 166–82

Takeuchi A 1992 Activities of small scale industries in Japan through inter-enterprise cooperation. *Report of Research* Nippon Institute of Technology 21: 283–92

Takeuchi A 1994 Location dynamics of industry in the Tokyo metropolitan region. *Report of Research* Nippon Institute of Technology 23: 195–220

Takeuchi A 1995 Globalization and localization of industry in Japan, a nation state, focus on the Tokyo region. *Report of Research* Nippon Institute of Technology 25: 241–54

Taylor A 1995 GM: why they might break up America's biggest company. *Fortune* 133 (May 13): 78–87

Taylor EG, Daysh GHJ, Fleure HJ and Smith W 1938 Discussion on the geographic distribution of industry. *Geographic Journal* 92: 22–32

Taylor MJ 1970 Location decisions of small firms. *Area* 2: 51–4

Taylor MJ 1975a Problems of minimum cost location: the Kuhn and Kuenne algorithm. *Occasional Paper 4*, Department of Geography, University of London, Queen Mary College

Taylor MJ 1975b Organizational growth, spatial interaction and locational decision-making. *Regional Studies* 9: 313–24

Taylor MJ 1984 *The geography of Australian corporate power* London: Croom Helm

Taylor MJ 1986 The product-cycle model: a critique. *Environment and Planning A* 18: 751–61

Taylor MJ 1986 (ed.) 1996 *Environmental change: industry, power and policy* Aldershot: Avebury

Taylor MJ and Thrift N 1981 British capital overseas: direct investment and corporate development in Australia. *Regional Studies* 15: 183–212

Taylor MJ and Thrift N 1982 Industrial linkage and the segmented economy 1. Some theoretical proposals. *Environment and Planning A* 14: 1601–13

Taylor MJ and Thift N 1983 Business organization, segmentation and location. *Regional Studies* 17: 445–65

Taylor MJ and Thrift N 1984 *The geography of multinationals* London: Croom Helm

Taylor MJ and Wood PA 1973 Industrial linkage and local agglomeration in the West Midlands metal industry. *Transactions, Institute of the British Geographers* 59: 129–54

Thirlwell AP 1982 Deindustrialization in the United Kingdom. *Lloyds Bank Review* 144: 22–37

Thomas MD 1964 The export base and development stages theories of regional economic growth: an appraisal. *Land Economics* 40: 421–32

Thomas MD 1969 Regional economic growth: some conceptual aspects. *Land Economics* 46: 43–51

Thornton CA and Koepke RL 1981 Federal legislation, clean air and local industry. *Geographical Review* 71: 324–39

Thwaites AT 1978 Technological change, mobile plants and regional development. *Regional Studies* 12: 445–61

Thwaites AT 1982 Some evidence of regional variations in the introduction and diffusion of industrial products and processes within British manufacturing industry. *Regional Studies* 16: 371–81

Tiano S 1994 *Patriarchy on the line: labor, gender, and ideology in the Mexican maquila industry* Philadelphia: Temple

Tickell A and Peck J 1992 Accumulation, regulation and the geographies of post-fordism: missing links in regulationist research. *Progress in Human Geography* 16: 190–218

Tiebout CM 1957 Location theory, empirical evidence and economic evolution. *Papers and Proceedings of the Regional Science Association* 3: 74–86

Tillman DA 1985 *Advanced technologies and economic analyses* Orlando: Academic Press

Todd D 1977 Regional intervention in Canada and the evolution of growth centre strategies. *Growth and Change* **8**: 29–34

Tornqvist G 1968 Flows of information and the location of economic activities. *Lund Studies in Geography* Series B35

Townroe P 1969 Locational choice and the individual firm. *Regional Studies* **3**: 15–24

Townroe P 1971 *Industrial location decisions* Occasional Paper No. 15, Centre for Urban and Regional Studies, University of Birmingham

Townroe P 1975 Branch plants and regional development. *Town Planning Review* **46**: 47–62

Townroe P 1976 *Planning industrial location* London: Leanard Hill Books

Townsend AR 1983 *The impact of recession* London: Croom Helm

Townsend AR 1993 The urban–rural cycle in the Thatcher growth years. *Transactions of the Institute of British Geographers* **18**: 207–221

Townsend H 1954 Economic theory and the cutlery trades. *Economica* **21**: 224–39

Trevor M 1983 *Japan's reluctant multinationals* London: Francis Pinter

Trist E 1979 New directions of hope: recent innovations connecting organisational, industrial, community and personal development. *Regional Studies* **13**: 439–51

Tweedale G 1986 Transatlantic speciality steels: Sheffield firms and the USA. In G Jones (ed.) *British multinationals: origins, management and performance* Aldershot: Gower, pp. 75–95

Tweedale G 1987a Metallurgy and technological change: a case study of Sheffield speciality steel and America, 1830–1930. *Technology and Culture* **27**: 189–222

Tweedale G 1987b *Sheffield steel and America: a century of commercial and technological interdependence, 1830–1930.* Oxford: Oxford University Press

UNCTNC 1985 *Transnational corporations in world development: third survey* New York: United Nations

UNCTC 1988 *Transnational corporations in world development: trends and prospects* New York: United Nations

UNIDO 1980 Export processing zones in developing countries. *UNIDO Working Paper on structural changes No. 19.* Vienna: UNIDO

UNIDO 1981 *Restructuring world industry in a period of crisis – the role of innovation: an analysis of recent developments in the semi-conductor industry.* Vienna: UNIDO

Ullman EL 1958 Amenities as a factor in regional growth. *Geographical Review* **44**: 119–32

Ullman EL 1958 Regional development and the geography of concentration. *Papers and Proceedings of the Regional Science Association* **4**: 179–98

Urry J 1986 Capitalist production, scientific management and the service class. In AJ Scott and M Storper (eds). *Production, work, territory: the geographical anatomy of industrial capitalism.* Boston: Allen and Unwin, pp. 43–66

US Congress 1984 *Technology innovation and regional economic development.* Background Paper 2, Congress of the United States Office of Technology Assessment, Washington: Government Printing Office

Utterbach JM and Abernathy WJ 1975 A dynamic model of process and product innovation. *Omega* **3**: 639–56

Vaessen P and Keeble D 1995 Growth-oriented SMEs in unfavourable regional environments. *Regional Studies* **29**: 489–505

Veblen T 1932 *The theory of business enterprise.* New York: Mentor Books

Vernon R 1966 International investment and international trade in the product life cycle. *Quarterly Journal of Economics* **80**: 190–207

Vernon R 1970 Organization as a scale factor in the growth of firms. In JW Markham and GF Papenek (eds) *Industrial organization and economic development: in honour of Professor Edward S. Mason.* Boston: Houghton Mifflin, pp. 47–66

Vernon R 1971 *Sovereign at bay: the multinational spread of US enterprises.* New York: Basic Books

Vernon R 1979 The product cycle hypothesis in a new international environment. *Oxford Bulletin of Economics and Statistics* **41**: 255–67

Vernon R 1985 Sovereignty at bay: ten years after. In TH Moran (ed.) *Multinational corporations: the political economy of foreign direct investment.* Lexington: DC Heath

Vernon R and Hoover EM 1959 *Anatomy of a metropolis.* Cambridge: Harvard University Press

Vianen J 1994 Small business developments in the long run and through the business cycle: lessons from the past in the Netherlands. In C Karlsson, B Johannisson and D Storey (eds) *Small business dynamics.* London: Routledge, pp. 18–39

Villeneuve P and Rose D 1988 Gender and the separation of employment from home in metropolitan Montreal 1971–1981. *Urban Geography* **9**: 155–79

Walby S 1990 *Theorizing patriarchy.* Oxford: Basil Blackwell

Walmsley DJ and Lewis GJ 1984 *Human geography: behavioural approaches.* London: Longman

Walker R 1989 A requiem for corporate geography: new directions in industrial organisation, the production of place and uneven development. *Geografisca Annaler* **71B**: 43–68

Wallace I 1974 The relationship between freight transport organization and industrial linkage in Britain. *Transactions of the Institute of British Geographers* **62**: 25–44

Wang J and Bradbury JNH 1990 The changing industrial geography of the Chinese special economic zones. *Economic Geography* **66**: 306–20

Warren K 1970 *The British iron and steel industry since 1840*. London: Methuen

Watkins MH 1963 A staple theory of economic growth. *Canadian Journal of Economics and Political Science* **29**: 141–8

Watkins Report 1968 *Foreign ownership and the structure of Canadian industry*. Ottawa: Privy Council Office

Watts HD 1971 The location of the sugar beet industry in England and Wales, 1912–36. *Transactions of the Institute of British Geographers* **53**: 95–116

Watts HD 1974 Locational adjustment in the British beet sugar industry. *Geography* **59**: 10–23

Watts HD 1977 Market areas and spatial rationalisation: the British brewing industry after 1945. *Tijdschrift voor Economische en Sociale Geografie* **68**: 224–40

Watts HD 1978 Interorganization relations and the location of industry. *Regional Studies* **12**: 215–25

Watts HD 1979 Large firms, multinationals, and regional development: some new evidence from the United Kingdom. *Environment and Planning A* **11**: 71–81

Watts HD 1980a Conflict and collusion in the British sugar beet industry, 1924 to 1928. *Journal of Historical Geography* **6**: 291–314

Watts HD 1980b *The large industrial enterprise: some spatial perspectives*. London: Croom Helm

Watts HD 1980c The location of European investment in the United Kingdom. *Tijdschrift voor Economische en Social Geografie* **71**: 3–14

Watts HD 1981 *The branch plant economy*. London: Longman

Watts HD 1982 The inter-regional distribution of West German multinationals in the United Kingdom. In MJ Taylor and N Thrift (eds). *A geography of multinationals*. London: Croom Helm, pp. 61–89

Watts HD 1987 *Industrial geography*. New York: Wiley

Watts HD 1991a Plant closures, multilocational firms, and the urban economy: Sheffield, UK. *Environment and Planning A* **23**: 37–58

Watts HD 1991b Plant closures in urban areas: towards a local policy response. *Urban Studies* **28**: 803–17

Watts HD and Stafford HA 1986 Plant closures and the multiplant firm: some conceptual issues. *Progress in Human Geography* **10**: 206–27

Webber MJ 1972 *Impact of uncertainty on location*. Cambridge, MA: MIT Press.

Webber MJ 1984 *Industrial location*. London: Sage.

Webber MJ 1986 Regional production and the production of regions: the case of steeltown. In AJ Scott and M Storper (eds) *Production, work, territory: the geographical anatomy of industrial capitalism*. Boston: Allen and Unwin, pp. 195–224

Webber MJ and Rigby DL 1986 The rate of profit in Canadian manufacturing, 1950–81. *Review of Radical Political Economics* **18**: 33–55

Weber A 1929 *Alfred Weber's theory of the location of industries* (translated by CJ Friedrich) Chicago: University of Chicago Press

Weiner MJ 1981 *English culture and the decline of the industrial spirit 1850–1980*. Cambridge: Cambridge University Press

Weinstein BL and Firestine RE 1978 *Regional growth and decline in the United States: the rise of the sunbelt and the decline of the northeast* New York: Praeger

Weinstein BL and Gross HT 1988 The rise and fall of sun, rust, and frost belts. *Economic Development Quarterly* **2**: 9–18

Weinstein BL, Gross HT and Rees J 1985 *Regional growth and decline in the United States*. New York: Praeger

Welch LS and Luostarinen R 1988 Internationalization: evolution of a concept. *Journal of General Management* **14**: 34–55

Wheeler JO 1986 Corporate spatial links with financial institutions: the role of the metropolitan hierarchy. *Annals of the Association of American Geographers* **76**: 262–74

Wheeler OJ 1987 Fortune firms and the fortune of their headquarters metropolises. *Geografisca Annaler B* **69**: 65–71

Wheeler OJ 1988 The corporate role of large metropolitan areas in the United States. *Annals of the Association of American Geographers* **79**: 523–43

Wheeler JO and Mitchelson RL 1989 Information flows among major metropolitan areas in the United States. *Annals of the Association of American Geographers* **79**: 523–43

Whitman ES and Schmidt WJ 1966 *Plant relocation*. New York: American Management Association

Whittington RC 1984 Regional bias in new firm formation in the UK. *Regional Studies* **18**: 253–5

Wild T and Jones P (eds) 1991 *Deindustrialisation and new industrialisation in Britain and Germany*. London: Anglo-German Foundation

Wilkins M 1970 *The emergence of multinational enterprise* Cambridge, MA: Harvard University Press

Wilkins M 1994 Comparative hosts. In G Jones (ed.) *The making of global enterprise*. London: Frank Cass, 18–50

Williams G 1983 *Not for export: towards a political economy of Canada's arrested industrialization.* Toronto: McClelland and Stewart

Williams K, Williams J and Thomas D 1983 *Why are the British bad at manufacturing?* London: Routledge and Kegan Paul

Williams K, Haslam C, Williams J and Wardlaw A 1987 *The breakdown of Austin Rover* Leamington Spa: Berg

Williams K, Williams J, Haslam C and Wardlow A 1989 Facing up to manufacturing failure. In P Hirst and J Zeitlin (eds) *Reversing industrial decline: industrial structure and policy in Britain and her competitors* New York: St Martin's Press, pp. 71–94

Williams K, Haslam C, Williams J, Johal S and Adcroft A 1994 *Cars: analysis, history and cases* Providence: Berghahn Books

Williamson OE 1975 *Markets and hierarchies, analysis and anti-trust implication: a study in the economics of industrial organization* New York: Free Press

Williamson OE 1985 *The economic institutions of capitalism* New York: The Free Press

Wilson P 1992 *Exports and local development: Mexico's new maquiladoras* Austin: University of Texas

Wise M 1949 On the evolution of the jewellery and gun quarters in Birmingham. *Transactions of the Institute of British Geographers* **15**: 57–72

Wolch J and Dear M 1989 *The power of geography: how territory shapes social life* Boston: Unwin Hyman

Womack JP, Jones DT and Roos D 1990 *The machine that changed the world* New York: Rawson Associates, Macmillan

Wright M, Thompson S, Chiploin B and Robbie K 1990 *Buy-outs and buy-ins: new strategies in corporate management* London: Graham and Trotman

Yaginuma H 1993 The keiretsu issue: a theoretical approach. *Japanese Economic Studies* **22**: 3–48

Yamazaki M 1980 *Japan's community-based industries: a case study of small industry* Tokyo: Asian Productivity Organization

Yeates M 1990 *The North American City* New York: Harper Row

Yoshihara K 1976 *Japanese investment in southeast Asia* Honolulu: The University Press of Hawaii

Yoshihara K 1978 Determinants of Japanese investment in south-east Asia. *International Social Science Journal* **30**: 363–76

Zeitlin J 1989 The third Italy. In R Murray (ed.) *Technology strategy and local economic intervention* Nottingham: Spokesman Books

Zuboff S 1988 *In the age of the smart machine* New York: Basic Books

Zukin S 1991 *Landscapes of power: from Detroit to Disney World* Berkeley: University of California Press.

Zurawicki L 1979 *Multinational enterprises in the west and east* Alphen aan den Rijn, Sijthoff and Noordhoff

INDEX